Recent Advances in Biotechnology

(*Volume 4*)

Progress in Food Biotechnology

Edited by

Ali Osman

DSM Food Specialties, Alexander Felminglaan 1, 2613 Delft, The Netherlands

General:

1. Any dispute or claim arising out of or in connection with this License Agreement or the Work (including non-contractual disputes or claims) will be governed by and construed in accordance with the laws of the U.A.E. as applied in the Emirate of Dubai. Each party agrees that the courts of the Emirate of Dubai shall have exclusive jurisdiction to settle any dispute or claim arising out of or in connection with this License Agreement or the Work (including non-contractual disputes or claims).

2. Your rights under this License Agreement will automatically terminate without notice and without the need for a court order if at any point you breach any terms of this License Agreement. In no event will any delay or failure by Bentham Science Publishers in enforcing your compliance with this License Agreement constitute a waiver of any of its rights.

3. You acknowledge that you have read this License Agreement, and agree to be bound by its terms and conditions. To the extent that any other terms and conditions presented on any website of Bentham Science Publishers conflict with, or are inconsistent with, the terms and conditions set out in this License Agreement, you acknowledge that the terms and conditions set out in this License Agreement shall prevail.

Bentham Science Publishers Ltd.
Executive Suite Y - 2
PO Box 7917, Saif Zone
Sharjah, U.A.E.
Email: subscriptions@benthamscience.org

BENTHAM SCIENCE

CONTENTS

PREFACE

Biotechnology is a dynamic, constantly-advancing, and multidisciplinary field which describes the application of any technology that uses biological systems, living organisms, and their products/derivatives to make or modify products and/or processes for specific uses. The advantages, advances, and relevant aspects of its disciplinary underpinnings in biology, chemistry, and engineering have covered various applied areas, such as medicine, pharmaceuticals, nutraceuticals, food and beverages, agriculture, environment, catalysis, *etc…* In the food processing sector, the recent advances in biotechnology have targeted the selection and manipulation of organisms and their products with the objectives to (1) improve process control, product quality, consistency, yield and process efficiency, and (2) produce novel opportunities in food preservation and in the production of value-added products (*i.e.* enzymes, flavour compounds, microbial cultures, food ingredients, *etc…*).

The global biotechnology industry is in the growth phase of its economic life cycle. The rapid increase in demand reflects the significant expansion in the numbers and volumes of industrial goods produced using biotechnology. Product lines increase as new technologies are developed, processes are learned, and products are commercialized. The contribution of biotechnology to the food production market was evaluated at about US$101.96bn in 2014 (Visiongain market research). Within food biotechnology, many segments are experiencing rapid growth rates. For example, the global food enzyme market is expected to grow with a CAGR (Compound Annual Growth rate) between 7 and 8.5 % from 2017 to 2023 (report linker, Mordor intelligence). Another example is the global starter culture market which is expected to grow with a CAGR of 5.5- 6.5 % between 2016 and 2024 (Credence Research, Markets and markets research).

The dynamics of the food biotechnology market and the fast pace with which the field develops are the main drivers behind the effort spent by the authors of this eBook to cover the most up-to-date advances taking place within food biotechnology. This current eBook 'Advances in Biotechnology: Progress in Food Biotechnology' has a pivotal role in describing and discussing snapshots of the most recent advances in the field of academic, industrial, and market perspectives. The eBook targets researchers, scholars, and particularly students and professionals from both academia and industry interested and working in this multidisciplinary subject. This eBook helps (i) students to strengthen their knowledge and expertise in the field, (ii) researchers and industry specialists to initiate and integrate new ideas and technologies in their product and process development, and (iii) specialists in governmental and non-governmental bodies in developing their policies and decision-making processes.

This eBook is written in a very simple and easy-to-understand language targeting a broad audience, and will be an interesting source of trustworthy information supported by graphs, tables, numbers, market trends, and stories of successful product launches. This eBook will not only cover the advances in food biotechnology from an academic perspective, but will also discuss in detail the current advances from an industrial perspective. The content of this eBook will also highlight, between the lines, the possible ways for creating the future of food biotechnology from a technological, market, industrial, and legal point of view.

This eBook is composed of 10 chapters. Chapter 1 presents the latest research on food protein-derived bioactive peptides and the impact of enzymatic hydrolysis on the functional and organoleptic properties of proteins, while chapter 2 discusses the classification, biosynthesis, functional properties, and various food and health applications of bacterial exopolysaccharides (EPS). In chapter 3, the recent advances in the enzymatic modification of phospholipids, the preparation of chemically defined structured phospholipids, and the diverse applications of these phospholipids are reviewed. Chapter 4 presents the current and future status of microbial culture research and production with emphasis on the main fermenting microorganisms, the mutual effects that microorganisms have on each other, and the resulting effects on food matrices. Furthermore, chapter 5 provides a comprehensive overview of the research carried out on probiotics, prebiotics, and synbiotics in the last 10 years as well as the new biotechnologies that contribute to the understanding of the host-microbiota interactions and the mechanisms of actions of pro-, pre- and synbiotics. In chapter 6, current food additives produced *via* biotechnological routes, such as amino acids, antimicrobial peptides, organic acids, vitamins, and sweeteners are discussed, whereas chapter 7 presents the latest results obtained with phenolic-based nanoparticles, showing the failures, achievements, and most promising routes for future work with these interesting nutraceuticals. Chapter 8 discusses (i) the recent advances in enzyme discovery approaches supported by few examples of relevance to the dairy industry, and (ii) the up-to-date developments in industrial dairy enzyme applications, with focus on lactose bioconversion, while chapter 9 highlights the role of biotechnology on the bioconversion of major industrial and agro-industrial by-products into various bio-products as examples of a future bio-based economy. In the last chapter of this eBook, plant epigenetics and future expectations that can employ epigenetics in order to improve crops and produce higher levels of *e.g.* vitamins and proteins are presented in detail.

Ali Osman
DSM Food Specialties
Alexander Felminglaan 1
2613 Delft
The Netherlands
E-mail: Ali.Osman@dsm.com

List of Contributors

Ali Osman	DSM Food Specialties, Alexander Felminglaan 1, 2613 Delft, The Netherlands
Amr R.A. Kataya	University of Stavanger, Centre for Organelle Research, Faculty of Science and Technology, N-4036 Stavanger, Norway
Andrea Monteagudo-Mera	Department of Food and Nutritional Sciences, University of Reading, Whiteknights, P.O. Box 226, Reading RG6 6AP, United Kingdom
Angela F. Jozala	Department of Technological and Environmental Processes, University of Sorocaba–UNISO, Sorocaba, SP, Brazil
Christina N. Economou	Laboratory of Biochemical Engineering & Environmental Technology (LBEET), Department of Chemical Engineering, University of Patras, 26504 Patras, Greece
Cid Ramón González-González	Instituto Tecnológico Superior de Acayucan, Carretera Costera del Golfo Km 216.4 Acayucan Veracruz, Mexico. CP. 96100
Clara Grosso	REQUIMTE/LAQV, Laboratório de Farmacognosia, Departamento de Química, Faculdade de Farmácia, Universidade do Porto, Rua de Jorge Viterbo Ferreira, no. 228, 4050-313 Porto, Portugal
Daniela A. V. Marques	Serra Talhada Campus, University of Pernambuco–UPE, Serra Talhada, PE, Brazil
Dimitris Sarris	Department of Food Science and Nutrition, School of the Environment, University of the Aegean, 81400 Myrina-Lemnos, Greece Laboratory of Food Microbiology and Biotechnology, Department of Food Science and Human Nutrition, Agricultural University of Athens, Athina 118 55, Greece
Dimitrios Charalampopoulos	Department of Food and Nutritional Sciences, University of Reading, Whiteknights, P.O. Box 226, Reading RG6 6AP, United Kingdom
Gurjot Deepika	Portob Biopharma Ltd., Manor Farm Road, Porton Down, Salisbury SP4 0JG, UK
Jan Kjølhede Vester	Novozymes A/S, Krogshøjvej 36, 2880 Bagsværd, Denmark
Jasmina Damnjanović	Laboratory of Molecular Biotechnology, Graduate School of Bioagricultural Sciences, Nagoya University, Furo-cho, Chikusa-ku, Nagoya 464-8601, Japan
Jeppe Wegener Tams	Novozymes A/S, Krogshøjvej 36, 2880 Bagsværd, Denmark
Jorge F. B. Pereira	Department of Bioprocesses and Biotechnology, School of Pharmaceutical Sciences, Univ. Estadual Paulista – UNESP, Araraquara, SP, Brazil
Patrícia Valentão	REQUIMTE/LAQV, Laboratório de Farmacognosia, Departamento de Química, Faculdade de Farmácia, Universidade do Porto, Rua de Jorge Viterbo Ferreira, no. 228, 4050-313 Porto, Portugal
Paula B. Andrade	REQUIMTE/LAQV, Laboratório de Farmacognosia, Departamento de Química, Faculdade de Farmácia, Universidade do Porto, Rua de Jorge Viterbo Ferreira, no. 228, 4050-313 Porto, Portugal

P. H. P. Prasanna Department of Animal & Food Sciences, Faculty of Agriculture, Rajarata University of Sri Lanka, Anuradhapura, Sri Lanka

Sander Sieuwerts Arla Innovation Centre, Agro Food Park 19, 8200 Aarhus, Denmark

Seraphim Papanikolaou Laboratory of Food Microbiology and Biotechnology, Department of Food Science and Human Nutrition, Agricultural University of Athens, Athina 118 55, Greece

Valker A. Feitosa Department of Biochemical and Pharmaceutical Technology, School of Pharmaceutical Sciences, University of São Paulo–USP, São Paulo, SP, Brazil
Bionanomanufacturing Center, Institute for Technological Research–IPT, São Paulo, SP, Brazil

Valéria C. Santos-Ebinuma Department of Bioprocesses and Biotechnology, School of Pharmaceutical Sciences, Univ. Estadual Paulista – UNESP, Araraquara, SP, Brazil

Yugo Iwasaki Laboratory of Molecular Biotechnology, Graduate School of Bioagricultural Sciences, Nagoya University, Furo-cho, Chikusa-ku, Nagoya 464-8601, Japan

Zied Khiari Cape Breton University - Verschuren Centre for Sustainability in Energy and the Environment, 1250 Grand Lake Road, Sydney, Nova Scotia, Canada, B1P 6L2

CHAPTER 1

Advances in Food Protein Biotechnology

Zied Khiari[1] and **Cid Ramón González-González**[2,*]

[1] *Cape Breton University - Verschuren Centre for Sustainability in Energy and the Environment. 1250 Grand Lake Road, Sydney, Nova Scotia, CanadaB1P 6L2*

[2] *Instituto Tecnológico Superior de Acayucan Carretera Costera del Golfo Km 216.4 Acayucan Veracruz, Mexico, CP 96100*

Abstract: The present book chapter deals with recent advances in food protein biotechnology. The latest research on food protein-derived bioactive peptides and the impact of enzymatic hydrolysis on the organoleptic properties of proteins is reviewed. Protein modifications, which have become the focus of many research studies during recent years, are also covered. Consideration is given to three different protein modification approaches (i) chemical modifications (glycation and disulfide cross-linking); (ii) physical modifications (high-pressure processing and ultrasound treatment); (iii) enzymatic modifications (transglutaminase cross-linking and proteolysis). Since the main purpose of protein modification is to enhance their functional properties, the effects of the chemical, physical, and enzymatic treatments on the solubility, emulsification, foamability, and rheological properties of food proteins are also discussed.

Keywords: Bioactivity, Enzyme, Emulsification, Functional Properties, Glycation, Hydrolysis, Protein, Peptide, Protein Modification, Solubility.

INTRODUCTION

Proteins are macromolecules composed of one or more polypeptides, which are made up of long chains of amino acids connected by peptide bonds. Proteins play a crucial role in both biological and food systems. In biology, proteins appear to have multiple functions, such as providing structural support, storing energy, chelating metal, and catalyzing biochemical reactions [1].

Proteins are also involved in immunity, transport, and regulation of cellular metabolism [1]. In food, proteins are considered to be valuable multifunctional biopolymers [2]. They are used in a variety of food formulations due to their natural origin, excellent nutritional value, and highly desirable physicochemical

* **Corresponding author Cid R. González-González:** Instituto Tecnológico Superior de Acayucan. Carretera Costera del Golfo Km 216.4 Acayucan Veracruz, Mexico. CP. 96100; Tel: +52-92424-50042; Fax: +52-92424-50042; Email: cidgonzalez@itsacayucan.edu.mx

Ali Osman (Ed.)

characteristics [2]. All these aspects positively contribute to the quality, safety, and stability of the food formulation [3].

The significance of proteins in the food industry lays on their ability to provide a wide range of functional properties, such as the formation of a gel network, emulsification, and foamability [3]. The protein physicochemical characteristics can further be enhanced through modifications. Innovative chemical, physical, and enzymatic modification approaches have recently been proposed. The protein modification can potentially lead to the creation of protein ingredients with enhanced and superior functional properties compared to the native proteins. This can, in turn, increase the industrial application of food proteins especially those from low-value sources (such as food processing by-products).

Traditionally, food proteins have been regarded as a source of essential and non-essential amino acids. However, the concept of functional foods that can reduce the risk of some chronic diseases and promote health in addition to providing nutrients has emerged in the last two decades [4]. Food proteins have now been recognized as valuable sources of bioactive peptides [5]. Both enzymatic hydrolysis and microbial fermentation have gained interest over the chemical hydrolysis for the production of peptides, due to safety and organoleptic issues associated with the latter method. Apart from producing bioactive peptides, the enzymatic hydrolysis, under controlled conditions, can improve the functional properties of the protein [6].

The objective of this chapter is to cover the recent advances in protein biotechnology. In the first section, the latest advances in the functionality of bioactive peptides released from food proteins and the impact of enzymatic hydrolysis on the organoleptic properties of proteins are reviewed. Recent research on chemical (glycation and disulfide cross-linking), physical (high-pressure processing and ultrasound treatment), and enzymatic (transglutaminase cross-linking and proteolysis) protein modifications are presented. The effects of these modifications on the protein functional properties (solubility, emulsification, foamability, and viscoelasticity) are also discussed.

ADVANCES IN FOOD PROTEIN-DERIVED BIOACTIVE PEPTIDES

The study of peptides with biological functions has gained importance in the last decades. Numerous applications have been found and investigated, such as antihypertensive, antioxidant, hypocholesterolemic, antiglycemic, and opioid like activities. The two most employed technologies to release bioactive peptides encrypted in food proteins are enzymatic hydrolysis and microbial fermentation, or a combination of both. One of the main advantages of bioactive peptides is that they come from food proteins and are, therefore, generally considered safe, as long as the enzymes or microbial agents used are also food grade and of appropriate quality [7]. The main challenges in the use of bioactive peptides are the elucidation of mechanisms of action in order to establish a cause-effect relationship, and their bioavailability and stability. The following section discusses these issues with a focus on the advances in the last decade.

Antihypertensive Activity

Angiotensin Converting Enzyme Inhibition

The antihypertensive activity of peptides has been largely investigated, mainly for the capability of particular peptide sequences to inhibit the metallopeptidase Angiotensin-I Converting Enzyme (ACE) – (EC 3.4.15.1), which plays a central role in modulating the Renin Angiotensin System (RAS) and the regulation of blood pressure (BP). The ACE catalyzes the hydrolysis of the decapeptide angiotensin I (Ang-I), releasing two amino acids at the C-terminal to produce the octapeptide angiotensin II (Ang-II). The latter upregulates blood pressure *via* vasoconstriction by targeting AT1 receptors in smooth muscle and blood vessels, and by promoting reabsorption of Na+ and water into the blood stream, which leads to an increase of blood volume and thus rising blood pressure [8]. Additionally, the ACE hydrolyzes bradykinin peptide, hindering its vasodilatory activity. Cheung *et al.* [9] designed a model of the ACE active site that allowed them to better understand the ACE inhibition–structure consisting of three subsites, S1, S1', and S2' interacting with the ACE-inhibitors (Fig. **1.1**) [9]. This led to the development of effective ACE inhibitory compounds, such as the synthetic peptide captopril, that was first used to alleviate the symptoms of hypertension [9]. In later years, food derived bioactive peptides with similar structure, *e.g.* containing proline at the C-terminal and a hydrophobic side chain amino acid at the antepenultimate position of the C-terminal, were also found to inhibit the ACE; however, they do it at a lower extent compared to synthetic modified peptides, such as captopril or lisinopril.

The most studied peptides with ACE inhibitory capabilities are Ile-Pro-Pro (IPP) and Val-Pro-Pro (VPP), derived from β-casein from bovine milk with a half maximal inhibitory concentration value (IC_{50}) of 5 and 9 μM, respectively; both have shown antihypertensive activity in spontaneous hypertensive rats (SHR) [11]. These peptides were originally released *via* fermentation by *Lactobacillus (Lb.) helveticus* and are sold under the commercial name of Calpis® in Japan. Additionally, Seppo *et al.* [12] reported a product containing the same IPP and VPP, fermented by *Lb. helveticus* LBK-16H [12]. Since then, other bioactive peptides derived from milk have been tested in animal and human models.

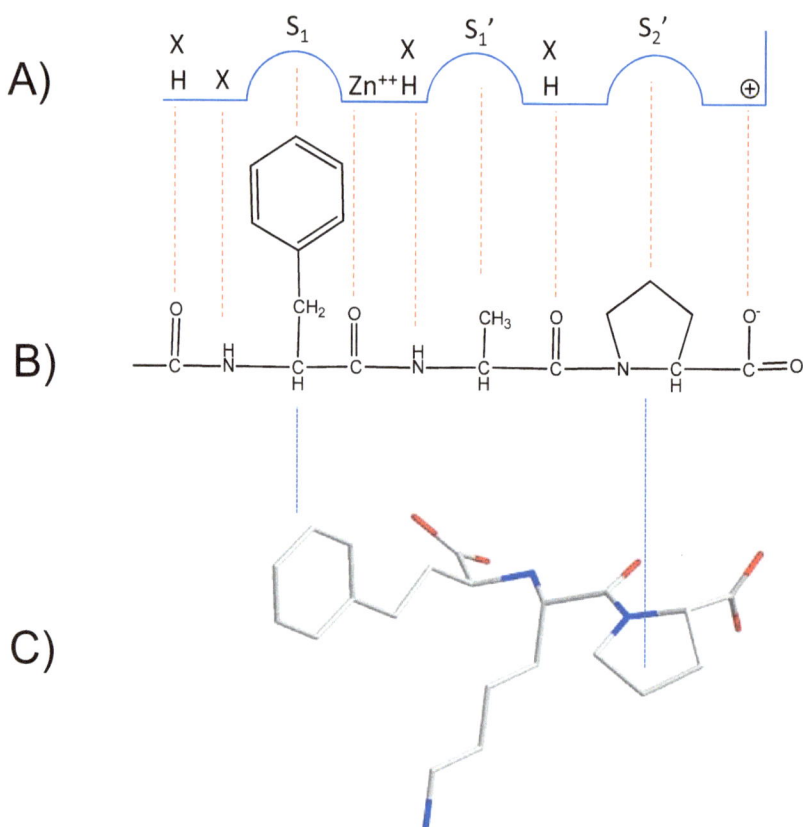

Fig. (1.1). A) Theoretical model of the ACE active site showing interactions with **B)** a venom peptide analogue. This led to the designing of the ACE-inhibitors, such as captopril and later the potent **C)** Lisinopril. This image was adapted from "The state of the ion channel research in 2004" [10].

In the last decade, a controversy regarding the actual mechanism of antihypertensive activity of these peptides has emerged. Wuezrner *et al.* [13] published a study with normotensive people where lactotripeptides (LTP) did not show ACE inhibitory activity *in vivo* in healthy subjects. However, it has been shown that LTP do not show antihypertensive activity in healthy subjects, but only in hypertensive, or even mild hypertensive people [13, 14]. Moreover, it was already reported that IPP and VPP did not show significant difference in lowering blood pressure against placebo, when LTP mixture was administered in hypertensive Dutch adults in a dosage of 4.8 ± 0.6 mg IPP and 5.4 ±0.4 mg VPP. They have also compared the effect of LTP obtained from different technologies: enzymatic hydrolysis, fermentation, and chemical synthesis, but no significant difference was found in any of the peptide formulations *vs* placebo. Additionally, Boelsma *et al.* [15] reported that IPP was the only peptide with detectable concentrations in blood after ingestion of tablets containing IPP and a mixture of MAP, LPP, and IPP, and that there was no evidence that the hypotensive activity observed could be due to ACE inhibition *in vivo* [15]. These studies are in contrast with research made by Hirota *et al.* [16], Jauhiainen *et al.* [17], Turpeinen *et al.* [18], Yoshizawa *et al.* [19], and Yoshizawa [20]; these researchers have reported antihypertensive activity in humans, mainly in systolic blood pressure (SBP) (-3.8 to -9 mmHg vs placebo) when administered in dosage ranges of 3.8 – 30 mg for IPP, and 2.4 – 22.5 mg for VPP. Interestingly, most of the studies showing considerable effect of LTP on BP are Japanese, whereas the studies performed with Caucasians in Europe have shown slight or no significant changes compared to placebo [21]. Based on the evidence of pharmacokinetics and pharmacodynamics with other antihypertensive drugs, it has been suggested that this variability in the outcome is more related to differences in diet rather than genetics [14].

Another important hypotensive peptide is Val-Tyr (VY), which has shown a strong ACE inhibitory activity *in vitro* (IC_{50} = 7.1µM comparable to that of IPP and VPP) and could be obtained from milk, sardine muscle, and sesame seed proteins [22 - 24]. It has hypotensive activity in SHR [25], and it is well absorbed into blood stream by normotensive and mild hypertensive people [26, 27]. Moreover, it has been observed that VY accumulates in higher concentrations in tissue than in blood stream; particularly in kidneys, lungs, and the abdominal aorta [28]. This observation coincides with the report of long term oral administration of peptides derived from jellyfish collagen with ACE inhibitory activity, where the concentration of Ang-II was reduced mainly in kidneys; meanwhile, the concentration in plasma was unaltered [29]. Therefore, it is important to consider that the presence of peptides in plasma after ingestion does not necessarily reflect a high bioactivity. Furthermore, using a novel model of organ bath to study *ex vivo* effect of VY on aortic rings, Vercruysse *et al.* [30] compared five different mechanisms [30], where the inhibition of vasoconstriction correlated to an Ang-I accumulation in tissue. This led to the conclusion that the mechanism of hypotensive activity for VY was ACE inhibition, which was similar to the result obtained by Kawasaki *et al.* [22], who observed an increase of Ang-I, and a decrease of Ang-II and aldosterone after an oral treatment of VY in mild-hypertensive people [22]. These two studies demonstrate that ACE inhibitory activity is the likely mechanism of hypotensive effect of VY *ex vivo* and *in vivo*. Interestingly, it was found that when the ACE inhibitor drug captopril was administered alongside VY in SHR, the hypotensive activity of the latter was hindered for an apparent competition on the membrane transport pathway; thus, suggesting that hypertensive subjects treated with ACE inhibitor drugs should avoid the consumption of small peptides [31]. However, in a recent experiment made with LTP, it was observed that the hindering effect on the hypotensive activity did not happen when the peptides were administrated 29 days after a treatment with the ACE inhibitor enalapril, indicating that the combination of BP with an ongoing treatment of ACE inhibitor does not alter its antihypertensive effect [32]. Thus, this statement deserves more research to validate the potential competitive interaction of bioactive sequences with ACE inhibitor drugs.

Renin Inhibition

An alternative studied mechanism related to the RAS is the inhibition of renin. Renin breaks down the angiotensinogen into Ang-I. He *et al.* [33] reported that the peptide Gly-His-Ser (GHS), obtained from rapeseed *via* enzymatic digestion with pepsin-pancreatin, was able to inhibit renin besides ACE activities *in vitro*, correlating with hypotensive activity in SHR (-17.29 ± 2.47 mmHg) after 6 h of treatment with a dosage of 30 mg/kg of synthetic peptide [33]. More recently, the peptide sequences Pro-Ser-Leu-Pro-Ala (PSLPA), Trp-Tyr-Thr (WYT), Ser-Val-Tyr-Thr (SVYT), and Ile-Pro-Ala-Gly-Val (IPAGV), contained in hemp seed proteins have been reported to exert hypotensive activity by ACE and renin inhibition. Among these, PSLPA and IPAGV showed the highest lowering of systolic blood pressure (SBP) (-40 and -36 mmHg, respectively) in SHR after 4 h of ingestion [34, 35]. A predicting and validated modelling study showed that small peptides formed by low molecular weight amino acids with hydrophobic chain at the N terminus and bulky side chains at C-terminus are desired for renin inhibition. Ile-Trp (IW) was the most bioactive sequence in this predictive model [36].

Other Hypotensive Mechanisms

Other antihypertensive mechanisms reported are the inhibition of endothelin-converting enzyme (ECE), the modulation of endothelin-1 (ET-1), the blocking of calcium channel, and the Ang-II receptor (AT1), as well as the upregulation of nitric oxide (NO) production [37]. The endothelin system (ES) has been less assessed for food derived peptides than the RAS. It represents, however, another key strategy to lower high blood pressure by bioactive peptides, provided that it is comprised by three peptide ligands and two activating peptidases, and plays a key role in the balance and development of disease in different organs, such as the kidneys, the lungs, and the heart [38]. For instance, it is reported that the β-lactoglobulin derived peptide Ala-Leu-Pro-Met-His-Ile-Arg (ALPMHIR) inhibits the release of ET-1 [39]. Moreover, the hypotensive peptides Gly-Ile-Leu- Arg-Pro-Tyr (GILRPY) and Arg-Glu-Pro-Tyr-Phe-Gly-Tyr (REPYFGY), derived from lactoferrin, showed ECE inhibitory activities *in vitro*, demonstrating that these sequences are able to inhibit vasoconstriction *ex vivo*, independently of ACE inhibition [40]. The same research group reported that the peptides Arg-Pro-Tyr-Leu (RPYL), Leu-Ile-Trp-Lys-Leu (LIWKL), and Arg-Arg-Trp-Gln-Trp-Arg (RRWQWR), also contained in lactoferrin, showed inhibition of vasoconstriction induced by angiotensin II, suggesting that these peptides exert a hypotensive effect by blocking the angiotensin AT1 receptors alongside ACE inhibitory activity [41]. Regarding the blocking of calcium channel, peptides such as His-Arg-Trp (HRW) showed a reduction in Ca^{2+} concentration in smooth muscle cells, which may indicate a potent suppressor of extracellular Ca^{2+} influx as a mechanism of hypotensive activity [42]. Wang *et al.* [44] showed that the dipeptide Trp-His (WH), derived from sardine muscle, is vaso-protective through the inhibition of voltage dependent *L*-type Ca2+ channel. This is also beneficial in order to reduce the inflammatory response in the distal colon [43, 44]. It is therefore important to pay attention to other mechanisms than ACE inhibition that may exert antihypertensive activity in humans, considering that until 2017 most of the studies on food derived antihypertensive peptides have been based on ACE inhibitory activity.

Endothelial Function Aid

There are reports on the beneficial effect of peptides on the endothelial function. Endothelial dysfunction conveys a diminution of the NO bioavailability, thus lowering the relaxing capability of the endothelium; this has been accepted as a key risk factor for cardiovascular disease [45]. One of the responses observed in the ingestion of bioactive peptides has been the decrease in the augmentation index (AI). Pal and Ellis [46] showed that whey proteins may reduce AI after 12 weeks of treatment in a group of overweight and obese individuals compared to another group fed with casein fraction and a third group fed with glucose as control suggesting that whey proteins might be the main factor for improving cardiovascular function of milk fractions, and that this is notorious when the consumption is chronic, rather than acute [46]. Moreover, a Finnish group reported that the consumption of LTP reduces the AI in mild hypertensive subjects [47, 48].

Ballard *et al.* [49] reported a study with a commercial formula of whey derived extract where they have observed a vascular dilation but no peptides were detected in the blood stream, signifying that the dilation effect is independent of the presence of bioactive peptides in the blood stream, and was not due to ACE inhibition *in vivo* [49]. Furthermore, Marcone *et al.* [50] demonstrated that hydrolysates of β-casein containing bioactive peptides showed inhibitory activity against the production of inflammatory proteins, such as MPC-1 and IL-8. They also demonstrated that β-casein hydrolysate inhibited, by more than 50%, the adhesion of monocytes to TNF-α activated human aortic endothelial cells at a concentration of 300 µg/mL, indicating the reduction of an important risk factor for developing atherosclerosis [50]. The sequence of the peptides responsible for this activity is to be reported. This endothelial protective effect is widely related to the antioxidant activity discussed below.

Antioxidant Activity

Oxidative stress caused by the presence of reactive oxygen species (ROS) may lead to dysfunction of endothelial physiology and to inflammatory processes and cell damage. An increase of ROS under certain conditions leads to an uncoupling of the endothelial nitric oxide synthase (eNOS) [51]. Scavenging of ROS is a route to diminish endothelial damage; this property can be found in peptides behaving as antioxidant agents [45]. Several studies have shown antioxidant activity of food-derived peptides *in vitro* and *ex vivo*. Power *et al.* [52] have made an excellent review on the antioxidant properties of peptides derived from milk proteins, analyzing different approaches to evaluate the antioxidant activity *in vitro*, *ex vivo,* and *in vivo* [52]. The main technique to study the antioxidant activity *in vitro* is the radical scavenging capability with the use of radicals, such as 2,2-Diphenil-1-picrylhydrazyl (DPPH), and 2,2'-azinobis (3-ethylbenzo-thiazoline-6-sulphonic acid (ABTS•+). The results are expressed in µM of Trolox equivalent antioxidant activity per mg of peptide (TEAC) or vitamin C antioxidant activity (VCEAC); both are antioxidant compounds with known radical scavenging capacity. Oxygen radical absorbance capacity assay (ORAC) is another assay to measure the protective capacity of a tested compound to inhibit the decomposition of 2,2'- azobis-2-methyl-propanimidamide dihydrochloride (AAPH) into peroxyl radicals.

Examples of potent antioxidative peptides (Table **1.1**) found in milk proteins are Val-Leu-Pro-Val-Pro-Gln-Lys (VLPVPQK), obtained from cow milk β-casein and, more recently, from buffalo milk β-casein [53, 54]. Trp-Tyr-Ser-Leu-Ala-Met-Ala-Ala-Ser-Asp-Ile (WYSLAMAASDI), Leu-Gln-Lys-Trp (LQLW), and Leu-Asp-Thr-Asp-Tyr-Lys-Lys (LDTDYKK), isolated from β-lactoglobulin A after enzymatic digestion, have been also identified for their antioxidant activity [55, 56].

Table 1.1. Examples of bioactive peptides, their source, and their mechanism of action.

Sequence	Protein source	Activity	Mechanism	References
IPA	β -lactoglobulin	Antiglycemic	DPP-IV inhibitor	[64]
RVPSLM	Egg white protein	Antiglycemic	α-glucosidase inhibitor	[65]

(Table 1.1) cont.....

Sequence	Protein source	Activity	Mechanism	References
IRW, IQW	Ovotransferrin	Anti-inflammatory	Inhibits TNF upregulation of ICAM-I	[66]
WH	Sardine Muscle	Anti-inflammatory, anti-atherosclerotic	Blockade of L-type Ca^{2+} channels	[43, 44]
WVYY, PSLPA	Hemp seed protein	Antioxidant	Radical scavenging, metal chelation	[33]
WYSLAMAASDI	β -lactoglobulin	Antioxidant	Radical scavenging	[55]
VLPVPQK	β-casein (Bovine & Buffalo)	Antioxidant	Radical scavenging	[53, 54]
IIAEK	β-lactoglobulin	Hypocholesterolemic	Inhibition of micellar solubility of cholesterol	[63]
FVVNATSN	Soy protein	Hypocholesterolemic	stimulate LDL-R transcription in human hepatocytes	[67]
VY	β-casein, sardine muscle	Hypotensive	ACE inhibition, Calcium channel blocker	[22 - 24]
PSLPA, IPAGV	Hemp seed protein	Hypotensive	ACE and renin inhibition	[34, 35]
IRLIIVLMPILMA	Red seaweed protein	Hypotensive	Renin inhibition	[68]
GHS	Rapeseed protein	Hypotensive	Renin inhibition	[33]
ALPMHIR	β-lactoglobulin	Hypotensive	Modulation of ET-1 synthesis	[39]
GILRPY, REPYFGY	lactoferrin	Hypotensive	ECE inhibition	[40]
RPYL, LIWKL	lactoferrin	Hypotensive	AT1 receptor blockage	[41]
HRW	Synthetic	Hypotensive	Suppressor of extracellular Ca^{2+} influx	[42]
MMP-2200 (lactomorphin)	β -lactoglobulin	Opioid, analgesic	μ and ∂ receptor agonism	[69]
HIRL	β-lactoglobulin	Opioid, anxiolytic	Antinociception mediated by NT2 and D1 receptors	[70, 71]

Other sources of peptides have been studied, such as egg and fish proteins. For example, Yousr and Howell [57] have identified Trp-Tyr-Gly-Pro-Asp (WYGPD), Lys-Gly-Leu-Trp-Glu (KGLWE), and Lys-Leu-Ser-Asp-Trp (KLSDW) as antioxidant peptides from egg yolk protein, showing radical scavenging and lipid oxidation inhibitory activity [57]. The antioxidative peptides Trp-Asn-Ile-Pro (WNIP) and Gly-Trp-Asn-Ile (GWNI) were obtained *via* enzymatic hydrolysis of ovotransferrin by thermolysin. Given that these two peptides showed the highest activity of the studied fractions, the antioxidant activity was attributed to the structure of Trp-Asn-Ile (WNI) [58]. Val-Cys-Ser-Val (VCSV) and Cys-Ala-Ala-Pro (CAAP) were discovered in flounder fish muscle (*Paralichthys olivaceus*); they have shown no toxicity in concentrations up to 200 µg/mL and scavenging activity in an *ex vivo* study using Vero cell lines [59]. Lunasin is a well characterized 43-amino acid peptide, with antioxidant and antitumor activity found in soy, amaranth, barley and wheat [60 - 62]. Its antioxidant capacity has been shown in Caco-2 cell lines, indicating a protective effect against damage of enterocytes caused by H_2O_2 and tert-butyl hydroperoxide (*t*-BOOH) oxidative stress, which indicates a potential protective effect of lunasin in the intestine [62]. One of the main limitations of lunasin is that it can only be administrated *via* parenteral as many other macropeptides that are likely hydrolyzed further into smaller inactive peptides when digested. More research is needed to investigate its toxicity and its correlation with radical scavenging activity *in vivo*.

Hypocholesterolemic and Antiglycemic Activities

The reduction of cholesterol has been observed after the consumption of peptides derived from casein, whey, and soy protein. Nagaoka *et al.* [63] reported a hypocholesterolemic peptide derived from β-lactoglobulin; Ile-Ile-Ala-Glu-Tyr (IIAEK) showed greater response to decrease low density lipoprotein in blood stream than β-sitosterol, a pharmaceutical used to control hypercholesterolemia [63].

In 2006, Pins and Keenan reported that a whey hydrolysate rich in bioactive peptides significantly lowered low density lipoprotein (LDL)-cholesterol in mild hypertensive subjects after 6 weeks of treatment [72]. In another study, Pal *et al.* [73] reported a diminution of total-cholesterol and LDL-cholesterol levels by 11% and 9.6%, respectively, after 12 weeks of treatment with whey protein isolate feed [73]. Moreover, Cho *et al.* [67] observed that soy protein hydrolysates enhance the transcription of LDL, and the peptide Phe-Val-Val-Asn-Ala-Thr-Ser-Asn (FVVNATSN) was identified in the most bioactive fraction of the soy protein hydrolysate [74]. Although these results indicate that food proteins and derived peptides may aid in reducing cholesterol levels in individuals with hypercholesterolemia, the mechanisms still remain unclear. The mechanisms that have been suggested so far are: 1) the effect of whey proteins on the synthesis of cholesterol in the liver [75], 2) the inhibitory activity of β-lactoglobulin, for instance, on the absorption of cholesterol in the intestine [76], 3) the inhibition of the expression of genes related to intestinal fatty acid and cholesterol absorption, and 4) peptides may reduce the solubility of micellar cholesterol, thus, easing its excretion [63].

Food derived peptides may also help to control type II diabetes. Soy, casein, and whey proteins induce insulin secretion after ingestion leading to a reduction of postprandial glucose [77]. The mechanisms reported are the inhibition of dipeptidyl-peptidase IV (DPP-IV) and α-glucosidase, which are enzymes involved in the glucose metabolism and digestion. DPP-IV inhibitors have been isolated from caseins and whey proteins: Ile-Pro-Ile (IPI: IC_{50} = 3.4 μM), Ile-Pro-Ile-Gln-Tyr (IPIQY: IC_{50} = 35.2 μM), Trp-Val (WV: IC_{50} = 65 μM), and Ile-Pro-Ala (IPA: IC_{50} = 49 μM) [64, 78, 79], indicating that proline at the penultimate position of the N-terminal is determinant for the inhibitory activity provided that the peptidase DPP-IV cleaves the dipeptide at the N-terminal from peptides with proline or alanine in the second position [80]. Moreover, the DPP-IV inhibitor peptide Leu-Pro-Gln-Asn-Ile-Pro-Pro-Leu (LPQNIPPL) was isolated from Gouda cheese; it showed postprandial glucose reduction in rats compared to placebo ($P<0.02$) [81]. Peptides that inhibit the α-glucosidase have been isolated from white egg proteins [65]. Also, hydrolysates produced during milk fermentation by probiotic strains have shown α-glucosidase inhibition [82]; but there are only few studies reporting this bioactivity, which represents an opportunity for more research to employ food-derived peptides in helping to control type 2 diabetes.

Opioid Activity

Peptides derived from food proteins have also shown opioid agonistic and antagonistic activities, such as modulation of social behaviour, constipation, and changes in the endocrinal system of the subjects [83]. The opioid activity of peptides is related to their structure similarity to those endogenous and exogenous ligands that possess affinity to opioid receptors, such as μ-(morphine), δ-(enkelaphine), and κ-(dinorphine) types [84]. Different opioid peptides derived from β-casein, such as Tyr-Pro-Phe-Pro-Gly (YPFPG: β-casomorphin-5) and Tyr-Pro-Phe-Pro-Gly-Pro-Ile (YPFPGPI: β-casomorphin-7), have been described in the literature as μ-type ligands, and are the most studied peptides with opioid activity, that have shown to exhibit anxiolytic and analgesic activity [83].

Glycosylated peptides derived from whey proteins have been studied for their potential use as analgesic. Animal models have shown social modulation behaviour and analgesic activity. For instance, lactomorphin (MMP-2200) showed benefits in reducing hyperkinesia in rat models of Parkinson's disease, where glycosylation of the peptide does not increase binding affinity for the receptors but instead facilitates the passage through the blood brain barrier and enhances its stability [69]. On the other hand, β-lactotensin (His-Ile-Arg-Leu (HIRL)) has been isolated from β-lactoglobulin, and has shown anxiolytic activity in mice, acting as an agonist of neurotensin NTS (2) receptor type [85]. No human trials with these peptides have been reported to date. There is an opportunity for more research on the opioid like activity of peptides from other food protein sources.

Future Perspectives in Food Protein Derived Peptides

Although several studies have been carried out showing bioactivity, such as lowering the blood pressure, enhancement of the endothelial function, antioxidant, opioid like, and anti-inflammatory activities, the cause-effect relationship is not yet well established in humans. Regarding the antihypertensive activity of peptides, the last trend showed the openness to elucidate mechanisms of hypotension other than ACE inhibition, given the poor evidence of ACE inhibitory activity *in vivo* and the low correlation of responses between *in vitro* and *in vivo* studies. Nevertheless, ACE inhibition should not be discarded as a possible mechanism, but other mechanisms should be considered as well. Provided the amount of data for these studies, and that most of them did not involve human subjects, there is a need for more, well-conducted, and properly-powered trials in order to establish the cause-effect relationship and to elucidate the mechanisms and the factors that influence the antihypertensive activity of food derived peptides. Potential benefits on the endothelial function deserve special attention for further work given the demonstrated antioxidant and anti-inflammatory capabilities of peptides.

ORGANOLEPTIC PROPERTIES OF FOOD PROTEIN HYDROLYSATES

One of the main methods for modifying protein organoleptic properties is proteolysis. Proteolysis, defined as the hydrolysis of proteins into smaller peptides or amino acids by breaking the peptide bond, could be carried out by physical, chemical, or biological processes. Proteolysis is one of the main occurring processes that brings about important changes in the food matrix during food processing, such as ripening, pasteurizing, or cooking. It has been widely recognized as a contributor to the development of taste and flavor through the production of low molecular weight peptides and amino acids [86].

Peptides comprising glutamine, asparagine, glutamic acid, and aspartic acid as well as free amino acids, like glutamic and aspartic acids, have been identified to possess a specific savory taste referred to as umami [87, 88]. According to the Japanese concept, umami means savory or delicious flavor (both taste and aroma) similar to that produced by glutamic or pyroglutamic acids [89].

Today, protein-based seasonings are commercially available. They are usually manufactured through an acid hydrolysis at elevated temperatures for different time periods up to 24 hours [86]. In addition to producing free amino acids and small peptides with the desired savory properties, the acid hydrolysis also generates a variety of carcinogenic chemical agents, such as mono and dichloropropanols and monochloropropanediols [86]. For this reason, the enzymatic hydrolysis of food proteins has been regarded as an alternative mild way to produce savoury peptides.

A large number of peptides obtained from the enzymatic hydrolysis of plant and animal proteins have been identified and characterized as possessing umami taste [90]. For example, Sonklin *et al.* [86] reported that the hydrolysis of mung bean meal protein isolate with 18% bromelain for 3 hours produced a hydrolysate with a combination of sensory characteristics described as bouillon, salty, sour taste, and umami [86]. The enzymatic hydrolysis also released several volatile compounds including benzaldehyde, 2-pentylfuran, and furfural. Koo *et al.* [89] found that hydrolyzing wheat gluten with alcalase for 24 hours decreased the bitterness and increased the umami taste as well as the overall acceptability of the hydrolysate [89]. Bagnasco *et al.* [91] prepared savoury hydrolysates from rice by-products using umamizyme and flavourzyme [91]. Their results indicated that both enzymes liberated peptides with intense umami and slightly bitter taste. In a comparative study, Su *et al.* [92] determined the physicochemical and sensory properties of defatted peanut meal hydrolysate prepared with crude protease extract from *Aspergillus oryzae* and three commercial proteolytic enzymes (alcalase, protamex and papain). Their findings showed that the crude protease extract from *Aspergillus oryzae* produced hydrolysates with better taste characteristics compared to those obtained from the commercial enzymes. They further separated the fractions with umami taste and identified two low molecular weight peptides (Ser-Ser-Arg-Asn-Glu-Gln-Ser-Arg (SSRNEQSR) and Glu-Gly-Ser-Glu-Ala-Pro-Asp-Gly-Ser-Ser-Arg (EGSEAPDGSSR)) as the active compounds [92].

Marine by-products have recently been investigated as natural sources for taste- and flavor-enhancing ingredients. In this respect, Cho and Kim [93] hydrolyzed sandy beach clam meat with a mixed protease blend comprising alcalase and flavourzyme [93]. The shellfish hydrolysate obtained under the optimum conditions (an enzyme concentration of 1%, a temperature of 54.7 °C, a pH of 5.9 and a hydrolysis period of 45 hours) had a large number of compounds (such as glutamic acid, lysine, glycine, adenosine diphosphate, adenosine monophosphate, and inosine) with taste effects. Similarly, Guo *et al.* [94] hydrolyzed Oriental shrimp (*Penaeus chinensis*) by-products using dispase. Under the optimum conditions (an enzyme concentration of 2%, a temperature of 57 °C, a pH of 6.5, and a hydrolysis period of 3 hours), the prepared hydrolysate possessed an intense shrimp flavor [94]. Using a different marine by-product, Laohakunjit *et al.* [95] studied the sensory characteristics of bromelain-derived protein hydrolysate from seaweed (*Gracilaria sp.*) by-products [95]. They reported that the seaweed protein hydrolysate, produced under optimal conditions (an enzyme concentration of 10%, a temperature of 50 °C, a pH of 6.0, and a hydrolysis period of 3 hours), elicited an umami taste and a seaweed odour. These three studies suggested that hydrolysates from marine by-products (*i.e.* clam meat, shrimp and seaweed) can be commercially marketed as natural ingredients for seasoning or for enhancing the flavor.

Several publications reported that the enzymatic hydrolysis of food proteins is an efficient way to release peptides with savory properties. However, it is important to note that bitterness is still a challenge in this area because it hinders their incorporation as functional ingredients in food products. The bitter taste intensity is correlated with the peptide chemical structure. The presence of hydrophobic peptides is also responsible for the bitterness of the protein hydrolysates. Despite its major contribution in the loss of valuable amino acids/peptides and lower recovery yields, activated carbon has long been used to reduce the bitterness of hydrolyzed food proteins through the removal of hydrophobic peptides [96]. One of the recent approaches to manage the bitterness include reducing the formation of bitter peptides and favoring the production of savoury peptides. This could be accomplished through an optimized enzymatic hydrolysis that achieves the desired organoleptic characteristics without compromising the taste. Factors that play a role in increasing the bitterness intensity (*i.e.* type of enzyme and the extent of the hydrolysis) have to be carefully controlled while the release of taste-active peptides should be targeted. It is well known that low molecular weight hydrophilic peptides are generally responsible for the flavour and taste of the entire protein hydrolysates. For instance, small oligopeptides with molecular masses lower than 3,000 Da have been identified as flavour/taste substances [97]. In addition, peptides rich in specific amino residues such as Gln, Asn, Glu, or Asp as well as the free amino acids Glu and Asp have a highly desirable sensory profile. Therefore, the use of proteases that are capable of releasing these particular peptides/amino acids is a promising approach for maximizing the value of food protein hydrolysates.

FUNCTIONAL PROPERTIES OF FOOD PROTEINS

Proteins play a major role in the formulation and stability of processed food products. In addition to their nutritional significance, food proteins possess specific characteristics which are usually referred to as functional properties. Protein functional properties are conventionally defined as "the physical and chemical aspects that influence the behavior of proteins in food systems during the entire production chain (*i.e.* processing, distribution, retailing, preparation, and consumption)" [98].

The functional properties of food proteins can be categorized into three major groups:

1. Hydration properties (such as solubility),
2. Surface properties (such as foaming, and emulsifying abilities),
3. Rheological properties (such as viscoelasticity, and gelation).

Both intrinsic and extrinsic parameters influence protein functional properties. Among the intrinsic factors, the molecular structure, the molecular size, the amino acid composition, and the amino acid sequence have been shown to affect the physical function of the proteins [98]. Ionic strength, pH, and temperature are among the most common extrinsic factors that alter the protein functionalities.

Solubility

By definition, the solubility of a protein is the equilibrium thermodynamic parameter that characterizes the solid-liquid interactions. In a practical sense, protein solubility, at constant temperature and pressure, represents the concentration of a protein in equilibrium with a solid phase in a crystalline or in an amorphous form [99].

The solubility is considered to be one of the most important physicochemical properties of food proteins. The determination of protein solubility, under various conditions, represents a useful tool in food biotechnology. For instance, the extraction, purification, and separation of proteins are mainly based on altering the solubility by varying the ionic strength and/or the pH of the medium. Increasing the ionic strength of the medium (generally by increasing the concentration of salts) results in the precipitation of proteins (*i.e.* salting-out effect) due to the preferential binding of ions and water to solutes [100]. Changing the pH of the medium modifies both the positive and negative charges of the hydrophobic and hydrophilic residues on the protein surface and, thus, their ability to interact with water. At a specific pH value, referred to as the isoelectric pH, proteins carry no net electrical charge which translates to a minimal hydration and a subsequent precipitation [101].

The isoelectric solubilization/precipitation (ISP) process, or more commonly known as the pH-shifting procedure, is an extraction method based on protein solubility. The ISP process (Fig. **1.2**), which has recently been successfully implemented on a large industrial scale [102, 103], is a selective pH-induced solubilization technique in which target proteins are isolated from lipids and other food components after solubilization at either acidic or alkaline pH values followed by precipitation at their isoelectric pH [104, 105]. The isoelectric solubilization/precipitation process represents an innovative processing technique to recover functional and nutritional proteins from food by-products [103].

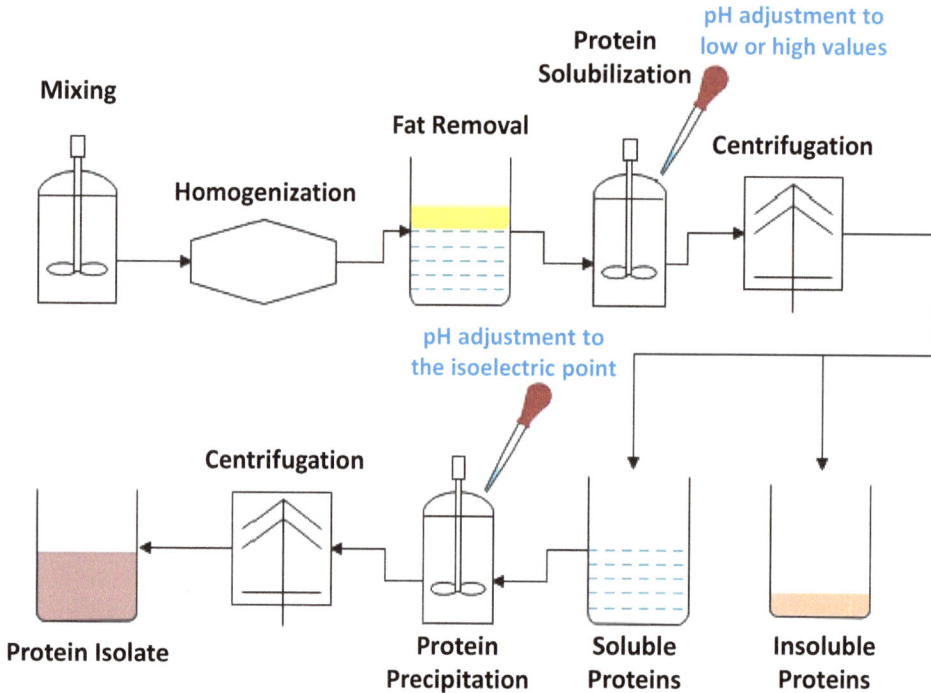

Fig. (1.2). Schematic representation of the industrial Isoelectric Solubilization/Precipitation (ISP) process for the extraction of protein isolate from meat processing by-products (Adapted from Khiari *et al.* [103]).

Emulsifying Properties

Emulsified foods constitute a wide category of products in the market [106, 107]. Technically, an emulsion is a system of two immiscible liquid phases, one of which is dispersed in the form of micro spherical droplets [106]. When the oil droplets are dispersed in the aqueous phase, the emulsion is termed oil-in-water (O/W). In contrast, water-in-oil (W/O) emulsion is characterized by water droplets distributed in the oily phase. The most commonly known examples for O/W emulsions are milk and cream while those for W/O emulsions are margarine and butter [107]. The formation of small dispersed droplets increases the interfacial area between the two liquids which, in turn, leads to a large interfacial positive energy. From a thermodynamic perspective, emulsions are considered to be unstable systems. Physical destabilization mechanisms of emulsions include oil droplet size variation processes (*i.e.* flocculation and coalescence), and particle migration phenomena (*i.e.* sedimentation and creaming) [108].

Synthetic chemical emulsifiers, such as polysorbates, have been widely used in the food and pharmaceutical industries to stabilize emulsions. Although synthetic emulsifiers are economical and effective, the growing consumer tendency to buy food products with "natural" ingredients created new challenges for the entire food sector. Some food proteins can act as emulsifiers due to their ability to reduce the interfacial tension between the oil and the water phases leading to the stabilization of the emulsion. This is mainly attributed to the formation of a protein protective coating around the oil droplets that prevents the coalescence phenomenon [109].

Functional and bioactive foods have long been believed to play a beneficial role in preventing some diseases and promoting health. Consumer interest in functional foods has resulted in an impressive market growth for this category of food products. However, recent knowledge indicated that several functional food ingredients cannot accomplish their health benefits due to their low bioavailability [110]. To overcome this issue, the food industry turned into nano-technology as a potential solution. In this regard, nano-emulsions have recently been investigated as a tool to enhance the delivery of bioactive lipophilic food components. Nano-emulsions are a new type of emulsions characterized by extremely small droplet sizes (diameter less than 100 nm) while most food emulsions today, consist of a dispersed phase with spherical droplets varying between 100 nm and 100 μm in diameter [106, 111]. The significance of nano-emulsions lays on their ability to improve the solubility, transport, and bioavailability of bioactive food ingredients [111]. These specific characteristics make food protein-based nano-emulsions promising encapsulating and carrier systems for bioactive food components [107]. Although several synthetic emulsifiers have been studied as nano-emulsifying agents, issues related to their safety and toxicity limit their potential industrial applications [111].

Various food proteins have recently been shown to act as emulsifiers in nano-emulsions with enhanced functions, such as the delivery of lipophilic bioactive compounds. For example, He *et al.* [112] investigated the safety and emulsifying properties of three food proteins (soybean protein isolate, whey protein isolate, and β-lactoglobulin) in nano-emulsions [112]. Their findings indicated that food protein nano-emulsions were more stable and biocompatible than traditional emulsifiers. When comparing the three proteins, β-lactoglobulin possessed better emulsifying capacity and biocompatibility than both soybean and whey protein isolates. Whey proteins have also been the focus of recent studies as emulsifiers in nano-emulsions. For instance, Relkin *et al.* [113] and Lee and McClements [114] prepared whey protein-based nanoemulsions, which have the potential to be useful delivery systems for functional food components [113, 114]. Although food proteins have attracted considerable interest as safe, natural, and low cost nano-emulsifiers, the thermodynamic instability of food protein-based nano-emulsions limits their practical application.

Foaming Properties

Foams are two-phase colloidal systems which contain a continuous liquid phase and a gas phase dispersed as bubbles or air cells. From a thermodynamic point of view, foams are unstable systems due to the large surface energy at the air-water interface and the significant differences between the densities of the two phases. The mechanisms of foam destabilization include drainage, coalescence, and coarsening. Gravity is the main cause of the drainage of the liquid. Coalescence refers to the breakage of films separating two bubbles leading to their merging into one bubble. Coarsening refers to the diffusion of gas from small bubbles to larger ones [115].

Due to their hydrophobicity and ability to rearrange and adsorb at the air-water interface, food proteins are considered to be excellent foam stabilizers in the food industry [116]. It is believed that proteins in foams spontaneously adsorb to the air/water interface. This is achieved through the unfolding of the proteins and the establishment of intermolecular interactions with surrounding proteins. This process results in the formation of an interfacial film of proteins [117]. From a practical point of view, proteins are less effective than synthetic surfactants in reducing the air/water interfacial tension. Although food proteins have the advantage of being natural and generally recognized as safe (GRAS), their use as surfactants in food systems can be a challenging task due to the presence of certain food components, such as salts, sugars, and lipids, which affect the foaming properties by altering both the physicochemical properties of the protein and the viscosity of the continuous phase [118].

Some examples of food proteins that have successfully been used as foaming agents include milk protein (whey protein, β-lactoglobulin), meat protein (gelatin), and egg white protein (albumen). β-lactoglobulin, which is milk-derived protein, has been shown to possess high foaming properties (foamability and foam stability), which can further be enhanced by pre-heating the protein [117]. Heating β-lactoglobulin leads to the formation of aggregates that produce stable foams. The enhanced foaming properties have been correlated to the increase in surface hydrophobicity which, in turn, stabilizes the foam through rapid formation of a viscoelastic film [117]. In a comparative study, Abirached *et al.* [119] assessed the foaming properties of soy and whey protein isolates and concluded that the foaming capacity and the stability of foams prepared with whey protein isolates (WPI) were better than those formulated with soy protein isolates (SPI) [119]. Meat proteins have also been evaluated as foaming agents. For example, Salvador *et al*. [120] analyzed the foaming properties of fresh and spray-dried porcine red cell protein concentrates [120]. The aim of their study was to determine the useful function of this protein as a functional food ingredient. Their finding indicated that both pH and the drying treatment affected the foaming property. Spray-dried porcine red cell protein concentrates were found to possess greater foaming capacity at neutral and acidic pH values. Another meat protein of great interest is gelatin. Gelatin (which is a heat-denatured form of collagen) is known to comprise both hydrophilic and hydrophobic portions. Due to this amphiphilic character, gelatin is regarded as an excellent protein surfactant. Until today and due to their unique functional properties, egg white proteins are considered the most widely used protein ingredients as foaming agents. Ovalbumin is the main protein constituent responsible for egg white functionality [121]. Their excellent foaming properties are thought to be the result of the interaction between the various constituent proteins [122].

Finding a food grade ingredient that can stabilize food foams is a challenge for the food industry. Protein fibrillization has recently emerged as a research area targeting the improvement of protein foaming properties. In this regard, Oboroceanu *et al*. [123] investigated the effect of protein fibrillization on their foaming properties. Their results were promising and showed that foams made from whey protein fibrils possessed significantly better foaming properties (foam formation and stability) compared to the non-fibrous whey proteins [123]. A more recent study by Wan *et al*. [124] reported that when heating soy glycinin (11S) at pH 2 for 20 h at 85 °C, a mixture consisting of long semi-flexible 11S fibrils and small peptides is formed. The study investigated the property of this mixture at different pH values [124]. Their results indicated that the fibril mixture at pH 5 and 7 provided better foam stability and can potentially represent a unique protein material for preparing stable foams. It seems, therefore, that the fibrillization of globular proteins can potentially improve their foaming properties.

Viscoelastic Properties and Gelation

"The study of the deformation and flow of matter" is the most commonly accepted definition of rheology. This interdisciplinary field of science aims at determining and understanding: 1- The geometrical changes induced by stress/forces and 2- The role of molecular structure on the rheological properties [125].

In rheology, ideal elastic solid materials deform under stress but return to their original shape when the stress is removed, while ideal viscous liquids flow and never recover to their initial state. Viscoelastic materials, on the other hand, are materials that exhibit both viscous and elastic properties under deformation.

Rheology has found many applications in food science and technology. For example, during the formulation and processing of food products fluid and semi-fluid foods are pumped and mixed. The determination and measurement of the viscoelastic properties allow the optimization of these unit operations.

The study of protein gelation is another key application of rheology in food industry. Protein gelation is an important functional property, which refers to the formation of a three-dimensional elastic network that entraps the liquid component. Protein gelation takes part in the processing and overall acceptability of several food products, such as meat, surimi, dairy, egg, and tofu. Protein gels are classified depending on their rheological and microstructural characteristics into three major categories: fine stranded, mixed, or particulate [3]. Although the gelation mechanisms of proteins are still not fully understood, most of the research studies indicate that proteins undergo denaturation and unfolding prior to rearrangement to an ordered state through protein-protein interactions and aggregation.

The complex sol-gel transition in heat-induced gelation involves a series of reactions starting with protein denaturation (loss of native structure with partial unfolding), followed by dissociation-association prior to aggregation [126]. The thermal-induced aggregation of globular proteins (such as soy and whey) can results in either random or ordered gels. It has been reported that the gelation of β-lactoglobulin is random and is characterized by a heterogeneous particulate network structure, while lysozyme undergoes a linear or stranded aggregation. Ovalbumin, on the other hand, can form either type of network structures [127, 128].

Protein gels can also be irreversible (*e.g.* egg albumen) or thermoreversible (*e.g.* gelatin). In the case of egg proteins, the formation of irreversible gel is ascribed to the formation of covalent bonds, disulphide linkages, as well as to hydrophobic interactions [122]. The formation of disulphide linkages is key for the stabilization of the egg protein gel matrix. Gelatin gels, on the other hand, represent a typical example for thermoreversible protein gels. The thermo-reversible gelling property of gelatin is the most important attribute for its successful usage in food industry as a multifunctional ingredient. Gelatin is a partially denatured form of collagen, a fibrous protein widely abundant in animal tissues. Collagens have several types (commonly termed type I, II, III, V, *etc.*), with type I being the major type which is abundant in connective tissues, bones and skin. Collagen is also distinctive by the presence of large amount of glycine and the two amino acids (proline and hydroxyproline) and the lack of tryptophan. Industrially, type I gelatin is extracted from collagen (type I) after an acid or an alkaline treatment, that partially breaks non-covalent linkages in collagen fibers, followed by a thermal solubilization in water [129]. Depending on the pre-treatment and the extraction conditions, the dissociation of collagen chains results in a heterogeneous polymeric mixture (Fig. **1.3**). Gelatin is hence characterized by the presence of separated α-chains (*i.e.* α-gelatin, monomeric form), two associated α-chains (*i.e.* β-gelatin, dimeric form), and three associated α-chains (*i.e.* γ-gelatin, trimeric form). The strength of gelatin gel (also known as Bloom strength or Bloom value) varies depending on gelatin molecular weight. The Bloom strength governs both the application and the price of gelatin with high Bloom gelatins being preferred.

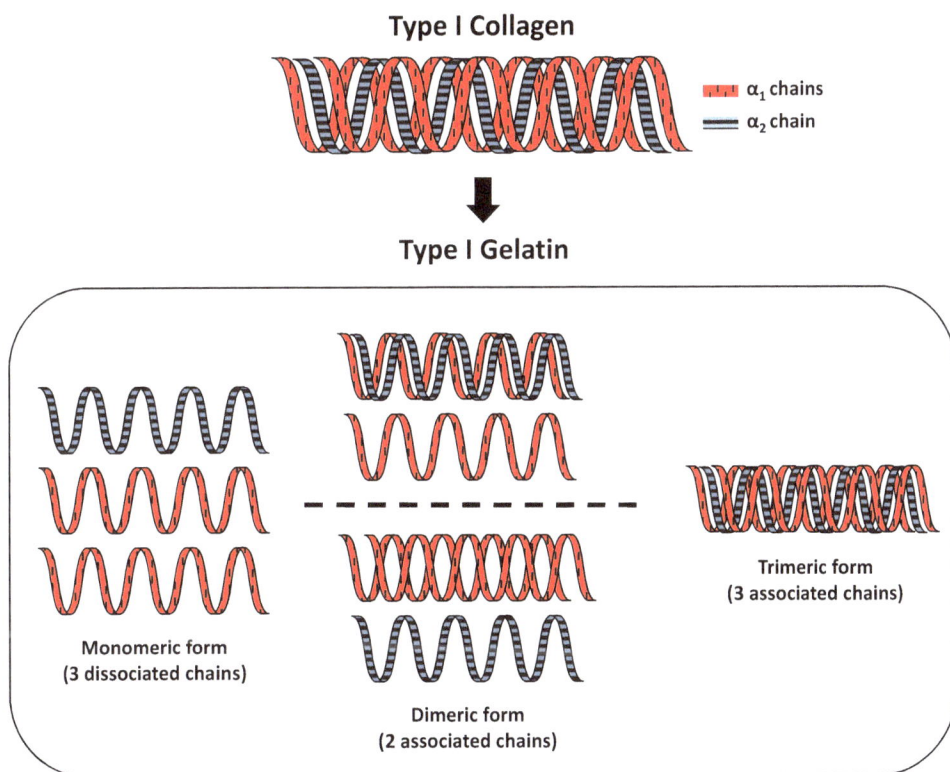

Type I Collagen

α₁ chains

α₂ chain

Type I Gelatin

Monomeric form
(3 dissociated chains)

Dimeric form
(2 associated chains)

Trimeric form
(3 associated chains)

Fig. (1.3). The conversion of type I collagen (ordered state) into gelatin (disordered state).

Many food products are based on protein colloidal gel structures. The study of the gelation mechanisms and the gel properties of food protein ingredients enables the prediction of the final texture of the food product. These rheological properties are usually assessed using highly sophisticated instruments called rheometers. With a rheometer running under oscillating mode, both elastic (G') and viscous (G") moduli of a material can be measured simultaneously. If the amplitude of the oscillation falls under the linear viscoelastic range (*i.e.* small strain dynamic oscillation), the sample material can be subjected to an oscillating shear stress without breakage of any molecular structure within the sample. The small strain dynamic oscillation mode is the method of choice for the evaluation of both viscoelastic and gelation properties of food proteins [130].

In gelatin, the mixture of polypeptides reversibly transforms into a gel. Gelatin shows a sol-to-gel transition when the temperature is lowered and a gel-to-sol transition as the temperature increases. The crosslinks in gelatin network are mainly based on hydrogen bonding and Van der Waals forces and thus very susceptible to changes of temperature, pH, and ionic strength [131].

The reversible gelation and melting of gelatin take place during the cooling and heating processes, respectively (Fig. **1.4**). During the gelation process, the elastic (or storage) modulus (G') of gelatin increases, representing the transition from a solution to gel state. Similar behavior is usually observed for the viscous (or loss) modulus (G"). However, the increase is mostly gradual. The sol-to-gel transition is also characterized by a sharp decrease in the phase angle (δ). The melting of gelatin is observed during the heating process and is marked by a significant decrease in the elastic modulus. It is also important to note that gelatin exhibits a thermal hysteresis in which the gelling temperature differs from the melting temperature.

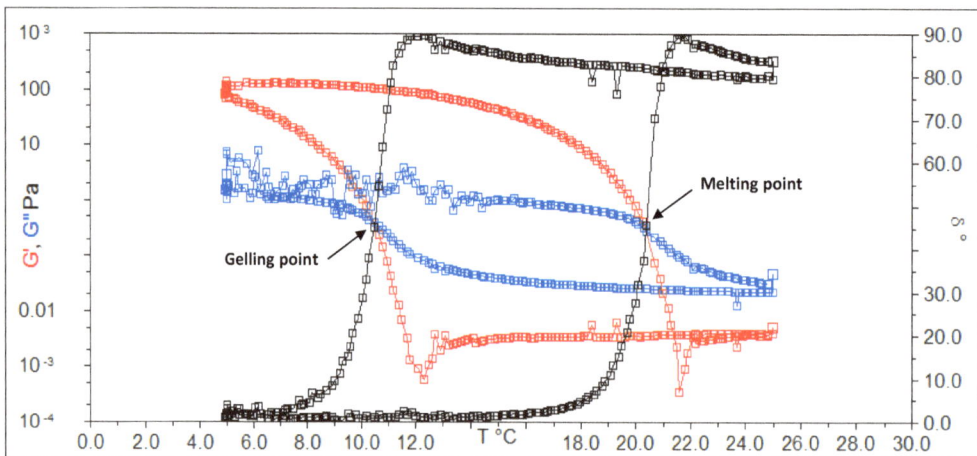

Fig. (1.4). A typical rheogram showing the viscoelastic properties of gelatin, including the elastic (G', in red) and viscous (G", in blue) moduli as well as the phase angle (δ, in black).

EFFECT OF PROTEIN MODIFICATIONS ON THE FUNCTIONAL PROPERTIES

In the last few years, research in food protein biotechnology focused on enhancing the protein functional properties. Chemical, physical, and enzymatic protein modification methods have been developed and proposed as tools to achieve this goal. Among these methods, glycation, disulfide and transglutaminase cross-linking, proteolysis, high-pressure processing, and ultrasonic treatment have recently been investigated.

Chemical Modification

Proteins can chemically be modified through the creation of cross-links either between the protein polypeptide chains (*i.e.* intramolecular cross-links), different proteins (*i.e.* intermolecular cross-links), or other food components (*i.e.* sugars, lipids, polyphenol, *etc.*). There are several types of chemical cross-links; however, glycation and disulfide cross-linking are by far the most studied cross-linking methods for food proteins.

Glycation

Glycation, or the Maillard reaction, is the non-enzymatic reaction of proteins with reducing sugars. The entire Maillard reaction occurs in three phases; early, advanced, and final stages, which are interrelated and can proceed at the same time [132]. The early stage is characterized by the formation of a covalent bond between a free amino group of either an amino acid, a peptide, or a protein and the carbonyl group of a reducing sugar (*i.e.* condensation reaction which forms a Schiff base, N-glycosylamine, and releases one molecule of water). N-glycosylamine is then irreversibly rearranged into the Amadori rearrangement product, 1-amino-1-deoxy-2-ketose [132, 133].

The conjugation of proteins with polysaccharides through the Maillard reaction has been shown to improve the functional properties of the conjugates. Recent studies on food protein conjugation with a wide range of sugars/polysaccharides (Table **1.2**) indicated that the glycation improves the solubility as well as the emulsifying and the foaming properties.

Table 1.2. Functionality of protein/sugar and protein/polysaccharide conjugates produced using Maillard reaction.

Protein	Sugar or Polysaccharide	Reaction Conditions	Effect on the Functional Properties	Reference
Whey protein isolate	Fenugreek gum	60 °C, 75% relative humidity for 3 days	Improved emulsion stability	[135]

(Table 1.2) cont.....

Protein	Sugar or Polysaccharide	Reaction Conditions	Effect on the Functional Properties	Reference
Peanut protein isolate	Dextran	60 °C, 79% relative humidity for 7 days	Improved thermal stability, solubility, emulsifying and foam properties	[136]
β-lactoglobulin and whey protein isolate	Corn fiber gum	75 °C, 79% relative humidity for up to 7 days	Improved emulsion stability	[137]
Egg white protein	Pectin	60 °C, 79% relative humidity for up to 48 h	Higher emulsion viscosity and stability	[138]
Soy β-conglycinin	Dextran	95 °C for 6 h in a liquid system	Increased protein solubility near the isoelectric point	[139]
Whey protein isolate	Dextran	60 °C for 48 h in a liquid system	Improved emulsifying ability and stability	[140]
Whey protein	Maltodextrin or corn syrup solids	90 °C for up to 24 h in a liquid system	Increased solubility and enhanced thermal stability	[141]
Oat protein isolate	Dextran	90 °C for up to 100 min in a liquid system	Improved emulsifying properties	[139]
Whey protein	Lactose or maltodextrin	130 °C, 79% relative humidity for up to 30 min	Enhanced thermal stability	[142]
Phosvitin from egg yolk	Dextran	100 °C for 6 h in a liquid system	Increased solubility and emulsion stability	[143]
Soy protein isolate	Maltodextrin	From 90 to 140 °C, 79% relative humidity for 2 h	Higher emulsifying stability	[144]
Shrimp hydrolysate	Xylose	65 °C for 4 h in a liquid system	Improved emulsifying properties	[145]
Wheat germ protein	Dextran	90 °C for 20 min in a liquid system	Better solubility and emulsifying properties	[146]
β-lactoglobulin	Galactose	50 °C for 48 h in a liquid system	Better foaming capacity and stability	[147]
Buckwheat protein	Dextran	60 °C for up to 48 h in a liquid system	Superior emulsifying activity	[148]
α-lactalbumin	Rhamnose and fucose	60 °C, 65% relative humidity for 24 h	Improved foam stability	[149]
Lysozyme	Potato galactan	60 °C, 65% relative humidity for up to 7 days	Higher solubility, thermal stability and emulsion stability	[150]

(Table 1.2) cont.....

Protein	Sugar or Polysaccharide	Reaction Conditions	Effect on the Functional Properties	Reference
Soy protein isolate	Glucose	50 °C, 65% relative humidity for up to 14 days	Improved solubility and emulsifying properties	[151]
Actomyosin	Glucosamine	40 °C for 8 h in a liquid system	Improved solubility and emulsifying properties	[152]

Among all the functional properties, the emulsifying capacity and stability of proteins are greatly enhanced by glycation with sugars/polysaccharides (Table **1.2**). This characteristic can potentially lead to the development of protein/polysaccharide-based emulsifiers which combine the emulsification properties of proteins and the stabilizing ability of polysaccharides [134]. These types of emulsifiers could potentially replace the synthetic industrial emulsifiers due to their natural origin, low cost, and safety.

It is important to note that although the cross-linking of proteins with sugars and polysaccharides seems to improve the protein functional properties, it can negatively affect the rheological and gel properties of the conjugates compared to the native protein, which has been attributed to the change in the secondary structure [153 - 155].

Disulfide Cross-linking

The disulfide cross-linking involves the binding of two cysteine residues covalently. In food science, the disulfide bonds have long been associated with increased heat stability (milk proteins) as well as with the formation of gel matrices (egg and meat proteins) and viscoelastic networks (wheat proteins) [156].

Due to their unique characteristics in enhancing the protein stability, novel disulfide bonds have recently been engineered in the aim of stabilizing enzymes [157]. The use of enzymes in food industry can potentially represent a sustainable alternative way to chemical processes [158]. However, for a successful industrial implementation, these biocatalysts have to be stable and maintain their activity during processing, which is not always achievable. Several protein engineering strategies have been proposed to stabilize enzymes including chemical modification, stabilization with the addition of additives, immobilization, crystallization, and genetic engineering [159]. Chemical cross-linking seems to be the easiest and the most efficient [159]. Among the chemical cross-linking approaches, the disulfide cross-linking of enzyme sub-units is the most efficient way to stabilize enzymes and prevent the dissociation of the sub-units [158].

The addition of new disulfide bonds to enhance the enzyme stability has recently been investigated. Le *et al.* [160] showed that the thermostability of lipase B from *Candida antarctica* was improved through the introduction of a new disulfide bridge [160]. Similarly, Wang *et al.* [161] and Han *et al.* [162] introduced a disulfide bond to xylanase from *Thermomyces lanuginosus* GH1 and lipase from *Rhizomucor miehei*, respectively, and they both were able to improve the thermal stability of the engineered enzymes compared to their native counterpart [161, 162]. Despite several reports indicating the potential of disulfide cross-linking in enhancing the enzyme thermal stability, this approach can, in some cases, result in destabilization of the enzyme [157].

Physical Modification

High hydrostatic pressure processing and ultrasound treatment have recently attracted a growing interest. These technologies have been widely accepted in the field of food science and technology as emerging technologies that can positively modify proteins and enhance the product characteristics.

High-Pressure Processing

High hydrostatic pressure (HHP) processing is a non-thermal technology that subjects food to pressures up to 1000 MPa [163]. HHP is not only used to inactivate enzymes and pathogens but also to create textured products through modifying the protein conformation (*i.e.* protein denaturation, aggregation, or gelation). The application of high-pressure on food proteins leads to the disruption of noncovalent interactions within the polypeptide chains followed by reformation of intra- and inter-molecular bonds within or between the polypeptide chains [164]. The changes of the protein properties under high-pressure processing are associated with the protein susceptibility, the magnitude of the pressure, the temperature as well as the duration of the treatment [164].

Muscle proteins (sarcoplasmic, myofibrillar, and stromal proteins) are perhaps the most widely studied proteins under high-pressure processing. Chan *et al.* [163] attempted to improve the functional and rheological properties of pale, soft, and exudative turkey meat using high-pressure processing [163]. Their research outcomes showed that the application of pressures of 50 and 100 MPa increased the water holding capacity of low pH meat and resulted in the formation of a better gel network. Various studies on gelatin (*i.e.* denatured collagen, a connective tissue protein) showed that the application of HHP induces changes in the molecular weight distribution and the viscoelastic properties [165]. Gekko and Fukamizu [166] and Shimada *et al.* [167] reported that HHP led to the improvement of gelatin stability as indicated by the increase of the melting temperature [166, 167].

Another recent application of high-pressure processing in food engineering is related to the preparation of multifunctional protein edible films. For instance, Condés *et al.* [168] developed a protein-based film using high-pressure modified amaranth proteins [168]. The produced amaranth protein film was uniform with better mechanical properties and lower water solubility as well as vapor permeability. These enhanced functional properties were associated with the structural changes due to the high-pressure treatment which unfolded the protein and promoted the formation of cross-links through both hydrogen and disulfide bonds [168].

Although high-pressure treatment effectively reduces bacterial spoilage and extends shelf-life, it can decrease some meat protein functionality and can negatively alter the sensory characteristics of the meat product [169 - 171]. Further studies on the impact of high-pressure processing on the functional properties of food proteins are still needed to better understand the effect of this technology on protein functionality.

Ultrasound Treatment

Ultrasound is one of the emerging technologies in food science due to its effectiveness as well as its minimal cost and energy consumption [172]. It applies high-frequency sound waves exceeding the audible limit of the human ear (*i.e.* frequencies greater than 16 kHz) [173]. At these frequency levels, changes in the physical, chemical, and biochemical properties of foods occur [172].

One of the applications of ultrasound in food processing technology consists of enhancing the extraction of bioactive and functional compounds from plant and animal materials [174]. Ultrasound-assisted extraction is believed to increase the extraction yield and rate and decrease the extraction duration. Recent research work investigated the effect of ultrasound technology on protein modification. It has been reported that ultrasound treatment can enhance the functional properties of proteins. For example, ultrasound treatment increased the solubility of whey proteins [175], soy protein concentrate [176], and α-lactalbumin [177]. Ultrasound seems also to improve the foaming and emulsifying properties of food proteins [175 - 177].

Arzeni *et al.* [121] applied ultrasound to egg white proteins. Their research findings indicated that emulsions prepared using sonicated samples had better stability to creaming and flocculation. Sonication also increased the surface hydrophobicity of egg white proteins which, in turn, resulted in a faster thermal aggregation [121]. In another study, Arzeni *et al.* [178] compared the effects of high-intensity ultrasound on whey protein concentrate, soy protein isolate, and egg white proteins [178]. Their results indicated that high-intensity sonication modified the gelation, viscosity, and solubility of these proteins. These changes were attributed to the increase in the hydrophobicity and alteration of the particle size.

Despite its beneficial processing aspects, it is important to note that the change induced by high-intensity ultrasound mainly depends on the nature of the protein and its degree of denaturation and aggregation [178]. Further studies need to be carried out and more knowledge has to be gathered regarding the effect of ultrasound technology on the characteristics and functionalities of food proteins.

Enzymatic Modification

The use of enzymes to modify food proteins has attracted interest over the chemical modification for many reasons including safety, improved reaction specificity, efficiency with less or no undesirable side-reactions, and the general consumer acceptance of enzymes [156]. There are several types of protein enzymatic modifications targeting the improvement of functional properties. Two of the most efficient approaches are transglutaminase cross-linking and proteolysis.

Transglutaminase Cross-linking

Protein-glutamine γ-glutamyltransferase (EC 2.3.2.13), more commonly known as transglutaminase, is a multifunctional enzyme involved in three types of post-translational protein modifications. For instance, transglutaminase catalyzes the acyl transfer reaction between the γ-carboxyamide group of a glutaminyl residue of a protein or a peptide (acyl donors) and a wide range of primary amines (acyl acceptors). If the ε-amino group of a lysine residue is the acyl-acceptor, peptide chains are covalently cross-linked through ε-(γ-glutamyl) lysine bonds. In addition to the cross-linking function, transglutaminase is able to hydrolyze the γ-carboxyamide group leading to the deamidation of the glutaminyl residue. This deamidation reaction is usually observed in the absence of amine substrates [179].

The protein cross-linking ability of transglutaminase is the main reason for its successful application in the food industry. Transglutaminase-induced covalent cross-linking significantly changes the physical and chemical properties of food proteins. The use of transglutaminase has been reported to enhance the textural and rheological properties of many protein-based food products, such as chicken myofibrillar protein isolate [180], ovalbumin [181], fish mince [182], milk [183], and rye dough [184].

Aside from improving the textural and viscoelastic properties of proteins through cross-linking, recent research in enzyme biotechnology investigated the transglutaminase glycosylation of food proteins and peptides with amino sugars (the acyl acceptor) as an alternative for the chemical glycation. For instance, the transglutaminase glycosylation of food proteins, such as casein [185] and actomyosin [186], with glucosamine at 37 °C produced glycoconjugates with improved emulsifying properties compared to the original protein. The positive effects of transglutaminase-mediated glycosylation on the emulsifying properties have been attributed to the superior water-binding ability of the glycoconjugates. When it comes to food-derived peptides, it seems that the transglutaminase-mediated glycosylation with glucosamine produces glycoconjugates with greater bioactive properties (*i.e.* antioxidant activity) than the native peptides [187, 188].

Proteolysis

The enzymatic hydrolysis of food proteins (*i.e.* proteolysis) is by far the most widely used approach to modify the protein physicochemical properties. Apart from producing bioactive peptides, the enzymatic hydrolysis *in vitro* is an effective approach for the improvement of functional and nutritional properties of food proteins [189].

Several parameters affect the composition, physicochemical, and functional properties of the hydrolysates. These factors include the type (serine-, cysteine-, aspartic- or metallo-) and origin (microbial, plant or animal) of the proteolytic enzymes, enzyme to substrate ratio, hydrolysis conditions (*i.e.* time, pH and temperature), and water to raw material ratio [189].

Table **1.3** details the effects of proteolysis on the functionality of food proteins.

Table 1.3. Functionality of food protein hydrolysates produced through proteolysis.

Protein	Enzyme	Hydrolysis Conditions	Effect on the Functional Properties	Reference
Corn glutelin	Protamex	pH 7.0, 50 °C, enzyme-to-substrate ratio 0.81% (w/w)	Improved solubility, foaming and emulsifying properties	[190]
Soy protein isolate	Papain	pH 7.0, 38 °C, enzyme-to-substrate ratio 1 and 2% (w/v)	Improved foaming capacity and foam stability	[190]
Pine nut protein isolate	Alcalase	pH 8.5, 55 °C, enzyme-to-substrate ratio 7% (w/w)	Improved solubility	[191]
Pea protein isolate	Papain	pH 8.0, 37 °C, enzyme-to-substrate ratio 0.5% (w/w)	Improved solubility, emulsifying properties and foaming capacity	[192]

(Table 1.3) cont.....

Protein	Enzyme	Hydrolysis Conditions	Effect on the Functional Properties	Reference
Egg white	Papain	pH 6.0, 50 °C, enzyme-to-substrate ratio 3% (w/w)	Improved solubility, emulsifying and foaming properties	[193]
Fish (Pink perch) muscle	Trypsin	pH 8.8, 37 °C, enzyme-to-substrate ratio 1% (w/w)	Improved solubility	[194]
Fish (Zebra blenny) muscle	Crude enzyme from zebra blenny	pH 8.0, 45 °C, enzyme-to-substrate ratio 3 U/mg protein	Improved solubility and foaming properties	[195]
Rice dreg protein	Alcalase	pH 8.5, 55 °C, enzyme-to-substrate ratio 1% (w/w)	Improved solubility and emulsifying capacity	[196]
Egg white	Trypsin	pH 7.5, 45 °C, enzyme-to-substrate ratio 3% (w/w)	Improved solubility	[193]
Cuttlefish skin and viscera	Proteases from *Bacillus licheniformis*	pH 10.0, 50 °C, enzyme-to-substrate ratio 0.5 U/mg protein	Improved solubility, foaming and emulsifying capacities	[197]
Peanut protein isolate	Alcalase	pH 7.0, 50 °C	Improved solubility	[198]
Chickpea protein isolate	Alcalase	pH 8.0, 50 °C, enzyme-to-substrate ratio 1.14 mg/g protein	Improved solubility and foaming properties	[199]
Fish (Ornate threadfin bream) muscle	Pepsin from skipjack tuna	pH 2.0, 50 °C	Improved solubility	[200]
Peanut protein isolate	Alcalase	pH 8.0, 50 °C, enzyme-to-substrate ratio 10% (w/w)	Improved solubility	[201]
Porcine blood plasma protein	Alcalase	pH 8.0, 55 °C, enzyme-to-substrate ratio 2% (w/w)	Improved solubility	[202]
Casein	Trypsin	pH 8.0, 37 °C, enzyme-to-substrate ratio 1% (w/w)	Improved solubility and emulsifying properties	[203]
Casein	Chymotrypsin	pH 8.0, 50 °C, enzyme-to-substrate ratio 1% (w/w)	Improved solubility and emulsifying properties	[203]
Heat-induced aggregates of whey protein isolate	Corolase	pH 8.0, 50 °C, enzyme-to-substrate ratio 1% (w/w)	Improved solubility	[204]
Turkey head collagen	Mixture of alcalase, flavourzyme and trypsin	pH 8.0, 50 °C, enzyme-to-substrate ratio 2% (w/w)	Improved solubility	[205]

The enzymatic hydrolysis modifies the molecular size of proteins, increases the number of ionizable groups, and exposes the hydrophobic groups [206]. These changes may result in the improvement of the functional properties. As illustrated in Table **1.3**, the solubility is the main functional property that is enhanced by proteolysis, followed by both emulsifying and foaming capacities. As the hydrolysis progresses, the structure of food protein is disrupted and small molecular weight peptides are generated. These peptides mostly contain polar residues that interact with water molecule through hydrogen bonds and lead to a better solubility [102]. As for the emulsifying and foaming properties, it is well known that good emulsifying and foaming agents must have an amphiphilic property and should be able to quickly adsorb at the oil/water or air/water interfaces, respectively (in such a way that the hydrophilic groups are oriented towards the aqueous phase while the lipophilic groups are aligned towards the non-polar phase). It was reported that hydrophobic amino acids control the adsorption of protein to the water interface [207]. The disruption of the protein structure and the exposure of hydrophobic residues have been correlated to the improvement of the emulsifying and foaming properties after proteolysis [102].

In the last few decades, a substantial research work has been done to add value to food processing by-products through proteolysis. Several microbial, plant, or animal proteolytic enzymes have been screened and used. Due to the high cost of commercial enzymes, another research segment focused on the use of digestive enzymes naturally present in fish viscera [208]. Proteases from fish guts include pepsin, chymotrypsin, trypsin, elastases, gastricsin, and collagenase [209]. Many fish endogenous enzymes have successfully been isolated from fish processing discards and used in protein hydrolysis [210]. However, differences among fish species, season, and the amount/activity of proteolytic enzymes constitute a major drawback for a potential industrial implementation of fish endogenous enzymes in hydrolyzing food proteins.

The enzymatic hydrolysis is generally monitored through the degree of hydrolysis (DH), which represents the ratio of the number of peptide bonds cleaved with respect to the total number of bonds per unit weight. The degree of hydrolysis has been shown to influence the functional properties of the food protein hydrolysate. High degrees of hydrolysis produce peptides with excellent solubility [211] but may negatively influence the rest of functional properties [193]. Therefore, it is recommended to target low degrees of hydrolysis (*i.e.* partial hydrolysis) in order to obtain the desired protein functionality.

Various methods have been developed to estimate the degree of hydrolysis of food protein hydrolysates. The main methods for the evaluation of the extent of hydrolysis are based on either the determination of liberated free α-amino groups, the titration of released protons, or the quantification of trichloroacetic acid soluble nitrogen (TCA-SN) [212]. The estimation of free amino groups can be achieved through various chemical methods, which are based on the reaction of primary amino groups with either o-phthaldialdehyde (OPA) or trinitro-benzen--sulfonic acid (TNBS) [213, 214]. The titration of released protons is commonly achieved through the pH static (pH-STAT) method. In this technique, the pH is maintained constant by the controlled addition of an acid or a base. The DH is subsequently calculated based on the molarity and the amount of the acid or base that has been consumed according to the formula proposed by Adler-Nissen [215]. The SN-TCA method is based on the determination of amount of TCA-soluble nitrogen. The addition of trichloroacetic acid induces the precipitation of undigested proteins while free amino acids and peptides remain soluble. The quantification of TCA-soluble nitrogen peptides can give an idea on the degree of hydrolysis. Up to now, there is no general agreement on the best method for monitoring the DH [216]. However, the pH-STAT method is frequently chosen because of its simplicity, wide applicability and practical convenience.

Both mesophilic and thermophilic enzymes have been widely used to hydrolyze proteins with the aim of improving their functional properties. These enzymes have optimal activities at temperatures ranging from 50 to 80 °C. Digesting food proteins at this temperature range is not practical due to safety risks (*i.e.* growth of pathogens) and unfavorable effects (*i.e.* oxidation) [217]. Cold-adapted enzymes, also known as psychrophilic enzymes, are active at temperature lower than 30 °C but are unstable at temperatures higher than 50 °C. Cold-adapted enzymes are very advantageous for the food industry because they can avoid spoilage and nutritional loss of the food products/ingredients [217]. Research in protein hydrolysis should move toward the discovery of novel enzymes with high activity at low temperatures for replacing mesophilic or thermophilic enzymes known to induce undesirable biological and chemical reactions at their optimum temperatures.

CONCLUDING REMARKS

The area of food protein biotechnology has been rapidly growing in the last few decades. Extensive research work has recently been done in preparing and identifying bioactive peptides and improving the physicochemical properties of food proteins. However, some challenges remain.

Firstly, more research is still needed in the bioactivity of food-derived peptides to tackle issues related to their poor bioavailability. Food scientists have to focus on developing methods for improving the stability of bioactive peptides against the gastric enzymatic degradation and enhancing their intestinal absorption.

Secondly, although effective protein modification methods for improving the functional properties of food proteins have been proposed and optimized, scaling-up these methods has not yet been achieved. Further research studies are needed in order to investigate the feasibility of protein modification methods at an industrial scale.

Finally, it is well known that the food industry generates significant amounts of low-value by-products rich in proteins. Converting these low-value by-products into functional and/or bioactive protein-based ingredients offers potential environmental and economic benefits. Economical and efficient processes capable of recovering food proteins from low-value by-products without altering their physicochemical and functional characteristics have to be discovered in the future.

CONSENT FOR PUBLICATION

Not applicable.

CONFLICT OF INTEREST

The authors declare no conflict of interest, financial or otherwise.

ACKNOWLEDGEMENTS

Declared none.

ABBREVIATIONS

AAPH	2,2'-azobis-2-methyl-propanimidamide dihydrochloride;
ABTS•+	2,2'-azinobis(3-ethylbenzothiazoline-6-sulphonic acid);
ACE	angiotensin-I converting enzyme;
AI	augmentation index;
Ang-I	angiotensin I;
Ang-II	angiotensin II;
AT-1 receptor	angiotensin II type 1 receptor;
BP	blood pressure;
DH	degree of hydrolysis;
DPP-IV	dipeptidyl-peptidase IV;
DPPH	2,2-Diphenil-1-picrylhydrazyl;
ECE	endothelin-converting enzyme; ET-1: endothelium-1;
ET-1	endothelium-1;
eNOS	endothelial nitric oxide synthase;
ES	endothelin system;
G'	elastic (or storage) modulus;

G"	viscous (or loss) modulus;
GRAS	generally recognized as safe;
HHP	high hydrostatic pressure;
ISP	isoclectric solubilization/precipitation;
LDL	low density lipoprotein;
LTP	lactotripeptides;
NO	nitric oxide;
NTS (2) receptor	neurotensin type 2 receptor;
O/W	oil-in-water; OPA: O-phthaldialdehyde;
OPA	O-phthaldialdehyde;
ORAC	oxygen radical absorbance capacity;
pH-STAT	pH static titration;
RAS	renin angiotensin system;
ROS	reactive oxygen species;
SBP	systolic blood pressure;
SHR	spontaneous hypertensive rats;
SPI	soy protein isolates;
t-BOOH	tert-butyl hydroperoxide;
TCA-SN	trichloroacetic acid soluble nitrogen;
TEAC	Trolox equivalent antioxidant capacity;
TNBS	trinitro-benzene-sulfonic acid;
VCEAC	vitamin C equivalent antioxidant capacity;
W/O	water-in-oil;
WPI	whey protein isolates;
δ	phase angle.

REFERENCES

[1]   Damodaran S. Food proteins: An overview. Food proteins and their applications. New York: Marcel Dekker, Inc. 1997; pp. 1-24.

[2]   Chen L, Remondetto GE, Subirade M. Food protein-based materials as nutraceutical delivery systems. Trends Food Sci Technol 2006; 17(5): 272-83.
[http://dx.doi.org/10.1016/j.tifs.2005.12.011]

[3]   Foegeding EA, Davis JP, Doucet D, McGuffey MK. Advances in modifying and understanding whey protein functionality. Trends Food Sci Technol 2002; 13(5): 151-9.
[http://dx.doi.org/10.1016/S0924-2244(02)00111-5]

[4]   Roberfroid MB. Concepts and strategy of functional food science: the European perspective. Am J Clin Nutr 2000; 71(6) (Suppl.): 1660S-4S.
[http://dx.doi.org/10.1093/ajcn/71.6.1660S] [PMID: 10837311]

[5]   Rutherfurd-Markwick KJ. Food proteins as a source of bioactive peptides with diverse functions. Br J Nutr 2012; 108(S2) (Suppl. 2): S149-57.
[http://dx.doi.org/10.1017/S000711451200253X] [PMID: 23107526]

[6]   Quaglia GB, Orban E. Influence of enzymatic hydrolysis on structure and emulsifying properties of sardine (*Sardina pilchardus*) protein hydrolysates. J Food Sci 1990; 55(6): 1571-3.
[http://dx.doi.org/10.1111/j.1365-2621.1990.tb03571.x]

[7]   Schaafsma G. Safety of protein hydrolysates, fractions thereof and bioactive peptides in human nutrition. Eur J Clin Nutr 2009; 63(10): 1161-8.
[http://dx.doi.org/10.1038/ejcn.2009.56] [PMID: 19623200]

[8]   Unger T. The role of the renin-angiotensin system in the development of cardiovascular disease. Am J Cardiol 2002; 89(2A): 3A-9A.
[http://dx.doi.org/10.1016/S0002-9149(01)02321-9] [PMID: 11835903]

[9]   Cheung HS, Wang FL, Ondetti MA, Sabo EF, Cushman DW. Binding of peptide substrates and inhibitors of angiotensin-converting enzyme. Importance of the COOH-terminal dipeptide sequence. J Biol Chem 1980; 255(2): 401-7.
[PMID: 6243277]

[10]   Ashcroft F, Benos D, Bezanilla F, *et al.* The state of ion channel research in 2004. Nat Rev Drug Discov 2004; 3(3): 239-78.
[http://dx.doi.org/10.1038/nrd1361]

[11]   Nakamura Y, Yamamoto N, Sakai K, Takano T. Antihypertensive effect of sour milk and peptides isolated from it that are inhibitors to angiotensin I-converting enzyme. J Dairy Sci 1995; 78(6): 1253-7.
[http://dx.doi.org/10.3168/jds.S0022-0302(95)76745-5] [PMID: 7673515]

[12]   Seppo L, Jauhiainen T, Poussa T, Korpela R. A fermented milk high in bioactive peptides has a blood pressure-lowering effect in hypertensive subjects. Am J Clin Nutr 2003; 77(2): 326-30.
[http://dx.doi.org/10.1093/ajcn/77.2.326] [PMID: 12540390]

[13]   Wuerzner G, Peyrard S, Blanchard A, Lalanne F, Azizi M. The lactotripeptides isoleucine-prolin--proline and valine-proline-proline do not inhibit the N-terminal or C-terminal angiotensin converting enzyme active sites in humans. J Hypertens 2009; 27(7): 1404-9.
[http://dx.doi.org/10.1097/HJH.0b013e32832b4759] [PMID: 19506528]

[14]   Boelsma E, Kloek J. Lactotripeptides and antihypertensive effects: a critical review. Br J Nutr 2009; 101(6): 776-86.
[http://dx.doi.org/10.1017/S0007114508137722] [PMID: 19061526]

[15]   Boelsma E, Kloek J. IPP-rich milk protein hydrolysate lowers blood pressure in subjects with stage 1 hypertension, a randomized controlled trial. Nutr J 2010; 9(1): 52.
[http://dx.doi.org/10.1186/1475-2891-9-52] [PMID: 21059213]

[16]   Hirota T, Ohki K, Kawagishi R, *et al.* Casein hydrolysate containing the antihypertensive tripeptides Val-Pro-Pro and Ile-Pro-Pro improves vascular endothelial function independent of blood pressure-lowering effects: contribution of the inhibitory action of angiotensin-converting enzyme. Hypertens Res 2007; 30(6): 489-96.
[http://dx.doi.org/10.1291/hypres.30.489] [PMID: 17664851]

[17]   Jauhiainen T, Rönnback M, Vapaatalo H, Wuolle K, Kautiainen H, Korpela R. *Lactobacillus helveticus* fermented milk reduces arterial stiffness in hypertensive subjects. Int Dairy J 2007; 17(10): 1209-11.
[http://dx.doi.org/10.1016/j.idairyj.2007.03.002]

[18]   Turpeinen AM, Kumpu M, Rönnback M, Seppo L, Kautiainen H, Jauhiainen T, *et al.* Antihypertensive and cholesterol-lowering effects of a spread containing bioactive peptides IPP and VPP and plant sterols. J Funct Foods 2009; 1(3): 260-5.

[http://dx.doi.org/10.1016/j.jff.2009.03.001]

[19]   Yoshizawa M, Maeda S, Miyaki A, *et al.* Additive beneficial effects of lactotripeptides and aerobic exercise on arterial compliance in postmenopausal women. Am J Physiol Heart Circ Physiol 2009; 297(5): H1899-903.
       [http://dx.doi.org/10.1152/ajpheart.00433.2009] [PMID: 19783777]

[20]   Yoshizawa M, Maeda S, Miyaki A, *et al.* Additive beneficial effects of lactotripeptides intake with regular exercise on endothelium-dependent dilatation in postmenopausal women. Am J Hypertens 2010; 23(4): 368-72.
       [http://dx.doi.org/10.1038/ajh.2009.270] [PMID: 20075849]

[21]   Pripp AH. Effect of peptides derived from food proteins on blood pressure: a meta-analysis of randomized controlled trials. Food Nutr Res 2008; 52: 1-9.
       [PMID: 19109662]

[22]   Kawasaki T, Seki E, Osajima K, *et al.* Antihypertensive effect of valyl-tyrosine, a short chain peptide derived from sardine muscle hydrolyzate, on mild hypertensive subjects. J Hum Hypertens 2000; 14(8): 519-23.
       [http://dx.doi.org/10.1038/sj.jhh.1001065] [PMID: 10962520]

[23]   Nakano D, Ogura K, Miyakoshi M, *et al.* Antihypertensive effect of angiotensin I-converting enzyme inhibitory peptides from a sesame protein hydrolysate in spontaneously hypertensive rats. Biosci Biotechnol Biochem 2006; 70(5): 1118-26.
       [http://dx.doi.org/10.1271/bbb.70.1118] [PMID: 16717411]

[24]   Gonzalez-Gonzalez C, Gibson T, Jauregi P. Novel probiotic-fermented milk with angiotensin I-converting enzyme inhibitory peptides produced by Bifidobacterium bifidum MF 20/5. Int J Food Microbiol 2013; 167(2): 131-7.
       [http://dx.doi.org/10.1016/j.ijfoodmicro.2013.09.002] [PMID: 24135669]

[25]   Matsufuji H, Matsui T, Ohshige S, Kawasaki T, Osajima K, Osajima Y. Antihypertensive effects of angiotensin fragments in SHR. Biosci Biotechnol Biochem 1995; 59(8): 1398-401.
       [http://dx.doi.org/10.1271/bbb.59.1398] [PMID: 7549089]

[26]   Matsui T, Tamaya K, Seki E, Osajima K, Matsumoto K, Kawasaki T. Val-Tyr as a natural antihypertensive dipeptide can be absorbed into the human circulatory blood system. Clin Exp Pharmacol Physiol 2002; 29(3): 204-8.
       [http://dx.doi.org/10.1046/j.1440-1681.2002.03628.x] [PMID: 11906484]

[27]   Matsui T, Tamaya K, Seki E, Osajima K, Matsumo K, Kawasaki T. Absorption of Val-Tyr with *in vitro* angiotensin I-converting enzyme inhibitory activity into the circulating blood system of mild hypertensive subjects. Biol Pharm Bull 2002; 25(9): 1228-30.
       [http://dx.doi.org/10.1248/bpb.25.1228] [PMID: 12230125]

[28]   Matsui T, Imamura M, Oka H, *et al.* Tissue distribution of antihypertensive dipeptide, Val-Tyr, after its single oral administration to spontaneously hypertensive rats. J Pept Sci 2004; 10(9): 535-45.
       [http://dx.doi.org/10.1002/psc.568] [PMID: 15473262]

[29]   Zhuang Y, Sun L, Zhang Y, Liu G. Antihypertensive effect of long-term oral administration of jellyfish (*Rhopilema esculentum*) collagen peptides on renovascular hypertension. Mar Drugs 2012; 10(2): 417-26.
       [http://dx.doi.org/10.3390/md10020417] [PMID: 22412809]

[30]   Vercruysse L, Morel N, Van Camp J, Szust J, Smagghe G. Antihypertensive mechanism of the dipeptide Val-Tyr in rat aorta. Peptides 2008; 29(2): 261-7.
       [http://dx.doi.org/10.1016/j.peptides.2007.09.023] [PMID: 18221823]

[31]   Matsui T, Zhu XL, Watanabe K, Tanaka K, Kusano Y, Matsumoto K. Combined administration of captopril with an antihypertensive Val-Tyr di-peptide to spontaneously hypertensive rats attenuates the blood pressure lowering effect. Life Sci 2006; 79(26): 2492-8.
       [http://dx.doi.org/10.1016/j.lfs.2006.08.013] [PMID: 16959271]

[32]   Watanabe M, Kurihara J, Suzuki S, Nagashima K, Hosono H, Itagaki F. The influence of dietary peptide inhibitors of angiotensin-converting enzyme on the hypotensive effects of enalapril. J Pharm Health Care Sci 2015; 1(1): 17.
[http://dx.doi.org/10.1186/s40780-015-0018-3] [PMID: 26819728]

[33]   He R, Malomo SA, Girgih AT, Ju X, Aluko RE. Glycinyl-histidinyl-serine (GHS), a novel rapeseed protein-derived peptide has blood pressure-lowering effect in spontaneously hypertensive rats. J Agric Food Chem 2013; 61(35): 8396-402.
[http://dx.doi.org/10.1021/jf400865m] [PMID: 23919612]

[34]   Girgih AT, Udenigwe CC, Li H, Adebiyi AP, Aluko RE. Kinetics of Enzyme Inhibition and Antihypertensive Effects of Hemp Seed (*Cannabis sativa* L.) Protein Hydrolysates. J Am Oil Chem Soc 2011; 88(11): 1767-74.
[http://dx.doi.org/10.1007/s11746-011-1841-9]

[35]   Girgih AT, He R, Malomo S, Offengenden M, Wu J, Aluko RE. Structural and functional characterization of hemp seed (Cannabis sativa L.) protein-derived antioxidant and antihypertensive peptides. J Funct Foods 2014; 6(1): 384-94.
[http://dx.doi.org/10.1016/j.jff.2013.11.005]

[36]   Udenigwe CC, Li H, Aluko RE. Quantitative structure-activity relationship modeling of renin-inhibiting dipeptides. Amino Acids 2012; 42(4): 1379-86.
[http://dx.doi.org/10.1007/s00726-011-0833-2] [PMID: 21246225]

[37]   Udenigwe CC, Mohan A. Mechanisms of food protein-derived antihypertensive peptides other than ACE inhibition. J Funct Foods 2014; 8(1): 45-52.
[http://dx.doi.org/10.1016/j.jff.2014.03.002]

[38]   Kedzierski RM, Yanagisawa M. Endothelin system: the double-edged sword in health and disease. Annu Rev Pharmacol Toxicol 2001; 41(3): 851-76.
[http://dx.doi.org/10.1146/annurev.pharmtox.41.1.851] [PMID: 11264479]

[39]   Maes W, Van Camp J, Vermeirssen V, *et al.* Influence of the lactokinin Ala-Leu-Pro-Met-His-Ile-Arg (ALPMHIR) on the release of endothelin-1 by endothelial cells. Regul Pept 2004; 118(1-2): 105-9.
[http://dx.doi.org/10.1016/j.regpep.2003.11.005] [PMID: 14759563]

[40]   Fernández-Musoles R, Salom JB, Martínez-Maqueda D, López-Díez JJ, Recio I, Manzanares P. Antihypertensive effects of lactoferrin hydrolyzates: Inhibition of angiotensin- and endothelin-converting enzymes. Food Chem 2013; 139(1-4): 994-1000.
[http://dx.doi.org/10.1016/j.foodchem.2012.12.049] [PMID: 23561201]

[41]   Fernández-Musoles R, Castelló-Ruiz M, Arce C, Manzanares P, Ivorra MD, Salom JB. Antihypertensive mechanism of lactoferrin-derived peptides: angiotensin receptor blocking effect. J Agric Food Chem 2014; 62(1): 173-81.
[http://dx.doi.org/10.1021/jf404616f] [PMID: 24354413]

[42]   Tanaka M, Watanabe S, Wang Z, Matsumoto K, Matsui T. His-Arg-Trp potently attenuates contracted tension of thoracic aorta of Sprague-Dawley rats through the suppression of extracellular Ca2+ influx. Peptides 2009; 30(8): 1502-7.
[http://dx.doi.org/10.1016/j.peptides.2009.05.012] [PMID: 19465074]

[43]   Kobayashi Y, Kovacs-Nolan J, Matsui T, Mine Y. The Anti-atherosclerotic Dipeptide, Trp-His, Reduces Intestinal Inflammation through the Blockade of L-Type Ca2+ Channels. J Agric Food Chem 2015; 63(26): 6041-50.
[http://dx.doi.org/10.1021/acs.jafc.5b01682] [PMID: 26079480]

[44]   Wang Z, Watanabe S, Kobayashi Y, Tanaka M, Matsui T. Trp-His, a vasorelaxant di-peptide, can inhibit extracellular Ca^{2+} entry to rat vascular smooth muscle cells through blockade of dihydropyridine-like L-type Ca^{2+} channels. Peptides 2010; 31(11): 2060-6.
[http://dx.doi.org/10.1016/j.peptides.2010.07.013] [PMID: 20688122]

[45]   Chakrabarti S, Wu J. Bioactive peptides on endothelial function. Food Sci Hum Wellness 2015; 5(1): 1-7.
[http://dx.doi.org/10.1016/j.fshw.2015.11.004]

[46]   Pal S, Ellis V. The chronic effects of whey proteins on blood pressure, vascular function, and inflammatory markers in overweight individuals. Obesity (Silver Spring) 2010; 18(7): 1354-9.
[http://dx.doi.org/10.1038/oby.2009.397] [PMID: 19893505]

[47]   Jauhiainen T, Rönnback M, Vapaatalo H, *et al.* Long-term intervention with Lactobacillus helveticus fermented milk reduces augmentation index in hypertensive subjects. Eur J Clin Nutr 2010; 64(4): 424-31.
[http://dx.doi.org/10.1038/ejcn.2010.3] [PMID: 20145666]

[48]   Jauhiainen T, Korpela R, Roennback M, Vapaatalo H. New use of therapeutically useful peptides. Patent WO2007132054A1 2007.

[49]   Ballard KD, Kupchak BR, Volk BM, *et al.* Acute effects of ingestion of a novel whey-derived extract on vascular endothelial function in overweight, middle-aged men and women. Br J Nutr 2013; 109(5): 882-93.
[http://dx.doi.org/10.1017/S0007114512002061] [PMID: 22691263]

[50]   Marcone S, Haughton K, Simpson PJ, Belton O, Fitzgerald DJ. Milk-derived bioactive peptides inhibit human endothelial-monocyte interactions via PPAR-γ dependent regulation of NF-κB. J Inflamm (Lond) 2015; 12(1): 1.
[http://dx.doi.org/10.1186/s12950-014-0044-1] [PMID: 25632270]

[51]   Karbach S, Wenzel P, Waisman A, Munzel T, Daiber A. eNOS uncoupling in cardiovascular diseases--the role of oxidative stress and inflammation. Curr Pharm Des 2014; 20(22): 3579-94.
[http://dx.doi.org/10.2174/13816128113196660748] [PMID: 24180381]

[52]   Power O, Jakeman P, FitzGerald RJ. Antioxidative peptides: enzymatic production, *in vitro* and *in vivo* antioxidant activity and potential applications of milk-derived antioxidative peptides. Amino Acids 2013; 44(3): 797-820.
[http://dx.doi.org/10.1007/s00726-012-1393-9] [PMID: 22968663]

[53]   Rival SG, Boeriu CG, Wichers HJ. Caseins and casein hydrolysates. 2. Antioxidative properties and relevance to lipoxygenase inhibition. J Agric Food Chem 2001; 49(1): 295-302.
[http://dx.doi.org/10.1021/jf0003911] [PMID: 11170591]

[54]   Shanmugam VP, Kapila S, Sonfack TK, Kapila R. Antioxidative peptide derived from enzymatic digestion of buffalo casein. Int Dairy J 2015; 42: 1-5.
[http://dx.doi.org/10.1016/j.idairyj.2014.11.001]

[55]   Hernández-Ledesma B, Dávalos A, Bartolomé B, Amigo L. Preparation of antioxidant enzymatic hydrolysates from alpha-lactalbumin and beta-lactoglobulin. Identification of active peptides by HPLC-MS/MS. J Agric Food Chem 2005; 53(3): 588-93.
[http://dx.doi.org/10.1021/jf048626m] [PMID: 15686406]

[56]   Contreras M del M, Hernàndez-Ledesma B, Amigo L, Martín-Álvarez PJ, Recio I. Production of antioxidant hydrolyzates from a whey protein concentrate with thermolysin: Optimization by response surface methodology. Lebensm Wiss Technol 2011; 44(1): 9-15.
[http://dx.doi.org/10.1016/j.lwt.2010.06.017]

[57]   Yousr M, Howell N. Antioxidant and ACE inhibitory bioactive peptides purified from egg yolk proteins. Int J Mol Sci 2015; 16(12): 29161-78.
[http://dx.doi.org/10.3390/ijms161226155] [PMID: 26690134]

[58]   Shen S, Chahal B, Majumder K, You SJ, Wu J. Identification of novel antioxidative peptides derived from a thermolytic hydrolysate of ovotransferrin by LC-MS/MS. J Agric Food Chem 2010; 58(13): 7664-72.
[http://dx.doi.org/10.1021/jf101323y] [PMID: 20568771]

[59]   Ko JY, Lee JH, Samarakoon K, Kim JS, Jeon YJ. Purification and determination of two novel antioxidant peptides from flounder fish (*Paralichthys olivaceus*) using digestive proteases. Food Chem Toxicol 2013; 52: 113-20.
[http://dx.doi.org/10.1016/j.fct.2012.10.058] [PMID: 23146692]

[60]   Hernández-Ledesma B, de Lumen BO. Lunasin: a novel cancer preventive seed Peptide. Perspect Medicin Chem 2008; 2: 75-80.
[http://dx.doi.org/10.4137/PMC.S372] [PMID: 19787099]

[61]   Galvez AF, Chen N, Macasieb J, de Lumen BO. Chemopreventive property of a soybean peptide (lunasin) that binds to deacetylated histones and inhibits acetylation. Cancer Res 2001; 61(20): 7473-8.
[PMID: 11606382]

[62]   García-Nebot MJ, Recio I, Hernández-Ledesma B. Antioxidant activity and protective effects of peptide lunasin against oxidative stress in intestinal Caco-2 cells. Food Chem Toxicol 2014; 65: 155-61.
[http://dx.doi.org/10.1016/j.fct.2013.12.021] [PMID: 24365261]

[63]   Nagaoka S, Futamura Y, Miwa K, *et al.* Identification of novel hypocholesterolemic peptides derived from bovine milk beta-lactoglobulin. Biochem Biophys Res Commun 2001; 281(1): 11-7.
[http://dx.doi.org/10.1006/bbrc.2001.4298] [PMID: 11178953]

[64]   Tulipano G, Sibilia V, Caroli AM, Cocchi D. Whey proteins as source of dipeptidyl dipeptidase IV (dipeptidyl peptidase-4) inhibitors. Peptides 2011; 32(4): 835-8.
[http://dx.doi.org/10.1016/j.peptides.2011.01.002] [PMID: 21256171]

[65]   Yu Z, Yin Y, Zhao W, *et al.* Novel peptides derived from egg white protein inhibiting alpha-glucosidase. Food Chem 2011; 129(4): 1376-82.
[http://dx.doi.org/10.1016/j.foodchem.2011.05.067]

[66]   Majumder K, Chakrabarti S, Davidge ST, Wu J. Structure and activity study of egg protein ovotransferrin derived peptides (IRW and IQW) on endothelial inflammatory response and oxidative stress. J Agric Food Chem 2013; 61(9): 2120-9.
[http://dx.doi.org/10.1021/jf3046076] [PMID: 23317476]

[67]   Cho SJ, Juillerat MA, Lee CH. Identification of LDL-receptor transcription stimulating peptides from soybean hydrolysate in human hepatocytes. J Agric Food Chem 2008; 56(12): 4372-6.
[http://dx.doi.org/10.1021/jf800676a] [PMID: 18500811]

[68]   Fitzgerald C, Mora-Soler L, Gallagher E, *et al.* Isolation and characterization of bioactive pro-peptides with *in vitro* renin inhibitory activities from the macroalga *Palmaria palmata*. J Agric Food Chem 2012; 60(30): 7421-7.
[http://dx.doi.org/10.1021/jf301361c] [PMID: 22747312]

[69]   Yue X, Falk T, Zuniga LA, *et al.* Effects of the novel glycopeptide opioid agonist MMP-2200 in preclinical models of Parkinson's disease. Brain Res 2011; 1413: 72-83.
[http://dx.doi.org/10.1016/j.brainres.2011.07.038] [PMID: 21840512]

[70]   Yamauchi R, Sonoda S, Jinsmaa Y, Yoshikawa M. Antinociception induced by β-lactotensin, a neurotensin agonist peptide derived from β-lactoglobulin, is mediated by NT2 and D1 receptors. Life Sci 2003; 73(15): 1917-23.
[http://dx.doi.org/10.1016/S0024-3205(03)00546-0] [PMID: 12899917]

[71]   Ohinata K, Sonoda S, Inoue N, Yamauchi R, Wada K, Yoshikawa M. beta-Lactotensin, a neurotensin agonist peptide derived from bovine beta-lactoglobulin, enhances memory consolidation in mice. Peptides 2007; 28(7): 1470-4.
[http://dx.doi.org/10.1016/j.peptides.2007.06.002] [PMID: 17629352]

[72]   Pins JJ, Keenan JM. Effects of whey peptides on cardiovascular disease risk factors. J Clin Hypertens (Greenwich) 2006; 8(11): 775-82.
[http://dx.doi.org/10.1111/j.1524-6175.2006.05667.x]

[73]   Pal S, Ellis V, Dhaliwal S. Effects of whey protein isolate on body composition, lipids, insulin and glucose in overweight and obese individuals. Br J Nutr 2010; 104(5): 716-23.
[http://dx.doi.org/10.1017/S0007114510000991] [PMID: 20377924]

[74]   Cho SJ, Juillerat MA, Lee CH. Cholesterol lowering mechanism of soybean protein hydrolysate. J Agric Food Chem 2007; 55(26): 10599-604.
[http://dx.doi.org/10.1021/jf071903f] [PMID: 18052124]

[75]   Zhang X, Beynen AC. Lowering effect of dietary milk-whey protein v. casein on plasma and liver cholesterol concentrations in rats. Br J Nutr 1993; 70(1): 139-46.
[http://dx.doi.org/10.1079/BJN19930111] [PMID: 8399095]

[76]   Nagaoka S, Kanamaru Y, Kuzuya Y, Kojima T, Kuwata T. Comparative studies on the serum cholesterol lowering action of whey protein and soybean protein in rats. Biosci Biotechnol Biochem 2014; 56(9): 1484-5.
[http://dx.doi.org/10.1271/bbb.56.1484]

[77]   Petersen BL, Ward LS, Bastian ED, Jenkins AL, Campbell J, Vuksan V. A whey protein supplement decreases post-prandial glycemia. Nutr J 2009; 8(1): 47.
[http://dx.doi.org/10.1186/1475-2891-8-47] [PMID: 19835582]

[78]   Nongonierma AB, FitzGerald RJ. Susceptibility of milk protein-derived peptides to dipeptidyl peptidase IV (DPP-IV) hydrolysis. Food Chem 2014; 145: 845-52.
[http://dx.doi.org/10.1016/j.foodchem.2013.08.097] [PMID: 24128555]

[79]   Nongonierma AB, FitzGerald RJ. Dipeptidyl peptidase IV inhibitory and antioxidative properties of milk protein-derived dipeptides and hydrolysates. Peptides 2013; 39(1): 157-63.
[http://dx.doi.org/10.1016/j.peptides.2012.11.016] [PMID: 23219487]

[80]   Lambeir A-M, Durinx C, Scharpé S, De Meester I. Dipeptidyl-peptidase IV from bench to bedside: an update on structural properties, functions, and clinical aspects of the enzyme DPP IV. Crit Rev Clin Lab Sci 2003; 40(3): 209-94.
[http://dx.doi.org/10.1080/713609354] [PMID: 12892317]

[81]   Uenishi H, Kabuki T, Seto Y, Serizawa A, Nakajima H. Isolation and identification of casein-derived dipeptidyl-peptidase 4 (DPP-4)-inhibitory peptide LPQNIPPL from gouda-type cheese and its effect on plasma glucose in rats. Int Dairy J 2012; 22(1): 24-30.
[http://dx.doi.org/10.1016/j.idairyj.2011.08.002]

[82]   Ramchandran L, Shah NP. Proteolytic profiles and angiotensin-I converting enzyme and alpha-glucosidase inhibitory activities of selected lactic acid bacteria. J Food Sci 2008; 73(2): M75-81.
[http://dx.doi.org/10.1111/j.1750-3841.2007.00643.x] [PMID: 18298740]

[83]   Meisel H. Biochemical properties of peptides encrypted in bovine milk proteins. Curr Med Chem 2005; 12(16): 1905-19.
[http://dx.doi.org/10.2174/0929867054546618] [PMID: 16101509]

[84]   Hayes M, Stanton C, Fitzgerald GF, Ross RP. Putting microbes to work: dairy fermentation, cell factories and bioactive peptides. Part II: bioactive peptide functions. Biotechnol J 2007; 2(4): 435-49.
[http://dx.doi.org/10.1002/biot.200700045] [PMID: 17407211]

[85]   Hou IC, Suzuki C, Kanegawa N, *et al.* β-Lactotensin derived from bovine β-lactoglobulin exhibits anxiolytic-like activity as an agonist for neurotensin NTS(2) receptor via activation of dopamine D(1) receptor in mice. J Neurochem 2011; 119(4): 785-90.
[http://dx.doi.org/10.1111/j.1471-4159.2011.07472.x] [PMID: 21895659]

[86]   Sonklin C, Laohakunjit N, Kerdchoechuen O. Physicochemical and flavor characteristics of flavoring agent from mungbean protein hydrolyzed by bromelain. J Agric Food Chem 2011; 59(15): 8475-83.
[http://dx.doi.org/10.1021/jf202006a] [PMID: 21739999]

[87]   Liao L, Qiu CY, Liu TX, Zhao MM, Ren JY, Zhao HF. Susceptibility of wheat gluten to enzymatic hydrolysis following deamidation with acetic acid and sensory characteristics of the resultant

hydrolysates. J Cereal Sci 2010; 52(3): 395-403.
[http://dx.doi.org/10.1016/j.jcs.2010.07.001]

[88] Dang Y, Gao X, Ma F, Wu X. Comparison of umami taste peptides in water-soluble extractions of Jinhua and Parma hams. Lebensm Wiss Technol 2015; 60(2): 1179-86.
[http://dx.doi.org/10.1016/j.lwt.2014.09.014]

[89] Koo SH, Bae IY, Lee S, Lee DH, Hur BS, Lee HG. Evaluation of wheat gluten hydrolysates as taste-active compounds with antioxidant activity. J Food Sci Technol 2014; 51(3): 535-42.
[http://dx.doi.org/10.1007/s13197-011-0515-9] [PMID: 24587529]

[90] Lioe HN, Takara Y, Kensaku M. Evaluation of peptide contribution to the intense umami taste of japanese soy sauces. J Food Sci 2005; 71(3): S277-83.
[http://dx.doi.org/10.1111/j.1365-2621.2006.tb15654.x]

[91] Bagnasco L, Pappalardo VM, Meregaglia A, *et al.* Use of food-grade proteases to recover umami protein-peptide mixtures from rice middlings. Food Res Int 2013; 50(1): 420-7.
[http://dx.doi.org/10.1016/j.foodres.2012.11.007]

[92] Su G, Cui C, Zheng L, Yang B, Ren J, Zhao M. Isolation and identification of two novel umami and umami-enhancing peptides from peanut hydrolysate by consecutive chromatography and MALDI-TOF/TOF MS. Food Chem 2012; 135(2): 479-85.
[http://dx.doi.org/10.1016/j.foodchem.2012.04.130] [PMID: 22868117]

[93] Cho W. Il, Kim SM. Taste compounds and biofunctional activities of the sandy beach clam hydrolysate for the shellfish flavoring condiment taste compounds and biofunctional activities of the sandy beach clam hydrolysate for the shellfish flavoring condiment. J Aquat Food Prod Technol 2016; 25(1): 24-34.
[http://dx.doi.org/10.1080/10498850.2013.822447]

[94] Guo X, Han X, He Y, Du H, Tan Z. Optimization of enzymatic hydrolysis for preparation of shrimp flavor precursor using response surface methodology. J Food Qual 2014; 37(4): 229-36.
[http://dx.doi.org/10.1111/jfq.12091]

[95] Laohakunjit N, Selamassakul O, Kerdchoechuen O. Seafood-like flavour obtained from the enzymatic hydrolysis of the protein by-products of seaweed (*Gracilaria* sp.). Food Chem 2014; 158: 162-70.
[http://dx.doi.org/10.1016/j.foodchem.2014.02.101] [PMID: 24731327]

[96] FitzGerald RJ, O'Cuinn G. Enzymatic debittering of food protein hydrolysates. Biotechnol Adv 2006; 24(2): 234-7.
[http://dx.doi.org/10.1016/j.biotechadv.2005.11.002] [PMID: 16386868]

[97] Fadda S, López C, Vignolo G. Role of lactic acid bacteria during meat conditioning and fermentation: peptides generated as sensorial and hygienic biomarkers. Meat Sci 2010; 86(1): 66-79.
[http://dx.doi.org/10.1016/j.meatsci.2010.04.023] [PMID: 20619799]

[98] Fox PF, Condon JJ, Eds. JE K. Relationship between structure and functional properties of food proteins. Food proteins. London: Applied Science Publishers Ltd 1982; pp. 51-103.

[99] Kramer RM, Shende VR, Motl N, Pace CN, Scholtz JM. Toward a molecular understanding of protein solubility: increased negative surface charge correlates with increased solubility. Biophys J 2012; 102(8): 1907-15.
[http://dx.doi.org/10.1016/j.bpj.2012.01.060] [PMID: 22768947]

[100] Li W, Zhou R, Mu Y. Salting effects on protein components in aqueous NaCl and urea solutions: toward understanding of urea-induced protein denaturation. J Phys Chem B 2012; 116(4): 1446-51.
[http://dx.doi.org/10.1021/jp210769q] [PMID: 22216970]

[101] Pihlasalo S, Auranen L, Hänninen P, Härmä H. Method for estimation of protein isoelectric point. Anal Chem 2012; 84(19): 8253-8.
[http://dx.doi.org/10.1021/ac301569b] [PMID: 22946671]

[102] Khiari Z, Omana DA, Pietrasik Z, Betti M. Evaluation of poultry protein isolate as a food ingredient:

physicochemical properties and sensory characteristics of marinated chicken breasts. J Food Sci 2013; 78(7): S1069-75.
[http://dx.doi.org/10.1111/1750-3841.12167] [PMID: 23772877]

[103] Khiari Z, Pietrasik Z, Gaudette NJ, Betti M. Poultry protein isolate prepared using an acid solubilization/precipitation extraction influences the microstructure, the functionality and the consumer acceptability of a processed meat product. Food Struct 2014; 2(1–2): 49-60.
[http://dx.doi.org/10.1016/j.foostr.2014.08.002]

[104] Gehring CK, Gigliotti JC, Moritz JS, Tou JC, Jaczynski J. Functional and nutritional characteristics of proteins and lipids recovered by isoelectric processing of fish by-products and low-value fish: A review. Food Chem 2011; 124(2): 422-31.
[http://dx.doi.org/10.1016/j.foodchem.2010.06.078]

[105] Tahergorabi R, Beamer SK, Matak KE, Jaczynski J. Isoelectric solubilization/precipitation as a means to recover protein isolate from striped bass (*Morone saxatilis*) and its physicochemical properties in a nutraceutical seafood product. J Agric Food Chem 2012; 60(23): 5979-87.
[http://dx.doi.org/10.1021/jf3001197] [PMID: 22624700]

[106] McClements DJ. Food emulsions: Principles, practices, and techniques. 3rd ed., Boca Raton: CRC Press 2015.
[http://dx.doi.org/10.1201/b18868]

[107] McClements DJ. Emulsion design to improve the delivery of functional lipophilic components. Annu Rev Food Sci Technol 2010; 1(1): 241-69.
[http://dx.doi.org/10.1146/annurev.food.080708.100722] [PMID: 22129337]

[108] Comas DI, Wagner JR, Tomás MC. Creaming stability of oil in water (O/W) emulsions: Influence of pH on soybean protein-lecithin interaction. Food Hydrocoll 2006; 20(7): 990-6.
[http://dx.doi.org/10.1016/j.foodhyd.2005.11.006]

[109] McClements DJ. Biopolymers in food emulsions.Modern biopolymer science. London: Academic Press 2009; pp. 129-66.
[http://dx.doi.org/10.1016/B978-0-12-374195-0.00004-5]

[110] Silva HD, Cerqueira MA, Vicente AA. Nanoemulsions for food applications: Development and characterization. Food Bioprocess Technol 2012; 5(3): 854-67.
[http://dx.doi.org/10.1007/s11947-011-0683-7]

[111] Adjonu R, Doran G, Torley P, Agboola S. Formation of whey protein isolate hydrolysate stabilised nanoemulsion. Food Hydrocoll 2014; 41: 169-77.
[http://dx.doi.org/10.1016/j.foodhyd.2014.04.007]

[112] He W, Tan Y, Tian Z, Chen L, Hu F, Wu W. Food protein-stabilized nanoemulsions as potential delivery systems for poorly water-soluble drugs: preparation, in vitro characterization, and pharmacokinetics in rats. Int J Nanomedicine 2011; 6: 521-33.
[PMID: 21468355]

[113] Relkin P, Shukat R, Bourgaux C, Meneau F. Nanostructures and polymorphisms in protein stabilised lipid nanoparticles, as food bioactive carriers: Contribution of particle size and adsorbed materials. Procedia Food Sci 2011; 1: 246-50.
[http://dx.doi.org/10.1016/j.profoo.2011.09.039]

[114] Lee SJ, McClements DJ. Fabrication of protein-stabilized nanoemulsions using a combined homogenization and amphiphilic solvent dissolution/evaporation approach. Food Hydrocoll 2010; 24(6–7): 560-9.
[http://dx.doi.org/10.1016/j.foodhyd.2010.02.002]

[115] Maestro A, Rio E, Drenckhan W, Langevin D, Salonen A. Foams stabilised by mixtures of nanoparticles and oppositely charged surfactants: relationship between bubble shrinkage and foam coarsening. Soft Matter 2014; 10(36): 6975-83.
[http://dx.doi.org/10.1039/C4SM00047A] [PMID: 24832218]

[116]   Schmidt I, Novales B, Boué F, Axelos MAV. Foaming properties of protein/pectin electrostatic complexes and foam structure at nanoscale. J Colloid Interface Sci 2010; 345(2): 316-24.
[http://dx.doi.org/10.1016/j.jcis.2010.01.016] [PMID: 20172532]

[117]   Moro A, Báez GD, Ballerini GA, Busti PA, Delorenzi NJ. Emulsifying and foaming properties of β-lactoglobulin modified by heat treatment. Food Res Int 2013; 51(1): 1-7.
[http://dx.doi.org/10.1016/j.foodres.2012.11.011]

[118]   Glaser LA, Paulson AT, Speers RA, Yada RY, Rousseau D. Foaming behavior of mixed bovine serum albumin-protamine systems. Food Hydrocoll 2007; 21(4): 495-506.
[http://dx.doi.org/10.1016/j.foodhyd.2006.05.008]

[119]   Abirached C, Medrano CA, Araujo AC, Moyna P, Añón MC, Panizzolo LA. Comparison of interfacial and foaming properties of soy and whey protein isolates. J Food Sci Eng 2012; 2: 376-81.

[120]   Salvador P, Saguer E, Parés D, Carretero C, Toldrà M. Foaming and emulsifying properties of porcine red cell protein concentrate. Food Sci Technol Int 2010; 16(4): 289-96.
[http://dx.doi.org/10.1177/1082013209353223] [PMID: 21339145]

[121]   Arzeni C, Pérez OE, Pilosof AMR. Functionality of egg white proteins as affected by high intensity ultrasound. Food Hydrocoll 2012; 29(2): 308-16.
[http://dx.doi.org/10.1016/j.foodhyd.2012.03.009]

[122]   Mine Y. Recent advances in the understanding of egg white protein functionality. Trends Food Sci Technol 1995; 6(7): 225-32.
[http://dx.doi.org/10.1016/S0924-2244(00)89083-4]

[123]   Oboroceanu D, Wang L, Magner E, Auty MAE. Fibrillization of whey proteins improves foaming capacity and foam stability at low protein concentrations. J Food Eng 2014; 121(1): 102-11.
[http://dx.doi.org/10.1016/j.jfoodeng.2013.08.023]

[124]   Wan Z, Yang X, Sagis LMC. Contribution of long fibrils and peptides to surface and foaming behavior of soy protein fibril system. Langmuir 2016; 32(32): 8092-101.
[http://dx.doi.org/10.1021/acs.langmuir.6b01511] [PMID: 27452662]

[125]   Malkin AY, Isayev AI. Rheology concepts, methods, and applications. 2nd ed., Toronto: ChemTec Publisher 2012.

[126]   Hermansson AM. Soy protein gelation. J Am Oil Chem Soc 1986; 63(5): 658-66.
[http://dx.doi.org/10.1007/BF02638232]

[127]   Cordobés F, Partal P, Guerrero A. Rheology and microstructure of heat-induced egg yolk gels. Rheol Acta 2004; 43(2): 184-95.
[http://dx.doi.org/10.1007/s00397-003-0338-3]

[128]   Perrechil FA, Braga ALM, Cunha RL. Acid gelation of native and heat-denatured soy proteins and locust bean gum. Int J Food Sci Technol 2013; 48(3): 620-7.
[http://dx.doi.org/10.1111/ijfs.12007]

[129]   Du L, Keplová L, Khiari Z, Betti M. Preparation and characterization of gelatin from collagen biomass obtained through a pH-shifting process of mechanically separated turkey meat. Poult Sci 2014; 93(4): 989-1000.
[http://dx.doi.org/10.3382/ps.2013-03609] [PMID: 24706977]

[130]   Clark AH, Ross-Murphy SB. Biopolymer network assembly. Measurement and theory.Modern biopolymer science. London: Academic Press 2009; pp. 1-27.
[http://dx.doi.org/10.1016/B978-0-12-374195-0.00001-X]

[131]   Carvalho W, Djabourov M. Physical gelation under shear for gelatin gels. Rheol Acta 1997; 36(6): 591-609.
[http://dx.doi.org/10.1007/BF00367355]

[132]   de Oliveira FC, Coimbra JSDR, de Oliveira EB, Zuñiga ADG, Rojas EEG. food protein-

polysaccharide conjugates obtained *via* the Maillard reaction: A review. Crit Rev Food Sci Nutr 2016; 56(7): 1108-25.
[http://dx.doi.org/10.1080/10408398.2012.755669] [PMID: 24824044]

[133] Ames JM. Maillard Reaction.Biochemistry of food proteins. London: Elsevier Science Publishers 1992; pp. 99-153.
[http://dx.doi.org/10.1007/978-1-4684-9895-0_4]

[134] Shepherd R, Robertson A, Ofman D. Dairy glycoconjugate emulsifiers: Casein-maltodextrins. Food Hydrocoll 2000; 14(4): 281-6.
[http://dx.doi.org/10.1016/S0268-005X(99)00067-3]

[135] Kasran M, Cui SW, Goff HD. Covalent attachment of fenugreek gum to soy whey protein isolate through natural Maillard reaction for improved emulsion stability. Food Hydrocoll 2013; 30(2): 552-8.
[http://dx.doi.org/10.1016/j.foodhyd.2012.08.004]

[136] Liu Y, Zhao G, Zhao M, Ren J, Yang B. Improvement of functional properties of peanut protein isolate by conjugation with dextran through Maillard reaction. Food Chem 2012; 131(3): 901-6.
[http://dx.doi.org/10.1016/j.foodchem.2011.09.074]

[137] Yadav MP, Parris N, Johnston DB, Onwulata CI, Hicks KB. Corn fiber gum and milk protein conjugates with improved emulsion stability. Carbohydr Polym 2010; 81(2): 476-83.
[http://dx.doi.org/10.1016/j.carbpol.2010.03.003]

[138] Al-Hakkak J, Al-Hakkak F. Functional egg white-pectin conjugates prepared by controlled Maillard reaction. J Food Eng 2010; 100(1): 152-9.
[http://dx.doi.org/10.1016/j.jfoodeng.2010.03.040]

[139] Zhang B, Guo X, Zhu K, Peng W, Zhou H. Improvement of emulsifying properties of oat protein isolate-dextran conjugates by glycation. Carbohydr Polym 2015; 127: 168-75.
[http://dx.doi.org/10.1016/j.carbpol.2015.03.072] [PMID: 25965470]

[140] Zhu D, Damodaran S, Lucey JA. Physicochemical and emulsifying properties of whey protein isolate (WPI)-dextran conjugates produced in aqueous solution. J Agric Food Chem 2010; 58(5): 2988-94.
[http://dx.doi.org/10.1021/jf903643p] [PMID: 20146423]

[141] Mulcahy EM, Mulvihill DM, O'Mahony JA. Physicochemical properties of whey protein conjugated with starch hydrolysis products of different dextrose equivalent values. Int Dairy J 2016; 53: 20-8.
[http://dx.doi.org/10.1016/j.idairyj.2015.09.009]

[142] Liu G, Zhong Q. High temperature-short time glycation to improve heat stability of whey protein and reduce color formation. Food Hydrocoll 2015; 44: 453-60.
[http://dx.doi.org/10.1016/j.foodhyd.2014.10.006]

[143] Chen H, Wang P, Wu F, *et al.* Preparation of phosvitin-dextran conjugates under high temperature in a liquid system. Int J Biol Macromol 2013; 55: 258-63.
[http://dx.doi.org/10.1016/j.ijbiomac.2013.01.018] [PMID: 23376557]

[144] Zhang J, Wu N, Lan T, Yang X. Improvement in emulsifying properties of soy protein isolate by conjugation with maltodextrin using high-temperature, short-time dry-heating Maillard reaction. Int J Food Sci Technol 2014; 49(2): 460-7.
[http://dx.doi.org/10.1111/ijfs.12323]

[145] Decourcelle N, Sabourin C, Dauer G, Guérard F. Effect of the Maillard reaction with xylose on the emulsifying properties of a shrimp hydrolysate (*Pandalus borealis*). Food Res Int 2010; 43(8): 2155-60.
[http://dx.doi.org/10.1016/j.foodres.2010.07.026]

[146] Niu L-Y, Jiang S-T, Pan L-J, Zhai Y-S. Characteristics and functional properties of wheat germ protein glycated with saccharides through Maillard reaction. Int J Food Sci Technol 2011; 46(10): 2197-203.
[http://dx.doi.org/10.1111/j.1365-2621.2011.02737.x]

[147] Corzo-Martínez M, Carrera Sánchez C, Moreno FJ, Rodríguez Patino JM, Villamiel M. Interfacial and foaming properties of bovine β-lactoglobulin: Galactose Maillard conjugates. Food Hydrocoll 2012; 27(2): 438-47.
[http://dx.doi.org/10.1016/j.foodhyd.2011.11.003]

[148] Guo X, Xiong YL. Characteristics and functional properties of buckwheat protein–sugar Schiff base complexes. Lebensm Wiss Technol 2013; 51(2): 397-404.
[http://dx.doi.org/10.1016/j.lwt.2012.12.003]

[149] Ter Haar R, Westphal Y, Wierenga PA, Schols HA, Gruppen H. Cross-linking behavior and foaming properties of bovine α-lactalbumin after glycation with various saccharides. J Agric Food Chem 2011; 59(23): 12460-6.
[http://dx.doi.org/10.1021/jf2032022] [PMID: 22010962]

[150] Seo S, Karboune S, L'Hocine L, Yaylayan V. Characterization of glycated lysozyme with galactose, galactooligosaccharides and galactan: Effect of glycation on structural and functional properties of conjugates. Lebensm Wiss Technol 2013; 53(1): 44-53.
[http://dx.doi.org/10.1016/j.lwt.2013.02.001]

[151] Tian S, Chen J, Small DM. Enhancement of solubility and emulsifying properties of soy protein isolates by glucose conjugation. J Food Process Preserv 2011; 35(1): 80-95.
[http://dx.doi.org/10.1111/j.1745-4549.2009.00456.x]

[152] Hrynets Y, Ndagijimana M, Betti M. Non-enzymatic glycation of natural actomyosin (NAM) with glucosamine in a liquid system at moderate temperatures. Food Chem 2013; 139(1-4): 1062-72.
[http://dx.doi.org/10.1016/j.foodchem.2013.02.026] [PMID: 23561210]

[153] Spotti MJ, Perduca MJ, Piagentini A, Santiago LG, Rubiolo AC, Carrara CR. Gel mechanical properties of milk whey protein-dextran conjugates obtained by Maillard reaction. Food Hydrocoll 2013; 31(1): 26-32.
[http://dx.doi.org/10.1016/j.foodhyd.2012.08.009]

[154] Spotti MJ, Martinez MJ, Pilosof AMR, Candioti M, Rubiolo AC, Carrara CR. Rheological properties of whey protein and dextran conjugates at different reaction times. Food Hydrocoll 2014; 38: 76-84.
[http://dx.doi.org/10.1016/j.foodhyd.2013.11.017]

[155] Sun WW, Yu SJ, Yang XQ, *et al.* Study on the rheological properties of heat-induced whey protein isolate-dextran conjugate gel. Food Res Int 2011; 44(10): 3259-63.
[http://dx.doi.org/10.1016/j.foodres.2011.09.019]

[156] Gerrard JA. Protein-protein crosslinking in food: Methods, consequences, applications. Trends Food Sci Technol 2002; 13(12): 389-97.
[http://dx.doi.org/10.1016/S0924-2244(02)00257-1]

[157] Dombkowski AA, Sultana KZ, Craig DB. Protein disulfide engineering. FEBS Lett 2014; 588(2): 206-12.
[http://dx.doi.org/10.1016/j.febslet.2013.11.024] [PMID: 24291258]

[158] Fernandez-Lafuente R. Stabilization of multimeric enzymes: Strategies to prevent subunit dissociation. Enzyme Microb Technol 2009; 45(6–7): 405-18.
[http://dx.doi.org/10.1016/j.enzmictec.2009.08.009]

[159] Hassani L. The effect of chemical modification with pyromellitic anhydride on structure, function, and thermal stability of horseradish peroxidase. Appl Biochem Biotechnol 2012; 167(3): 489-97.
[http://dx.doi.org/10.1007/s12010-012-9671-2] [PMID: 22562551]

[160] Le QAT, Joo JC, Yoo YJ, Kim YH. Development of thermostable *Candida antarctica* lipase B through novel *in silico* design of disulfide bridge. Biotechnol Bioeng 2012; 109(4): 867-76.
[http://dx.doi.org/10.1002/bit.24371] [PMID: 22095554]

[161] Wang Y, Fu Z, Huang H, *et al.* Improved thermal performance of *Thermomyces lanuginosus* GH11 xylanase by engineering of an N-terminal disulfide bridge. Bioresour Technol 2012; 112: 275-9.

[http://dx.doi.org/10.1016/j.biortech.2012.02.092] [PMID: 22425398]

[162] Han ZL, Han SY, Zheng SP, Lin Y. Enhancing thermostability of a *Rhizomucor miehei* lipase by engineering a disulfide bond and displaying on the yeast cell surface. Appl Microbiol Biotechnol 2009; 85(1): 117-26.
[http://dx.doi.org/10.1007/s00253-009-2067-8] [PMID: 19533118]

[163] Chan JTY, Omana DA, Betti M. Application of high pressure processing to improve the functional properties of pale, soft, and exudative (PSE)-like turkey meat. Innov Food Sci Emerg Technol 2011; 12(3): 216-25.
[http://dx.doi.org/10.1016/j.ifset.2011.03.004]

[164] Sun XD, Holley RA. High hydrostatic pressure effects on the texture of meat and meat products. J Food Sci 2010; 75(1): R17-23.
[http://dx.doi.org/10.1111/j.1750-3841.2009.01449.x] [PMID: 20492191]

[165] Gómez-Guillén MC, Giménez B, Montero P. Extraction of gelatin from fish skins by high pressure treatment. Food Hydrocoll 2005; 19(5): 923-8.
[http://dx.doi.org/10.1016/j.foodhyd.2004.12.011]

[166] Gekko K, Fukamizu M. Effect of pressure on the sol-gel transition of gelatin. Int J Biol Macromol 1991; 13(5): 295-300.
[http://dx.doi.org/10.1016/0141-8130(91)90030-X] [PMID: 1801903]

[167] Shimada K, Sakai Y, Nagamatsu K, Hori T, Hayashi R. Gel-setting and gel-melting temperatures of aqueous gelatin solutions under high pressure measured by a hot-wire method. Biosci Biotechnol Biochem 1996; 60(8): 1349-50.
[http://dx.doi.org/10.1271/bbb.60.1349]

[168] Condés MC, Añón MC, Mauri AN. Amaranth protein films prepared with high-pressure treated proteins. J Food Eng 2015; 166: 38-44.
[http://dx.doi.org/10.1016/j.jfoodeng.2015.05.005]

[169] Kruk ZA, Yun H, Rutley DL, Lee EJ, Kim YJ, Jo C. The effect of high pressure on microbial population, meat quality and sensory characteristics of chicken breast fillet. Food Control 2011; 22(1): 6-12.
[http://dx.doi.org/10.1016/j.foodcont.2010.06.003]

[170] Souza CM, Boler DD, Clark DL, *et al.* The effects of high pressure processing on pork quality, palatability, and further processed products. Meat Sci 2011; 87(4): 419-27.
[http://dx.doi.org/10.1016/j.meatsci.2010.11.023] [PMID: 21172731]

[171] Lowder AC, Mireles Dewitt CA. Impact of high pressure processing on the functional aspects of beef muscle injected with salt and/or sodium phosphates. J Food Process Preserv 2014; 38(4): 1840-8.
[http://dx.doi.org/10.1111/jfpp.12155]

[172] Awad TS, Moharram HA, Shaltout OE, Asker D, Youssef MM. Applications of ultrasound in analysis, processing and quality control of food: A review. Food Res Int 2012; 48(2): 410-27.
[http://dx.doi.org/10.1016/j.foodres.2012.05.004]

[173] Soria AC, Villamiel M. Effect of ultrasound on the technological properties and bioactivity of food: A review. Trends Food Sci Technol 2010; 21(7): 323-31.
[http://dx.doi.org/10.1016/j.tifs.2010.04.003]

[174] Vilkhu K, Mawson R, Simons L, Bates D. Applications and opportunities for ultrasound assisted extraction in the food industry - A review. Innov Food Sci Emerg Technol 2008; 9(2): 161-9.
[http://dx.doi.org/10.1016/j.ifset.2007.04.014]

[175] Jambrak AR, Mason TJ, Lelas V, Herceg Z, Herceg IL. Effect of ultrasound treatment on solubility and foaming properties of whey protein suspensions. J Food Eng 2008; 86(2): 281-7.
[http://dx.doi.org/10.1016/j.jfoodeng.2007.10.004]

[176] Jambrak AR, Lelas V, Mason TJ, Krešić G, Badanjak M. Physical properties of ultrasound treated soy

proteins. J Food Eng 2009; 93(4): 386-93.
[http://dx.doi.org/10.1016/j.jfoodeng.2009.02.001]

[177] Jambrak AR, Mason TJ, Lelas V, Paniwnyk L, Herceg Z. Effect of ultrasound treatment on particle size and molecular weight of whey proteins. J Food Eng 2014; 121(1): 15-23.
[http://dx.doi.org/10.1016/j.jfoodeng.2013.08.012]

[178] Arzeni C, Martínez K, Zema P, Arias A, Pérez OE, Pilosof AMR. Comparative study of high intensity ultrasound effects on food proteins functionality. J Food Eng 2012; 108(3): 463-72.
[http://dx.doi.org/10.1016/j.jfoodeng.2011.08.018]

[179] Yokoyama K, Nio N, Kikuchi Y. Properties and applications of microbial transglutaminase. Appl Microbiol Biotechnol 2004; 64(4): 447-54.
[http://dx.doi.org/10.1007/s00253-003-1539-5] [PMID: 14740191]

[180] Sun XD, Arntfield SD. Gelation properties of chicken myofibrillar protein induced by transglutaminase crosslinking. J Food Eng 2011; 107(2): 226-33.
[http://dx.doi.org/10.1016/j.jfoodeng.2011.06.019]

[181] Giosafatto CVL, Rigby NM, Wellner N, Ridout M, Husband F, Mackie AR. Microbial transglutaminase-mediated modification of ovalbumin. Food Hydrocoll 2012; 26(1): 261-7.
[http://dx.doi.org/10.1016/j.foodhyd.2011.06.003]

[182] Binsi PK, Shamasundar BA. Purification and characterisation of transglutaminase from four fish species: Effect of added transglutaminase on the viscoelastic behaviour of fish mince. Food Chem 2012; 132(4): 1922-9.
[http://dx.doi.org/10.1016/j.foodchem.2011.12.027]

[183] De Sá EMF, Bordignon-Luiz MT. The effect of transglutaminase on the properties of milk gels and processed cheese. Int J Dairy Technol 2010; 63(2): 243-51.
[http://dx.doi.org/10.1111/j.1471-0307.2010.00568.x]

[184] Beck M, Jekle M, Selmair PL, Koehler P, Becker T. Rheological properties and baking performance of rye dough as affected by transglutaminase. J Cereal Sci 2011; 54(1): 29-36.
[http://dx.doi.org/10.1016/j.jcs.2011.01.012]

[185] Jiang S-J, Zhao X-H. Cross-linking and glucosamine conjugation of casein by transglutaminase and the emulsifying property and digestibility *in vitro* of the modified product. Int J Food Prop 2012; 15(6): 1286-99.
[http://dx.doi.org/10.1080/10942912.2010.521274]

[186] Hrynets Y, Ndagijimana M, Betti M. Transglutaminase-catalyzed glycosylation of natural actomyosin (NAM) using glucosamine as amine donor: Functionality and gel microstructure. Food Hydrocoll 2014; 36: 26-36.
[http://dx.doi.org/10.1016/j.foodhyd.2013.09.001]

[187] Hong PK, Gottardi D, Ndagijimana M, Betti M. Glycation and transglutaminase mediated glycosylation of fish gelatin peptides with glucosamine enhance bioactivity. Food Chem 2014; 142: 285-93.
[http://dx.doi.org/10.1016/j.foodchem.2013.07.045] [PMID: 24001843]

[188] Gottardi D, Hong PK, Ndagijimana M, Betti M. Conjugation of gluten hydrolysates with glucosamine at mild temperatures enhances antioxidant and antimicrobial properties. Lebensm Wiss Technol 2014; 57(1): 181-7.
[http://dx.doi.org/10.1016/j.lwt.2014.01.013]

[189] Rustad T, Storrø I, Slizyte R. Possibilities for the utilisation of marine by-products. Int J Food Sci Technol 2011; 46(10): 2001-14.
[http://dx.doi.org/10.1111/j.1365-2621.2011.02736.x]

[190] Zheng XQ, Wang JT, Liu XL, *et al.* Effect of hydrolysis time on the physicochemical and functional properties of corn glutelin by Protamex hydrolysis. Food Chem 2015; 172: 407-15.

[http://dx.doi.org/10.1016/j.foodchem.2014.09.080] [PMID: 25442571]

[191]   Cai L, Xiao L, Liu C, Ying T. Functional Properties and Bioactivities of Pine Nut (*Pinus gerardiana*) Protein Isolates and Its Enzymatic Hydrolysates. Food Bioprocess Technol 2013; 6(8): 2109-17.
[http://dx.doi.org/10.1007/s11947-012-0885-7]

[192]   Barac M, Cabrilo S, Stanojevic S, Pesic M, Pavlicevic M, Zlatkovic B, *et al.* Functional properties of protein hydrolysates from pea (*Pisum sativum*, L) seeds. Int J Food Sci Technol 2012; 47(7): 1457-67.
[http://dx.doi.org/10.1111/j.1365-2621.2012.02993.x]

[193]   Chen C, Chi YJ. Antioxidant, ace inhibitory activities and functional properties of egg white protein hydrolysate. J Food Biochem 2012; 36(4): 383-94.
[http://dx.doi.org/10.1111/j.1745-4514.2011.00555.x]

[194]   Naqash SY, Nazeer RA. Antioxidant and functional properties of protein hydrolysates from pink perch (*Nemipterus japonicus*) muscle. J Food Sci Technol 2013; 50(5): 972-8.
[http://dx.doi.org/10.1007/s13197-011-0416-y] [PMID: 24426005]

[195]   Ktari N, Jridi M, Bkhairia I, Sayari N, Ben Salah R, Nasri M. Functionalities and antioxidant properties of protein hydrolysates from muscle of zebra blenny (*Salaria basilisca*) obtained with different crude protease extracts. Food Res Int 2012; 49(2): 747-56.
[http://dx.doi.org/10.1016/j.foodres.2012.09.024]

[196]   Zhao Q, Xiong H, Selomulya C, *et al.* Enzymatic hydrolysis of rice dreg protein: effects of enzyme type on the functional properties and antioxidant activities of recovered proteins. Food Chem 2012; 134(3): 1360-7.
[http://dx.doi.org/10.1016/j.foodchem.2012.03.033] [PMID: 25005954]

[197]   Balti R, Bougatef A, El-Hadj Ali N, Zekri D, Barkia A, Nasri M. Influence of degree of hydrolysis on functional properties and angiotensin I-converting enzyme-inhibitory activity of protein hydrolysates from cuttlefish (*Sepia officinalis*) by-products. J Sci Food Agric 2010; 90(12): 2006-14.
[PMID: 20583200]

[198]   Su G, Ren J, Yang B, Cui C, Zhao M. Comparison of hydrolysis characteristics on defatted peanut meal proteins between a protease extract from *Aspergillus oryzae* and commercial proteases. Food Chem 2011; 126(3): 1306-11.
[http://dx.doi.org/10.1016/j.foodchem.2010.11.083]

[199]   Yust M del M, Pedroche J, Millán-Linares M del C, Alcaide-Hidalgo JM, Millán F. Improvement of functional properties of chickpea proteins by hydrolysis with immobilised Alcalase. Food Chem 2010; 122(4): 1212-7.
[http://dx.doi.org/10.1016/j.foodchem.2010.03.121]

[200]   Nalinanon S, Benjakul S, Kishimura H, Shahidi F. Functionalities and antioxidant properties of protein hydrolysates from the muscle of ornate threadfin bream treated with pepsin from skipjack tuna. Food Chem 2011; 124(4): 1354-62.
[http://dx.doi.org/10.1016/j.foodchem.2010.07.089]

[201]   Jamdar SN, Rajalakshmi V, Pednekar MD, Juan F, Yardi V, Sharma A. Influence of degree of hydrolysis on functional properties, antioxidant activity and ACE inhibitory activity of peanut protein hydrolysate. Food Chem 2010; 121(1): 178-84.
[http://dx.doi.org/10.1016/j.foodchem.2009.12.027]

[202]   Liu Q, Kong B, Xiong YL, Xia X. Antioxidant activity and functional properties of porcine plasma protein hydrolysate as influenced by the degree of hydrolysis. Food Chem 2010; 118(2): 403-10.
[http://dx.doi.org/10.1016/j.foodchem.2009.05.013]

[203]   Banach JC, Lin Z, Lamsal BP. Enzymatic modification of milk protein concentrate and characterization of resulting functional properties. Lebensm Wiss Technol 2013; 54(2): 397-403.
[http://dx.doi.org/10.1016/j.lwt.2013.06.023]

[204]   O'Loughlin IB, Murray BA, Kelly PM, FitzGerald RJ, Brodkorb A. Enzymatic hydrolysis of heat-

induced aggregates of whey protein isolate. J Agric Food Chem 2012; 60(19): 4895-904.
[http://dx.doi.org/10.1021/jf205213n] [PMID: 22533541]

[205] Khiari Z, Ndagijimana M, Betti M. Low molecular weight bioactive peptides derived from the enzymatic hydrolysis of collagen after isoelectric solubilization/precipitation process of turkey by-products. Poult Sci 2014; 93(9): 2347-62.
[http://dx.doi.org/10.3382/ps.2014-03953] [PMID: 24931971]

[206] Zhao G, Liu Y, Zhao M, Ren J, Yang B. Enzymatic hydrolysis and their effects on conformational and functional properties of peanut protein isolate. Food Chem 2011; 127(4): 1438-43.
[http://dx.doi.org/10.1016/j.foodchem.2011.01.046]

[207] Foegeding EA, Luck PJ, Davis JP. Factors determining the physical properties of protein foams. Food Hydrocoll 2006; 20(2–3): 284-92.
[http://dx.doi.org/10.1016/j.foodhyd.2005.03.014]

[208] Shahidi F, Janak Kamil YVA. Enzymes from fish and aquatic invertebrates and their application in the food industry. Trends Food Sci Technol 2001; 12(12): 435-64.
[http://dx.doi.org/10.1016/S0924-2244(02)00021-3]

[209] Ferraro V, Cruz IB, Jorge RF, Malcata FX, Pintado ME, Castro PML. Valorisation of natural extracts from marine source focused on marine by-products: A review. Food Res Int 2010; 43(9): 2221-33.
[http://dx.doi.org/10.1016/j.foodres.2010.07.034]

[210] Venugopal V, Shahidi F. Value-added products from underutilized fish species. Crit Rev Food Sci Nutr 1995; 35(5): 431-53.
[http://dx.doi.org/10.1080/10408399509527708] [PMID: 8573282]

[211] Shahidi F, Han X. Production and characteristics of protein hydrolysates from capelin (*Mallotus villosus*). Food Chem 1995; 53: 285-93.
[http://dx.doi.org/10.1016/0308-8146(95)93934-J]

[212] Aspevik T, Egede-Nissen H, Oterhals Å. A systematic approach to comparison of the cost efficiency of endopeptidases to hydrolyze Atlantic salmon (*Salmo salar*) by-products. Food Technol Biotechnol 2016; 54(3): 1-37.
[http://dx.doi.org/10.17113/ftb.54.04.16.4553] [PMID: 27904386]

[213] Gonzalez-Gonzalez CR, Tuohy KM, Jauregi P. Production of angiotensin-I-converting enzyme (ACE) inhibitory activity in milk fermented with probiotic strains: Effects of calcium, pH and peptides on the ACE-inhibitory activity. Int Dairy J 2011; 21(9): 615-22.
[http://dx.doi.org/10.1016/j.idairyj.2011.04.001]

[214] Adler-Nissen J. Determination of the degree of hydrolysis of food protein hydrolysates by trinitrobenzenesulfonic acid. J Agric Food Chem 1979; 27(6): 1256-62.
[http://dx.doi.org/10.1021/jf60226a042] [PMID: 544653]

[215] Adler-Nissen J. Methods in food protein hydrolysis. Enzymic hydrolysis of food proteins. New York: Elsevier Applied Science Publishers 1986; pp. 116-24.

[216] Bah CSF, Carne A, McConnell MA, Mros S, Bekhit Ael-D. Production of bioactive peptide hydrolysates from deer, sheep, pig and cattle red blood cell fractions using plant and fungal protease preparations. Food Chem 2016; 202: 458-66.
[http://dx.doi.org/10.1016/j.foodchem.2016.02.020] [PMID: 26920319]

[217] Feller G, Gerday C. Psychrophilic enzymes: hot topics in cold adaptation. Nat Rev Microbiol 2003; 1(3): 200-8.
[http://dx.doi.org/10.1038/nrmicro773] [PMID: 15035024]

CHAPTER 2

Bacterial Exopolysaccharides and their Applications in the Food Industry

P.H.P. Prasanna[*]

Department of Animal & Food Sciences, Faculty of Agriculture, Rajarata University of Sri Lanka, Anuradhapura, Sri Lanka

Abstract: Exopolysaccharide (EPS) production has been reported in both prokaryotes and eukaryotes. Bacterial EPSs have been used in both food and non-food applications. The demand for these EPSs has increased in the last two decades due to various advantages associated with these biopolymers compared to other plant and animal based hydrocolloids. Synthesis mechanism, structure, and characteristics of EPS vary with species of bacteria. Different methods of isolation and characterization of bacterial EPS have been reported. EPSs of bacteria have various technological properties, which suit their application in various industries. This chapter discusses the classification of EPS, EPS-producing bacterial species, biosynthesis of homopolysaccharides and heteropolysaccharides, extraction and purification of EPS, characterization of EPS, functional properties and various food applications of EPSs, and health benefits of EPSs.

Keywords: Animal, Bacteria, Characterization, Exopolysaccharide, Food Applications, Health Benefits, Heteropolysaccharides, Homopolysaccharides, Hydrocolloids, Isolation, Synthesis, Plant.

INTRODUCTION

Polysaccharides are commonly used as major ingredients in the food industry. They are basically used as stabilizers, thickeners, and gelling agents in different food products [1]. They are natural substances and mainly derived from plants and microorganisms. The main sources for these polysaccharides include bacteria, fungi, algae, and plants; the last two are commercially used to fulfil the demand of the food industry. Bacteria produce different polysaccharides, which can be defined based on the location of the cell. Fig. (**2.1**) shows the classification of bacterial polysaccharides. Bacteria use cytosolic polysaccharides to fulfil their carbon and energy requirements. Cell wall polysaccharides, such as

---

[*] **Corresponding author P.H.P. Prasanna:** Department of Animal & Food Sciences, Faculty of Agriculture, Rajarata University of Sri Lanka, Anuradhapura, Sri Lanka; Tel: 94714458656; Fax: 94252221614; Email: phpprasanna@yahoo.com

Ali Osman (Ed.)

peptidoglycans and teichoic acids are considered as constituents of cell walls. There are two types of exo-cellular polymers namely capsular polysaccharide (CPS) and exopolysaccharide (EPS). CPSs form capsules while EPSs are secreted into the environment [2].

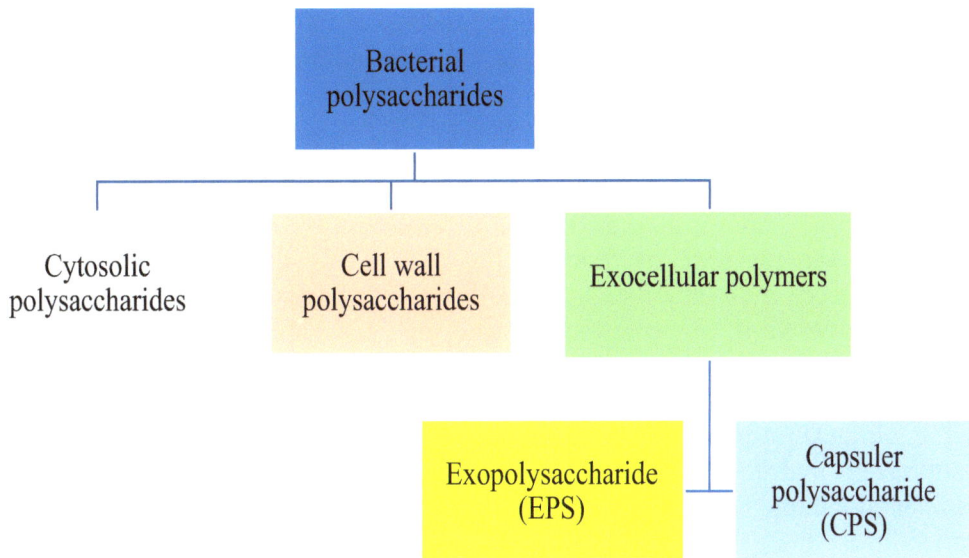

Fig. (2.1). Classification of bacterial polysaccharides.

EPSs have many physiological roles associated with their producers. In general, EPSs are produced by these bacteria to address adverse environmental conditions, such as temperature, pressure, and light intensity [3]. Bacteria do not use EPSs as energy reserves and they have no ability to catabolize EPSs to fulfil their energy requirements [4]. EPSs are highly hydrated, which can protect bacteria from desiccation, phagocytosis, predation by protozoa and phage attack, antibiotics or toxic compounds, and osmotic stress. In addition, EPSs are used to capture essential nutrients and minerals from the environment. Some EPSs have been shown to degrade some metals, which is due to their anionic characteristics. Furthermore, they have a role in cell recognition and adhesion to surfaces [5]. In addition, some EPS-producing bacteria produce biofilms that create a hydrated scaffolding for the producers. The biofilm can give a protection for them from biocides and antimicrobial agents [6].

EPSs of many different bacteria have been used in various food and non-food applications. These include food products, such as dairy products, pharmaceuticals, bio-emulsifiers, bio-flocculants, and chemical products [7]. The demand for EPSs from different bacteria has been increasing in the last decade due to their generally recognized as safe (GRAS) status and environmentally friendly production. In addition, EPSs of many bacterial species have been shown to have high activity in various applications at a low concentration compared to commercially available polymers [8]. The demand for EPSs increases further due to the reported health benefits associated with the consumption of EPSs containing food products, and these health benefits include immunomodulation, immunostimulation, antitumor properties, anti-inflammatory properties, and antioxidant activities [9].

EPSs from bacteria have some advantages over plant or animal hydrocolloids in food and other applications [10, 11]. They show lower toxicity and higher biodegradability. In addition, they are more environmentally compatible than other materials. In the case of food applications, EPSs show higher foaming activity and selectivity over other hydrocolloids. Furthermore, EPSs have been shown to be effective in extreme temperature, pH, and salinity conditions [12]. Sometimes the demand for hydrocolloids cannot be fulfilled by components of animal and plant origin [13] and furthermore, these hydrocolloids cannot fulfil the desired rheological properties in food products. In addition, prices and availability of plant-based hydrocolloids change with time due to drought and other environmental factors in areas where these plants are grown [12]. In the context of vegetarian lifestyles and also due to cultural considerations, some consumers refrain from eating food containing animal-based hydrocolloids; thus, the demand for bacterial EPSs increases [14]. EPSs have no taste and, therefore, EPSs can be used to develop food products without masking their original flavour characteristics [15].

EPS produced by LAB can be considered as natural additives, which are preferred by some consumers over stabilizers of plant or animal origin [16, 17]. EPSs from many species of bacteria have been shown to have good physicochemical, rheological, and emulsifying properties. Moreover, EPSs produced by different bacteria have been shown to have potential application in food products, including dairy products, sauces, soup mixtures, and desserts, by improving their rheological and textural properties [18].

EPS PRODUCING BACTERIA

EPS production has been reported in both prokaryotic and eukaryotic microorganisms. The common EPS-producing eukaryotic organisms include yeasts (*Cryptococcus* sp., *Hansenula* sp., *Rhodotorula* sp., *Lipomyces* sp., *Bullera* sp., *Aureobasidium* sp., and *Sporobolomyces* sp.) [19], filamentous fungi (*Aureobasidium pullulans*, *Slerotium* sp., *Schizophyllum commune*, *Sporobolo-*

myces sp., *Rhodoturula* sp., *Tremella fusiformes, Cryptptococcus* sp., and *Tremella aurantia*), Microalgae (*Porphyridium cruentum* and *Botryococcus braunii*), and some diatoms (*Amphora holsatica, Navicula directa* and *Melosira nummuloïdes*) [20].

The EPS-producing prokaryotes include bacteria and cyanobacteria. Most of EPS producing bacteria belong to the genera *Streptococcus* (*S. thermophilus* and *S. mutans*), *Lactobacillus* (*Lb. delbrueckii* subsp. *bulgaricus, Lb. casei, Lb. plantarum, Lb. fermentum, Lb. reuteri, Lb. sanfranciscensis, Lb. fermentum, Lb. brevis, Lb. rhamnosus, Lb. kefiranofaciens, Lb. helveticus, Lb. acidophilus, Lb. johnsonii, Lb.* curvatus, and *Lb. pentosus*), *Lactococcus* (*L. lactis*), *Leuconostoc* (*Leu. mesenteroides, Leu. citreum, Leu. lactis, Leu. Pseudomesenteroides,* and *Leu. dextranicum*), *Pediococcus* (*P. damnosus, P.parvulus, P. acidilactici,* and *P. pentosaceus*), *Pseudomonas* (*Pseu. aeruginosa, Pseu. Solanacearum,* and *Pseu. putida*), *Xanthomonas* (*Xan.campestris* and *Xan.oryzae pv. oryzae*), *Erwinia* (*E. amylovora, E. pyrifoliae,* and *E. stewartii*), *Bacillus* (*B. thermoantarcticus, B. licheniformis, B. polymyxa, B.subtilis, B. amyloliquefaciens,* and *B. cereus*) and *Weissella* (*W. confuse, W. hellenica, W. confuse,* and *W. cibaria*). In addition, there are some other bacteria associated with fermented foods and various parts of human body, such as propionibacteria and bifidobacteria, which may produce EPSs. However, compared to other genera, it is a much rarer phenomenon in bifidobacteria [21, 22].

Table **2.1** shows the major EPSs produced by bacteria that have been extensively studied and used for various applications.

Table 2.1. Bacterial exopolysaccharides and their applications.

Bacterial Strain/s	Exopolysaccharide	Monomer Composition	Applications	Reference/s
Xanthomonas campestris	Xanthan	Glucose, mannose, glucuronic acid, acetate and pyruvate	Foods, toiletries, oil recovery, cosmetics, and water-based paints	Rosalam and England [23]
Pseudomonas aeruginosa and Azotobacter sp.	Alginate	Mannuronic acid and guluronic acid	Textile printing, paper and board treatments, water treatment, can sealing, pharmaceutical, and biotechnology industry	Pawar and Edgar [24]; Rehm and Valla [25]

(Table 2.1) cont.....

Bacterial Strain/s	Exopolysaccharide	Monomer Composition	Applications	Reference/s
Leuconostoc sp., *Streptococcus* sp., *Lactobacillus* sp. and *Gluconobacter* sp.	Dextran	Glucose	Therapeutic agent, a component of Sephadex derivatives, alleviation of iron deficiency anaemia and in environmental clean-up of heavy metal	Naessens, Cerdobbel [26]
Sphingomonas sp.	Gellan	Glucose, glucuronic acid and rhamnose	Food ingredient (gelling, stabilizing, and suspending agent), biomedical industries, personal care and production processes	Fialho, Moreira [27]
Agrobacterium sp.	Curdlan	Glucose	Biomedical applications and food applications	Zhan, Lin [28]
Acetobacter xylinum	Cellulose	Glucose	Medical and food applications	Czaja, Krystynowicz [29]
Bacillus sp., Zymomonas sp., Erwinia sp. and *Lactobacillus sp.*	Levan	Fructose	Food, biotechnological and medical applications	Czaja, Krystynowicz [29]
Streptococcus zooepidemicus	Hyaluronan	N-acetyl-D-glucosamine and glucuronic acid	Biomedical applications	Chong, Blank [30]
Sinorhizobium meliloti	Succinoglycan	Galactose and glucose	Food applications	Becker, Ruberg [31]

BACTERIAL SYNTHESIS OF EPS

EPSs are produced either throughout growth during the late logarithmic phase or during the stationary phase, and the rate of production may depend on stress conditions, such as nutrient imbalance, salt, pH, and temperature [32]. There are two types of EPSs; heteropolysaccharides (HePSs), which contain different types of monosaccharide, and homopolysaccharides (HoPSs), which are composed of one type of monosaccharide (Fig. **2.2**). HePSs contain two to eight repeating units of monosaccharide, whereas HoPSs are mostly composed of either glucose or fructose [33].

Two major EPS synthesis pathways have been reported in bacteria; extracellular and intracellular production [20]. In general, HoPSs production is carried out either outside the cell or within the cell wall, and they include β-D-glucans, α-D-glucans, fructans, and polygalactan [5]. All HoPSs are synthesized by extracellular specific enzymes, such as glycansucrases, which belongs to glycosyltransferases; the required energy for this synthesis comes from the hydrolysis of sucrose. Glucansucrases convert sucrose to glucans, while sucrose is converted to fructans by fructansucrases [34].

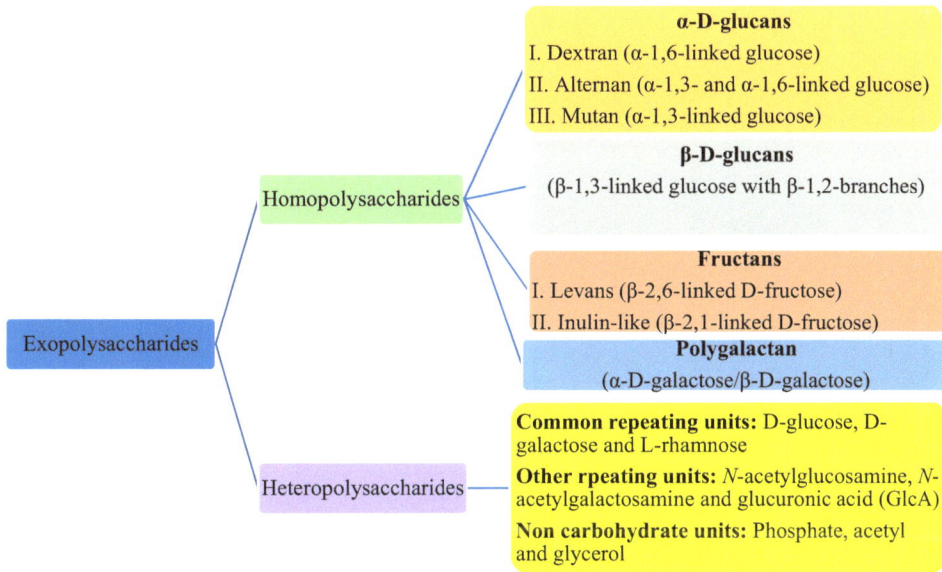

α-D-glucans
I. Dextran (α-1,6-linked glucose)
II. Alternan (α-1,3- and α-1,6-linked glucose)
III. Mutan (α-1,3-linked glucose)

β-D-glucans
(β-1,3-linked glucose with β-1,2-branches)

Homopolysaccharides

Fructans
I. Levans (β-2,6-linked D-fructose)
II. Inulin-like (β-2,1-linked D-fructose)

Polygalactan
(α-D-galactose/β-D-galactose)

Exopolysaccharides

Common repeating units: D-glucose, D-galactose and L-rhamnose

Other rpeating units: *N*-acetylglucosamine, *N*-acetylgalactosamine and glucuronic acid (GlcA)

Non carbohydrate units: Phosphate, acetyl and glycerol

Heteropolysaccharides

Fig. (2.2). Classification of bacterial exopolysaccharides.

HePS synthesis happens intracellularly. It has more complex mechanisms than that of HoPSs (Fig. **2.3**). In HePS synthesis, the precursor repeating units are formed intracellularly and they are translocated across the membrane for polymerization extracellularly [5]. According to Sutherland [35], HePSs synthesis has distinctive phases; up-taking of sugar subunits into the cytoplasm, the synthesis of sugar-1-phospahates, assembling of repeating monosaccharide units, the addition of any acyl groups, polymerization of repeating units, and secretion of HePSs to the external environment.

In the synthesis of HePSs, monosaccharides and disaccharides are transported into the cytoplasm and this process is regulated by some specific proteins. However, active or passive transport systems can be involved in the process based on the substrate [36]. The sugars in the cytoplasm are activated by phosphorylation using the Phospho Enol Pyruvate-sugar PhosphoTransferase System (PEP-PTS), that results in sugar-6-phosphates [13, 34]. Phosphorylated sugars are used to synthesize energy-rich nucleoside diphosphate sugars (NDP-sugars), which are essential for the production of HePSs. The main sugar nucleotides used for the synthesis of HEPSs are UDP-Glucose, UDP-Galactose, GDP-Mannose, and TDP-Rhamnose. These can be interconverted by reactions, such as epimerization, rearrangement, oxidation, reduction, and decarboxylation [35].

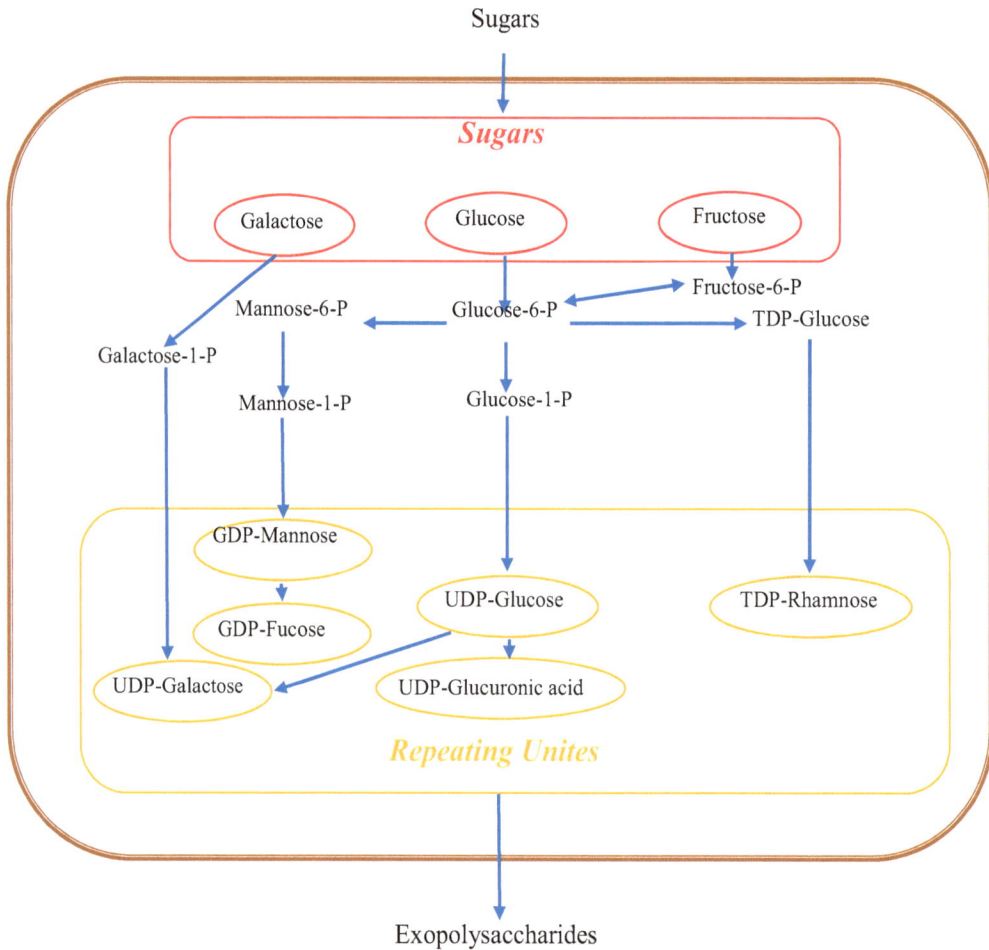

Fig. (2.3). Simple schematic representation of bacterial HePSs biosynthesis. Adopted and modified from Mishra and Jha [32], Rehm [38], and Laws, Gu [13]. GDP, guanosine diphosphate; TDP, tyrosine diphosphate; UDP, uridine diphosphate.

In the case of glucose in the cytoplasm, glucokinase converts glucose to glucose-6-phosphate (Glu-6-P) and it is then converted to glucose-1-phosphate (Glc-1-P) by phosphoglucomutase. Glucose-1-phosphate is transformed into UDP-glucose (UDP-Glc) or deoxythymidine diphosphate (dTDP)-glucose, dTDP-mannose, and dTDP-rhamnose. Galactose can be converted to galactose-1-phophate and to UDP-galactose (UDP-Gal). UDP-Gal may be converted to UDP-glucuronic acid and UDP-Gal can be also interconverted to UDP-Glc and *vice versa* using epimerases. Fructose can be used to produce Glc-6-P. Mannose is converted to mannose-1-P and GDP-glucuronic acid. Fructose-6-phosphate can be interconverted through several steps to UDP-N-acetylglucosamine. Glycosyltransferases can transfer sugar nucleotides produced from the above reactions to a repeating unit attached to a glycosyl carrier lipid. The exported repeating units through the cell membrane are polymerized by enzymes outside the cell membrane [37].

Selection of strains of a particular bacteria and optimization of growth conditions are the common strategies which are used to enhance the production of EPS [39]. EPS production by bacteria normally starts during the exponential phase of the growth and may reach the maximum in the stationary phase [40]. EPS production and their composition depend on culture and fermentation conditions. It was reported that production of EPS varied with temperature, pH of medium, and composition of the medium in terms of nitrogen source, mineral and vitamin content [5, 41].

The EPS production of bacteria was shown to be growth associated and therefore, a maximum production could be seen under conditions optimal for their growth [42]. Most of the studies have shown that optimal EPS production in mesophilic strains occurs at low temperatures, around 15 to 25 °C with most bacterial strains, whereas in thermophilic strains, EPS production is favoured at the higher temperature around 42 °C [43]. The optimum pH level for EPS production of LAB has been reported to be in the range of 5 to 7 [40]. In terms of growth media composition, EPS synthesis in some species of LAB was reported to be favoured by high carbon and low nitrogen substrate ratio. The proportion of low and high molecular weight EPS production of *Streptococcus thermophilus* LY03 was shown to be strongly dependent on the carbon/nitrogen ratio of the fermentation broth [44]. A shift from a high-molecular-mass exopolysaccharide to a low-molecular-mass EPS production of *S. thermophilus* LY03 with increasing initial complex nitrogen concentrations was observed. Regarding the effect of the carbon source, it was reported that addition of glucose in excess had a stimulatory effect on the rate of EPS production of *Lactobacillus delbrueckii* subsp. *bulgaricus* [40]. Changes in EPS production of bacteria have been observed with changing the source of sugar. In one study, *Lactobacillus delbrueckii* was shown to produce 25 mg L-1 with fructose, while the same strain yielded 80 mg L^{-1} with a mixture of fructose and glucose [45]. In addition, Looijesteijn *et al.* [46] reported that *Lactococcus lactis* produced more exopolysaccharide using glucose than fructose as the sugar substrate. Overall, complex media support the production of EPS which indicates the necessity of rich nutrients, such as peptone, beef extract, and yeast extract for the biomass development prior to EPS production [47]. The concentration of yeast extract was shown to have an influence on growth and exopolysaccharide production by *Propionibacterium acidi-propionici* DSM [48]. However, the presence of mannans, other carbohydrates, and yeast extract may lead to problems in isolation and purification of EPS from such complex media. Furthermore, supplementation with vitamins and amino acids were shown to improve bacterial growth and EPS production of some bacteria [49, 50].

Selection of strains that produce high amounts of EPS is the other common practice used to enhance EPS synthesis in a particular product [36]. However, both strain selection and optimization of culture conditions are considered as traditional techniques, as they can increase the production of EPS up to a certain level due to the physiological limits of the particular bacterium. Therefore, metabolic engineering has been used as a novel strategy to improve EPS production of bacteria [36]. This can be done by manipulation of the genes, which encode the enzymes which catalyse reactions in the EPS-producing pathway, or by altering the regulatory pathways that affect gene expression and enzyme activity [51].

EXTRACTION AND PURIFICATION OF EPS

In general, bacteria produce EPSs in complex media, which contain different biopolymers. These biopolymers may interact with EPSs. This becomes more complex in media; such as milk since EPS-producing bacteria may be entrapped in milk protein network. Therefore, it is complex to isolate and purify EPSs from the complex medium. Furthermore, some EPSs are attached to the bacterial cells while some are freely released into the growth medium.

There are different techniques, which have been tested to isolate and purify EPSs from bacteria. Each method has its own advantages and disadvantages [49, 52 - 54]. In general, EPS isolation and purification from bacteria involves: (i) cell and protein removal by centrifugation or trichloroacetic acid (TCA) precipitation followed by centrifugation, (ii) precipitation of EPS from cell-free supernatant by addition of ethanol followed by centrifugation, and (iii) drying of precipitated polysaccharide by lyophilisation (laboratory scale) or drum drying (commercial scale). However, there are variations on the above steps between different studies. All three steps have been used mostly with complex media and milk [55 - 59].

In some procedures, heating (90 to 100 °C) of the medium has been used at the end of the fermentation process, before protein and cell removal [8, 60 - 62]. Heating may help to inactivate the enzymes which may lead to the degradation of EPS [36]. In addition, membrane filtration has been used to remove some impurities (proteins, sugars, and cell debris) from the EPS fraction. Sometimes, ion exchange or size exclusion is used to purify the EPS fractions further [63, 64]. Taking all the above into account, the selection of the most suitable method for the isolation and purification of EPS depends on the media used for microbial growth, the required purity of the final product, and the impact on the properties of the polymer and the operation costs [36].

CHARACTERIZATION OF BACTERIAL EPS

There are several steps to be followed for the full characterization of EPS. These start from the determination of molecular weight of EPS followed by the identification of monosaccharide composition, the linkage pattern of monomers, and the absolute configuration of the monomers.

Determination of EPS Yield

The simplest method for the quantification of EPS is measuring the weight of lyophilized samples after the end of cleaning up procedures [59, 65, 66]. However, this is not an accurate method when other impurities, such as proteins and other carbohydrates are present in the EPS [2]. The colorimetric procedure explained by Dubois *et al.* [67] has been commonly used in many studies to quantify the isolated EPS [42, 68 - 73]. However, this procedure may lead to overestimation if other low molecular weight carbohydrates are present in the samples. Therefore, EPS can be more precisely quantified using gel permeation chromatography (GPC) or size exclusion chromatography with HPLC system coupled to a refractive detector (RI) [74 - 77].

Determination of Average Molecular Weight

The molecular mass of EPS is widely determined by using GPC with refractive index (RI) detection [34]. In this technique, the molecular weight of EPS is determined by comparison with control molecules. Dextran standards of different molar mass are commonly used to compare the EPS peaks. In addition, high-performance GPC with multi-angle laser light scatter detection can be used to evaluate the absolute molecular weight without the need of standards or column calibration [36].

Determination of Chemical Composition

This includes the identification of sugars, repeating units, and chain groups, such as acyl and phosphate groups. The common technique involves acid hydrolysis of EPS followed by derivatization to alditol acetates, and these released monomers are determined by gas chromatography [34]. In addition, high-performance anion exchange chromatography (HPAEC) with pulsed amperometric detection (PAD) is commonly used, and it is straightforward since it does not need a derivatisation step. Furthermore, capillary electrophoresis has been used to characterize carbohydrates; the method does not involve significant carbohydrate modification after hydrolysis [78 - 81].

Determination of Absolute Configuration

This usually starts with the analysis of monosaccharide composition as mentioned in the above section and is followed by the analysis of the linkage pattern of the monosaccharides [82]. For identification of linkage pattern of the monosaccharide, all free hydroxyl groups of EPS are methylated and hydrolysed to monosaccharides. Then, they are reduced to form partially methylated alditol acetates. This is called a derivatisation process and it results in products containing both methoxy (-OMe) and acetyl (-OAc) groups. The methoxy groups represent the location of free hydroxyl group, while acetylated sites indicate the location of glycosidic linkage. These partially methylated deuterated alditol acetates could be identified using GCMS [2, 58, 74, 83].

The repeating units of EPS are determined by using nuclear magnetic resonance (NMR) [2]. There are techniques, such as 2D correlated spectroscopy, 2D nuclear overhauser effect enhancement spectroscopy, 2D total correlation spectroscopy, and 2D heteronuclear single quantum coherence to determine the structure. These techniques can provide information about the interaction of each carbon/hydrogen with adjacent atoms and chemical groups. In addition, they provide their relative position in the structure of EPS [36].

FUNCTIONAL PROPERTIES AND FOOD APPLICATIONS OF EPS

In general, bacterial EPSs have been used in various food products, medical products, cosmetic products, and pharmaceuticals. In addition, their effectiveness as bio-flocculants, bio-absorbents, heavy metal removal agents, and drug delivery agents have been reported [7]. In the case of food application, EPSs are primarily used as thickening, stabilizing, and structure modification agents due to their non-Newtonian behaviour and high viscosity in aqueous media [36]. However, they are also used in various industries due to their rheological properties since EPSs have the ability to form viscous solutions at very low concentrations and they can protect their stability over a wide range of temperature and pH [84].

EPSs show very interesting interactions with proteins in food products that result in various desirable properties. These interactions lead to the desirable rheological properties of EPS containing foods. However, these interactions may depend on the type of EPS and the type of interactions between EPS and proteins in the food product. There are some common industrially important microbial EPS, such as dextran, xanthan, gellan, pullulan, and bacterial alginates [7].

Xanthan from *Xanthomonas campestris* is widely used in many food formulations due to its unique characteristics, such as high-water retention, texturizing, modification of viscosity, and flavour releasing properties. Xanthan is used to increase water binding during baking and storage of baked goods. In addition, it has been shown to improve quality characteristics of cakes, muffins, and biscuits [85]. Furthermore, it has been shown to improve appearance and texture of fruit drinks containing particles of fruit pulp. In the case of dried fruit powders, xanthan is mixed with the fruit powders to improve the quality parameters of rehydrated fruit powders. In addition, the incorporation of xanthan can improve viscosity of ice cream, yoghurt and cottage cheese [86]. Mandala and Bayas [87] studied the effect of xanthan on swelling power (SP) and solubility index (SOL) of wheat starch. Addition of xanthan (0.09%, w/w) increased SP of the xanthan-wheat starch mixture from 10.33 g/g to 12.88 g/g. Similarly, the incorporation of xanthan increased SOL of the mixture from 5.01% to 5.44%. In another study, rheological properties of xanthan-rice starch mixture at different xanthan gum concentration (0%, 0.2%, 0.4%, 0.6%, and 0.8% w/w) were evaluated [88]. The magnitudes of apparent viscosity of the mixture increased by 0.21 Pa s with increase in xanthan concentration from 0% to 0.8%.

Dextran is produced from sucrose by bacteria using dextransucrase [26]. Dextran has been shown to be effective as conditioners, stabilizers, and bodying agents in various food products. Dextran can improve viscosity, moisture retention, and inhibition of sugar crystallization in confectionaries. It is also used as a gelling agent in gums and jelly candies. In addition, bacterial dextran can inhibit sugar crystallization in ice cream and improve body and mouth feel of pudding mixtures [89]. EPS-positive *Weissella confusa* sourdough bread was shown to yield bread volume 10% higher than EPS-negative *W. confusa* sourdough bread [90]. In another study, sourdough containing dextran was shown to improve the volume (up to 12%) and softness of rye breads and rye mixed breads [91].

Gellan is a high molecular mass polysaccharide, which is industrially produced by *Sphingomonas paucimobilis* ATCC 31461 [92]. Incorporation of gellan has been shown to improve texture, physical stability, flavour release, and water binding capacity in different food products. It is used to have desired structure and texture of confectionaries. In addition, it has the ability to minimize the moisture fluctuation in sugary products, icings, and toppings. Furthermore, gellan has the ability to reduce the oil content in fried foods. In addition, it can improve textural and rheological characteristics of puddings, beverages, dairy products, fruit spreads, bakery fillings, sauces, and batters [89, 92]. Gellan-based (0.5% w/v) coating formulations was shown to improve water vapour resistance on fresh-cut papaya pieces compared to that of alginate based coating material [93]. In another study, Bajaj and Singhal [94] showed that addition of gellan gum at 0.25% (w/w) reduced the oil content in the deep fat fried product from 37.02% in the control to 27.91%.

Curdlan is a neutral water insoluble EPS produced by *Agrobacterium* sp [28]. It is tasteless, colourless and odourless and it can form gels with different properties (textual qualities, physical stabilities, and water holding capacities) depending on the heat treatment [95]. It is commonly used as a bio-thickening and gelling agent in different foods. It has been used to improve textural properties of dairy, meat, and baking products. In addition, curdlan can improve quality characteristics of batter systems and mixing of curdlan with noddle mixture can improve elasticity and strength of noddle [89]. Hydration and heating of curdlan can improve the mouth feel of fatty foods. Thixotropic properties were also reported with hydrated curdlan, which makes it suitable to be incorporated in low-fat dairy products and sauces [95].

Bacterial cellulose is commercially produced by using *Acetobacter xylinus* and it can form hydrogels. This property of bacterial cellulose is beneficial in many food products as a thickener, stabilizer, and texture modifier [96]. In addition, bacterial cellulose is tasteless, stable over the wide pH range, temperature, and freezing conditions [97]. *Nata* is a bacterial cellulose gel, which is a traditional dessert in Southeast Asia, and it has been popular around the world [96]. In addition, bacterial cellulose has been used in different sauces, gravies, icings, sour cream, whipped toppings, aerated desserts, and cultured dairy products. It has been shown to be effective as a bulking agent and a fat replacer and, therefore, there is a potential to be used as a low-calorie food ingredient [96].

Levan is a HoPS and, for the food applications, it is produced by using LAB, such as *Streptococcus salivarius*, *Streptococcus mutants*, *Leuconostoc mesenteroides* NRRL B-512F, *Lactobacillus sanfranciscensis* LTH 2590, and *Lactobacillus reuteri* LB121. It has been shown to be effective as a bio-thickener [98]. In addition, it has a potential as an emulsifier, a stabilizer, an encapsulating agent, and a carrier of flavours. It was also used to improve the water-holding capacity of food products. In addition, levan is considered as a prebiotic and is useful in the modulation of colonic microorganisms [99].

Lipid emulsifying effect of EPSs from different bacteria has been reported [7, 84]. The demand for bio-emulsifiers has increased due to their biodegradability and low toxicity [100]. Bacterial EPSs have been used in various food applications as emulsifiers [101]. In addition, EPS from different bacterial species has been shown to be effective as a bio-emulsifier in other industries, such as petroleum, metal processing, detergent manufacturing, paints, pulp, and agriculture [102]. Emulsions prepared with EPS of *Enterobacter cloaceae* were shown to be superior or comparable with that of Arabic gum, tragacanth, and karaya. The emulsion with this EPS showed a good shelf life under standard conditions and under stress conditions, such as low pH and high salinity [12].

Flocculants have been widely used in waste water treatment, drinking water treatment, downstream processing, fermentation and food processing [7]. Flocculants are classified into synthetic organic flocculants (polyacrylamide, polyethylimine), synthetic inorganic flocculants (polyaluminium chloride and aluminium sulphate), and bio-flocculants [103]. Although the synthetic organic flocculants are common, they have less biodegradability and can cause environmental problems [104]. Hence, the development of safe and biodegradable bio-flocculants, such as EPS, produced by bacteria have been investigated as the alternative flocculants [105]. Higher flocculation activity of EPS isolated from *Lactobacillus kefiranofaciens* ZW3 and *Streptococcus phocae* P180 was revealed compared to that of organic flocculants [7, 8]. In addition, EPS isolated from *Weissella confuse* AJ53 was shown to have higher flocculating activity than xanthan gum, guar gum, arabic gum, tragacanth gum, and ghatti gum [106].

Fermented milk products are the most common food type, which has been extensively researched with regards to the *in-situ* production of EPSs by different LAB. In the case of fermented milk products, EPSs of different bacterial species have been shown to improve the texture [107] and have been used commercially to substitute the fat in low-fat products and improve the mouth feel [108]. However, EPS of different bacteria varies greatly in composition, charge, spatial arrangement, rigidity, and ability to interact with proteins, although a clear effect of EPS concentration on viscosity of the products has not been reported [15]. EPS-producing LAB including thermophilic *Streptococcus thermophilus,* and *Lactobacillus delbrueckii* subsp. *bulgaricus* and mesophilic *Lactococcus lactis* subsp. *cremoris* strains have been widely used in fermented milk products and have been investigated for their properties [5, 34]. Prasanna *et al.* [11] and Audy *et al.* [57] reported that EPS produced by some bifidobacteria could enhance the viscosity of fermented dairy products. When EPS are secreted into the milk medium, they may lead to a ropy appearance of the milk gels [109]. The effect of EPS on protein matrix and structure formation depends on their concentration, interactions with proteins and rheological characteristics [110]. Furthermore, it was reported that the effect of EPS on yoghurt texture depends on molecular weight, the degree of branching and chain flexibility [111]. In addition, some studies have shown that EPS could positively interact with milk proteins and lead to higher viscoelastic moduli and firmness. This may be due to electrostatic interactions between casein and EPS as suggested by Ayala-Hernandez *et al.*

[112], Girard and Schaffer-Lequart [113] and Gentès *et al.* [114]. This interaction could strengthen the casein gel structure. EPS production *in situ* can result in fermented milk products with higher water holding capacity and lower syneresis. This is due to the ability of EPS to bind free water in fermented milk products [34]. Furthermore, fermentation has the main role in determining the functionalities of *in situ* EPS. EPS produced after gelation is limited to pockets around the bacteria [109]. Therefore, it is difficult to study the real interactions between milk proteins and EPSs, as other stabilizers which are directly mixed with milk increase rheological properties [110].

EPS have been successfully used in different cheese manufacturing process to improve various properties including texture and water content. Mozzarella cheese produced with EPS producing *Streptococcus thermophilus* MR-1C was shown to have 2% higher moisture content than that of the control, which was produced with EPS negative *S. thermophilus* TA061 [115]. In another study, ropy *Lactococcus lactis* ssp. *cremoris* was shown to be more effective at increasing moisture content and improving textural properties of reduced fat Cheddar cheese compared to those of the control [116]. Furthermore, EPS producing *Streptococcus thermophilus* ST446 positively affected texture of low-fat Caciotta-type cheese [117].

EPS are sometimes mixed with other polymers to enhance their polymeric structure, properties, and functionalities. Xanthan was mixed with locust bean gum and shown to have improved quality parameters [118]. In addition, EPS have been used to develop encapsulation materials for some probiotics and prebiotics [36]. Some EPSs can form the film, which could be used to develop edible coating materials [119, 120]. In addition, EPS has been used to deliver some drugs due to their ability to form polymeric matrices [121].

HEALTH BENEFITS OF EPS

Consumption of EPS containing food products has shown to have various health benefits [21, 122, 123]. Fig. (**2.4**) shows the different health benefits reported with consumption of bacterial EPSs, such as immunostimulatory, immunomodulatory, antitumor, antiviral, anti-inflammatory, and antioxidant effects [9, 43, 124, 125].

Modification of Intestinal Microbial Population

There are reports that EPSs help probiotic bacteria for passing through severe gastrointestinal conditions and colonization to intestinal mucosa [126]. The protective coat of EPS could help probiotics to live in high acid level and bile salt concentration of the upper part of the intestinal tract of the human. Therefore, EPS production of probiotics can ensure safe delivery of probiotics from mouth to the colon with minimum effect on their viability [16]. EPS production of some probiotic bacteria has been shown to modify the environment of the human gut which can change interactions among the intestinal microbial population in a beneficial way [127]. Furthermore, it was reported that EPS production of some probiotic bacteria including bifidobacteria interfere with the adhesion of pathogens to human intestinal mucus [128].

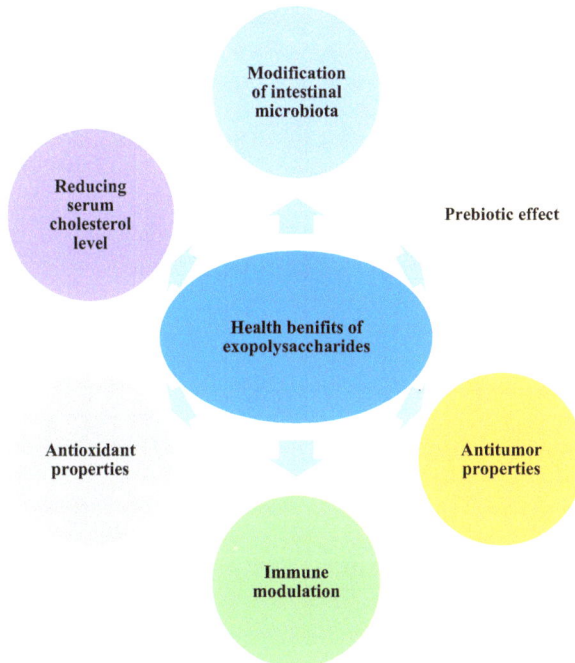

Fig. (2.4). Health benefits associated with exopolysaccharides.

Prebiotic Effects

Prebiotics are defined as "nondigestible food ingredients that beneficially affect the host by selectively stimulating the growth and/or activity of one or a limited number of bacterial species already resident in the colon, and thus attempt to improve host health" [129]. Prebiotic properties of EPSs have been reported; in one study the prebiotic property of a fructan produced by *Lactobacillus sanfranciscensis* was shown *in vitro* [130], while EPS of *Weissella cibaria* showed the long-lasting prebiotic effect in the colon model [131]. In another study, Mårtensson *et al.* [132] observed a significant increase in the population of bifidobacteria in faecal samples from human volunteers who were served with oat-based product co-fermented with EPS-producing *Pediococcus damnosus*. In addition, a relationship between digestibility of EPS with their physicochemical characteristics has been reported. Mozzi *et al.* [133] observed that digestion of

EPS from *Streptococcus thermophiles* did not change the monosaccharide composition or molecular mass of EPS.

Antitumor Activity

The reported positive effects of EPSs in controlling tumour has increased research activities on EPS-producing bacteria. EPS of *Lactobacillus plantarum* 70810 was shown to have antitumor properties against HepG-2, BGC-823 and HT-29 tumour cells *in vitro* [123]. In another study, a levan-type EPS from *Paenibacillus polymyxa* EJS-3 was shown to act against tumour cells *in vitro* [134]. In a separate *in vitro* antitumor activity test, a levan from *Microbacterium laevaniformans*, *Rahnella aquatilis,* and *Zymomonas mobilis* was shown to be effective against eight different tumour cell lines [135]. In addition, antitumor activity was observed with EPS of *Rhizobium* sp. N613 *in vitro* [136]. Kim *et al.* [137] conducted an *in vitro* study with EPS of *Lactobacillus acidophilus* 606 to evaluate the anti-tumour activity against HT-29 colon cancer cells and observed effectiveness of EPS in controlling the tumour cells. Furthermore, *Lactobacillus bulgaricus* 878R was shown to produce EPS with anti-tumour activity *in vitro* [138].

Immune Modulation Activity

There are two immune responses; innate immune response and adaptive immune responses, where the first type plays an important role in the development of the second type. Bacterial strains can enhance both immune responses through the induction of dendritic cell (DC) maturation. In addition, they stimulate immune cells to release pro-inflammatory cytokines; interferon-gamma (IFN-γ), tumour necrosis factor-alpha (TNF-α), and interleukin-12 (IL-12) [139]. EPSs from different bacteria have been shown to have immune responses. Hosono *et al.* [140] conducted an *in vitro* study with a water-soluble polysaccharide fraction of *B. adolescentis* and observed immune modulating activity. In another study, EPS produced by *Lactobacillus delbrueckii* subsp. *bulgaricus* OLL 1073R-1 was evaluated for immunological functions with mice and an increase in the immune responses was reported [141]. Oral administration of EPS-producing *Bifidobacterium breve* UCC2003 was shown to modulate immune responses of mice. This study further suggests that EPS of this strain can facilitate their colonization in the host through evasion of potentially damaging immune responses and they can reduce pathogen colonization [142]. In another *in vitro* study, EPS of *Bifidobacterium longum* was shown to have mild immune modulating properties for macrophages, which increased the ability of the bacterium against gastrointestinal infections [143]. Furthermore, oral administration of selenium EPS-from *Enterobacter cloacae* Z0206 was shown to improve cellular and humoral immune responses of mice [144]. Feeding of EPS from *Lactobacillus kefiranofaciens* was observed to induce the gut mucosal response of mice and EPS could maintain intestinal homeostasis and enhance the IgA production at both the small and large intestine level [145]. In addition, there are many other studies that reported immune modulation activities of EPS of different bacteria. EPSs of *Lactobacillus rhamnosus* RW- 9595M [146], *Bifidobacterium lactis* Bb12 [147], *Lactobacillus rhamnosus* KF5 [148], *Lactobacillus brevis* KB290 [149], and *Bifidobacterium bifidum* PRL2010 [150] were shown to exert immune modulation activities.

Antioxidant Properties of EPS

Oxidative stress aids in the development of various diseases, such as Alzheimer's disease, Parkinson's disease, inflammation, and cancer. Both synthetic and natural antioxidants have been shown to play an important role in controlling the condition. Many EPS-producing bacteria have been shown to have good antioxidant properties, which can control the oxidative stress [151]. *In vitro* antioxidant testing of EPS of *Lactococcus lactis* subsp. *lactis* was shown to reduce peroxidation, increase anti-superoxide anion, and increase anti-hydroxyl radical scavenging activity [152]. In another *in vitro* study, EPS of *Bacillus cereus* was observed to have scavenging activity against superoxide and hydroxyl radical. In addition, this EPS could protect oxidative DNA damage and PC12 cells [153]. Meng *et al.* [154] used mice to evaluate antioxidant activity of EPS of *Morchella esculenta* SO-01. They reported that EPS can increase the antioxidant activity by improving superoxide dismutase (SOD) and glutathione peroxidase (GSH-Px) activities, and malondialdehyde (MDA) content in blood, liver, heart, spleen, and kidney. They proposed a mechanism that activity of these enzymes is increased by EPS, and MDA production could be inactivated by direct activation of enzymes or binding of metal ions, which is necessary in the production of free radical. In another study, Sun *et al.* [155] reported the antioxidant activity of EPS from *Zunongwangia profunda*. They recommended applying this EPS as an antioxidant in foods to control oxidative damage to lipids and proteins. The antioxidant activity of EPS of some other bacteria has been reported. They include *Lactobacillus plantarum* R315 [156], *Pseudomonas fluorescens* [157], *Lactobacillus rhamnosus* [158], *Bifidobacterium bifidum* WBIN03 [156], *Lactobacillus helveticus* MB2-1 [159], *Pseudomonas* PT-8 [160], *Lactococcus lactis* subsp. *lactis* [161], *Enterococcus faecium* [162], and *Lactobacillus planterum* LP6 [163].

Reduction of Serum Cholesterol Level

Changes in the serum cholesterol level with consumption of EPSs from bacteria have been reported. In one study, oral administration of EPS-producing *Lactobacillus paracasei* was observed to modulate lipid metabolism in a mouse model with atherosclerosis. They further observed a reduction of serum cholesterol level and hepatic lipid concentrations of mice, which were fed with the EPS-producing bacterium [164]. In addition, it was revealed that slime materials produced by *Lactococcus lactis* subsp. *cremoris* SBT 0495 have a favorable effect on rat cholesterol metabolism [165]. However, *Lindström et al.* [166] did not observe any changes in cholesterol level of mice which were fed with EPS-producing *Pediococcus parvulus*.

CONCLUSION

Application of EPSs originated from bacteria in various industries has become popular around the world due to many advantages associated with them compared to the traditional materials, such as plants and animal hydrocolloids. EPS-producing bacterial species have been isolated from various ecosystems and isolation of new potential species continues. Understanding of the mechanism involved in EPS synthesis of a particular species is important to maximize the potential applications of EPS in various industries. Isolation and characterization of EPS are complex and huge variation has been reported. Application of EPS of bacteria in various food application has been popular due to improved quality characteristics and health benefits associated with EPS containing foods. Potential health benefits associated with EPS of bacteria have been mostly reported with *in vitro* experiments and, therefore, these EPSs should be tested with *in vivo* model to have a clear understanding of their activities.

CONSENT FOR PUBLICATION

Not applicable.

CONFLICT OF INTEREST

The author declares no conflict of interest, financial or otherwise.

ACKNOWLEDGEMENTS

Declared none.

REFERENCES

[1]   Varela P, Fiszman S. Hydrocolloids in fried foods. A review. Food Hydrocoll 2011; 25(8): 1801-12. [http://dx.doi.org/10.1016/j.foodhyd.2011.01.016]

[2]   Ruas-Madiedo P, de los Reyes-Gavilán CG. Invited review: methods for the screening, isolation, and characterization of exopolysaccharides produced by lactic acid bacteria. J Dairy Sci 2005; 88(3): 843-56.

[http://dx.doi.org/10.3168/jds.S0022-0302(05)72750-8] [PMID: 15738217]

[3]   Looijesteijn PJ, Trapet L, de Vries E, Abee T, Hugenholtz J. Physiological function of exopolysaccharides produced by Lactococcus lactis. Int J Food Microbiol 2001; 64(1-2): 71-80.
      [http://dx.doi.org/10.1016/S0168-1605(00)00437-2] [PMID: 11252513]

[4]   Cerning J, Renard CMGC, Thibault JF, *et al.* Carbon source requirements for exopolysaccharide production by *Lactobacillus casei* CG11 and partial structure analysis of the polymer. Appl Environ Microbiol 1994; 60(11): 3914-9.
      [PMID: 16349427]

[5]   Ruas-Madiedo P, Hugenholtz J, Zoon P. An overview of the functionality of exopolysaccharides produced by lactic acid bacteria. Int Dairy J 2002; 12(2-3): 163-71.
      [http://dx.doi.org/10.1016/S0958-6946(01)00160-1]

[6]   Jackson KD, Starkey M, Kremer S, Parsek MR, Wozniak DJ. Identification of psl, a locus encoding a potential exopolysaccharide that is essential for *Pseudomonas aeruginosa* PAO1 biofilm formation. J Bacteriol 2004; 186(14): 4466-75.
      [http://dx.doi.org/10.1128/JB.186.14.4466-4475.2004] [PMID: 15231778]

[7]   Wang Y, Ahmed Z, Feng W, Li C, Song S. Physicochemical properties of exopolysaccharide produced by *Lactobacillus kefiranofaciens* ZW3 isolated from Tibet kefir. Int J Biol Macromol 2008; 43(3): 283-8.
      [http://dx.doi.org/10.1016/j.ijbiomac.2008.06.011] [PMID: 18662712]

[8]   Kanmani P, Satish kumar R, Yuvaraj N, Paari KA, Pattukumar V, Arul V. Production and purification of a novel exopolysaccharide from lactic acid bacterium Streptococcus phocae PI80 and its functional characteristics activity in vitro. Bioresour Technol 2011; 102(7): 4827-33.
      [http://dx.doi.org/10.1016/j.biortech.2010.12.118] [PMID: 21300540]

[9]   Liu C, Lu J, Lu L, Liu Y, Wang F, Xiao M. Isolation, structural characterization and immunological activity of an exopolysaccharide produced by *Bacillus licheniformis* 8-37-0-1. Bioresour Technol 2010; 101(14): 5528-33.
      [http://dx.doi.org/10.1016/j.biortech.2010.01.151] [PMID: 20199860]

[10] Prasanna PHP, Bell A, Grandison AS, Charalampopoulos D. Emulsifying, rheological and physicochemical properties of exopolysaccharide produced by *Bifidobacterium longum* subsp. *infantis* CCUG 52486 and *Bifidobacterium infantis* NCIMB 702205. Carbohydr Polym 2012; 90(1): 533-40.
      [http://dx.doi.org/10.1016/j.carbpol.2012.05.075] [PMID: 24751074]

[11] Prasanna PHP, Grandison AS, Charalampopoulos D. Microbiological, chemical and rheological properties of low fat set yoghurt produced with exopolysaccharide (EPS) producing *Bifidobacterium* strains. Food Res Int 2013; 51(1): 15-22.
      [http://dx.doi.org/10.1016/j.foodres.2012.11.016]

[12] Iyer A, Mody K, Jha B. Emulsifying properties of a marine bacterial exopolysaccharide. Enzyme Microb Technol 2006; 38(1): 220-2.
      [http://dx.doi.org/10.1016/j.enzmictec.2005.06.007]

[13] Laws A, Gu Y, Marshall V. Biosynthesis, characterisation, and design of bacterial exopolysaccharides from lactic acid bacteria. Biotechnol Adv 2001; 19(8): 597-625.
      [http://dx.doi.org/10.1016/S0734-9750(01)00084-2] [PMID: 14550013]

[14] Karim AA, Bhat R. Gelatin alternatives for the food industry: recent developments, challenges and prospects. Trends Food Sci Technol 2008; 19(12): 644-56.
      [http://dx.doi.org/10.1016/j.tifs.2008.08.001]

[15] Jolly L, Vincent SJF, Duboc P, Neeser JR. Exploiting exopolysaccharides from lactic acid bacteria. Antonie van Leeuwenhoek 2002; 82(1-4): 367-74.
      [http://dx.doi.org/10.1023/A:1020668523541] [PMID: 12369204]

[16] Alp G, Aslim B. Relationship between the resistance to bile salts and low pH with exopolysaccharide

(EPS) production of *Bifidobacterium* spp. isolated from infants feces and breast milk. Anaerobe 2010; 16(2): 101-5.
[http://dx.doi.org/10.1016/j.anaerobe.2009.06.006] [PMID: 19576995]

[17]   Salazar N, Binetti A, Gueimonde M, *et al.* Safety and intestinal microbiota modulation by the exopolysaccharide-producing strains *Bifidobacterium animalis* IPLA R1 and *Bifidobacterium longum* IPLA E44 orally administered to Wistar rats. Int J Food Microbiol 2011; 144(3): 342-51.
[http://dx.doi.org/10.1016/j.ijfoodmicro.2010.10.016] [PMID: 21078530]

[18]   Garai-Ibabe G, Dueñas MT, Irastorza A, *et al.* Naturally occurring 2-substituted (1,3)-beta-D-glucan producing *Lactobacillus suebicus* and *Pediococcus parvulus* strains with potential utility in the production of functional foods. Bioresour Technol 2010; 101(23): 9254-63.
[http://dx.doi.org/10.1016/j.biortech.2010.07.050] [PMID: 20691585]

[19]   Pavlova K, Koleva L, Kratchanova M, Panchev I. Production and characterization of an exopolysaccharide by yeast. World J Microbiol Biotechnol 2004; 20(4): 435-9.
[http://dx.doi.org/10.1023/B:WIBI.0000033068.45655.2a]

[20]   Donot F, Fontana A, Baccou JC, Schorr-Galindo S. Microbial exopolysaccharides: main examples of synthesis, excretion, genetics and extraction. Carbohydr Polym 2012; 87(2): 951-62.
[http://dx.doi.org/10.1016/j.carbpol.2011.08.083]

[21]   Xu R, Shang N, Li P. *In vitro* and *in vivo* antioxidant activity of exopolysaccharide fractions from *Bifidobacterium animalis* RH. Anaerobe 2011; 17(5): 226-31.
[http://dx.doi.org/10.1016/j.anaerobe.2011.07.010] [PMID: 21875680]

[22]   Prasanna PHP, Grandison AS, Charalampopoulos D. Screening human intestinal *Bifidobacterium* strains for growth, acidification, EPS production and viscosity potential in low-fat milk. Int Dairy J 2012; 23(1): 36-44.
[http://dx.doi.org/10.1016/j.idairyj.2011.09.008]

[23]   Rosalam S, England R. Review of xanthan gum production from unmodified starches by *Xanthomonas comprestris* sp. Enzyme Microb Technol 2006; 39(2): 197-207.
[http://dx.doi.org/10.1016/j.enzmictec.2005.10.019]

[24]   Pawar SN, Edgar KJ. Alginate derivatization: a review of chemistry, properties and applications. Biomaterials 2012; 33(11): 3279-305.
[http://dx.doi.org/10.1016/j.biomaterials.2012.01.007] [PMID: 22281421]

[25]   Rehm BHA, Valla S. Bacterial alginates: biosynthesis and applications. Appl Microbiol Biotechnol 1997; 48(3): 281-8.
[http://dx.doi.org/10.1007/s002530051051] [PMID: 9352672]

[26]   Naessens M, Cerdobbel A, Soetaert W, Vandamme EJ. Leuconostoc dextransucrase and dextran: production, properties and applications. J Chem Technol Biotechnol 2005; 80(8): 845-60.
[http://dx.doi.org/10.1002/jctb.1322]

[27]   Fialho AM, Moreira LM, Granja AT, Popescu AO, Hoffmann K, Sá-Correia I. Occurrence, production, and applications of gellan: current state and perspectives. Appl Microbiol Biotechnol 2008; 79(6): 889-900.
[http://dx.doi.org/10.1007/s00253-008-1496-0] [PMID: 18506441]

[28]   Zhan XB, Lin CC, Zhang HT. Recent advances in curdlan biosynthesis, biotechnological production, and applications. Appl Microbiol Biotechnol 2012; 93(2): 525-31.
[http://dx.doi.org/10.1007/s00253-011-3740-2] [PMID: 22124723]

[29]   Czaja W, Krystynowicz A, Bielecki S, Brown RM Jr. Microbial cellulose--the natural power to heal wounds. Biomaterials 2006; 27(2): 145-51.
[http://dx.doi.org/10.1016/j.biomaterials.2005.07.035] [PMID: 16099034]

[30]   Chong BF, Blank LM, Mclaughlin R, Nielsen LK. Microbial hyaluronic acid production. Appl Microbiol Biotechnol 2005; 66(4): 341-51.

[http://dx.doi.org/10.1007/s00253-004-1774-4] [PMID: 15599518]

[31]   Becker A, Rüberg S, Baumgarth B, Bertram-Drogatz PA, Quester I, Pühler A. Regulation of succinoglycan and galactoglucan biosynthesis in *Sinorhizobium meliloti.* J Mol Microbiol Biotechnol 2002; 4(3): 187-90.
[PMID: 11931545]

[32]   Mishra A, Jha B. Microbial exopolysaccharides.The Prokaryotes. Berlin, Heidelberg: Springer 2013; pp. 179-92.
[http://dx.doi.org/10.1007/978-3-642-31331-8_25]

[33]   Wang Y, Gänzle MG, Schwab C. Exopolysaccharide synthesized by *Lactobacillus reuteri* decreases the ability of enterotoxigenic *Escherichia coli* to bind to porcine erythrocytes. Appl Environ Microbiol 2010; 76(14): 4863-6.
[http://dx.doi.org/10.1128/AEM.03137-09] [PMID: 20472719]

[34]   Badel S, Bernardi T, Michaud P. New perspectives for Lactobacilli exopolysaccharides. Biotechnol Adv 2011; 29(1): 54-66.
[http://dx.doi.org/10.1016/j.biotechadv.2010.08.011] [PMID: 20807563]

[35]   Sutherland IW. Bacterial exopolysaccharides.Comprehensive glycoscience. Oxford: Elsevier 2007; pp. 521-57.
[http://dx.doi.org/10.1016/B978-044451967-2/00133-1]

[36]   Freitas F, Alves VD, Reis MAM. Advances in bacterial exopolysaccharides: from production to biotechnological applications. Trends Biotechnol 2011; 29(8): 388-98.
[http://dx.doi.org/10.1016/j.tibtech.2011.03.008] [PMID: 21561675]

[37]   De Vuyst L, De Vin F, Vaningelgem F, Degeest B. Recent developments in the biosynthesis and applications of heteropolysaccharides from lactic acid bacteria. Int Dairy J 2001; 11(9): 687-707.
[http://dx.doi.org/10.1016/S0958-6946(01)00114-5]

[38]   Rehm BHA. Bacterial polymers: biosynthesis, modifications and applications. Nat Rev Microbiol 2010; 8(8): 578-92.
[http://dx.doi.org/10.1038/nrmicro2354] [PMID: 20581859]

[39]   Looijesteijn PJ, van Casteren WHM, Tuinier R, Doeswijk-Voragen CHL, Hugenholtz J. Influence of different substrate limitations on the yield, composition and molecular mass of exopolysaccharides produced by *Lactococcus lactis* subsp. *cremoris* in continuous cultures. J Appl Microbiol 2000; 89(1): 116-22.
[http://dx.doi.org/10.1046/j.1365-2672.2000.01082.x] [PMID: 10945787]

[40]   Petry S, Furlan S, Crepeau MJ, Cerning J, Desmazeaud M. Factors affecting exocellular polysaccharide production by *Lactobacillus delbrueckii* subsp. *bulgaricus* grown in a chemically defined medium. Appl Environ Microbiol 2000; 66(8): 3427-31.
[http://dx.doi.org/10.1128/AEM.66.8.3427-3431.2000] [PMID: 10919802]

[41]   Younghoon Kim YK, Ji Uk Kim JUK, Sejong Oh SO, Young Jun Kim YJK, Myunghee Kim MK, Sae Hun Kim SHK. Technical optimization of culture conditions for the production of exopolysaccharide (EPS) by *Lactobacillus rhamnosus* ATCC 9595. Food Sci Biotechnol 2008; 17(3): 587-93.

[42]   Zisu B, Shah NP. Effects of pH, temperature, supplementation with whey protein concentrate, and adjunct cultures on the production of exopolysaccharides by *Streptococcus thermophilus* 1275. J Dairy Sci 2003; 86(11): 3405-15.
[http://dx.doi.org/10.3168/jds.S0022-0302(03)73944-7] [PMID: 14672169]

[43]   De Vuyst L, Degeest B. Heteropolysaccharides from lactic acid bacteria. FEMS Microbiol Rev 1999; 23(2): 153-77.
[http://dx.doi.org/10.1111/j.1574-6976.1999.tb00395.x] [PMID: 10234843]

[44]   Degeest B, De Vuyst L. Indication that the nitrogen source influences both amount and size of exopolysaccharides produced by streptococcus thermophilus LY03 and modelling of the bacterial

growth and exopolysaccharide production in a complex medium. Appl Environ Microbiol 1999; 65(7): 2863-70.
[PMID: 10388677]

[45]   Grobben JG, Smith RM, Sikkema J, de Bont MJA. Influence of fructose and glucose on the production of exopolysaccharides and the activities of enzymes involved in the sugar metabolism and the synthesis of sugar nucleotides in *Lactobacillus delbrueckii* subsp. *bulgaricus* NCFB 2772. Appl Microbiol Biotechnol 1996; 46(3): 279-84.
[http://dx.doi.org/10.1007/s002530050817]

[46]   Looijesteijn PJ, Boels IC, Kleerebezem M, Hugenholtz J. Regulation of exopolysaccharide production by *Lactococcus lactis* subsp. *cremoris* By the sugar source. Appl Environ Microbiol 1999; 65(11): 5003-8.
[PMID: 10543815]

[47]   Patel AK, Michaud P, Singhania RR, Soccol CR, Pandey A. Polysaccharides from probiotics: new developments as food additives. Food Technol Biotechnol 2010; 48(4): 451-63.

[48]   Gorret N, Maubois JL, Engasser JM, Ghoul M. Study of the effects of temperature, pH and yeast extract on growth and exopolysaccharides production by *Propionibacterium acidi-propionici* on milk microfiltrate using a response surface methodology. J Appl Microbiol 2001; 90(5): 788-96.
[http://dx.doi.org/10.1046/j.1365-2672.2001.01310.x] [PMID: 11348440]

[49]   Bajaj IB, Saudagar PS, Singhal RS, Pandey A. Statistical approach to optimization of fermentative production of gellan gum from *Sphingomonas paucimobilis* ATCC 31461. J Biosci Bioeng 2006; 102(3): 150-6.
[http://dx.doi.org/10.1263/jbb.102.150] [PMID: 17046526]

[50]   Li H, Xu H, Xu H, Li S, Ouyang PK. Biosynthetic pathway of sugar nucleotides essential for welan gum production in *Alcaligenes* sp. CGMCC2428. Appl Microbiol Biotechnol 2010; 86(1): 295-303.
[http://dx.doi.org/10.1007/s00253-009-2298-8] [PMID: 19838696]

[51]   Martin Lo Y, Argin-Soysal S, Chia-Hua H. Bioconversion of whey lactose into microbial exopolysaccharides.Bioprocessing for value-added products from renewable resources: new technologies and applications. Oxford: Elsevier Science 2007; pp. 559-85.

[52]   Freitas F, Alves VD, Pais J, *et al.* Characterization of an extracellular polysaccharide produced by a *Pseudomonas* strain grown on glycerol. Bioresour Technol 2009; 100(2): 859-65.
[http://dx.doi.org/10.1016/j.biortech.2008.07.002] [PMID: 18713662]

[53]   Grobben GJ, Van Casteren WHM, Schols HA, Oosterveld A, Sala G, Smith MR, *et al.* Analysis of the exopolysaccharides produced by *Lactobacillus delbrueckii* subsp. *bulgaricus* NCFB 2772 grown in continuous culture on glucose and fructose. Appl Microbiol Biotechnol 1997; 48(4): 516-21.
[http://dx.doi.org/10.1007/s002530051089]

[54]   Pham PL, Dupont I, Roy D, Lapointe G, Cerning J. Production of exopolysaccharide by *Lactobacillus rhamnosus* R and analysis of its enzymatic degradation during prolonged fermentation. Appl Environ Microbiol 2000; 66(6): 2302-10.
[http://dx.doi.org/10.1128/AEM.66.6.2302-2310.2000] [PMID: 10831403]

[55]   Cerning J, Renard CM, Thibault JF, *et al.* Carbon source requirements for exopolysaccharide production by *Lactobacillus casei* CG11 and partial structure analysis of the polymer. Appl Environ Microbiol 1994; 60(11): 3914-9.
[PMID: 16349427]

[56]   Amatayakul T, Halmos AL, Sherkat F, Shah NP. Physical characteristics of yoghurts made using exopolysaccharide-producing starter cultures and varying casein to whey protein ratios. Int Dairy J 2006; 16(1): 40-51.
[http://dx.doi.org/10.1016/j.idairyj.2005.01.004]

[57]   Audy J, Labrie S, Roy D, Lapointe G. Sugar source modulates exopolysaccharide biosynthesis in *Bifidobacterium longum* subsp. longum CRC 002. Microbiology 2010; 156(Pt 3): 653-64.

[http://dx.doi.org/10.1099/mic.0.033720-0] [PMID: 19850611]

[58]   Harding LP, Marshall VM, Elvin M, Gu Y, Laws AP. Structural characterisation of a perdeuteriomethylated exopolysaccharide by NMR spectroscopy: characterisation of the novel exopolysaccharide produced by *Lactobacillus delbrueckii* subsp. *bulgaricus* EU23. Carbohydr Res 2003; 338(1): 61-7.
[http://dx.doi.org/10.1016/S0008-6215(02)00354-3] [PMID: 12504382]

[59]   Frengova GI, Simova ED, Beshkova DM, Simov ZI. Production and monomer composition of exopolysaccharides by yogurt starter cultures. Can J Microbiol 2000; 46(12): 1123-7.
[http://dx.doi.org/10.1139/w00-103] [PMID: 11142402]

[60]   Ai L, Guo B, Wang Y, Chen W, Zhang H, Wu Z. Extraction of exopolysaccharides from Lactobacillus Casei LC 2 W. China Dairy Industry 2008; 36(7): 14-8.

[61]   Cerning J, Bouillanne C, Landon M, Desmazeaud M. Isolation and characterization of exopolysaccharides from slime-forming mesophilic lactic acid bacteria. J Dairy Sci 1992; 75(3): 692-9.
[http://dx.doi.org/10.3168/jds.S0022-0302(92)77805-9]

[62]   Degeest B, Mozzi F, De Vuyst L. Effect of medium composition and temperature and pH changes on exopolysaccharide yields and stability during *Streptococcus thermophilus* LY03 fermentations. Int J Food Microbiol 2002; 79(3): 161-74.
[http://dx.doi.org/10.1016/S0168-1605(02)00116-2] [PMID: 12371651]

[63]   Kumar AS, Mody K, Jha B. Bacterial exopolysaccharides-a perception. J Basic Microbiol 2007; 47(2): 103-17.
[http://dx.doi.org/10.1002/jobm.200610203] [PMID: 17440912]

[64]   Nicolaus B, Kambourova M, Oner ET. Exopolysaccharides from extremophiles: from fundamentals to biotechnology. Environ Technol 2010; 31(10): 1145-58.
[http://dx.doi.org/10.1080/09593330903552094] [PMID: 20718297]

[65]   Vijayendra SVN, Palanivel G, Mahadevamma S, Tharanathan RN. Physico-chemical characterization of an exopolysaccharide produced by a non-ropy strain of *Leuconostoc* sp. CFR 2181 isolated from dahi, an Indian traditional lactic fermented milk product. Carbohydr Polym 2008; 72(2): 300-7.
[http://dx.doi.org/10.1016/j.carbpol.2007.08.016]

[66]   Bramhachari PV, Dubey SK. Isolation and characterization of exopolysaccharide produced by *Vibrio harveyi* strain VB23. Lett Appl Microbiol 2006; 43(5): 571-7.
[http://dx.doi.org/10.1111/j.1472-765X.2006.01967.x] [PMID: 17032234]

[67]   Dubois M, Gilles KA, Hamilton JK, Rebers PA, Smith F. Colorimetric method for determination of sugars and related substances. Anal Chem 1956; 28(3): 350-6.
[http://dx.doi.org/10.1021/ac60111a017]

[68]   Petry S, Furlan S, Waghorne E, Saulnier L, Cerning dagger J, Maguin E. Comparison of the thickening properties of four Lactobacillus delbrueckii subsp. bulgaricus strains and physicochemical characterization of their exopolysaccharides. FEMS Microbiol Lett 2003; 221(2): 285-91.
[http://dx.doi.org/10.1016/S0378-1097(03)00214-3] [PMID: 12725940]

[69]   Macedo MG, Lacroix C, Gardner NJ, Champagne CP. Effect of medium supplementation on exopolysaccharide production by *Lactobacillus rhamnosus* RW-9595M in whey permeate. Int Dairy J 2002; 12(5): 419-26.
[http://dx.doi.org/10.1016/S0958-6946(01)00173-X]

[70]   Kimmel SA, Roberts RF, Ziegler GR. Optimization of exopolysaccharide production by *Lactobacillus delbrueckii* subsp. *bulgaricus* RR grown in a semidefined medium. Appl Environ Microbiol 1998; 64(2): 659-64.
[PMID: 9464404]

[71]   Ta-Chen L, Chang JS, Young CC. Exopolysaccharides produced by *Gordonia alkanivorans* enhance

bacterial degradation activity for diesel. Biotechnol Lett 2008; 30(7): 1201-6.
[http://dx.doi.org/10.1007/s10529-008-9667-8] [PMID: 18286235]

[72]   Saija N, Welman AD, Bennett RJ. Development of a dairy-based exopolysaccharide bioingredient. Int Dairy J 2010; 20(9): 603-8.
[http://dx.doi.org/10.1016/j.idairyj.2010.03.011]

[73]   Dong ZH, Liu T, Liu HF. Influence of EPS isolated from thermophilic sulphate-reducing bacteria on carbon steel corrosion. Biofouling 2011; 27(5): 487-95.
[http://dx.doi.org/10.1080/08927014.2011.584369] [PMID: 21604218]

[74]   Faber EJ, Zoon P, Kamerling JP, Vliegenthart JFG. The exopolysaccharides produced by *Streptococcus thermophilus* Rs and Sts have the same repeating unit but differ in viscosity of their milk cultures. Carbohydr Res 1998; 310(4): 269-76.
[http://dx.doi.org/10.1016/S0008-6215(98)00189-X] [PMID: 9821263]

[75]   Ruas-Madiedo P, Alting AC, Zoon P. Effect of exopolysaccharides and proteolytic activity of *Lactococcus lactis* subsp. *cremoris* strains on the viscosity and structure of fermented milks. Int Dairy J 2005; 15(2): 155-64.
[http://dx.doi.org/10.1016/j.idairyj.2004.05.009]

[76]   Hwang HJ, Kim SW, Xu CP, Choi JW, Yun JW. Production and molecular characteristics of four groups of exopolysaccharides from submerged culture of *Phellinus gilvus*. J Appl Microbiol 2003; 94(4): 708-19.
[http://dx.doi.org/10.1046/j.1365-2672.2003.01903.x] [PMID: 12631207]

[77]   Ruas-Madiedo P, Tuinier R, Kanning M, Zoon P. Role of exopolysaccharides produced by *Lactococcus lactis* subsp. *cremoris* on the viscosity of fermented milks. Int Dairy J 2002; 12(8): 689-95.
[http://dx.doi.org/10.1016/S0958-6946(01)00161-3]

[78]   Levander F, Svensson M, Rådström P. Small-scale analysis of exopolysaccharides from *Streptococcus thermophilus* grown in a semi-defined medium. BMC Microbiol 2001; 1(1): 23.
[http://dx.doi.org/10.1186/1471-2180-1-23] [PMID: 11602017]

[79]   Vaningelgem F, Zamfir M, Mozzi F, *et al.* Biodiversity of exopolysaccharides produced by *Streptococcus thermophilus* strains is reflected in their production and their molecular and functional characteristics. Appl Environ Microbiol 2004; 70(2): 900-12.
[http://dx.doi.org/10.1128/AEM.70.2.900-912.2004] [PMID: 14766570]

[80]   Verhoef R, de Waard P, Schols HA, Rättö M, Siika-aho M, Voragen AGJ. Structural elucidation of the EPS of slime producing *Brevundimonas vesicularis* sp. isolated from a paper machine. Carbohydr Res 2002; 337(20): 1821-31.
[http://dx.doi.org/10.1016/S0008-6215(02)00280-X] [PMID: 12431884]

[81]   Tieking M, Kaditzky S, Valcheva R, Korakli M, Vogel RF, Gänzle MG. Extracellular homopolysaccharides and oligosaccharides from intestinal lactobacilli. J Appl Microbiol 2005; 99(3): 692-702.
[http://dx.doi.org/10.1111/j.1365-2672.2005.02638.x] [PMID: 16108811]

[82]   Welman AD. Exploitation of exopolysaccharides from lactic acid bacteria: Nutritional and functional benefits.Bacterial polysaccharides: Current innovations and future trends. Norfolk: Caister Academic Press 2009; pp. 331-44.

[83]   Hallack LF, Passos DS, Mattos KA, *et al.* Structural elucidation of the repeat unit in highly branched acidic exopolysaccharides produced by nitrogen fixing Burkholderia. Glycobiology 2010; 20(3): 338-47.
[http://dx.doi.org/10.1093/glycob/cwp181] [PMID: 19933228]

[84]   Freitas F, Alves VD, Carvalheira M, Costa N, Oliveira R, Reis MAM. Emulsifying behaviour and rheological properties of the extracellular polysaccharide produced by *Pseudomonas oleovorans* grown on glycerol byproduct. Carbohydr Polym 2009; 78(3): 549-56.

[http://dx.doi.org/10.1016/j.carbpol.2009.05.016]

[85]   Sharma BR, Naresh L, Dhuldhoya NC, Merchant SU, Merchant UC. Xanthan gum-A boon to food industry. Food Promotion Chronicle 2006; 1(5): 27-30.

[86]   Palaniraj A, Jayaraman V. Production, recovery and applications of xanthan gum by *Xanthomonas campestris*. J Food Eng 2011; 106(1): 1-12.
       [http://dx.doi.org/10.1016/j.jfoodeng.2011.03.035]

[87]   Mandala I, Bayas E. Xanthan effect on swelling, solubility and viscosity of wheat starch dispersions. Food Hydrocoll 2004; 18(2): 191-201.
       [http://dx.doi.org/10.1016/S0268-005X(03)00064-X]

[88]   Kim C, Yoo B. Rheological properties of rice starch–xanthan gum mixtures. J Food Eng 2006; 75(1): 120-8.
       [http://dx.doi.org/10.1016/j.jfoodeng.2005.04.002]

[89]   Khan T, Park JK, Kwon JH. Functional biopolymers produced by biochemical technology considering applications in food engineering. Korean J Chem Eng 2007; 24(5): 816-26.
       [http://dx.doi.org/10.1007/s11814-007-0047-1]

[90]   Katina K, Maina NH, Juvonen R, *et al. In situ* production and analysis of *Weissella confusa* dextran in wheat sourdough. Food Microbiol 2009; 26(7): 734-43.
       [http://dx.doi.org/10.1016/j.fm.2009.07.008] [PMID: 19747607]

[91]   Lacaze G, Wick M, Cappelle S. Emerging fermentation technologies: development of novel sourdoughs. Food Microbiol 2007; 24(2): 155-60.
       [http://dx.doi.org/10.1016/j.fm.2006.07.015] [PMID: 17008159]

[92]   Bajaj IB, Survase SA, Saudagar PS, Singhal RS. Gellan gum: fermentative production, downstream processing and applications. Food Technol Biotechnol 2007; 45(4): 341.

[93]   Tapia M, Rojas-Graü M, Carmona A, Rodriguez F, Soliva-Fortuny R, Martin-Belloso O. Use of alginate-and gellan-based coatings for improving barrier, texture and nutritional properties of fresh-cut papaya. Food Hydrocoll 2008; 22(8): 1493-503.
       [http://dx.doi.org/10.1016/j.foodhyd.2007.10.004]

[94]   Bajaj I, Singhal R. Gellan gum for reducing oil uptake in sev, a legume based product during deep-fat frying. Food Chem 2007; 104(4): 1472-7.
       [http://dx.doi.org/10.1016/j.foodchem.2007.02.011]

[95]   Zhang R, Edgar KJ. Properties, chemistry, and applications of the bioactive polysaccharide curdlan. Biomacromolecules 2014; 15(4): 1079-96.
       [http://dx.doi.org/10.1021/bm500038g] [PMID: 24552241]

[96]   Shi Z, Zhang Y, Phillips GO, Yang GJ. Utilization of bacterial cellulose in food. Food Hydrocoll 2014; 35: 539-45.
       [http://dx.doi.org/10.1016/j.foodhyd.2013.07.012]

[97]   Lin SP, Calvar IL, Catchmark JM, Liu JR, Demirci A, Cheng KC. Biosynthesis, production and applications of bacterial cellulose. Cellulose 2013; 20(5): 2191-219.
       [http://dx.doi.org/10.1007/s10570-013-9994-3]

[98]   Srikanth R, Reddy CHSSS, Siddartha G, Ramaiah MJ, Uppuluri KB. Review on production, characterization and applications of microbial levan. Carbohydr Polym 2015; 120: 102-14.
       [http://dx.doi.org/10.1016/j.carbpol.2014.12.003] [PMID: 25662693]

[99]   Zannini E, Waters DM, Coffey A, Arendt EK. Production, properties, and industrial food application of lactic acid bacteria-derived exopolysaccharides. Appl Microbiol Biotechnol 2016; 100(3): 1121-35.
       [http://dx.doi.org/10.1007/s00253-015-7172-2] [PMID: 26621802]

[100] Banat IM, Makkar RS, Cameotra SS. Potential commercial applications of microbial surfactants. Appl Microbiol Biotechnol 2000; 53(5): 495-508.

[http://dx.doi.org/10.1007/s002530051648] [PMID: 10855707]

[101] Degeest B, Vaningelgem F, De Vuyst L. Microbial physiology, fermentation kinetics, and process engineering of heteropolysaccharide production by lactic acid bacteria. Int Dairy J 2001; 11(9): 747-57.
[http://dx.doi.org/10.1016/S0958-6946(01)00118-2]

[102] Rosenberg E, Ron EZ. High- and low-molecular-mass microbial surfactants. Appl Microbiol Biotechnol 1999; 52(2): 154-62.
[http://dx.doi.org/10.1007/s002530051502] [PMID: 10499255]

[103] Yu GH, He PJ, Shao LM. Characteristics of extracellular polymeric substances (EPS) fractions from excess sludges and their effects on bioflocculability. Bioresour Technol 2009; 100(13): 3193-8.
[http://dx.doi.org/10.1016/j.biortech.2009.02.009] [PMID: 19269815]

[104] Huang XW, Cheng W, Hu YY. Screening of flocculant-producing strains by NTG mutagenesis. J Environ Sci (China) 2005; 17(3): 494-8.
[PMID: 16083133]

[105] Li WW, Zhou WZ, Zhang YZ, Wang J, Zhu XB. Flocculation behavior and mechanism of an exopolysaccharide from the deep-sea psychrophilic bacterium *Pseudoalteromonas* sp. SM9913. Bioresour Technol 2008; 99(15): 6893-9.
[http://dx.doi.org/10.1016/j.biortech.2008.01.050] [PMID: 18353634]

[106] Dubey AK, Jeevaratnam K. Structural characterization and functional evaluation of an exopolysaccharide produced by *Weissella confusa* AJ53, an isolate from fermented Uttapam batter supplemented with *Piper betle* L. leaves. Food Sci Biotechnol 2015; 24(6): 2117-24.
[http://dx.doi.org/10.1007/s10068-015-0281-y]

[107] Welman AD, Maddox IS. Exopolysaccharides from lactic acid bacteria: perspectives and challenges. Trends Biotechnol 2003; 21(6): 269-74.
[http://dx.doi.org/10.1016/S0167-7799(03)00107-0] [PMID: 12788547]

[108] Kodali VP, Das S, Sen R. An exopolysaccharide from a probiotic: Biosynthesis dynamics, composition and emulsifying activity. Food Res Int 2009; 42(5-6): 695-9.
[http://dx.doi.org/10.1016/j.foodres.2009.02.007]

[109] Kristo E, Miao Z, Corredig M. The role of exopolysaccharide produced by *Lactococcus lactis* subsp. *cremoris* in structure formation and recovery of acid milk gels. Int Dairy J 2011; 21: 656-62.
[http://dx.doi.org/10.1016/j.idairyj.2011.02.002]

[110] Hassan AN. ADSA Foundation Scholar Award: Possibilities and challenges of exopolysaccharide-producing lactic cultures in dairy foods. J Dairy Sci 2008; 91(4): 1282-98.
[http://dx.doi.org/10.3168/jds.2007-0558] [PMID: 18349221]

[111] Folkenberg DM, Dejmek P, Skriver A, Skov Guldager H, Ipsen R. Sensory and rheological screening of exopolysaccharide producing strains of bacterial yoghurt cultures. Int Dairy J 2006; 16(2): 111-8.
[http://dx.doi.org/10.1016/j.idairyj.2004.10.013]

[112] Ayala-Hernandez I, Hassan AN, Goff HD, Corredig M. Effect of protein supplementation on the rheological characteristics of milk permeates fermented with exopolysaccharide-producing *Lactococcus lactis* subsp. *cremoris*. Food Hydrocoll 2009; 23(5): 1299-304.
[http://dx.doi.org/10.1016/j.foodhyd.2008.11.004]

[113] Girard M, Schaffer-Lequart C. Attractive interactions between selected anionic exopolysaccharides and milk proteins. Food Hydrocoll 2008; 22(8): 1425-34.
[http://dx.doi.org/10.1016/j.foodhyd.2007.09.001]

[114] Gentès MC, St-Gelais D, Turgeon SL. Gel formation and rheological properties of fermented milk with *in situ* exopolysaccharide production by lactic acid bacteria. Dairy Sci Technol 2012; 91(5): 1-17.

[115] Perry DB, McMahon DJ, Oberg CJ. Manufacture of low fat mozzarella cheese using exopolysaccharide-producing starter cultures1. J Dairy Sci 1998; 81(2): 563-6.

[http://dx.doi.org/10.3168/jds.S0022-0302(98)75608-5]

[116] Dabour N, Kheadr E, Benhamou N, Fliss I, LaPointe G. Improvement of texture and structure of reduced-fat Cheddar cheese by exopolysaccharide-producing lactococci. J Dairy Sci 2006; 89(1): 95-110.
       [http://dx.doi.org/10.3168/jds.S0022-0302(06)72073-2] [PMID: 16357272]

[117] Di Cagno R, De Pasquale I, De Angelis M, Buchin S, Rizzello CG, Gobbetti M. Use of microparticulated whey protein concentrate, exopolysaccharide-producing *Streptococcus thermophilus*, and adjunct cultures for making low-fat Italian Caciotta-type cheese. J Dairy Sci 2014; 97(1): 72-84.
       [http://dx.doi.org/10.3168/jds.2013-7078] [PMID: 24183686]

[118] Shobha MS, Tharanathan RN. Rheological behaviour of pullulanase-treated guar galactomannan on co-gelation with xanthan. Food Hydrocoll 2009; 23(3): 749-54.
       [http://dx.doi.org/10.1016/j.foodhyd.2008.04.006]

[119] Oms-Oliu G, Soliva-Fortuny R, Martín-Belloso O. Edible coatings with antibrowning agents to maintain sensory quality and antioxidant properties of fresh-cut pears. Postharvest Biol Technol 2008; 50(1): 87-94.
       [http://dx.doi.org/10.1016/j.postharvbio.2008.03.005]

[120] Nguyen VT, Gidley MJ, Dykes GA. Potential of a nisin-containing bacterial cellulose film to inhibit *Listeria monocytogenes* on processed meats. Food Microbiol 2008; 25(3): 471-8.
       [http://dx.doi.org/10.1016/j.fm.2008.01.004] [PMID: 18355672]

[121] Rodríguez-Carmona E, Villaverde A. Nanostructured bacterial materials for innovative medicines. Trends Microbiol 2010; 18(9): 423-30.
       [http://dx.doi.org/10.1016/j.tim.2010.06.007] [PMID: 20674365]

[122] O'Connor EB, Barrett E, Fitzgerald G, Hill C, Stanton C, Ross RP. Production of vitamins, exopolysaccharides and bacteriocins by probiotic bacteria.Probiotic dairy products. Oxford: Blackwell Publishing Ltd 2005; pp. 167-95.

[123] Wang K, Li W, Rui X, Chen X, Jiang M, Dong M. Characterization of a novel exopolysaccharide with antitumor activity from *Lactobacillus plantarum* 70810. Int J Biol Macromol 2014; 63: 133-9.
       [http://dx.doi.org/10.1016/j.ijbiomac.2013.10.036] [PMID: 24189393]

[124] Doleyres Y, Schaub L, Lacroix C. Comparison of the functionality of exopolysaccharides produced *in situ* or added as bioingredients on yogurt properties. J Dairy Sci 2005; 88(12): 4146-56.
       [http://dx.doi.org/10.3168/jds.S0022-0302(05)73100-3] [PMID: 16291605]

[125] Pan D, Mei X. Antioxidant activity of an exopolysaccharide purified from *Lactococcus lactis* subsp. *lactis* 12. Carbohydr Polym 2010; 80(3): 908-14.
       [http://dx.doi.org/10.1016/j.carbpol.2010.01.005]

[126] Górska S, Jachymek W, Rybka J, Strus M, Heczko PB, Gamian A. Structural and immunochemical studies of neutral exopolysaccharide produced by *Lactobacillus johnsonii* 142. Carbohydr Res 2010; 345(1): 108-14.
       [http://dx.doi.org/10.1016/j.carres.2009.09.015] [PMID: 19897181]

[127] Salazar N, Gueimonde M, Hernández-Barranco AM, Ruas-Madiedo P, de los Reyes-Gavilán CG. Exopolysaccharides produced by intestinal *Bifidobacterium* strains act as fermentable substrates for human intestinal bacteria. Appl Environ Microbiol 2008; 74(15): 4737-45.
       [http://dx.doi.org/10.1128/AEM.00325-08] [PMID: 18539803]

[128] Ruas-Madiedo P, Moreno JA, Salazar N, *et al.* Screening of exopolysaccharide-producing *Lactobacillus* and *Bifidobacterium* strains isolated from the human intestinal microbiota. Appl Environ Microbiol 2007; 73(13): 4385-8.
       [http://dx.doi.org/10.1128/AEM.02470-06] [PMID: 17483284]

[129] Gibson GR, Roberfroid MB. Dietary modulation of the human colonic microbiota: introducing the

concept of prebiotics. J Nutr 1995; 125(6): 1401-12.
[PMID: 7782892]

[130] Bello FD, Walter J, Hertel C, Hammes WP. In vitro study of prebiotic properties of levan-type exopolysaccharides from *Lactobacilli* and non-digestible carbohydrates using denaturing gradient gel electrophoresis. Syst Appl Microbiol 2001; 24(2): 232-7.
[http://dx.doi.org/10.1078/0723-2020-00033] [PMID: 11518326]

[131] Hongpattarakere T, Cherntong N, Wichienchot S, Kolida S, Rastall RA. *In vitro* prebiotic evaluation of exopolysaccharides produced by marine isolated lactic acid bacteria. Carbohydr Polym 2012; 87(1): 846-52.
[http://dx.doi.org/10.1016/j.carbpol.2011.08.085]

[132] Mårtensson O, Biörklund M, Lambo AM, *et al.* Fermented, ropy, oat-based products reduce cholesterol levels and stimulate the bifidobacteria flora in humans. Nutr Res 2005; 25(5): 429-42.
[http://dx.doi.org/10.1016/j.nutres.2005.03.004]

[133] Mozzi F, Gerbino E, Font de Valdez G, Torino MI. Functionality of exopolysaccharides produced by lactic acid bacteria in an *in vitro* gastric system. J Appl Microbiol 2009; 107(1): 56-64.
[http://dx.doi.org/10.1111/j.1365-2672.2009.04182.x] [PMID: 19291238]

[134] Liu J, Luo J, Ye H, Zeng X. Preparation, antioxidant and antitumor activities in vitro of different derivatives of levan from endophytic bacterium *Paenibacillus polymyxa* EJS-3. Food Chem Toxicol 2012; 50(3-4): 767-72.
[http://dx.doi.org/10.1016/j.fct.2011.11.016] [PMID: 22142695]

[135] Yoo SH, Yoon EJ, Cha J, Lee HG. Antitumor activity of levan polysaccharides from selected microorganisms. Int J Biol Macromol 2004; 34(1-2): 37-41.
[http://dx.doi.org/10.1016/j.ijbiomac.2004.01.002] [PMID: 15178007]

[136] Zhao L, Chen Y, Ren S, Han Y, Cheng H. Studies on the chemical structure and antitumor activity of an exopolysaccharide from *Rhizobium* sp. N613. Carbohydr Res 2010; 345(5): 637-43.
[http://dx.doi.org/10.1016/j.carres.2009.11.017] [PMID: 20100608]

[137] Kim Y, Oh S, Yun HS, Oh S, Kim SH. Cell-bound exopolysaccharide from probiotic bacteria induces autophagic cell death of tumour cells. Lett Appl Microbiol 2010; 51(2): 123-30.
[PMID: 20536712]

[138] Ebina T, Ogata N, Murata K. Antitumor effect of *Lactobacillus bulgaricus* 878R. Biotherapy 1995; 9: 65-70.

[139] Tsai YT, Cheng PC, Pan TM. The immunomodulatory effects of lactic acid bacteria for improving immune functions and benefits. Appl Microbiol Biotechnol 2012; 96(4): 853-62.
[http://dx.doi.org/10.1007/s00253-012-4407-3] [PMID: 23001058]

[140] Hosono A, Lee J, Ametani A, *et al.* Characterization of a water-soluble polysaccharide fraction with immunopotentiating activity from *Bifidobacterium adolescentis* M101-4. Biosci Biotechnol Biochem 1997; 61(2): 312-6.
[http://dx.doi.org/10.1271/bbb.61.312] [PMID: 9058970]

[141] Kitazawa H, Harata T, Uemura J, Saito T, Kaneko T, Itoh T. Phosphate group requirement for mitogenic activation of lymphocytes by an extracellular phosphopolysaccharide from *Lactobacillus delbrueckii* ssp. *bulgaricus*. Int J Food Microbiol 1998; 40(3): 169-75.
[http://dx.doi.org/10.1016/S0168-1605(98)00030-0] [PMID: 9620124]

[142] Fanning S, Hall LJ, Cronin M, *et al.* Bifidobacterial surface-exopolysaccharide facilitates commensal-host interaction through immune modulation and pathogen protection. Proc Natl Acad Sci USA 2012; 109(6): 2108-13.
[http://dx.doi.org/10.1073/pnas.1115621109] [PMID: 22308390]

[143] Wu MH, Pan TM, Wu YJ, Chang SJ, Chang MS, Hu CY. Exopolysaccharide activities from probiotic *bifidobacterium*: Immunomodulatory effects (on J774A.1 macrophages) and antimicrobial properties.

Int J Food Microbiol 2010; 144(1): 104-10.
[http://dx.doi.org/10.1016/j.ijfoodmicro.2010.09.003] [PMID: 20884069]

[144] Xu CL, Wang YZ, Jin ML, Yang XQ. Preparation, characterization and immunomodulatory activity of selenium-enriched exopolysaccharide produced by bacterium *Enterobacter cloacae* Z0206. Bioresour Technol 2009; 100(6): 2095-7.
[http://dx.doi.org/10.1016/j.biortech.2008.10.037] [PMID: 19056259]

[145] Vinderola G, Perdigón G, Duarte J, Farnworth E, Matar C. Effects of the oral administration of the exopolysaccharide produced by *Lactobacillus kefiranofaciens* on the gut mucosal immunity. Cytokine 2006; 36(5-6): 254-60.
[http://dx.doi.org/10.1016/j.cyto.2007.01.003] [PMID: 17363262]

[146] Chabot S, Yu HL, De Léséleuc L, Cloutier D, Van Calsteren MR, Lessard M, *et al.* Exopolysaccharides from *Lactobacillus rhamnosus* RW-9595M stimulate TNF, IL-6 and IL-12 in human and mouse cultured immunocompetent cells, and IFN-$gamma in mouse splenocytes. Lait 2001; 81(6): 683-97.
[http://dx.doi.org/10.1051/lait:2001157]

[147] Amrouche T, Boutin Y, Prioult G, Fliss I. Effects of bifidobacterial cytoplasm, cell wall and exopolysaccharide on mouse lymphocyte proliferation and cytokine production. Int Dairy J 2006; 16(1): 70-80.
[http://dx.doi.org/10.1016/j.idairyj.2005.01.008]

[148] Shao L, Wu Z, Zhang H, Chen W, Ai L, Guo B. Partial characterization and immunostimulatory activity of exopolysaccharides from *Lactobacillus rhamnosus* KF5. Carbohydr Polym 2014; 107: 51-6.
[http://dx.doi.org/10.1016/j.carbpol.2014.02.037] [PMID: 24702917]

[149] Sasaki E, Suzuki S, Fukui Y, Yajima N. Cell-bound exopolysaccharides of *Lactobacillus brevis* KB290 enhance cytotoxic activity of mouse splenocytes. J Appl Microbiol 2015; 118(2): 506-14.
[http://dx.doi.org/10.1111/jam.12686] [PMID: 25376258]

[150] Turroni F, Serafini F, Foroni E, *et al.* Role of sortase-dependent pili of *Bifidobacterium bifidum* PRL2010 in modulating bacterium-host interactions. Proc Natl Acad Sci USA 2013; 110(27): 11151-6.
[http://dx.doi.org/10.1073/pnas.1303897110] [PMID: 23776216]

[151] Zhang L, Liu C, Li D, *et al.* Antioxidant activity of an exopolysaccharide isolated from *Lactobacillus plantarum* C88. Int J Biol Macromol 2013; 54: 270-5.
[http://dx.doi.org/10.1016/j.ijbiomac.2012.12.037] [PMID: 23274679]

[152] Guo Y, Pan D, Li H, Sun Y, Zeng X, Yan B. Antioxidant and immunomodulatory activity of selenium exopolysaccharide produced by *Lactococcus lactis* subsp. *lactis*. Food Chem 2013; 138(1): 84-9.
[http://dx.doi.org/10.1016/j.foodchem.2012.10.029] [PMID: 23265459]

[153] Zheng LP, Zou T, Ma YJ, Wang JW, Zhang YQ. Antioxidant and DNA damage protecting activity of exopolysaccharides from the endophytic bacterium *Bacillus cereus* SZ1. Molecules 2016; 21(2): 174.
[http://dx.doi.org/10.3390/molecules21020174] [PMID: 26861269]

[154] Meng F, Zhou B, Lin R, *et al.* Extraction optimization and *in vivo* antioxidant activities of exopolysaccharide by Morchella esculenta SO-01. Bioresour Technol 2010; 101(12): 4564-9.
[http://dx.doi.org/10.1016/j.biortech.2010.01.113] [PMID: 20153962]

[155] Sun ML, Liu SB, Qiao LP, *et al.* A novel exopolysaccharide from deep-sea bacterium *Zunongwangia profunda* SM-A87: low-cost fermentation, moisture retention, and antioxidant activities. Appl Microbiol Biotechnol 2014; 98(17): 7437-45.
[http://dx.doi.org/10.1007/s00253-014-5839-8] [PMID: 24872221]

[156] Li S, Huang R, Shah NP, Tao X, Xiong Y, Wei H. Antioxidant and antibacterial activities of exopolysaccharides from *Bifidobacterium bifidum* WBIN03 and *Lactobacillus plantarum* R315. J Dairy Sci 2014; 97(12): 7334-43.
[http://dx.doi.org/10.3168/jds.2014-7912] [PMID: 25282420]

[157] Sirajunnisa AR, Vijayagopal V, Sivaprakash B, Viruthagiri T, Surendhiran D. Optimization, kinetics and antioxidant activity of exopolysaccharide produced from rhizosphere isolate, *Pseudomonas fluorescens* CrN6. Carbohydr Polym 2016; 135: 35-43.
[http://dx.doi.org/10.1016/j.carbpol.2015.08.080] [PMID: 26453848]

[158] Polak-Berecka M, Waśko A, Szwajgier D, Chomaz A. Bifidogenic and antioxidant activity of exopolysaccharides produced by *Lactobacillus rhamnosus* E/N cultivated on different carbon sources. Pol J Microbiol 2013; 62(2): 181-8.
[PMID: 24053021]

[159] Li W, Ji J, Chen X, Jiang M, Rui X, Dong M. Structural elucidation and antioxidant activities of exopolysaccharides from *Lactobacillus helveticus* MB2-1. Carbohydr Polym 2014; 102: 351-9.
[http://dx.doi.org/10.1016/j.carbpol.2013.11.053] [PMID: 24507291]

[160] Ye S, Zhang J, Liu Z, Zhang Y, Li J, Li YO. Biosynthesis of selenium rich exopolysaccharide (Se-EPS) by *Pseudomonas* PT-8 and characterization of its antioxidant activities. Carbohydr Polym 2016; 142: 230-9.
[http://dx.doi.org/10.1016/j.carbpol.2016.01.058] [PMID: 26917395]

[161] Guo Y, Pan D, Sun Y, Xin L, Li H, Zeng X. Antioxidant activity of phosphorylated exopolysaccharide produced by *Lactococcus lactis* subsp. *lactis*. Carbohydr Polym 2013; 97(2): 849-54.
[http://dx.doi.org/10.1016/j.carbpol.2013.06.024] [PMID: 23911523]

[162] Abdhul K, Ganesh M, Shanmughapriya S, Kanagavel M, Anbarasu K, Natarajaseenivasan K. Antioxidant activity of exopolysaccharide from probiotic strain *Enterococcus faecium* (BDU7) from Ngari. Int J Biol Macromol 2014; 70: 450-4.
[http://dx.doi.org/10.1016/j.ijbiomac.2014.07.026] [PMID: 25062992]

[163] Li JY, Jin MM, Meng J, Gao SM, Lu RR. Exopolysaccharide from *Lactobacillus planterum* LP6: antioxidation and the effect on oxidative stress. Carbohydr Polym 2013; 98(1): 1147-52.
[http://dx.doi.org/10.1016/j.carbpol.2013.07.027] [PMID: 23987456]

[164] London LEE, Kumar AHS, Wall R, *et al.* Exopolysaccharide-producing probiotic Lactobacilli reduce serum cholesterol and modify enteric microbiota in ApoE-deficient mice. J Nutr 2014; 144(12): 1956-62.
[http://dx.doi.org/10.3945/jn.114.191627] [PMID: 25320181]

[165] Nakajima H, Suzuki Y, Hirota T. Cholesterol lowering activity of ropy fermented milk. J Food Sci 1992; 57(6): 1327-9.
[http://dx.doi.org/10.1111/j.1365-2621.1992.tb06848.x]

[166] Lindström C, Holst O, Nilsson L, Öste R, Andersson KE. Effects of *Pediococcus parvulus* 2.6 and its exopolysaccharide on plasma cholesterol levels and inflammatory markers in mice. AMB Express 2012; 2(1): 66.
[http://dx.doi.org/10.1186/2191-0855-2-66] [PMID: 23234432]

CHAPTER 3

Enzymatic Modification of Phospholipids

Jasmina Damnjanović and **Yugo Iwasaki**[*]

Laboratory of Molecular Biotechnology, Graduate School of Bioagricultural Sciences, Nagoya University, Furo-cho, Chikusa-ku, Nagoya 464-8601, Japan

Abstract: Phospholipids (PLs) are naturally-occurring amphiphilic compounds, which can be used as emulsifiers, liposome components, and nutritional supplements in food, cosmetic, and pharmaceutical industries. Enzymatic modifications of natural PLs are of importance to enhance the quality of PLs as value-added products. This chapter reviews the advances in the enzymatic modification of PLs, including acyl group modification and head group modification. Acyl group modification can be performed using lipase- or phospholipase A-mediated transesterification or ester synthesis to introduce particular fatty acid residues. Head group modification is carried out by phospholipase D-catalyzed transphosphatidylation. This reaction enables introduction of a wide range of natural and unnatural functional molecules into the phospholipids, making it possible to prepare chemically defined structured PLs for diverse applications.

Keywords: Emulsifier, Esterification, Phospholipid, Phospholipase, Transesterification.

INTRODUCTION

Phospholipids (PLs) are lipids containing phosphorus. PLs are classified into two classes, glycerophospholipids and sphingophospholipids. The major representatives of natural glycerophospholipids are phosphatidylcholine (PC), phosphatidylethanolamine (PE), phosphatidylglycerol (PG), phosphatidylinositol (PI), phosphatidylserine (PS), phosphatidic acid (PA), and cardiolipin (CL), while the representative of sphingophospholipids is sphingomyelin (SM). They contain two or more hydrophobic acyl or alkyl moieties bound by ester or ether bonds to a glycerol or a sphingosine backbone and a polar head group bound *via* phosphodiester bond.

The hydrophobicity of the acyl or alkyl moieties and the hydrophilicity of the head groups make PLs amphiphilic compounds. Therefore, one of the most

---

[*] **Corresponding author Yugo Iwasaki:** Laboratory of Molecular Biotechnology, Graduate School of Bioagricultural Sciences, Nagoya University, Furo-cho, Chikusa-ku, Nagoya 464-8601, Japan; Tel: +81-52-789-4145; Fax: +81-5--789-4143; Email: iwasaki@agr.nagoya-u.ac.jp

important characteristics of PLs is that, in aqueous environment, they can form several different structures, such as monomers, micelles, reverse micelles, and bilayer vesicles. Due to the PLs' characteristics, *i.e.,* physical properties, biocompatibility and nutritional functions, PLs are useful compounds in many industrial fields, such as food, cosmetics and pharmaceuticals.

For example, in food and cosmetic applications, lecithin and lyso-lecithin have been traditionally used as emulsifiers, aerating agents, viscosity modifiers, dispersing agents, lubricants, mold-releasing agents, emollients, moisturizers, *etc.*

As more specific examples of particular PL classes, PA suppresses the taste response to bitter substances; and thereby it is used as a masking agent of bitterness in food and medicines [1]. PS shows therapeutic effects on human health including memory-related disorders, and it is used as an over-the-counter food supplement [2]. PI is an attractive nutritional supplement for food or pharmaceutical industries due to its biological effects, mainly related to improvement of lipid metabolism [3]. As fine chemicals, PLs of defined chemical structures are used for liposome formulations, which are applied in drug delivery systems [4]. A formulation consisting of bovine lung extract containing PLs and surfactant-associated proteins is used in surfactant therapy for neonatal respiratory distress syndrome, a condition caused by developmental insufficiency of the lung surfactant [5].

To make better use of PLs, the chemical structures and purity are of critical importance. The primary sources of PLs for industrial applications are usually of plant origin, *e.g.,* soy lecithin, which is obtained upon degumming of edible oil. However, not all the PL classes are easily available from such natural sources, which are usually heterogeneous mixtures of various PL species with different acyl groups and head groups. For this reason, it is important to modify the structures of available PLs by means of chemical, enzymatic, or chemo-enzymatic reactions. Enzymatic reactions generally proceed under mild conditions with desirable specificity (chemo-, regio-, and stereospecificity).

Phospholipases (PLases) are enzymes that hydrolyze PLs. Based on their modes of action, PLases are classified into several classes, PLases A_1 (PLA$_1$), A_2 (PLA$_2$), B (PLB), C (PLC), and D (PLD) (Fig. **3.1**). PLA$_1$ and PLA$_2$ hydrolyze the acyl ester bond at 1- and 2-position, respectively, while PLB hydrolyzes both. PLC and PLD act on the phosphodiester bond, where PLC cleaves the linkage between the glycerol and the phosphate, whereas PLD attacks the bond between the phosphate and the alcohol moiety of the polar head.

This chapter reviews the advances of enzymatic modification of PLs with respect to two categories, *i.e.*, acyl group modification and head group modification.

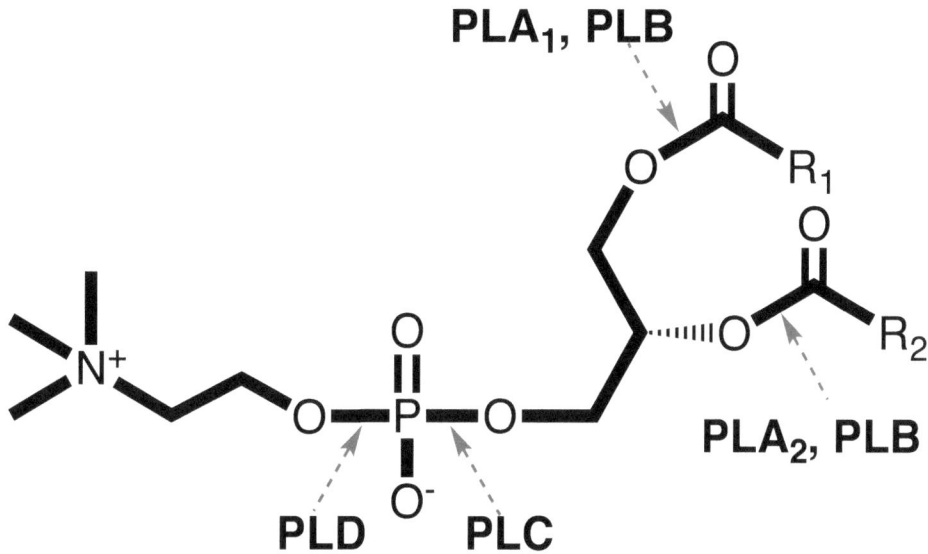

PLA₁, PLB
PLA₂, PLB
PLD
PLC

Fig. (3.1). Catalytic activities of different phospholipases on the example of PC. The arrows indicate the bonds cleaved by different phospholipases.

ACYL GROUP MODIFICATION

Importance of the Acyl Group Modification of PLs

PLs from natural sources are heterogeneous mixtures consisting of various molecular species with different fatty acid (FA) residues and polar head groups. As different PL species have different physical, chemical, and biological properties, it is often necessary to use PLs with defined chemical structure, which are referred to as structured PLs (sPLs). For example, in liposome preparations, such sPL species are used to control the properties of the vesicles. Also, in the field of lipid biochemistry, sPLs are required as laboratory reagents for evaluation of PL's bioactivities.

Incorporating particular FAs into the PLs seems especially beneficial for the utilization of polyunsaturated fatty acids (PUFAs). PUFAs, such as eicosapentaenoic acid (EPA), docosahexaenoic acid (DHA), and conjugated linoleic acid (CLA) are known to have various bioactivities [6, 7]. Therefore, the use of these FAs as functional foods is of special interest. However, sensitivity to oxidation is often a serious problem with using PUFAs. It was reported that DHA residues in the form of PLs show higher stability against oxidation than the ones in the form of free acids, ethyl esters, and triacylglycerols (TAGs) [8, 9]. Therefore, incorporation of PUFAs into PLs is advantageous for their protection against oxidation.

Another practical advantage to design sPLs is the use of PLs as carriers of PUFAs to the cells or the tissues of living animals. Brenna's group demonstrated that arachidonic acid (ARA)- and DHA-containing PCs were more efficiently delivered into the brain of neonatal baboons and piglets, respectively, than the corresponding PUFA-containing TAG [10, 11]. Rossmeisl *et al.* [12] demonstrated that DHA/EPA administered in PL form were superior in preserving a healthy metabolic profile under obesogenic conditions, compared to the form of TAGs in mice [12].

In addition, it was reported that unsaturated FAs [oleic acid (OA), linoleic acid (LA) and ARA] were more efficiently delivered into the developing rat brain when administered in the form of 1-lyso-2-acyl-PC (1-LPC) than unesterified FAs [13], or 1-acyl-2-lyso-PC (2-LPC) [14]. Furthermore, because 1-lyso-2-DHA-PC (an efficient transporter of DHA to the brain) easily isomerizes to 1-DHA-2-lyso-PC by spontaneous acyl migration, Lagarde *et al.* [15] synthesized 1-acetyl-2-DHA-PC (AceDoPC) to avoid the isomerization by blocking the free hydroxyl group at the *sn*-1 position. The synthesized AceDoPC showed similar behavior to that of 1-lyso-2-DHA-PC [15].

Still another benefit for the preparation of sPLs is the case of medium-chain FAs (MCFA)-containing PLs. Vikbjerg *et al.* [16] showed that MCFA-PL had superior emulsifying properties compared to the natural soybean lecithin [16]. There are several reports about the preparation of MCFA-PL by enzymatic reactions [17 - 20].

Chemical Synthesis of Diacyl-PLs

PCs with the same FA residues both at *sn*-1 and *sn*-2 positions (mono-acid PC) can be synthesized chemically (Fig. **3.2A**). Natural PC is first deacylated with alkaline, such as tetrabutylammonium hydroxide to obtain glycerophosphorylcholine (GPC), followed by chemical acylation with an appropriate FA donor (*e.g.*, FA-anhydride) in the presence of a chemical catalyst (*e.g.*, dimethylaminopyridine, DMAP) [21]. GPC is often isolated as a $CdCl_2$ complex to facilitate its preparation and handling [22], but the use of such toxic metals should be avoided. An alternative method was reported, in which GPC adsorbed on Kieselguhr was directly esterified without using $CdCl_2$ [23].

PC species with different FA residues at the *sn*-1 and *sn*-2 positions (mixed-acid PC) are prepared by hydrolysis of the corresponding mono-acid PC by PLA_2 to afford 2-LPC, followed by further chemical acylation at the *sn*-2 position [24].

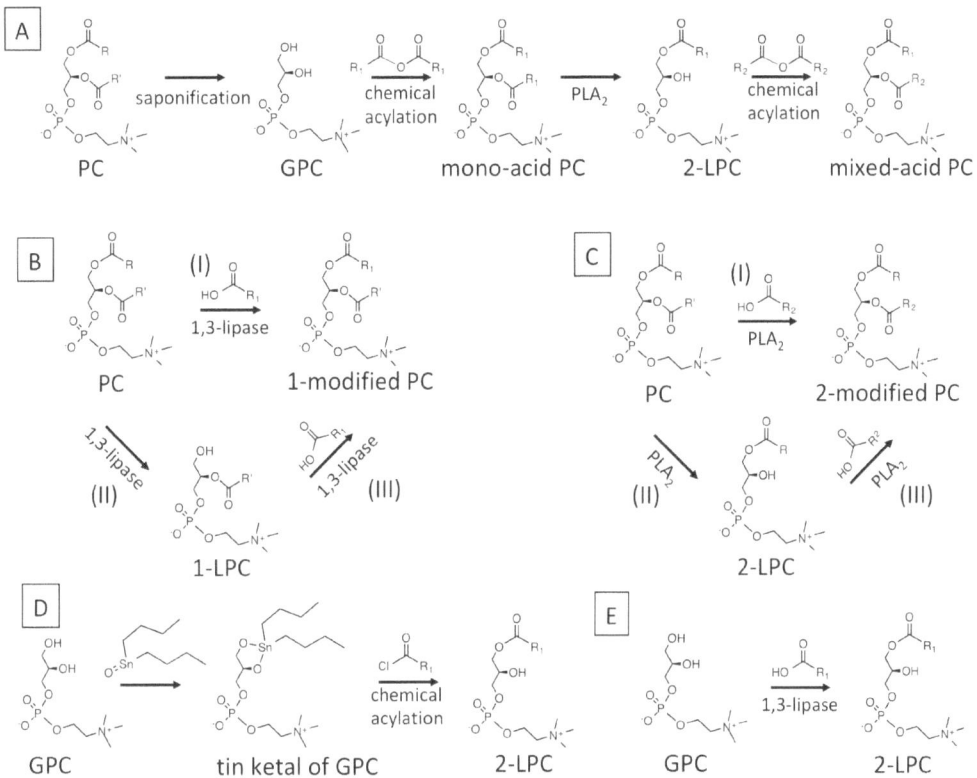

Fig. (3.2). Modification of acyl groups of PLs. **(A)** Chemical and chemo-enzymatic modification of mono- and mixed-acid PC. **(B)***sn*-1,3-specific lipase-mediated *sn*-1 specific modification; (I) direct acidolysis, (II and III) enzymatic deacylation followed by esterification. **(C)** PLA₂-mediated *sn*-2 specific modification; (I) direct acidolysis, (II and III) enzymatic deacylation followed by esterification. **(D)** Tin-mediated regio-selective chemical acylation of GPC. **(E)***sn*-1,3-specific lipase-mediated acylation of GPC.

Enzymatic Synthesis of Diacyl-PLs

Use of Lipases and PLA₂

The chemical and chemo-enzymatic methods are useful for the preparation of sPLs especially as fine chemicals. Still, there are drawbacks in the use of chemical methods, i) the requirements for activated acyl donors, such as FA chloride and chemical catalysts, such as DMAP, and ii) the requirements for protection/deprotection steps of certain head groups (*e.g.*, amino groups of PE or PS, hydroxyl groups of PI or PG) during the deacylation and re-acylation [25]. More importantly, these chemically synthesized PLs are not suitable for applications in food due to the safety issues related to the use of chemical reagents, and their relatively high cost. For these reasons, many attempts have been made to enzymatically incorporate particular FAs into natural PLs (*e.g.*, soy lecithin) using enzymes including lipase and PLA_2. Introducing functional FAs into natural PLs without the use of toxic chemical reagents is especially important.

Lipase is an enzyme that catalyzes the hydrolysis of esters including acylglycerols and PLs, but it also catalyzes ester synthesis, transesterification (alcoholysis and acidolysis), and interesterificaiton (Fig. **3.3**). An important characteristic of lipases is positional specificity towards the glycerol backbone of lipids. With respect to the positional specificity, lipases are classified as either 1,3-specific or non-specific. A 1,3-specific lipase acts on the ester bonds of glycerides at the *sn*-1,3 positions, while a non-specific lipase acts on all positions without discrimination. The positional specificity of a lipase depends on its origin; some are strictly 1,3-specific, some are almost completely non-specific, and others can be classified in between the two [26]. Therefore, the choice of lipase is very important for position-specific modification of PLs. Note that there are no "strictly" *sn*-2-specific lipases, although some are reported to show "preference" towards *sn*-2 position over the *sn*-1, 3 positions [27 - 29].

Hydrolysis

Ester synthesis

Transesterification

(1) alcoholysis

(2) acidolysis

(3) interesterification

Fig. (3.3). Reactions catalyzed by lipases.

A great number of research articles about the enzyme-catalyzed modification of PLs are currently available (Table **3.1**). According to the positional specificities of the modification, the modifications are categorized into: *sn*-1 selective, *sn*-2 selective, and non-specific modification.

Table 3.1. Enzymatic preparation of structured diacyl-PLs

Enzyme[a]	Type[b]	Substrate[c]	Solvent System	Results[d]	Note	Ref.
RDL	A	Egg PE + OA	Buffer	18% of OA in PE, < 20% PE rec.		[30]
RDL	A	DPPC + OA	Hexane-Buffer	25% inc., 25% PC rec.		[31]
MML	A	Egg PC+ HDA	Water-sat. Toluene	~100% inc. at *sn*-1, 40% PC rec.		[32]
RAL	A	Egg PC+ HDA	Toluene	~100% inc. at *sn*-1, 60% PC rec.	a_w controlled	[33]
ROL	A (E)	DPPC (LPC) + C6:0	Toluene	78% modified PC	a_w controlled	[34]
ROL	I	Soy PC + C12:0-ME	Solvent-Free	48% inc., 28% PC rec.		[35]
RML	A	DPPC + EPA	Solvent-Free	58% EPA inc., 39% PC rec.		[36]
RML	A	DMPC + OA	Solvent-Free	35% oleoyl-myristoy--PC		[37]
RML	A	PC + EPA	Hexane	17.7% EPA inc.		[38]
RML	A	Soy PC + CLA	Solvent-Free	16% CLA inc.		[39]
RML	I	PC+PUFAEE	Hexane	56.8% PUFA inc.	Mg^{2+}, urea addition	[40]
RML	I	Soy PL+PUFA-EE	Hexane	12.3% PUFA inc.	Mg^{2+} addition	[41]
CALB	I	Egg PC + Linseed oil	Hexane	34% linolenic acid inc. at *sn*-1		[42]
snPLA$_2$	E	LPC + OA	Toluene	6.5% diacyl-PC		[43]
poPLA$_2$	E	LPC+PUFA	Isooctane-AOT microemulsion	~6% diacyl-PC	RSM optimization	[44]
poPLA$_2$	E	LPC + OA	Toluene	60% diacyl-PC	a_w controlled	[45]
poPLA$_2$	E	LPC + EPA	Glycerol	60% diacyl-PC	Formamide addition	[46]
poPLA$_2$	E	LPC+ DHA	Glycerol	90% diacyl-PC	Formamide addition, Vacuum	[47]

(Table 3.1) cont.....

Enzyme[a]	Type[b]	Substrate[c]	Solvent System	Results[d]	Note	Ref.
poPLA$_2$	E	LPC + EPA	Glycerol	49% diacyl-PC	Gly, Ala addition, Vacuum	[48]
poPLA$_2$	A	PC + C8:0	Solvent-Free	36% inc., 29% PC rec.	RSM optimization	[49]
poPLA2	A	Soy PC + C8:0	Hexane	45% ML-PC		[50]
CRL	I	Soy PC + PUFA-EE	Hexane	47.1% PUFA inc. at *sn*-2		[51]
CRL	I	Soy PC+ MG or DG	Hexane	3~60% targeted FA inc.		[52]
CCL	I	Soy PC+ Sardine oil	Hexane-Glycerol	32% inc., 47% PC rec.		[53]
CCL	A	DPPC + EPA	Water-sat. Benzene	3.5% PC rec. with EPA		[54]
PLA$_1$	A	Soy PC + CLA	Solvent-Free	85.8% CLA inc.	RSM optimization	[55]
PLA$_1$	A	Soy PC + CLA	Solvent-Free	90% CLA inc.		[56]
PLA$_1$	A	Soy PC + PUFA	Solvent-Free	35% PUFA inc.		[57]
PLA$_1$	A	PC+PUFA	Solvent-Free	28% PUFA inc.		[58]
PLA$_1$	I	PC+PUFA-EE	Solvent-Free	30.7% PUFA inc., 16.5% PC Rec.		[59]
PLA$_1$	A	PC+PUFA	Solvent-Free	57.4% PUFA inc., 16.7% PC rec.		[60]
TLL	A	Soy PL + C8:0	Solvent-Free	39% inc.	RSM optimization	[17]
TLL	A	Soy PC + C8:0	Hexane	46% inc., 60% PC rec.	RSM optimization	[19]

[a] RDL, *Rhizopus delemar* lipase; MML, *Mucor miehei* lipase; RAL, *Rhizopus arrhizus* lipase; ROL, *Rhizopus oryzae* lipase; RML, *Rhizomucor miehei* lipase (Lipozyme® RM IM); CALB, *Candida antarctica* lipase B (Novozym 435); snPLA$_2$, snake venom PLA$_2$; poPLA$_2$, porcine pancreatic PLA$_2$; CRL, *Candida rugosa* lipase; CCL, *Candida cylindracea* lipase; PLA$_1$, a protein-engineered enzyme obtained by combination of *Thermomyces lanuginosus* lipase *and Fusarium oxysporum* PLA$_1$ (Lecitase Ultra); TLL, *Thermomyces lanuginosa* lipase (Lipozyme® TLIM).

[b] Reaction type: A, acidolysis; E, esterification; I, interesterification.

[c] Represented as "PL + acyl donor". OA, oleic acid; HDA, heptadecanoic acid; DPPC, dipalmitoyl-PC; C6:0, hexanoic acid; C12:0-ME, lauric acid methylester; DMPC, dimyristoyl-PC; PUFA-EE, PUFA ethylester; C8:0, octanoic acid;

[d] For acidolysis and interesterificaiton reactions, the incorporation (inc.) of the targeted FAs acyl residues and the recovery yield (rec.) of diacyl-PL are shown as well as the positional distribution. For esterification reactions, the yields of diacyl-PL are shown. ML-PC, PC with one medium and one long chain acyl residues.

Sn-1 Selective Modification

The FAs of interest can be introduced into the *sn*-1 position of PLs using 1,3-

specific lipases (Fig. **3.2B**). Brockerhoff *et al.* [30] reported the first example of such a reaction using a lipase from *Rhizopus delemar*, a filamentous fungus, for the acidolysis of egg PE with ^{14}C-radiolabelled OA to obtain *sn*-1-radiolabeled PLs in an aqueous reaction system [30]. Yagi *et al.* [31] employed *R. delemar* lipase for acidolysis of PC in hexane-buffer biphasic system [31], while Svensson *et al.* [32] used immobilized *Rhizomucor miehei* lipase (Lipozyme RMIM) in water-saturated toluene [32]. Owing to the strict positional specificity of the catalysts, the reaction occurred selectively at the *sn*-1 position.

A problem in the lipase-catalyzed reaction is hydrolysis of the substrate and the target product, which results in the loss of product (= diacyl-PL) yields. For example, in the early works [30 - 32], although the incorporation of the target FAs was reasonable, the recovery ratio of diacyl-PLs was in the range of 20-40%, indicating that more than half of the initial PLs were lost by hydrolysis to lyso PC (LPC) or further to GPC. One may think that performing the reaction under completely anhydrous conditions will overcome the problem of hydrolysis; however, in anhydrous environment, enzymes often lose activity. In addition, considering the reaction mechanism of acidolysis, a certain degree of hydrolysis must occur to form LPC as an intermediate, which is then re-esterified with the target acyl groups. Therefore, control of the water content in the reaction system is very important. Svensson *et al.* [33] studied this topic on acidolysis of PC, and performed the reaction in toluene using immobilized *Rhizopus arrhizus* lipase [33]. Prior to the reaction, all reactants and the catalyst were equilibrated over saturated salt solutions to obtain defined initial water activity (a_w). By optimizing a_w in the reaction system, almost 100% incorporation in the *sn*-1 position with PC recovery of 60% was achieved. Adlercreutz *et al.* [34] compared acidolysis (Fig. **3.2B**, route I) of PC and esterification of 1-lyso-2-acyl-PC (1-LPC) (Fig. **3.2B**, route III) for incorporation of hexanoic acid into the *sn*-1 position with *Rhizopus oryzae* lipase in toluene [34]. In both acidolysis and esterification, the yield of more than 70% of the desired PC (1-hexanoyl-2-palmitoyl-PC) was achieved.

Sn-2 Selective Modification

Introduction of FA into *sn*-2 position is performed using PLA$_2$. For this purpose, condensation reaction between particular FA and 2-LPC by PLA$_2$ is often used (Fig. **3.2C**, route III). The first example of PLA$_2$-mediated ester synthesis was demonstrated by Pernas *et al.* [43]. They introduced OA into 2-LPC using lyophilized snake venom PLA$_2$ in dry toluene with a low yield (6.5%) [43]. Also, Na *et al.* [44] demonstrated condensation between PUFA and 2-LPC in a microemulsion system consisting of isooctane, sodium *bis*-(2-ethylhexyl)-sulfosuccinate and low water content, but the yield was still very low (~6%) [44]. Later on, many research groups worked on the yield improvement. Egger *et al.*

[45] performed condensation of 2-LPC and OA with porcine pancreatic PLA_2 (Lecitase 10L) in toluene. By careful control of moisture content, they successfully achieved the yield of 60% at a_w of 0.11 [45]. Hosokawa *et al.* [46] employed formamide as a water mimic for the condensation of 2-LPC with PUFA in glycerol to afford 60% yield [46]. The same research group further improved the yield by performing the reaction under reduced pressure for the removal of the condensed water (and excess formamide) to achieve as high as ~90% yield [47]. Because formamide is not food-compatible, Tanaka *et al.* [48] devised an alternative method compatible for food production. They found that addition of amino acids, such as glycine and alanine into the reaction mixture of glycerol, 2-LPC, FA and porcine pancreatic PLA_2, all of which were of food grade, was effective to give a yield of 63% for OA and 49% for EPA [48].

As 2-LPC is usually prepared by PLA_2-hydrolysis, the introduction of FAs into *sn*-2 position by condensation reactions should include hydrolysis of PLs by PLA_2, recovery of the 2-LPC and re-esterification of the 2-LPC by PLA_2 (Fig. **3.2C**, routes II and III). Besides, direct acidolysis of diacyl-PLs with FAs by PLA_2 is also possible (Fig. **3.2C**, route I) [49]. This is practically advantageous, as the whole process is performed in one step. Yamamoto *et al.* [51] reported incorporation of PUFA into PL using *Candida cylindracea* lipase. Interestingly, the incorporated PUFA residues were located selectively at the *sn*-2 positions, although the enzyme is known as a non-specific lipase [51].

Non-Specific Modification

If the sole purpose is enrichment of particular FAs with no preference for the position (*sn*-1 or -2) to be modified, lipases with no or low positional specificity are usually used. Early works often employed non-specific lipase from *Candida rugosa* (*C. cylindracea*; Lipase OF). Yoshimoto *et al.* [54] reported acidolysis of PC with EPA using polyethyleneglycol-modified Lipase OF (with increased solubility in solvents) in benzene with relatively low degree of conversion [54]. Other works employed Lipase OF powder for interesterifcation of PC with TAG, diacylglycerol (DAG), or monoacylglycerol (MAG) in hexane, with reasonable degrees of conversion and PC recoveries [52, 53].

Indeed, 1,3-specific lipase and PLA_1 can also be used for non-specific modification. A commercially available PLA_1, such as Lecitase™ Ultra, an enzyme engineered by fusing two lipases, from *Thermomyces lanuginosa* and *Fusarium oxysporum* [61], is often used for this purpose. One notable example is the work by Baeza-Jiménez *et al.* [56], who performed acidolysis of PC with CLA in a solvent-free system. Optimizing the conditions resulted in 90% CLA incorporation, which is the highest ever reported, although the recovery of PC has not been discussed.

Process Optimization for Acyl Group Modification

For efficient conversion, many parameters should be considered as the operation variables that affect the yield of the target product. These parameters include temperature, time, enzyme amount, substrate ratio, and water content. In the case of acidolysis, for example, it is ideal to achieve both high degree of FA incorporation and at the same time high degree of PL recovery. However, high incorporation often results in low recovery of diacyl-PL and *vice versa*. The lowered recovery of diacyl-PL is mainly due to the hydrolysis during the reaction, but it is indeed necessary for acidolysis to proceed. Hence, there should be a compromise between incorporation and recovery. In addition, for more practical production, other factors, such as production cost and feasibility of scale-up should be taken into account.

Response surface methodology (RSM) is a practical option for optimizing multiple parameters of a production process. A number of studies are dedicated to optimization of enzyme-mediated PL modification by using RSM [17, 19, 20, 49, 53, 57]. Among them, an excellent example is the one by Vikbjerg *et al.* [19]. They studied acidolysis of soy PC with caprylic acid using *Thermomyces lanuginosa* lipase as the catalyst. By optimizing parameters (enzyme dose, temperature, solvent amount, time, and substrate ratio) for both high incorporation and high recovery, incorporation of 46% and PC recovery of 60% have been achieved.

In the production of modified diacyl-PC, the hydrolytic byproducts LPC and GPC are contaminants, which should be removed after the reaction. Doig and Diks [35] demonstrated a simple way for removal of the contaminants by solvent extraction with hexane/2-propanol/water, all of which are food-compatible solvents.

Synthesis of Lyso PLs

Compared to diacyl-PLs, lyso PLs (LPLs, also called lysolecithin), obtained by hydrolysis of lecithin with PLA_2 (Fig. **3.2C**, route III), have better water solubility and emulsifying properties. For this reason, lysolecithin has been produced by hydrolysis with PLA_2 and used as an industrial food emulsifier [62, 63]. Besides this traditional use as emulsifiers, LPLs of defined chemical structures have been attracting or getting special attention due to their biological effects related to cell signaling [64 - 66].

A simple way for the synthesis of 1-acyl-2-lyso-PL (2-LPL) of a particular structure is the PLA_2-mediated hydrolysis of diacyl-PL, of which the acyl-groups/

at least in the *sn*-1 position, is the desired one (Fig. **3.2C**, route III). Alternatively, it can be prepared by deacylation of diacyl-PL with 1,3-specific lipase to generate 1-lyso-2-acyl-PL (1-LPL), followed by acyl migration from the *sn*-2 to *sn*-1 position by exposure to alkaline conditions, such as alkaline buffer [67] or ammonia vapor [68].

In addition to deacylation, another way for preparation of 2-LPC is direct acylation of GPC. As shown in Fig. (**3.2A**), chemical acylation of GPC, at both *sn*-1 and *sn*-2 positions, is possible. In contrast, it is usually difficult to specifically esterify only the *sn*-1 position, due to the non-selective nature of the chemical acylation. Although an excellent method for selective chemical mono-acylation of GPC at *sn*-1 position was developed [69], the method requires dibutyltin oxide, a toxic tin compound (Fig. **3.2D**).

Alternatively, 1,3-specific lipase can be used for *sn*-1 selective direct esterification of GPC, which was first reported by Mazur *et al.* [70]. They performed the enzymatic acylation of GPC with octanoic anhydride using immobilized *Mucor miehei* lipase (Fig. **3.2E**). The overall yield of the product was 73% with high isomeric purity. Kim and Kim [71] employed immobilized *M. miehei* lipase and FA to give a yield of up to 90%, while Virto *et al.* [72] used an immobilized *Candida antarctica* lipase B (Novozym™ 435) and FA-vinylester with the yield of >95%. By carrying out the condensation reaction under vacuum to remove the condensed water, and thereby shift the equilibrium towards the ester formation, Hong *et al.* [73] demonstrated Novozym™ 435-mediated synthesis of CLA-containing 2-LPC at 70% yield [73].

1-LPL is prepared by *sn*-1-specific deacylation of diacyl-PC with 1,3-specific lipase. Sarney *et al.* [68] performed alcoholysis of lecithin in ethanol (therefore the reaction is ethanolysis) using immobilized *M. miehei* lipase [68]. The reaction proceeded almost quantitatively, and the isomeric purity of the product indicated it was almost completely the desired 1-lyso-2-acyl-form, meaning that no acyl migration occurred.

HEAD GROUP MODIFICATION

Basics of PLD

Enzymatic modification of head group of PLs, also called transphosphatidylation, is a reaction of polar head group exchange between the substrate PL and a corresponding alcohol acceptor. Catalyzed by a PLD, this reaction can be performed in a single step and under mild conditions, which makes it highly competitive to the traditional methods relying on extraction from natural sources or chemical synthesis. While extraction suffers from inefficiency and health risks, chemical synthesis requires many steps and harsh conditions.

PLD is a naturally occurring enzyme that primarily catalyzes the hydrolysis of phosphodiester bond of PLs (Fig. **3.4A**) to produce PA functioning as a signal molecule and, a corresponding alcohol. The majority of PLDs also catalyze transphosphatidylation (Fig. **3.4A**), which, in particular, has a very high industrial relevance as it enables production of less available or unnatural PLs from the ones highly abundant in nature, *e.g.*, PC.

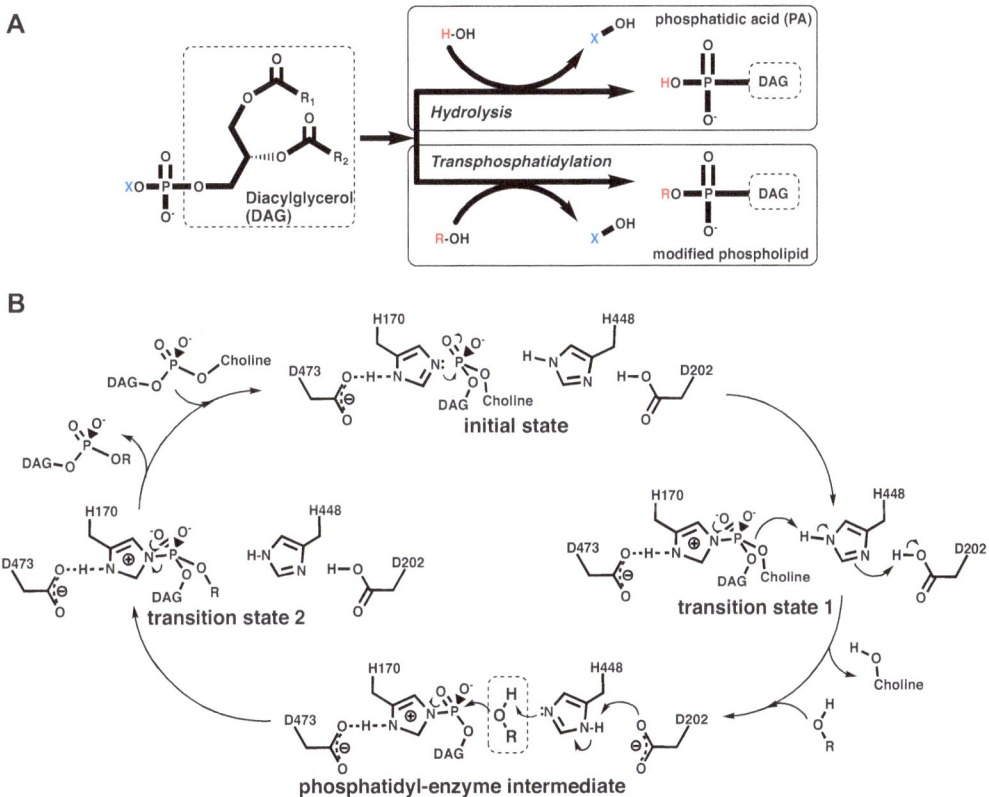

Fig. (3.4). (A) Reactions catalyzed by PLD; (B) Catalytic mechanism of the HKD PLD shown with residue numbers of *Streptomyces* PMF PLD.

PLD activity has been observed in viruses, prokaryotic (bacteria, *e.g.*, *Streptomyces*) and eukaryotic (unicellular and multicellular) organisms, as well as in Archaea [74]. The majority of these enzymes belong to the same PLD superfamily and catalyze the above-mentioned reactions with the same catalytic mechanism (Fig. **3.4B**). Water acts as a nucleophile in hydrolysis, while in transphosphatidylation the role of a nucleophile is played by an alcohol [75 - 77]. Some enzymes showing PLD activity belong to different protein families and follow different reaction mechanisms. Generally, PLDs are considered to have wide substrate specificities [77]. Distinction between natural 1,2-diacyl and artificial 1,3-diacyl-PC stereoisomers is pronounced, and activity is observed only towards 1,2-diacyl-PC [78]. Most of the PLDs show higher activity towards aggregated substrates (*e.g.* micelles, liposomes) compared to the soluble ones. The geometry of aggregated substrates plays a critical role in catalysis [79].

Origin of PLD Enzymes Used in Phospholipid Modification

PLD activity was first observed in the extracts of carrots and cabbage leaves, according to Hanahan and Chaikoff in the late 1940's [80, 81]. Discovery of PLD activity in animal tissues came much later, in 1975, in the report of Saito and Kanfer describing PLD activity in the rat brain [82]. However, interest in mammalian PLDs began only after revealing its rapid activation in response to extracellular stimuli [83 - 85]. Until today, PLD activity has been observed in almost all forms of life, with around 8000 genes registered in the NCBI GeneBank. Still, not all correspond to the genuine PLD enzymes, which will be explained in the following parts of this chapter.

Even though the early studies on PL modification employed PLD enzymes from cabbage leaves [86], soon after, their role was taken over by microbial enzymes, mainly PLDs from Actinomycetes, in particular the genera *Streptomyces, Streptoverticillium*, and *Actinomadura* isolated from soil samples [87 - 94], due to their high transphosphatidylation activity compared to PLDs of any other origin. Further studies on catalytic properties of these enzymes pointed out that they belong to different protein families and significantly differ in their catalytic mechanism. The difference was first observed in the work of Juneja *et al.* [95] who compared PLDs of different origin including *Streptomyces antibioticus* (SaPLD), *Streptomyces chromofuscus* (ScPLD), and cabbage in synthesis of PE [95]. SaPLD showed the highest transphosphatidylation activity while the cabbage PLD came in as second in line. ScPLD showed significantly lower activity compared to the other two. This indicated that the two *Streptomyces* PLDs (*i.e.,* SaPLD and ScPLD) are quite different in their enzymatic properties, despite the closely related origin. It was later confirmed that ScPLD is a member of phosphodiesterase/alkaline phosphatase family, along with purple acid phosphatase, protein phosphatase, and nucleotide phosphodiesterase [96], while SaPLD belongs to the PLD superfamily.

Reaction System

While hydrolysis readily proceeds in aqueous systems, transphosphatidylation is usually favored in biphasic systems consisting of a water-immiscible organic solvent containing PLs, such as diethylether and ethylacetate, and an aqueous phase containing the enzyme and an alcohol acceptor [86]. In biphasic systems, the reaction can be favored by simple introduction of salt, such as NaCl, at saturating concentrations, which affects the re-localization of the enzyme from the water phase to the interface by strengthening the hydrophobic interactions between the enzyme and acyl chains of PL [97]. Although the high salt concentrations inhibit the enzyme, the apparent activation seems to outweigh the salt-induced inhibitory effect. Transphosphatidylation can also occur in a purely aqueous environment under the optimized conditions, even though PLD is intrinsically a hydrolytic enzyme. This eliminates the need to control water content in the reaction system or dehydrate the enzyme prior to the reaction; these steps are usually required in lipase-catalyzed transesterifications. An advantage of the biphasic system is that the target product is soluble in organic phase and can be separated easily from the aqueous phase.

It is preferable to avoid the use of organic solvents, especially when the product should be used as a food additive or pharmaceutical. Such organic solvent-free system was utilized for the synthesis of PS, by performing the reaction using lecithin adsorbed on calcium sulfate powder in an aqueous buffer, affording PS in high yield [98].

PLD - Mediated Synthesis of Natural Phospholipids

PLD-mediated synthesis of natural PLus was reported as early as in the late 1980's and early 1990's. The synthesis of natural PLs, such as PG [99 - 101], PE [95], or PS [102, 103] from lecithin or PC proceeded with the yield of nearly 100% under the optimized conditions. Transphosphatidylation with L- and D-serine gave phosphatidyl-L- and -D-serine, respectively. Interestingly, bacterial PLD showed a 2-fold higher reaction rate with D-serine than with L-serine, while cabbage PLD reacted only with L-serine [102]. In the presence of high PC concentration, the efficiency of the reaction decreased due to the inhibitory effect of released choline. The efficiency was restored by introduction of choline oxidase and catalase [103]. Alternatively, choline can be effectively removed by a cation-exchange resin, as reported by Rich and Khmelnitsky, for the transphos-phatidylation of PC with various alcohol acceptors in anhydrous chloroform when yields of over 80% have been achieved [104]. The reaction of egg lecithin containing 75% PC and 25% PE with choline gave PC with almost 100% purity [105]. While the first reports utilized other *Streptomyces* PLDs, PS production has also been recently demonstrated using ScPLD displayed on the surface of the yeast with the yield of 67.5% [106].

CL is a type of natural PL having two phosphatidyl moieties linked *via* a glycerol group. Formation of CL during the synthesis of PG from PC and glycerol has been reported, especially at low glycerol concentrations [107]. The suggested mechanism is that PG, formed from PC and glycerol, becomes another phosphatidyl acceptor to compete with the remaining glycerol. However, the final CL yield remains low even at prolonged reaction times. Higher reaction efficiency has been achieved when PG was used as the only substrate [108]. The lack of positive charge in polar head group of PG is reported to be the limiting step towards CL formation in the system containing PC and glycerol.

PI is present mainly in eukaryotic cellular membranes where it plays important roles in lipid metabolism, such as stimulation of reverse cholesterol transport [109, 110], increase of high-density lipoprotein cholesterol levels [111], and decrease of TAG levels in the serum and liver [3, 112]. Recent reports claim that dietary PI prevents development of nonalcoholic fatty liver disease [113] and influences immune functions to prevent induced pathogenesis and development of liver injury [114]. For these reasons, PI holds a great potential as a supplemental compound. Since PI cannot be readily obtained by enzymatic means, significant protein engineering efforts have been made to finally yield an engineered SaPLD enzyme with an ability to synthesize 1-PI, a naturally occurring positional isomer of PI having various acyl chains, with >90% specificity and in reasonable yields of 30% on average as per total PLs [97, 115 - 117]. The optimized reaction proceeded with high yields and retained specificity at low temperatures in a biphasic system containing salt-saturated acetate buffer.

Cyclic phosphatidic acid (cPA) is naturally occurring PL identified from human serum, analogous to the lyso phosphatidic acid (LPA) with an acyl group at *sn*-1 and a cyclic phosphate group at *sn*-2 and *sn*-3 positions of the glycerol backbone. cPA was shown to participate in regulation of many cellular functions, interestingly, with an effect opposite to that of the LPA [118]. Serum cPA is produced in a transphosphatidylation reaction catalyzed by serum lysoPLD, autotaxin, along with its homolog LPA [119]. Other PLD enzymes have also been reported to catalyze formation of cPA and LPA. As confirmed by ^{31}P NMR, ScPLD catalyzes formation of cPA and LPA where cPA can be detected only as a transient product which is soon hydrolyzed to LPA [120]. Exclusive formation of cPA from LPC has been reported for PLD from *Actinomadura* sp. 362 (AcPLD) [118], similar to the *Loxosceles* PLD-like sphingomyelinase (SMase) and PLDs from pathogenic actinobacteria (*Arcanobactirium haemoliticum*) and ascomycete fungi (*Coccidioides posadasii*), which catalyze exclusive formation of cPA from LPC as well as cyclic ceramide-1,3-phosphate from SM [121, 122].

PLD-Mediated Synthesis of Unnatural Phospholipids

The PLD-mediated transphosphatidylation has a great potential in the synthesis of

PLs with unnatural polar head groups. With the development of enzymatic processes, the bioactive natural PLs became or are expected to become readily available for industrial applications. However, limited diversity of their polar heads is a bottleneck for the increasing need of PLs with novel or improved properties for application in food, pharmaceutical, and cosmetic industries. With rapid development of pharmaceuticals, drug delivery systems, food supplements, and stabilizers, PLs having distinct and diversified properties are in high demand. Notable is the generation of efficient liposome-based drug delivery systems, where it is often necessary to fine-tune the properties of the phospholipid layer in terms of responsiveness to different triggers, stability, and possibility to decorate the liposomes with bioactive molecules by attaching them to the PL backbone. These requests can be met by incorporation of unnatural PLs into the liposome formulations. As an example, incorporation of phosphatidylascorbic acid or unnatural phosphatidylsaccharides increases stability of liposomes against oxidation, aggregation, and dehydration. In the form of liposomes, PLs containing drug molecules as their polar heads can be used as drug carriers of high bioavailability.

Until now, enzymatic synthesis of a variety of artificial PLs has been demonstrated. Structures of the representative ones are listed in Table **3.2**. However, there are several considerations to whether a hydroxyl compound would be a good or poor acceptor of the phosphatidyl group.

Table 3.2. List of representative unnatural PLs synthesized in PLD-catalyzed reaction systems.

Phospholipid	Function	Enzyme	Reaction system	Yield (%)	Ref.
6-phosphatidyl-D-glucose	Liposome stabilizer	*Streptomyces* PLD	Biphasic/ ethyl acetate	95	[124]
		AcPLD	Biphasic/ diethyl ether	86	[125]
phosphatidylserinol	Pharmaceutical	PLDP	Biphasic/ chloroform	95 (4/1) *R/S*	[126]
diphosphatidylated serinol	Liposome stabilizer	PLDP	Biphasic/ diethyl ether	62	[108]

(Table 3.2) cont.....

Phospholipid	Function	Enzyme	Reaction system	Yield (%)	Ref.
phosphatidyl- 2-methyl-1-phen-1-2-propanol	Antioxidant	*Streptomyces* PLD	Biphasic/ ethyl acetate	< 20	[128]
phosphatidyl-*p*-methoxyphenol	Bioactive compound	*Streptomyces* PLD	Biphasic/ benzene	48	[132]
phosphatidyl geraniol	Pharmaceutical	*Streptomyces* PLD	terpene-water	90	[130]
phosphatidylascorbic acid	Antioxidant	*Streptomyces* PLD	Biphasic/ diethyl ether	>80	[133, 134]
phosphatidylchromanol	Antioxidant	*Streptomyces lydicus* PLD	Biphasic/ diethyl ether	94	[135, 136]
phosphatidylarbutin	Tyrosinase inhibitor	*Streptomyces* PLD	Biphasic/ ethyl acetate	18	[138]
phosphatidylkojic acid	Tyrosinase inhibitor	*Streptomyces* PLD	Biphasic/ diethyl ether	60	[138]
phosphatidyl-5-fluorouridine	Anti-cancer agent	PLDP	Biphasic/ chloroform	68	[139, 140]

(Table 3.2) cont.....

Phospholipid	Function	Enzyme	Reaction system	Yield (%)	Ref.
phosphatidyl-*N*-acetylneuraminic acid	Anti-viral agent	*Streptomyces* PLD, chemical synthesis	Chemo-enzymatic process	25	[141]
phosphatidylpeptide	Inhibitor against fibronectin adhesion to integrin	PLDP	Biphasic/chloroform	15	[126]
phosphatidyl dihydrohyacetone	Tanning agent for cosmetics	*Streptomyces* PLD	Biphasic/ethyl acetate	<80	[142]
phosphatidyltyrosol	Antioxidant	*Streptomyces* PLD	Biphasic/ethyl acetate	87	[128]
phosphatidylthiamine	Vitamin B1	*Streptomyces* PLD	Biphasic/ethyl acetate	95	[144]
phosphatidylpanthothenic acid	Vitamin B5	*Streptomyces* PLD	Biphasic/ethyl acetate	1.6	[144]
phosphatidylgenipin	Chinese medicine	*Streptomyces* PLD	Biphasic/ethyl acetate	>80	[145]

(Table 3.2) cont.....

Phospholipid	Function	Enzyme	Reaction system	Yield (%)	Ref.
phosphatidylsitosterol	Inhibitor of cholesterol absorption	*Streptomyces* PLD	Biphasic/ chloroform	~30	[146]

Type of the hydroxyl group and molecular size: Primary hydroxyl groups are preferred over the secondary ones, making aliphatic primary alcohols the best acceptors. Some secondary alcohols, such as the straight chain aliphatic alcohols and dihydroxycyclohexanols, can also be transphosphatidylated, although with lower efficiency than the primary ones [123]. While cyclohexanol and cyclohexanediols (1,2- 1,3- and 1,4-) are good PLD substrates, further hydroxylation introduces significant steric hindrances that obstruct the acceptor binding and proceeding of the reaction. This is the case with 1,3,5- or 1,2,3-cyclohexanetriols, inositol, its *ortho* ester, and secondary hydroxyls of monosaccharaides [115, 123]. In compounds with multiple hydroxyl groups, the primary ones are always selectively reacted. For example, the transphosphatidylation of PC with D-glucose gave predominantly 6-phosphatidy--D- glucose (6-PGlc), in which the 6th primary hydroxyl group of glucose is linked to the phosphatidyl group [124, 125]. PG prepared from PC and glycerol is a mixture of diastereomers, in which the phosphatidyl group is linked to either *sn*-1 or *sn*-3 primary hydroxyl group of glycerol, but not to the *sn*-2 hydroxyl group [126, 127]. Tertiary hydroxyl groups are very poor acceptors. Still, there is an example where a tertiary alcohol, 2-methyl-1-phenyl-2-propanol, was transphosphatidylated, although in a low yield [128].

In addition, the length of the sugar chain also introduces certain steric obstacles and hinders the reaction. While mono- and disaccharides are good acceptors, reactivity towards trimers or longer sugar polymers is either reduced or abolished [125, 129]. A similar effect has been observed in the synthesis of phosphatidylpeptides and phosphatidylterpenes. While Ser-methylester is reacted with efficiency of 95%, tripeptide Ser-Gly-Val-methylester was a fairly good substrate providing a 31% product yield. The hexapeptide Ser-Gly-Arg-Gly-Asp-Val-methylester proved to be a poor substrate providing only 15% product yield [126]. In the case of terpenes, the shortest used substrate, geraniol, can be reacted with a yield of 53 mol%, which gradually decreased towards longer substrates, such as geranylgeraniol and phytol to reach 17 and 14 mol% respectively [130].

PLDs can also discriminate between the two primary hydroxyl groups of prochiral diols, such as serinol and 2-amino-1,3-propanediol, or triols, such as glycerol. PLD from *Streptomyces* sp. (PLDP) shows a clear discrimination between two primary hydroxyl groups of prochiral serinol with preference for pro-*R* over pro-*S* hydroxyl group [126]. Generated phosphatidylserinol with *R*-configuration (with respect to the C2' atom) was obtained in four times excess compared to the one with *S*-configuration. However, in PG synthesis, the same enzyme showed no distinction between *sn*-1 and *sn*-3 position of glycerol. In another study on enzymatic PG synthesis, the authors observed that the recognition of two primary hydroxyl groups of prochiral glycerol was influenced largely by the reaction temperature [127], where the stereoselectivity became less stringent at higher temperatures if PLD of *Streptomyces septatus* TH-2 (TH-2 PLD) and AcPLD were used, and unaffected if *Streptomyces halstedii* K6 PLD or cabbage PLD were used. Since all used enzymes belong to the same PLD superfamily and share the same catalytic mechanism, the differences in stereoselectivity likely arise from the structure around the catalytic site.

In transphosphatidylation reactions yielding CL and related diphosphatidyl lipids, referred to as CL analogs, the authors used ethanolamine and its derivatives as well as glycerol as substrates and concluded that the steric and charge effects strongly influence the formation of diphosphatidyl lipids [108]. While transphosphatidylation of serinol by PLDP provides phosphatidylserinol as the main product in the early stage of the reaction, longer reaction times result in accumulation of CL analog. The formation of CL analogs proceeds well only in the reaction with the *Streptomyces* PLD, while cabbage PLD is recommended if monophosphatidyl products are preferred [131].

PLDs can accept phenolic substrates as well; however, the efficiency of transphosphatidylation will significantly depend on the type of the substituent. Among the phenols, *para*-substituted ones were the best substrates, especially with more electron-donating substituents (lower σ_p) [132]. The study considering phosphatidylation of tyrosol and its derivatives drew similar conclusions stating that *para*-amino and *para*-hydroxyl groups do not inhibit the transphosphatidylation, while *meta*-hydroxyl, *meta*- and *ortho*-methyl groups have mild inhibitory effect [128].

Solubility of the acceptor: The concentration of the acceptor should be high to ensure high efficiency and suppress the hydrolysis, which competes with transphosphatidylation. Acceptors can either be soluble in water or in the organic solvent phase. Even with approx. 0.1% solubility in water, 2-naphthol can be a good acceptor for transphosphatidylation providing the yield of 41% of the phosphatidylated product [132]. To avoid the use of organic solvents, acceptors can be used as a reaction medium, or one of the reaction phases, provided that they are liquid under the experimental conditions, water-immiscible and can dissolve PC. This kind of system has been successfully tested in synthesis of phosphatidylterpenes, namely phosphatidylgeraniol, phosphatidylfarnesol, phosphatidylgeranylgeraniol and phosphatidylphytol, where achieved molar yields of products reached much higher levels compared to the conventional biphasic system [130].

The ability of PLD to react on a wide range of acceptors has been extensively used to introduce biofunctional moieties into the PL backbone as it can serve as a non-toxic and biocompatible carrier within different liposome formulations.

One special interest is antioxidants. In the early 1990's, phosphatidylascorbic acid (PAsA) [133, 134] and phosphatidylchromanol (PCh) [135, 136] were successfully synthesized and characterized as liposome components. PAsA was present on the water-lipid interface of the liposome, enhancing the effective concentration of ascorbic acid moiety for scavenging aqueous peroxy radicals and suppressing oxidation of PC in multilamelar liposomes. Similarly, PCh containing 2,5,7,8-tetramethyl-6-hydroxy-2-(hydroxyethyl)chroman, the bioactive part of α-tocopherol, suppressed autooxidation of lard even more effectively than the original compound by formation of the reverse micelles in oil, which trapped the residual water dissolving trace amounts of metal ions, such as iron, known to initiate the oxidation. In a later study, tyrosol, hydroxytyrosol, and derivatives of known antioxidant and anti-cancer compounds have been phosphatidylated in relatively high yields, depending on the nature and size of the substituents on the benzene ring [128]. Yet another recent example is synthesis of phosphatidylmethylferrocene using PLD from peanuts [137].

In the same time, introduction of arbutin and kojic acid, known tyrosinase inhibitors, into the PL backbone was reported [138]. While arbutin is a competitive tyrosinase inhibitor, kojic acid inhibits the enzyme as a chelator and an antioxidant, which ultimately results in prevention of overproduction of melanin in epidermal cells. Phosphatidylation improved the stability of both compounds, while conserving the inhibitory effects toward tyrosinase.

Nucleoside analogues, primarily anti-cancer agents, such as 5-fluorouridine have also been successfully turned into PL derivatives [139]. Some of the phosphatidyl nucleosides even had stronger anti-tumor effects than the original compounds [140].

A novel sialylphospholipid, an anti-viral inhibitor, was synthesized from *N*-acetylneuraminic acid (NeuAc) and PC by a chemo-enzymatic method [141]. Inhibitory effect of the resultant liposome containing phosphatidylated NeuAc for rotavirus infection was 10^3-10^4-fold higher than that of NeuAc. The improvement in inhibitory strength is attributed to the fact that the PL derivative forms bilayers with multivalent NeuAc moieties, displayed on the surface and interacting with the virus in a multivalent manner.

Recent development of liposomes and drug delivery systems raises the need for introduction of stabilizing components into the PL backbone to improve their structural stability, for example, upon freeze-drying process. Saccharides have been shown to convey such function, thus efficient production of phosphatidylsaccharides is getting more attention. Among the first examples, 6-PGlc has been synthesized by *Streptomyces* PLD in a biphasic system with 95 mol% yield [124]. In further work, the same group of authors enzymatically introduced di- and tri-saccharides (sucrose and raffinose) into the PL backbone by the same PLD. Liposomes prepared using the phosphatidylsaccharides indeed showed increased stability compared to the ones made of PC [129].

Besides, phosphatidyl derivatives of peptide-based inhibitor against fibronectin adhesion to integrin [126], dihydroxyacetone (tanning agent for cosmetics) [142], geraniol (with anti-proliferative effect to cancer cells) [130, 143], thiamine (vitamin B1) [144], panthothenic acid (vitamin B5) [144], genipin (component of a Chinese medicine) [145], and sitosterol (inhibitor of cholesterol absorption) [146] are also interesting examples.

Enzymology and Protein Engineering of PLD

Structure

The majority of PLDs belong to the PLD superfamily, encompassing PLD enzymes, CL synthases, PS synthases (PSS), poxvirus envelope proteins, *Yersinia* murine toxin (Ymt), tyrosil-DNA phosphodiesterase, and several nucleases. Superfamily members have been found to exist across prokaryotic, eukaryotic, and archaeal world. PFam (http://pfam.sanger.ac.uk) database classified PLD superfamily (clan) in 8 families with total of 11654 domains and 86 structures from 2005 species. Within the superfamily, commercially significant bacterial PLD enzymes belong to the PLDc 2 family (*as of March 2016*: 6935 sequences from 1815 species; although this number may also include proteins with other functions, since the members are affiliated according to the conservation of the catalytic domain; for some family members the function is not yet known).

The structural feature of PLD superfamily, HxKxxxxD (HKD) motif, is present in a single or double copy in the primary structure [147, 148]. Regardless of the protein function, His residue of the motif is critically important for the catalysis [149 - 153]. Conservation of the HKD motif implies that the superfamily members share a similar reaction mechanism towards different substrates. Besides the structural core which is relevant for the catalytic mechanism, eukaryotic PLDs share other conserved regions, such as phox-homologous (PX), pleckstrin-homologous (PH), and calcium-dependent PL-binding (C2) domains, which are involved in regulation of the enzymes inside the cells [74, 154 - 156].

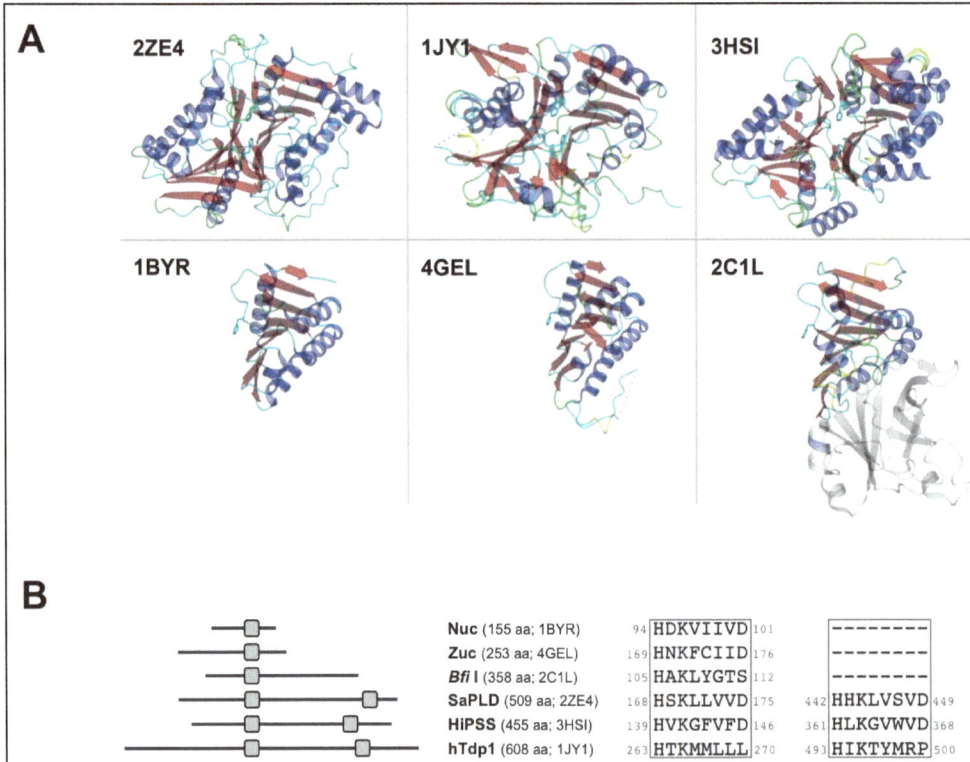

Fig. (3.5). (A) Structural comparison between homodimeric 1BYR (Nuc), 4GEL (Zuc), 2C1L (*Bfi* I) and dual domain PLD superfamily members 2ZE4 (SaPLD), 1JY1 (Tdp 1) and 3HSI (*Hi* PSS); (B) Schematic presentation of the primary structures of PLD superfamily members with HKD domain sequences encircled on the right. Gray boxes on the left scheme represent HKD motifs and are placed according to the appearance in the primary sequence. Numbers in parenthesis in the middle represent the length in amino acid number and PDB entry.

The crystal structures have been resolved for a limited number of superfamily members. The most studied are the enzymes from PLDc 2 family (30 structures from 7 species) and tyrosyl-DNA phosphodiesterase (43 structures from 2 species). Typical representatives are *Salmonella typhimurium* endonuclease (Nuc, 1BYR) [153], *Drosophila* zucchini ribonuclease (Zuc, 4GEL) [157], *Bacillus firmus* endonuclease (*Bfi* I, 2C1L) [158], human tyrosyl-DNA phosphodiesterase (Tdp1, 1JY1) [159], *Streptomyces* sp. PMF PLD (PMF-PLD, 1F0I) [152], SaPLD

(2ZE4), *Escherichia coli* polyphosphate kinase (PKK, 1XDO) [160], *Haemophilus influenzae* PSS (3HSI), and *Pyrococcus furiosus* global transcriptional regulator (TrmB, 3QPH) [161]. Members having one HKD motif in the primary structure (*i.e.*, Nuc, Zuc and *Bfi* I) exist as homodimers with each of the two motifs located at the subunit interface to form one catalytic site (Fig. **3.5A**). PLD, Tdp1, and PSS, on the other hand, contain two copies of HKD motif in the primary sequence and form a dual domain structure consisting of two (N-terminal and C-terminal) half-domains of a similar fold (Fig. **3.5A**). It is considered that these dual domain enzymes evolved *via* gene duplication event [147, 148]. To support this theory, a dual domain PLD enzyme was successfully reconstituted from two fragments, its N- and C-terminal half-domains, expressed and purified independently [162, 163]. The core structure of these half-domains is similar to the one of Nuc, Zuc, or *Bfi* I, confirming that all PLD superfamily members share a common structural pattern. All of these findings point to the common ancestry of superfamily members.

Catalytic Mechanism

PLD superfamily members share not only the common structural core but also a common catalytic mechanism based on the nucleophilic substitution at the phosphorus atom of phosphate esters or phosphate anhydride (Fig. **3.4B**). After two independent research groups reported that cabbage PLD in addition to hydrolysis also catalyzes head group exchange in the presence of primary alcohols, the hypothesis on the two-step reaction mechanism involving formation of phosphatidyl-enzyme intermediate was born [164, 165]. One of the pioneering experiments studied changes in the stereo configuration of the phosphorus before and after the reaction with cabbage PLD using a PL substrate chirally labeled with oxygen isotopes at phosphorus [166, 167]. The authors concluded that the reaction follows two-step S_N-2 mechanism since the ^{31}P-NMR showed overall retention of the phosphorus configuration, meaning that the reaction proceeds in two steps with two Walden inversions. If the reaction would follow a one-step mechanism *via* direct attack of the phosphorus by activated water or alcohol, the phosphorus configuration should be inverted due to one Walden inversion.

It took more than a decade to prove the formation of a phosphatidyl-enzyme intermediate and identify the catalytic residue/s in the work of Gottlin *et al.* [150]. The authors labeled endonuclease Nuc with ^{32}P-phosphate, and identified phospho-histidine in the protein hydrolyzate. This was the first clear proof of the hypothesis. Additional experiments using mutated Nuc variants confirmed that His 94 of the HKD motif was the nucleophile that attacks the phosphorus. After obtaining the crystal structure of Nuc, the first crystal structure of a PLD superfamily member, the role of His 94 in the formation of the phospho-histidine intermediate was confirmed. Crystallization of Nuc with inhibitor tungstate revealed that His 94, Lys 96, and Asn 111 directly interact with tungstate ion, while Ser 109 and Glu 122 help maintaining the hydrogen bond network in tungstate-enzyme complex [153]. It was further proposed that in homodimeric, Nuc His 94 of one monomer acts as a nucleophile and His 94 of the other monomer as a general acid (GA)/general base (GB) that protonates oxygen of the leaving group and deprotonates the incoming water. Still, since the two monomers cannot be readily distinguished, it remained unclear which of the two HKD motifs in dual domain PLDs carries the nucleophile, and which carries the residue with GA/GB function.

The first report on distribution of roles of the catalytic His in dual domain PLDs came soon after the work of Gottlin *et al.* [150] from the Iwasaki's group. Our labeling experiment using a fragmentary PLD reconstituted from N- and C-terminal domains pointed at the C-terminal half-domain of SaPLD to contain the nucleophile [163]. However, this was rebutted a year later, after the resolution of the first crystal structure of dual domain PLD, PMF PLD [152], followed by the proposal of the catalytic mechanism [151]. The work of Leiros *et al.* [151] revealed that two His of the catalytic site have specific, different roles (Fig. **3.4B**). Initially, the protonation states of the catalytic Histidines at participating/ designated nitrogen atoms are different and believed to be regulated by the nearby Asp 202 and Asp 473. While His 170 in the N-terminal half-domain is not protonated, His 448 in the C-terminal half domain is. The catalysis is initiated by the nucleophilic attack of the His 170 to the phosphorus, and subsequent formation of a negatively charged pentavalent transition state, which is stabilized by the Lysines of the HKD motifs. In the following, His 448 (as GA) protonates the leaving group to generate alcohol (*e.g.*, choline) and phosphatidyl-enzyme intermediate. In the second step, His 448 (as GB) deprotonates the incoming water or alcohol molecule, which then acts as the second nucleophile attacking the phosphorus of the phosphatidyl-enzyme intermediate, to produce PA or phosphatidylalcohol, leaving the catalytic residues regenerated. Using kinetic measurements and NMR, another research group suggested that formation of the phosphatidyl-enzyme intermediate is the rate-limiting step, while deprotonation of water or alcohol and release of PA or phosphatidylalcohol proceeds rapidly [168].

The next solid proof of the two-step reaction mechanism with the formation of phosphatidyl-enzyme intermediate came by 2010, when Orth *et al.* [169] studied the phosphodiester hydrolysis catalyzed by two imidazoles, one functioning as a nucleophile and the other as GA, using an intramolecular hydrolysis of a model compound, bis(2-(1-methyl-1H-imidazolyl)phenyl)phosphate [169]. The work concluded that the two-step mechanism is thermodynamically favored over the one-step mechanism. The authors also confirmed the formation of unstable phosphatidyl-imidazole intermediate.

Besides the roles of Histidines and Lysines of the HKD motif/s, following studies identified other residues important for substrate binding and recognition. Starting from highly homologous TH-2PLD and PLDP, Uesugi *et al.* [170] used chimeragenesis approach to probe the residues [79, 170]. As a result, two regions, 188-203 ('G188 loop') and 425-442 ('K438 loop'), were found to be important in catalysis and substrate recognition. These loop structures are positioned between β7 and α7 (188-203) and between β13 and β14 (425-442), and form an entrance to the active site. Within the loops, G188, D191, A426, and K438 were found critical for substrate recognition [79, 170]. Additional two loops, 123-130 ('Y126 loop') and 372-414 ('G381 loop'), have been recently suggested to form a gate-like structure near the active site of SaPLD and control PL acceptance and entrapment in the active site [171]. Based on the different position of the loops in the structure with and without the substrate, it is believed that they act to "grasp" the substrate upon binding. Since PLD mostly acts on PL-water interface, the enzyme needs to pull out one PL molecule from the aggregated substrate, and keep it within the active site against the intrinsic property of PLs to spontaneously aggregate. This concept, although reasonable, still requires further verification.

Protein Engineering of PLD

Improving the catalytic performance by protein engineering has been mainly done for bacterial PLDs. There is, however, one notable protein engineering study on a plant PLD, cabbage PLD2, in which the ratio of transphosphatidylation *vs.* hydrolysis has been improved by the mutations near the HKD motif (C310S and C625S) [172, 173]. The lack of tertiary structure of this enzyme hinders further detailed analysis of structure-function relationship. In contrast, the availability of the tertiary structures as well as the ease of preparation of recombinant enzymes [174, 175] facilitates the mutational studies in bacterial PLDs.

Improving the Catalytic Activity

To synthesize structured PLs using PLDs, we need to deal with an intrinsic property of the enzyme to hydrolyze the substrate PL and produce PA. An attempt to change the transphosphatidylation vs. hydrolysis ratio has been reported by Ogino *et al.* [176]. In the study, *Streptoverticillium cinnamoneum* PLD was used as target for site-directed mutagenesis of conserved glycine-glycine (GG) and glycine-serine (GS) motifs located near the HKD motifs, and suggested to maintain local conformation of the active site by positioning the catalytic His through the hydrogen bond network [176]. The mutations revealed significance of the GG/GS motifs on both enzymatic activity and stability. Remarkable 9 to 27-fold enhancement in transphosphatidylation activity was observed in G215S,

G216S and G216S/S489G variants.

Enhancing Thermostability

The first study to tackle the issue of thermostability in PLDs identified the key residues by chimeragenesis between thermolabile *S. halstedii* K1 (K1PLD) and thermostable TH-2PLD [177]. Residue 346 of K1PLD was highlighted, since E346D mutation significantly enhanced enzyme's thermostability. In the following, the authors constructed chimera gene library by *in vivo* DNA shuffling between K1PLD and TH-2PLD and identified G188 as another target for thermostability improvement [178]. The G188F mutation enhanced thermostability without activity loss.

Altering Head Group Specificity

As mentioned previously, most of the natural and considerable number of unnatural PLs can be synthesized from PC and the corresponding alcohols in PLD-mediated transphosphatidylation. However, PI is an exception, due to the low affinity of the enzyme towards *myo*-inositol caused by its size and shape, which sterically hindered the entry to the active site.

However, PI has a great potential as a supplement in foods or pharmaceutical products due to its biological effects related to regulation and improvement of lipid metabolism in animals and humans [3, 109 - 112]. Unlike PC, the abundance of PI in rich natural sources, such as soybeans is only 287 mg/100 g. Combined with unsatisfactory extraction efficiency, the price of such product is a critical limitation for its intended use.

Aiming to obtain a PI-synthesizing PLD, our group created SaPLD variants able to catalyze transphosphatidylation of secondary hydroxyls of *myo*-inositol yielding PI, by site-directed saturation mutagenesis of three residues lining the entry to the active site, W187, Y191, and Y385, and believed to affect acceptor accommodation and binding. High-throughput screening of approx. 30,000 clones resulted in isolation of approx. 90 PI-synthesizing PLD mutants [115]. This was the first time to change natural acceptor specificity of a PLD enzyme.

The issue with the PI-synthesizing PLDs of the first generation was the positional specificity of the PI. While only 1-PI is naturally occurring, and known to be bioactive, the PLD variants synthesized mixture of PI positional isomers with phosphatidyl moiety linked to either one of the six non-equivalent hydroxyl groups of *myo*-inositol. Several were found to predominantly produce mixture of 1-PI and 3-PI over the other PI isomers (187D/191Y/385R: DYR, 187A/191Y/385R: AYR and 187M/191Y/385R: MYR) [116]. These were used as templates for the second generation of PI-synthesizing variants aiming for exclusive specificity of the enzyme for 1-OH of *myo*-inositol. Residue 187 was randomized while keeping the 191Y and 385R unchanged. The resultant variants differed in their PI-synthesizing activity and markedly in 1-PI/3-PI ratio, which confirmed importance of the selected position in PI isomer specificity. Among the "XYR" variants, NYR and HYR showed the highest 1- over 3-PI specificity, while another variant, TYR, acquired very high 3- over 1-PI specificity [116, 179]. Interestingly, 3- over 1-PI specificity proved easier to acquire. It seems that the reaction with 3-OH group of *myo*-inositol is less sterically hindered than the reaction with 1-OH group. This inspired us to test the possibility of using other inositol stereoisomers as substrates in PI synthesis. Among the isomers, *allo*-inositol was the most preferred while *scyllo*-inositol was the least preferred substrate [180]. In addition, specificity of the most 1-PI specific variant, NYR, has been further improved by site-directed mutagenesis targeting another four residues of the acceptor-binding site. The resultant best variant, NYR-186T, synthesized 93% of 1-PI accompanied by 7% of 3-PI in total PI yield [117]. Structure model analyses pointed at G186T mutation to increase rigidity of the acceptor-binding site, thus restricting the possible orientations of inositol. Further rigidification achieved by lowering the reaction temperature from 37 to 20°C resulted in 97% of 1-PI *vs.* 3% of 3-PI in total PI yield. Total PI yield showed no significant change between the two temperatures.

Once the specificity of the enzyme was optimized towards exclusive production of 1-PI, our group tackled the next challenge, optimization of the PI yield, which initially amounted between 10 and 15 mol % of all PLs even with the most productive PLD variants. The first idea was to increase the yield by addition of excess *myo*-inositol to the reaction mixture. Since the solubility of *myo*-inositol is highly temperature-dependent, elevating the reaction temperature should improve its solubility and ensure its excess in the reaction system, thus promoting the PI synthesis. For the reaction at high temperatures (60-70°C), we have developed engineered enzymes by two approaches, site-directed saturation mutagenesis and rational design. Saturation mutagenesis of the seven sites known to have high B-factor yielded only moderate improvement, while rational strategy based on deletion of the flexible loop on the enzyme's surface yielded an enzyme with 11 times extended activity half-life at 70°C [171, 181]. However, after realizing that the positional specificity of the engineered enzyme is not maintained at high temperatures, we started testing different approaches to improve the PI yield and retain the positional specificity. A simple and elegant solution to the problem proved to be conducting the reaction in a biphasic system with 1-PI specific PLD dissolved in salt-saturated aqueous phase at 20°C [97, 117]. In the presence of 4.3 M of NaCl in the aqueous phase, PI yield of 35 mol % of all PLs was achieved for *sn*-1,2-dioleoyl-PI. Although we have observed that high concentrations of salt inhibit the enzyme, improvement of PI synthesis under the high salt concentration mainly occurs due to the accumulation of the enzyme at the interface by strengthened hydrophobic interactions with acyl chains of PC, which outweigh the inhibitory effect. Since the reaction proceeds well at low temperatures, tight control of the PI positional specificity is now possible in addition to the yield improvement. Using the new system, other natural PI species such as *sn*- 1-palmitoyl-2-oleoyl-PI, and *sn*-1-stearoyl-2-arachidonoyl-PI, were synthesized with total yields of 25 and 37 mol% respectively, and isomeric purities of 91 and 96%.

Mode of the Substrate Binding in Streptomyces PLD

Currently available data provide sufficient evidence to propose a hypothesis about a mechanism of recognition and binding of PLs by PLD. The first contact between the enzyme and PL occurs at the interface, where polar head groups are oriented towards the aqueous phase and thus can easily come in contact with the enzyme. The entrance formed by the "G188 and K438 loops" [79, 170] recognizes the polar heads of the substrate. As the substrate enters the active site, further accommodation of the polar head is made by the residues of the acceptor-binding site. Subsequently, the gate-like structure made of "Y126 and G381 loops" [182] pulls one PL molecule into the active site, and keeps it enclosed during the catalysis, while the hydrophobic pocket accommodates the acyl chains. As the polar heads are cleaved, the formed alcohol leaves the active site while the acceptor comes in a way suitable for the protonation/deprotonation by the His in the role of GA/GB. After one catalytic cycle, the gate will open to release the modified PL, allowing the next cycle to begin.

In cPA synthesis, which is an intramolecular cyclization, the acceptor is the *sn*- 2-hydroxyl group of the LPL substrate. For the *sn*-2-hydroxyl to be deprotonated by the His during the cyclization, it should be in a proper position and orientation. However, judging by the position of the corresponding *sn*-2 oxygen atom of the phosphatidyl-enzyme complex, such positioning seems impossible; the *sn*- 2-hydroxyl of LPC is far from and opposite to the ideal position with respect to the phosphorus. This is the reason why we speculate there might be a unique mechanism that enables this cyclization, which may involve different catalytic residue/s, distribution of roles among the two catalytic Histidines or rotation of the lysophosphatidyl moiety so as to properly orient the *sn*-2-hydroxyl group. Efforts are currently being made to explain this mechanism.

Non-HKD PLD

Enzymes showing PLD activity *via* different catalytic mechanisms from the one described for PLD superfamily are referred as non-HKD PLDs, to distinguish them from PLD superfamily characterized by the catalytic HKD motif. From the application viewpoint, most relevant representatives include PLD from *Streptomyces chromofuscus* (ScPLD), *Loxosceles* PLD and autotaxin.

PLD from S. chromofuscus

According to the PFam classification, ScPLD (57 kDa, monomer) is a member of Calcineurin superfamily and phosphodiesterase/alkaline phosphatase family (PhoD), along with purple acid phosphatase and PhoD of *Bacillus subtilis* to which it has 52% sequence homology [96]. The enzyme shows metal ion dependence (ferric ion is an essential cofactor) and activation [183]. While manganese (II) helps proper substrate binding, calcium ion stimulates activity by directly binding to the enzyme or by binding the PA, an allosteric activator of ScPLD [74]. Proteolytically cleaved ScPLD shows higher activity than the full-length enzyme, however it loses the ability of allosteric activation by PA, which indicates that the N-terminal domain is catalytic while C-terminal domain (residues 351-510) is a regulatory part of the enzyme [96, 183, 184]. For transphosphatidylation, ScPLD requires very high alcohol concentration (8-10 M) to reach the maximum activity, while the reaction practically does not proceed in the presence of 1-2 M methanol or ethanol, which is the concentration that enables the generation of 95% of the transphosphatidylation product by HKD PLDs [96]. ScPLD has been also shown to catalyze hydrolysis of LPC and lyso phosphatidylethanolamine (LPE) with subsequent liberation of LPA *via* an intermediate cPA product [120]. Hydrolysis of cPA occurred at a 4-fold lower rate than its formation in neutral aqueous solution. Accumulation of cPA was stimulated in the presence of phosphatase competitive inhibitor, sodium vanadate, in concentration of 5-10 mM.

Loxosceles PLD

Spiders from the *Loxosceles* genus are known to produce enzymes with PLD-like activity, mainly in hydrolysis of SM releasing the choline and cyclic ceramide-1,3-phosphate. These enzymes are the main component of the spider venom and are responsible for damaging effects to the skin and blood upon the spider bite. In addition to SMase activity, some of the *Loxosceles* venom enzymes also catalyze hydrolysis of LPC or LPE to produce cPA. The intramolecular transphospha-tidylation seems to be exclusively dominant over hydrolysis yielding LPA [121], with preference towards SM over LPLs and charged over neutral head groups [185]. While the enzyme from *Sicaris terrosus* prefers ethanolamine over choline head group, the opposite was observed for *Loxosceles arizonica.*

Autotaxin

Autotaxin is a lysophospholipase D belonging to the nucleotide pyrophosphatase/ phosphodiesterase 2 family, present in the human and animal serum [119, 186]. Its physiological target is LPC while the activity is inhibited by LPA and sphingosine-1-phosphate [187]. Autotaxin catalyzes both hydrolysis and transphosphatidylation of LPC yielding LPA and cPA. Interestingly, the LPA/cPA ratio can change under different conditions, such as nature and concentration of cations. While Co^{2+} has been reported to facilitate the hydrolysis of cPA to LPA, high concentration of NaCl (1 M) stimulates the formation of cPA rather than LPA [119].

Interfacial Kinetics of PLD

As previously explained, PLDs prefer aggregated substrates where the catalysis proceeds on the interface and, thus is affected by the interface properties, namely physical structure and chemical properties. Until now, a significant number of models describing catalysis at the interface has been made available. One of the first reports by Verger *et al.* [188] in 1973 combined Michaelis-Menten kinetic model with interfacial enzyme activation [188]. Other widely recognized models include "surface dilution kinetics" developed to explain the action of cobra venom PLA_2 on mixed micelles [189 - 192] and "hopping *vs.* scooting model" developed to describe two extreme modes of PLA_2 kinetics [193, 194]. Although these models have been used to describe interfacial kinetic of mammalian and plant PLDs [195 - 200], being developed for PLA_2, the models assume that the catalytic product diffuses from the interface and remains dissolved in the water phase, which is not the case for products of the catalysis by PLD. A recent review suggested that "hopping vs. scooting model" can be applied to distinguish different behaviors of ScPLD and HKD PLDs on the interface [74]. While hopping mode was used to describe catalysis of ScPLD, scooting mode was attributed to HKD PLDs. Based on the study utilizing PLD from *Streptomyces* sp. NA684, an HKD enzyme, another article claims that HKD PLDs can acquire different modes to recognize and bind the substrate, depending on the physical properties imposed by different molecule shape and head groups of the PLs [201]. In addition, Mayer's group recently proposed a novel model for interfacial catalysis by PLD in lipid bilayers containing long-chain PLs, which is an extension of the classic model described by Verger *et al.* [188] including the effects of activation by the product (PA) and substrate depletion [202]. The experimental system used relies on the planar lipid bilayers with chemical and electrical conductance to both sides of the bilayer, enabled by embedded gramicidin A channel. The model introduces two redefined kinetic constants, the specificity constant and interfacial quality constant. Accumulation of the products at the interface is mainly considered as it contributes to the substrate dilution, inhibition/activation and altered binding of the enzyme due to the changes of the bilayer curvature.

Activity Assays for PLD

Until today, various assays have been developed for effective and selective PLD detection, based on the hydrolysis of natural or synthetic substrates followed by direct or indirect detection of the products, PA, or released head groups.

Direct detection of PLD activity *via* released head group is routinely done by spectrophotometric quantification of yellow-colored *p*-nitrophenol released from phosphatidyl-*p*-nitrophenol [203], or alternatively, by qualitative colorimetric detection of a colored azo dye formed by coupling of diazonium salt with 2-naphtol released from a synthetic substrate, phosphatidyl-2-naphthol [204], which is especially suitable for detection of activity on solid materials. Recent availability of fluorescently labeled PLs enabled the development of simple assays for direct quantification of the released head groups by fluorescence. One example is PLD biosensor based on phospholipid-graphene nano-assembly composed of graphene oxide sheet placed between hydrophobic parts of the bilayer consisting of PE with fluorescein-labeled ethanolamine and unlabeled PC [205]. Indirect methods rely on the use of PC as a substrate and subsequent liberation of the choline which is then converted into betaine and hydrogen peroxide by choline oxidase, followed by formation of a colored product by peroxidase in a hydrogen peroxide-dependent oxidation [206, 207]. Commercial kits employing this principle are currently available.

Measurement of PLD activity by detection and quantification of the hydrolytic product, PA, can be done by a simple spectrophotometric measurement [208]. The produced PA is reacted with purple ferric salicylate (FeSal) where PA competitively binds ferric ions causing decrease in the absorbance at 490 nm due to deterioration of colored FeSal.

Transphosphatidylation activity is difficult to detect by a simple, straightforward method. For this purpose, transphosphatidylation product is usually detected after separation by thin layer chromatography (TLC) and staining, followed by quantification of the product. One such assay employs densitometric quantification of phosphatidyl-1-butanol after reaction of 1-butanol and PC in a biphasic system, HPTLC separation and staining [209].

CONCLUDING REMARKS

PLases and lipases have become powerful tools for enzymatic modification of

PLs. Many of the reactions described here are very simple, and indeed, it has now become possible to prepare most of the natural-type PLs having particular acyl groups and head groups from readily available lecithin, fatty acids and head group alcohols.

Lowering the production costs is important for practical applications, and the prices of the enzymes often account for the majority of the production costs. Immobilization of enzymes will facilitate continuous operation or recycling of the catalysts, leading to the overall cost reduction. Effective production of enzymes by recombinant techniques is also promising.

In addition, discovery of new enzymes with novel or improved properties (activity, stability and specificity) by isolation from environment or by protein engineering is of great interest.

CONSENT FOR PUBLICATION

Not applicable.

CONFLICT OF INTEREST

The authors declare no conflict of interest, financial or otherwise.

ACKNOWLEDGEMENTS

Declared none.

ABBREVIATIONS

1-LPC	1-lyso-2-acyl-phosphatidylcholine
1-LPL	1-lyso-2-acyl-phospholipid
2-LPC	1-acyl-2-lyso-phosphatidylcholine
2-LPL	1-acyl-2-lyso-phospholipid
6-PGlc	6-phosphatidylglucose
AceDoPC	1-acetyl-2-docosahexaenoyl-phosphatidylcholine
AcPLD	PLD from *Actinomadura* sp. 362
ARA	arachidonic acid
***Bfi* I**	*Bacillus firmus* endonuclease
CL	cardiolipin
CLA	conjugated linoleic acid
cPA	cyclic phosphatidic acid
DAG	diacylglycerol docosahexaenoic acid

DMAP	dimethylaminopyridine
EPA	eicosapentaenoic acid
FA	fatty acid
FeSal	Ferric salicylate
GA	general acid
GB	general base
GPC	glycerophosphorylcholine
K1PLD	PLD from *Streptomyces halstedii* K1
LA	linoleic acid
LPA	lyso phosphatidic acid
LPC	lyso phosphatidylcholine
LPE	lysophosphatidylethanolamine
LPL	lysophospholipid
MAG	monoacylglycerol
MCFA	medium-chain fatty acid
NeuAc	*N*-acetylneuraminic acid
Nuc	*Salmonella typhimurium* endonuclease
OA	oleic acid
PA	phosphatidic acid
PAsA	phosphatidylascorbic acid
PC	phosphatidylcholine
PCh	phosphatidylchromanol
PE	phosphatidylethanolamine
PG	phosphatidylglycerol
PH	pleckstrin-homologous
PhoD	phosphatase D from *Bacillus subtilis*
PI	phosphatidylinositol
PKK	*Escherichia coli* polynucleotide kinase
PL	phospholipid
PLA$_1$	phospholipase A$_1$
PLA$_2$	phospholipase A$_2$
PLase	phospholipase
PLC	phospholipase C
PLD	phospholipase D
PLDP	PLD from *Streptomyces* sp.

PMF-PLD	PLD from *Streptomyces* sp. PMF
PS	phosphatidylserine
PSS	phosphatidylserine synthase
PUFA	polyunsaturated fatty acid
PX	phox-homologous
RSM	response surface methodology
SaPLD	PLD from *Streptomyces antibioticus*
ScPLD	PLD from *Streptomyces chromofuscus*
SM	sphingomyelin
SMase	sphingomyelinase
sPL	structured phospholipid triacylglycerol
Tdp1	human tyrosyl-DNA phosphodiesterase
TH-2 PLD	PLD from *Streptomyces septatus* TH-2
TLC	thin layer chromatography
TrmB	*Pyrococcus furiosus* global transcriptional regulator
Zuc	*Drosophila* zucchini ribonuclease

REFERENCES

[1]   Katsuragi Y, Kurihara K. Specific inhibitor for bitter taste. Nature 1993; 365(6443): 213-4.
      [http://dx.doi.org/10.1038/365213b0] [PMID: 8371778]

[2]   Liu X, Shiihara M, Taniwaki N, Shirasaka N, Atsumi Y, Shiojiri M. Phosphatidylserine: Biology, technologies, and applications. In: Ahmad MU, Xu X, Eds. Polar lipids Biology, chemistry, and technology. Urbana, Illinois: AOCS Press 2015; pp. 145-84.

[3]   Yanagita T, Nagao K. Functional lipids and the prevention of the metabolic syndrome. Asia Pac J Clin Nutr 2008; 17 (Suppl. 1): 189-91.
      [PMID: 18296334]

[4]   Li J, Wang X, Zhang T, *et al.* A review on phospholipids and their main applications in drug delivery systems. Asian J Pharm Sci 2015; 10(2): 81-98.
      [http://dx.doi.org/10.1016/j.ajps.2014.09.004]

[5]   Fujiwara T, Maeta H, Chida S, Morita T, Watabe Y, Abe T. Artificial surfactant therapy in hyaline-membrane disease. Lancet 1980; 1(8159): 55-9.
      [http://dx.doi.org/10.1016/S0140-6736(80)90489-4] [PMID: 6101413]

[6]   Dyall SC. Long-chain omega-3 fatty acids and the brain: a review of the independent and shared effects of EPA, DPA and DHA. Front Aging Neurosci 2015; 7: 52.
      [http://dx.doi.org/10.3389/fnagi.2015.00052] [PMID: 25954194]

[7]   Pariza MW. Perspective on the safety and effectiveness of conjugated linoleic acid. Am J Clin Nutr 2004; 79(6) (Suppl.): 1132S-6S.
      [http://dx.doi.org/10.1093/ajcn/79.6.1132S] [PMID: 15159246]

[8]   Lyberg A-M, Fasoli E, Adlercreutz P. Monitoring the oxidation of docosahexaenoic acid in lipids. Lipids 2005; 40(9): 969-79.
      [http://dx.doi.org/10.1007/s11745-005-1458-1] [PMID: 16329470]

[9]   Song J-H, Inoue Y, Miyazawa T. Oxidative stability of docosahexaenoic acid-containing oils in the form of phospholipids, triacylglycerols, and ethyl esters. Biosci Biotechnol Biochem 1997; 61(12): 2085-8.
[http://dx.doi.org/10.1271/bbb.61.2085] [PMID: 9438988]

[10]   Wijendran V, Huang M-C, Diau G-Y, Boehm G, Nathanielsz PW, Brenna JT. Efficacy of dietary arachidonic acid provided as triglyceride or phospholipid as substrates for brain arachidonic acid accretion in baboon neonates. Pediatr Res 2002; 51(3): 265-72.
[http://dx.doi.org/10.1203/00006450-200203000-00002] [PMID: 11861929]

[11]   Liu L, Bartke N, Van Daele H, *et al.* Higher efficacy of dietary DHA provided as a phospholipid than as a triglyceride for brain DHA accretion in neonatal piglets. J Lipid Res 2014; 55(3): 531-9.
[http://dx.doi.org/10.1194/jlr.M045930] [PMID: 24470588]

[12]   Rossmeisl M, Jilkova ZM, Kuda O, *et al.* Metabolic effects of *n*-3 PUFA as phospholipids are superior to triglycerides in mice fed a high-fat diet: possible role of endocannabinoids. PLoS One 2012; 7(6): e38834.
[http://dx.doi.org/10.1371/journal.pone.0038834] [PMID: 22701720]

[13]   Thiés F, Delachambre MC, Bentejac M, Lagarde M, Lecerf J. Unsaturated fatty acids esterified in 2-acyl-l-lysophosphatidylcholine bound to albumin are more efficiently taken up by the young rat brain than the unesterified form. J Neurochem 1992; 59(3): 1110-6.
[http://dx.doi.org/10.1111/j.1471-4159.1992.tb08353.x] [PMID: 1494901]

[14]   Thiès F, Pillon C, Moliere P, Lagarde M, Lecerf J. Preferential incorporation of *sn*-2 lysoPC DHA over unesterified DHA in the young rat brain. Am J Physiol 1994; 267(5 Pt 2): R1273-9.
[PMID: 7977854]

[15]   Lagarde M, Hachem M, Bernoud-Hubac N, Picq M, Véricel E, Guichardant M. Biological properties of a DHA-containing structured phospholipid (AceDoPC) to target the brain. Prostaglandins Leukot Essent Fatty Acids 2015; 92: 63-5.
[http://dx.doi.org/10.1016/j.plefa.2014.01.005] [PMID: 24582148]

[16]   Vikbjerg AF, Rusig J-Y, Jonsson G, Mu H, Xu X. Comparative evaluation of the emulsifying properties of phosphatidylcholine after enzymatic acyl modification. J Agric Food Chem 2006; 54(9): 3310-6.
[http://dx.doi.org/10.1021/jf052665w] [PMID: 16637690]

[17]   Peng L, Xu X, Mu H, Høy C-E, Adler-Nissen J. Production of structured phospholipids by lipase-catalyzed acidolysis: Optimization using response surface methodology. Enzyme Microb Technol 2002; 31(4): 523-32.
[http://dx.doi.org/10.1016/S0141-0229(02)00147-3]

[18]   Vikbjerg AF, Mu H, Xu X. Monitoring of monooctanoylphosphatidylcholine synthesis by enzymatic acidolysis between soybean phosphatidylcholine and caprylic acid by thin-layer chromatography with a flame ionization detector. J Agric Food Chem 2005; 53(10): 3937-42.
[http://dx.doi.org/10.1021/jf0480389] [PMID: 15884820]

[19]   Vikbjerg AF, Mu H, Xu X. Lipase-catalyzed acyl exchange of soybean phosphatidylcholine in n-Hexane: a critical evaluation of both acyl incorporation and product recovery. Biotechnol Prog 2005; 21(2): 397-404.
[http://dx.doi.org/10.1021/bp049633y] [PMID: 15801777]

[20]   Vikbjerg AF, Mu H, Xu X. Elucidation of acyl migration during lipase-catalyzed production of structured phospholipids. J Am Oil Chem Soc 2006; 83(7): 609-14.
[http://dx.doi.org/10.1007/s11746-006-1246-3]

[21]   Gupta CM, Radhakrishnan R, Khorana HG. Glycerophospholipid synthesis: improved general method and new analogs containing photoactivable groups. Proc Natl Acad Sci USA 1977; 74(10): 4315-9.
[http://dx.doi.org/10.1073/pnas.74.10.4315] [PMID: 270675]

[22]   Chadha JS. Preparation of crystalline l-α-glycerophosphoryl-choline-cadmium chloride adduct from commercial egg lecithin. Chem Phys Lipids 1970; 4(1): 104-8.
[http://dx.doi.org/10.1016/0009-3084(70)90067-8]

[23]   Ichihara K, Iwasaki H, Ueda K, Takizawa R, Naito H, Tomosugi M. Synthesis of phosphatidylcholine: an improved method without using the cadmium chloride complex of *sn*-glycero-3-phosphocholine. Chem Phys Lipids 2005; 137(1-2): 94-9.
[http://dx.doi.org/10.1016/j.chemphyslip.2005.06.001] [PMID: 16054615]

[24]   Mason JT, Broccoli AV, Huang C. A method for the synthesis of isomerically pure saturated mixed-chain phosphatidylcholines. Anal Biochem 1981; 113(1): 96-101.
[http://dx.doi.org/10.1016/0003-2697(81)90049-X] [PMID: 6895006]

[25]   Aneja R, Chadha JS, Cubero Robles E, van Daal R. Partial synthesis of phosphatidylethanolamines. Biochim Biophys Acta 1969; 187(3): 439-41.
[http://dx.doi.org/10.1016/0005-2760(69)90018-6] [PMID: 5391259]

[26]   Matori M, Asahara T, Ota Y. Positional specificity of microbial lipases. J Ferment Bioeng 1991; 72(5): 397-8.
[http://dx.doi.org/10.1016/0922-338X(91)90094-W]

[27]   Sugihara A, Shimada Y, Tominaga Y. A novel *Geotrichum candidum* lipase with some preference for the 2-position on a triglyceride molecule. Appl Microbiol Biotechnol 1991; 35(6): 738-40.
[http://dx.doi.org/10.1007/BF00169887]

[28]   Rogalska E, Cudrey C, Ferrato F, Verger R. Stereoselective hydrolysis of triglycerides by animal and microbial lipases. Chirality 1993; 5(1): 24-30.
[http://dx.doi.org/10.1002/chir.530050106] [PMID: 8448074]

[29]   Asahara T, Matori M, Ikemoto M, Ota Y. Production of two types of lipases with opposite positional specificity by *Geotrichum* sp. FO401b. Biosci Biotechnol Biochem 1993; 57(3): 390-4.
[http://dx.doi.org/10.1271/bbb.57.390]

[30]   Brockerhoff H, Schmidt PC, Fong JW, Tirri LJ. Letter: Introduction of labeled fatty acid in position 1 of phosphoglycerides. Lipids 1976; 11(5): 421-2.
[http://dx.doi.org/10.1007/BF02532851] [PMID: 1271981]

[31]   Yagi T, Nakanishi T, Yoshizawa Y, Fukui F. The enzymatic acyl exchange of phospholipids with lipases. J Ferment Bioeng 1990; 69(1): 23-5.
[http://dx.doi.org/10.1016/0922-338X(90)90158-S]

[32]   Svensson I, Adlercreutz P, Mattiasson B. Interesterification of phosphatidylcholine with lipases in organic media. Appl Microbiol Biotechnol 1990; 33(3): 255-8.
[http://dx.doi.org/10.1007/BF00164517] [PMID: 1366637]

[33]   Svensson I, Adlercreutz P, Mattiasson B. Lipase-catalyzed transesterification of phosphatidylcholine at controlled water activity. J Am Oil Chem Soc 1992; 69(10): 986-91.
[http://dx.doi.org/10.1007/BF02541063]

[34]   Adlercreutz D, Budde H, Wehtje E. Synthesis of phosphatidylcholine with defined fatty acid in the *sn*-1 position by lipase-catalyzed esterification and transesterification reaction. Biotechnol Bioeng 2002; 78(4): 403-11.
[http://dx.doi.org/10.1002/bit.10225] [PMID: 11948447]

[35]   Doig SD, Diks RMM. Toolbox for exchanging constituent fatty acids in lecithins. Eur J Lipid Sci Technol 2003; 105(7): 359-67.
[http://dx.doi.org/10.1002/ejlt.200390074]

[36]   Haraldsson GG, Thorarensen A. Preparation of phospholipids highly enriched with n-3 polyunsaturated fatty acids by lipase. J Am Oil Chem Soc 1999; 76(10): 1143-9.
[http://dx.doi.org/10.1007/s11746-999-0087-2]

[37] Mustranta A, Suortti T, Poutanen K. Transesterification of phospholipids in different reaction conditions. J Am Oil Chem Soc 1994; 71(12): 1415-9.
[http://dx.doi.org/10.1007/BF02541365]

[38] Mutua LN, Akoh CC. Lipase-catalyzed modification of phospholipids: Incorporation of n-3 fatty acids into biosurfactants. J Am Oil Chem Soc 1993; 70(2): 125-8.
[http://dx.doi.org/10.1007/BF02542613]

[39] Hossen M, Hernandez E. Enzyme-catalyzed synthesis of structured phospholipids with conjugated linoleic acid. Eur J Lipid Sci Technol 2005; 107(10): 730-6.
[http://dx.doi.org/10.1002/ejlt.200501190]

[40] Marsaoui N, Naghmouchi K, Baah J, Raies A, Laplante S. Incorporation of ethyl esters of EPA and DHA in soybean lecithin using *Rhizomucor miehei* lipase: Effect of additives and solvent-free conditions. Appl Biochem Biotechnol 2015; 176(3): 938-46.
[http://dx.doi.org/10.1007/s12010-015-1621-3] [PMID: 25894950]

[41] Marsaoui N, Laplante S, Raies A, Naghmouchi K. Incorporation of omega-3 polyunsaturated fatty acids into soybean lecithin: effect of amines and divalent cations on transesterification by lipases. World J Microbiol Biotechnol 2013; 29(12): 2233-8.
[http://dx.doi.org/10.1007/s11274-013-1388-z] [PMID: 23749246]

[42] Chojnacka A, Gładkowski W, Kiełbowicz G, Gliszczyńska A, Niezgoda N, Wawrzeńczyk C. Lipase-catalyzed interesterification of egg-yolk phosphatidylcholine and plant oils. Grasas Aceites 2014; 65(4): e053.
[http://dx.doi.org/10.3989/gya.0585141]

[43] Pernas P, Olivier JL, Legoy MD, Bereziat G. Phospholipid synthesis by extracellular phospholipase A_2 in organic solvents. Biochem Biophys Res Commun 1990; 168(2): 644-50.
[http://dx.doi.org/10.1016/0006-291X(90)92369-B] [PMID: 2334428]

[44] Na A, Eriksson C, Eriksson S-G, Österberg E, Holmberg K. Synthesis of phosphatidylcholine with (n-3) fatty acids by phospholipase A_2 in microemulsion. J Am Oil Chem Soc 1990; 67(11): 766-70.
[http://dx.doi.org/10.1007/BF02540488]

[45] Egger D, Wehtje E, Adlercreutz P. Characterization and optimization of phospholipase A_2 catalyzed synthesis of phosphatidylcholine. Biochim Biophys Acta 1997; 1343(1): 76-84.
[http://dx.doi.org/10.1016/S0167-4838(97)00115-5] [PMID: 9428661]

[46] Hosokawa M, Takahashi K, Kikuchi Y, Hatano M. Preparation of therapeutic phospholipids through porcine pancreatic phospholipase A_2-mediated esterification and Lipozyme-mediated acidolysis. J Am Oil Chem Soc 1995; 72(11): 1287-91.
[http://dx.doi.org/10.1007/BF02546201]

[47] Awano S, Miyamoto K, Hosokawa M, Mankura M, Takahashi K. Production of docosahexaenoic acid bounded phospholipids via phospholipase A_2 mediated bioconversion. Fish Sci 2006; 72(4): 909-11.
[http://dx.doi.org/10.1111/j.1444-2906.2006.01236.x]

[48] Tanaka T, Isezaki T, Nakano H, Iwasaki Y. Synthesis of phospholipids containing polyunsaturated fatty acids by phospholipase $A_{(2)}$-mediated esterification with food-compatible reagents. J Oleo Sci 2010; 59(7): 375-80.
[http://dx.doi.org/10.5650/jos.59.375] [PMID: 20513971]

[49] Vikbjerg AF, Mu H, Xu X. Synthesis of structured phospholipids by immobilized phospholipase A_2 catalyzed acidolysis. J Biotechnol 2007; 128(3): 545-54.
[http://dx.doi.org/10.1016/j.jbiotec.2006.11.006] [PMID: 17150274]

[50] More HT, Pandit AB. Enzymatic acyl modification of phosphatidylcholine using immobilized lipase and phospholipase A_2. Eur J Lipid Sci Technol 2010; 112(4): 428-33.
[http://dx.doi.org/10.1002/ejlt.200900150]

[51] Yamamoto Y, Mizuta E, Ito M, Harata M, Hiramoto S, Hara S. Lipase-catalyzed preparation of

phospholipids containing n-3 polyunsaturated fatty acids from soy phospholipids. J Oleo Sci 2014; 63(12): 1275-81.
[http://dx.doi.org/10.5650/jos.ess14125] [PMID: 25452265]

[52]   Hara S, Hasuo H, Nakasato M, Higaki Y, Totani Y. Modification of soybean phospholipids by enzymatic transacylation. J Oleo Sci 2002; 51(6): 417-21.
[http://dx.doi.org/10.5650/jos.51.417]

[53]   Totani Y, Hara S. Preparation of polyunsaturated phospholipids by lipase-catalyzed transesterification. J Am Oil Chem Soc 1991; 68(11): 848-51.
[http://dx.doi.org/10.1007/BF02660600]

[54]   Yoshimoto T, Nakata M, Yamaguchi S, Funada T, Saito Y, Inada Y. Synthesis of eicosapentaenoyl phosphatidylcholines by polyethylene glycol-modified lipase in benzene. Biotechnol Lett 1986; 8(11): 771-6.
[http://dx.doi.org/10.1007/BF01020820]

[55]   Baeza-Jiménez R, Noriega-Rodríguez JA, García HS, Otero C. Structured phosphatidylcholine with elevated content of conjugated linoleic acid: Optimization by response surface methodology. Eur J Lipid Sci Technol 2012; 114(11): 1261-7.
[http://dx.doi.org/10.1002/ejlt.201200038]

[56]   Baeza-Jiménez R, González-Rodríguez J, Kim I-H, García HS, Otero C. Use of immobilized phospholipase A$_1$-catalyzed acidolysis for the production of structured phosphatidylcholine with an elevated conjugated linoleic acid content. Grasas Aceites 2012; 63(1): 44-52.
[http://dx.doi.org/10.3989/gya.045211]

[57]   Garcia HS, Kim I-H, López-Hernández A, Hill CG Jr. Enrichment of lecithin with n-3 fatty acids by acidolysis using immobilized phospholipase A$_1$. Grasas Aceites 2008; 59(4): 368-74.
[http://dx.doi.org/10.3989/gya.2008.v59.i4.531]

[58]   Kim I-H, Garcia HS, Hill CG Jr. Phospholipase A$_1$-catalyzed synthesis of phospholipids enriched in n-3 polyunsaturated fatty acid residues. Enzyme Microb Technol 2007; 40(5): 1130-5.
[http://dx.doi.org/10.1016/j.enzmictec.2006.08.018]

[59]   Li X, Chen J-F, Yang B, Li D-M, Wang Y-H, Wang W-F. Production of structured phosphatidylcholine with high content of DHA/EPA by immobilized phospholipase A$_1$-catalyzed transesterification. Int J Mol Sci 2014; 15(9): 15244-58.
[http://dx.doi.org/10.3390/ijms150915244] [PMID: 25170810]

[60]   Zhao T, No S, Kim BH, Garcia HS, Kim Y, Kim I-H. Immobilized phospholipase A$_1$-catalyzed modification of phosphatidylcholine with n-3 polyunsaturated fatty acid. Food Chem 2014; 157: 132-40.
[http://dx.doi.org/10.1016/j.foodchem.2014.02.024] [PMID: 24679762]

[61]   Fernandez-Lorente G, Filice M, Terreni M, Guisan JM, Fernandez-Lafuente R, Palomo JM. Lecitase® Ultra as regioselective biocatalyst in the hydrolysis of fully protected carbohydrates: Strong modulation by using different immobilization protocols. J Mol Catal, B Enzym 2008; 51(3-4): 110-7.
[http://dx.doi.org/10.1016/j.molcatb.2007.11.017]

[62]   Cabezas DM, Madoery R, Diehl BWK, Tomás MC. Application of enzymatic hydrolysis on sunflower lecithin using a pancreatic PLA$_2$. J Am Oil Chem Soc 2011; 88(3): 443-6.
[http://dx.doi.org/10.1007/s11746-010-1684-9]

[63]   Morgado MAP, Cabral JMS, Prazeres DMF. Hydrolysis of lecithin by phospholipase A$_2$ in mixed reversed micelles of lecithin and sodium dioctyl sulphosuccinate. J Chem Technol Biotechnol 1995; 63(2): 181-9.
[http://dx.doi.org/10.1002/jctb.280630214]

[64]   Ishii I, Fukushima N, Ye X, Chun J. Lysophospholipid receptors: signaling and biology. Annu Rev Biochem 2004; 73(1): 321-54.
[http://dx.doi.org/10.1146/annurev.biochem.73.011303.073731] [PMID: 15189145]

[65]   Chun J, Rosen H. Lysophospholipid receptors as potential drug targets in tissue transplantation and autoimmune diseases. Curr Pharm Des 2006; 12(2): 161-71.
[http://dx.doi.org/10.2174/138161206775193109] [PMID: 16454733]

[66]   D'Arrigo P, Servi S. Synthesis of lysophospholipids. Molecules 2010; 15(3): 1354-77.
[http://dx.doi.org/10.3390/molecules15031354] [PMID: 20335986]

[67]   VanMiddlesworth F, Lopez M, Zweerink M, Edison AM, Wilson K. Chemicoenzymic synthesis of lysofungin. J Org Chem 1992; 57(17): 4753-4.
[http://dx.doi.org/10.1021/jo00043a040]

[68]   Sarney DB, Fregapane G, Vulfson EN. Lipase-catalyzed synthesis of lysophospholipids in a continuous bioreactor. J Am Oil Chem Soc 1994; 71(1): 93-6.
[http://dx.doi.org/10.1007/BF02541478]

[69]   Fasoli E, Arnone A, Caligiuri A, D'Arrigo P, de Ferra L, Servi S. Tin-mediated synthesis of lyso-phospholipids. Org Biomol Chem 2006; 4(15): 2974-8.
[http://dx.doi.org/10.1039/b604636c] [PMID: 16855747]

[70]   Mazur AW, Hiler GD II, Lee SSC, Armstrong MP, Wendel JD. Regio- and stereoselective enzymatic esterification of glycerol and its derivatives. Chem Phys Lipids 1991; 60(2): 189-99.
[http://dx.doi.org/10.1016/0009-3084(91)90041-9] [PMID: 1814641]

[71]   Kim J, Kim B-G. Lipase-catalyzed synthesis of lysophosphatidylcholine using organic cosolvent for *in situ* water activity control. J Am Oil Chem Soc 2000; 77(7): 791-7.
[http://dx.doi.org/10.1007/s11746-000-0126-1]

[72]   Virto C, Adlercreutz P. Lysophosphatidylcholine synthesis with *Candida antarctica* lipase B (Novozym 435). Enzyme Microb Technol 2000; 26(8): 630-5.
[http://dx.doi.org/10.1016/S0141-0229(00)00147-2] [PMID: 10793211]

[73]   Hong SI, Kim Y, Kim C-T, Kim I-H. Enzymatic synthesis of lysophosphatidylcholine containing CLA from *sn*-glycero-3-phosphatidylcholine (GPC) under vacuum. Food Chem 2011; 129(1): 1-6.
[http://dx.doi.org/10.1016/j.foodchem.2011.04.038]

[74]   Selvy PE, Lavieri RR, Lindsley CW, Brown HA, Phospholipase D. Phospholipase D: enzymology, functionality, and chemical modulation. Chem Rev 2011; 111(10): 6064-119.
[http://dx.doi.org/10.1021/cr200296t] [PMID: 21936578]

[75]   Ulbrich-Hofmann R, Lerchner A, Oblozinsky M, Bezakova L. Phospholipase D and its application in biocatalysis. Biotechnol Lett 2005; 27(8): 535-44.
[http://dx.doi.org/10.1007/s10529-005-3251-2] [PMID: 15973486]

[76]   Jenkins GM, Frohman MA. Phospholipase D: a lipid centric review. Cell Mol Life Sci 2005; 62(19-20): 2305-16.
[http://dx.doi.org/10.1007/s00018-005-5195-z] [PMID: 16143829]

[77]   Song J, Han J, Rhee J. Phospholipases: Occurrence and production in microorganisms, assay for high-throughput screening, and gene discovery from natural and man-made diversity. J Am Oil Chem Soc 2005; 82(10): 691-705.
[http://dx.doi.org/10.1007/s11746-005-1131-0]

[78]   Mansfeld J, Brandt W, Haftendorn R, Schöps R, Ulbrich-Hofmann R. Discrimination between the regioisomeric 1,2- and 1,3-diacylglycerophosphocholines by phospholipases. Chem Phys Lipids 2011; 164(3): 196-204.
[http://dx.doi.org/10.1016/j.chemphyslip.2010.12.009] [PMID: 21195068]

[79]   Uesugi Y, Hatanaka T. Phospholipase D mechanism using *Streptomyces* PLD. Biochim Biophys Acta 2009; 1791(9): 962-9.
[http://dx.doi.org/10.1016/j.bbalip.2009.01.020] [PMID: 19416643]

[80]   Hanahan DJ, Chaikoff IL. A new phospholipide-splitting enzyme specific for the ester linkage

between the nitrogenous base and the phosphoric acid grouping. J Biol Chem 1947; 169(3): 699-705.
[PMID: 20259103]

[81]   Hanahan DJ, Chaikoff IL. On the nature of the phosphorus-containing lipides of cabbage leaves and their relation to a phospholipide-splitting enzyme contained in these leaves. J Biol Chem 1948; 172(1): 191-8.
[PMID: 18920784]

[82]   Saito M, Kanfer J. Phosphatidohydrolase activity in a solubilized preparation from rat brain particulate fraction. Arch Biochem Biophys 1975; 169(1): 318-23.
[http://dx.doi.org/10.1016/0003-9861(75)90346-X] [PMID: 239638]

[83]   Bocckino SB, Blackmore PF, Wilson PB, Exton JH. Phosphatidate accumulation in hormone-treated hepatocytes via a phospholipase D mechanism. J Biol Chem 1987; 262(31): 15309-15.
[PMID: 3117799]

[84]   Bocckino SB, Wilson PB, Exton JH. Ca^{2+}-mobilizing hormones elicit phosphatidylethanol accumulation via phospholipase D activation. FEBS Lett 1987; 225(1-2): 201-4.
[http://dx.doi.org/10.1016/0014-5793(87)81157-2] [PMID: 3319693]

[85]   Liscovitch M, Blusztajn JK, Freese A, Wurtman RJ. Stimulation of choline release from NG108-15 cells by 12-O-tetradecanoylphorbol 13-acetate. Biochem J 1987; 241(1): 81-6.
[http://dx.doi.org/10.1042/bj2410081] [PMID: 3566713]

[86]   Eibl H, Kovatchev S. Preparation of phospholipids and their analogs by phospholipase D.Methods in Enzymology. Academic Press 1981; Vol. 72: pp. 632-9.
[http://dx.doi.org/10.1016/S0076-6879(81)72055-X]

[87]   Imamura S, Horiuti Y. Purification of *Streptomyces chromofuscus* phospholipase D by hydrophobic affinity chromatography on palmitoyl cellulose. J Biochem 1979; 85(1): 79-95.
[http://dx.doi.org/10.1093/oxfordjournals.jbchem.a132334] [PMID: 762053]

[88]   Kokusho Y, Kato S, Machida H, Iwasaki S. Purification and properties of phospholipase D from *Actinomadura* sp. strain no. 362. Agric Biol Chem 1987; 51(9): 2515-24.

[89]   Shimbo K, Iwasaki Y, Yamane T, Ina K. Purification and properties of phospholipase D from *Streptomyces antibioticus*. Biosci Biotechnol Biochem 1993; 57(11): 1946-8.
[http://dx.doi.org/10.1271/bbb.57.1946]

[90]   Shimbo K, Yano H, Miyamoto Y. Two *Streptomyces* strains that produce phospholipase D with high transphosphatidylation activity. Agric Biol Chem 1989; 53(11): 3083-5.

[91]   Shimbo K, Yano H, Miyamoto Y. Purification and properties of phospholipase D from *Streptomyces lydicus*. Agric Biol Chem 1990; 54(5): 1189-93.

[92]   Carrea G, D'Arrigo P, Piergianni V, Roncaglio S, Secundo F, Servi S. Purification and properties of two phospholipases D from *Streptomyces* sp. Biochim Biophys Acta 1995; 1255(3): 273-9.
[http://dx.doi.org/10.1016/0005-2760(94)00241-P] [PMID: 7734443]

[93]   Ogino C, Negi Y, Matsumiya T, *et al*. Purification, characterization, and sequence determination of phospholipase D secreted by *Streptoverticillium cinnamoneum*. J Biochem 1999; 125(2): 263-9.
[http://dx.doi.org/10.1093/oxfordjournals.jbchem.a022282] [PMID: 9990122]

[94]   Hatanaka T, Negishi T, Kubota-Akizawa M, Hagishita T. Purification, characterization, cloning and sequencing of phospholipase D from *Streptomyces septatus* TH-2. Enzyme Microb Technol 2002; 31(3): 233-41.
[http://dx.doi.org/10.1016/S0141-0229(02)00121-7]

[95]   Juneja LR, Kazuoka T, Yamane T, Shimizu S. Kinetic evaluation of conversion of phosphatidylcholine to phosphatidylethanolamine by phospholipase D from different sources. Biochim Biophys Acta 1988; 960(3): 334-41.
[http://dx.doi.org/10.1016/0005-2760(88)90041-0] [PMID: 3382679]

[96]   Yang H, Roberts MF. Expression and characterization of a heterodimer of *Streptomyces chromofuscus* phospholipase D. Biochim Biophys Acta 2004; 1703(1): 43-51.
[http://dx.doi.org/10.1016/j.bbapap.2004.09.014] [PMID: 15588701]

[97]   Muraki M, Damnjanović J, Nakano H, Iwasaki Y. Salt-induced increase in the yield of enzymatically synthesized phosphatidylinositol and the underlying mechanism. J Biosci Bioeng 2016; 122(3): 276-82.
[http://dx.doi.org/10.1016/j.jbiosc.2016.02.011] [PMID: 27009527]

[98]   Iwasaki Y, Mizumoto Y, Okada T, Yamamoto T, Tsutsumi K, Yamane T. An aqueous suspension system for phospholipase D-mediated synthesis of PS without toxic organic solvent. J Am Oil Chem Soc 2003; 80(7): 653-7.
[http://dx.doi.org/10.1007/s11746-003-0754-5]

[99]   Juneja L, Hibi N, Yamane T, Shimizu S. Repeated batch and continuous operations for phosphatidylglycerol synthesis from phosphatidylcholine with immobilized phospholipase D. Appl Microbiol Biotechnol 1987; 27(2): 146-51.
[http://dx.doi.org/10.1007/BF00251937]

[100]   Juneja LR, Hibi N, Inagaki N, Yamane T, Shimizu S. Comparative study on conversion of phosphatidylcholine to phosphatidylglycerol by cabbage phospholipase D in micelle and emulsion systems. Enzyme Microb Technol 1987; 9(6): 350-4.
[http://dx.doi.org/10.1016/0141-0229(87)90058-5]

[101]   Juneja LR, Hibi N, Yamane T, Shimizu S. Studies on enzymatic conversion of phospholipids. 3. Repeated batch and continuous operations for phosphatidylglycerol synthesis from phosphatidylcholine with immobilized phospholipase D. Appl Microbiol Biotechnol 1987; 27(2): 146-51.
[http://dx.doi.org/10.1007/BF00251937]

[102]   Juneja LR, Kazuoka T, Goto N, Yamane T, Shimizu S. Conversion of phosphatidylcholine to phosphatidylserine by various phospholipases D in the presence of L- or D-serine. Biochim Biophys Acta 1989; 1003(3): 277-83.
[http://dx.doi.org/10.1016/0005-2760(89)90233-6]

[103]   Juneja LR, Taniguchi E, Shimizu S, Yamane T. Increasing productivity by removing choline in conversion of phosphatidylcholine to phosphatidylserine by phospholipase D. J Ferment Bioeng 1992; 73(5): 357-61.
[http://dx.doi.org/10.1016/0922-338X(92)90278-3]

[104]   Rich JO, Khmelnitsky YL. Phospholipase D-catalyzed transphosphatidylation in anhydrous organic solvents. Biotechnol Bioeng 2001; 72(3): 374-7.
[http://dx.doi.org/10.1002/1097-0290(20010205)72:3<374::AID-BIT16>3.0.CO;2-E]   [PMID: 11135209]

[105]   Juneja L, Yamane T, Shimizu S. Enzymatic method of increasing phosphatidylcholine content of lecithin. J Am Oil Chem Soc 1989; 66(5): 714-7.
[http://dx.doi.org/10.1007/BF02669959]

[106]   Liu Y, Zhang T, Qiao J, *et al.* High-yield phosphatidylserine production via yeast surface display of phospholipase D from *Streptomyces chromofuscus* on *Pichia pastoris.* J Agric Food Chem 2014; 62(23): 5354-60.
[http://dx.doi.org/10.1021/jf405836x] [PMID: 24841277]

[107]   Piazza GJ, Marmer WN. Conversion of phosphatidylcholine to phosphatidylglycerol with phospholipase D and glycerol. J Am Oil Chem Soc 2007; 84(7): 645-51.
[http://dx.doi.org/10.1007/s11746-007-1081-1]

[108]   Müller AO, Mrestani-Klaus C, Schmidt J, Ulbrich-Hofmann R, Dippe M. New cardiolipin analogs synthesized by phospholipase D-catalyzed transphosphatidylation. Chem Phys Lipids 2012; 165(7): 787-93.

[http://dx.doi.org/10.1016/j.chemphyslip.2012.09.005] [PMID: 23059117]

[109]   Stamler CJ, Breznan D, Neville TA-M, Viau FJ, Camlioglu E, Sparks DL. Phosphatidylinositol promotes cholesterol transport *in vivo*. J Lipid Res 2000; 41(8): 1214-21.
[PMID: 10946008]

[110]   Burgess JW, Boucher J, Neville TA, *et al.* Phosphatidylinositol promotes cholesterol transport and excretion. J Lipid Res 2003; 44(7): 1355-63.
[http://dx.doi.org/10.1194/jlr.M300062-JLR200] [PMID: 12700341]

[111]   Burgess JW, Neville TA, Rouillard P, Harder Z, Beanlands DS, Sparks DL. Phosphatidylinositol increases HDL-C levels in humans. J Lipid Res 2005; 46(2): 350-5.
[http://dx.doi.org/10.1194/jlr.M400438-JLR200] [PMID: 15576836]

[112]   Yanagita T. Nutritional functions of dietary phosphatidylinositol. Inform 2003; 14(2): 64-6.

[113]   Shirouchi B, Nagao K, Inoue N, *et al.* Dietary phosphatidylinositol prevents the development of nonalcoholic fatty liver disease in Zucker (*fa/fa*) rats. J Agric Food Chem 2008; 56(7): 2375-9.
[http://dx.doi.org/10.1021/jf703578d] [PMID: 18324772]

[114]   Inafuku M, Nagao K, Inafuku A, *et al.* Dietary phosphatidylinositol protects C57BL/6 mice from concanavalin A-induced liver injury by modulating immune cell functions. Mol Nutr Food Res 2013; 57(9): 1671-9.
[http://dx.doi.org/10.1002/mnfr.201200607] [PMID: 23653180]

[115]   Masayama A, Takahashi T, Tsukada K, *et al. Streptomyces* phospholipase D mutants with altered substrate specificity capable of phosphatidylinositol synthesis. ChemBioChem 2008; 9(6): 974-81.
[http://dx.doi.org/10.1002/cbic.200700528] [PMID: 18338352]

[116]   Masayama A, Tsukada K, Ikeda C, Nakano H, Iwasaki Y. Isolation of phospholipase D mutants having phosphatidylinositol-synthesizing activity with positional specificity on *myo*-inositol. ChemBioChem 2009; 10(3): 559-64.
[http://dx.doi.org/10.1002/cbic.200800651] [PMID: 19123198]

[117]   Damnjanović J, Kuroiwa C, Tanaka H, Ishida K, Nakano H, Iwasaki Y. Directing positional specificity in enzymatic synthesis of bioactive 1-phosphatidylinositol by protein engineering of a phospholipase D. Biotechnol Bioeng 2016; 113(1): 62-71.
[http://dx.doi.org/10.1002/bit.25697] [PMID: 26154602]

[118]   Murakami-Murofushi K, Uchiyama A, Fujiwara Y, *et al.* Biological functions of a novel lipid mediator, cyclic phosphatidic acid. Biochim Biophys Acta 2002; 1582(1-3): 1-7.
[http://dx.doi.org/10.1016/S1388-1981(02)00131-2] [PMID: 12069804]

[119]   Tsuda S, Okudaira S, Moriya-Ito K, *et al.* Cyclic phosphatidic acid is produced by autotaxin in blood. J Biol Chem 2006; 281(36): 26081-8.
[http://dx.doi.org/10.1074/jbc.M602925200] [PMID: 16837466]

[120]   Friedman P, Haimovitz R, Markman O, Roberts MF, Shinitzky M. Conversion of lysophospholipids to cyclic lysophosphatidic acid by phospholipase D. J Biol Chem 1996; 271(2): 953-7.
[http://dx.doi.org/10.1074/jbc.271.2.953] [PMID: 8557710]

[121]   Lajoie DM, Zobel-Thropp PA, Kumirov VK, Bandarian V, Binford GJ, Cordes MHJ. Phospholipase D toxins of brown spider venom convert lysophosphatidylcholine and sphingomyelin to cyclic phosphates. PLoS One 2013; 8(8): e72372.
[http://dx.doi.org/10.1371/journal.pone.0072372] [PMID: 24009677]

[122]   Lajoie DM, Cordes MHJ. Spider, bacterial and fungal phospholipase D toxins make cyclic phosphate products. Toxicon 2015; 108: 176-80.
[http://dx.doi.org/10.1016/j.toxicon.2015.10.008] [PMID: 26482933]

[123]   D'Arrigo P, de Ferra L, Piergianni V, Ricci A, Scarcelli D, Servi S. Phospholipase D from Streptomyces catalyses the transfer of secondary alcohols. J Chem Soc, Chem Comm 1994; 1(14): 1709-0.

[124] Song S, Cheong L-Z, Guo Z, *et al.* Phospholipase D (PLD) catalyzed synthesis of phosphatidyl-glucose in biphasic reaction system. Food Chem 2012; 135(2): 373-9.
[http://dx.doi.org/10.1016/j.foodchem.2012.05.020] [PMID: 22868102]

[125] Kokusho Y, Tsunoda A, Kato S, Machida H, Iwasaki S. Production of various phosphatidyl-saccharides by phospholipase D from *Actinomadura* sp. strain no. 362. Biosci Biotechnol Biochem 1993; 57(8): 1302-5.
[http://dx.doi.org/10.1271/bbb.57.1302]

[126] Wang P, Schuster M, Wang YF, Wong CH. Synthesis of phospholipid-inhibitor conjugates by enzymic transphosphatidylation with phospholipase D. J Am Chem Soc 1993; 115(23): 10487-91.
[http://dx.doi.org/10.1021/ja00076a004]

[127] Sato R, Itabashi Y, Suzuki A, Hatanaka T, Kuksis A. Effect of temperature on the stereoselectivity of phospholipase D toward glycerol in the transphosphatidylation of phosphatidylcholine to phosphatidylglycerol. Lipids 2004; 39(10): 1019-23.
[http://dx.doi.org/10.1007/s11745-004-1325-0] [PMID: 15691025]

[128] Yamamoto Y, Kurihara H, Miyashita K, Hosokawa M. Synthesis of novel phospholipids that bind phenylalkanols and hydroquinone via phospholipase D-catalyzed transphosphatidylation. N Biotechnol 2011; 28(1): 1-6.
[http://dx.doi.org/10.1016/j.nbt.2010.06.014] [PMID: 20601268]

[129] Song S, Cheong L-Z, Falkeborg M, *et al.* Facile synthesis of phosphatidyl saccharides for preparation of anionic nanoliposomes with enhanced stability. PLoS One 2013; 8(9): e73891.
[http://dx.doi.org/10.1371/journal.pone.0073891] [PMID: 24069243]

[130] Yamamoto Y, Hosokawa M, Kurihara H, Miyashita K. Preparation of phosphatidylated terpenes via phospholipase D-mediated transphosphatidylation. J Am Oil Chem Soc 2008; 85(4): 313-20.
[http://dx.doi.org/10.1007/s11746-008-1206-1]

[131] Dippe M, Mrestani-Klaus C, Schierhorn A, Ulbrich-Hofmann R. Phospholipase D-catalyzed synthesis of new phospholipids with polar head groups. Chem Phys Lipids 2008; 152(2): 71-7.
[http://dx.doi.org/10.1016/j.chemphyslip.2008.01.005] [PMID: 18302938]

[132] Takami M, Hidaka N, Suzuki Y. Phospholipase D-catalyzed synthesis of phosphatidyl aromatic compounds. Biosci Biotechnol Biochem 1994; 58(12): 2140-4.
[http://dx.doi.org/10.1271/bbb.58.2140]

[133] Nagao A, Ishida N, Terao J. Synthesis of 6-phosphatidyl-L-ascorbic acid by phospholipase D. Lipids 1991; 26(5): 390-4.
[http://dx.doi.org/10.1007/BF02537205] [PMID: 1895887]

[134] Nagao A, Terao J. Antioxidant activity of 6-phosphatidyl-L-ascorbic acid. Biochem Biophys Res Commun 1990; 172(2): 385-9.
[http://dx.doi.org/10.1016/0006-291X(90)90684-F] [PMID: 2241939]

[135] Koga T, Nagao A, Terao J, Sawada K, Mukai K. Synthesis of a phosphatidyl derivative of vitamin E and its antioxidant activity in phospholipid bilayers. Lipids 1994; 29(2): 83-9.
[http://dx.doi.org/10.1007/BF02537147] [PMID: 8152350]

[136] Koga T, Terao J. Antioxidant activity of a novel phosphatidyl derivative of vitamin E in lard and its model system. J Agric Food Chem 1994; 42(6): 1291-4.
[http://dx.doi.org/10.1021/jf00042a007]

[137] Correia Ledo D, Mauzeroll J. Synthesis of metal complex modified phospholipids by phospholipase D-catalyzed transphosphatidylation. ECS Trans 2009; 19(33): 1-10.

[138] Takami M, Hidaka N, Miki S, Suzuki Y. Enzymatic synthesis of novel phosphatidylkojic acid and phosphatidylarbutin, and their inhibitory effects on tyrosinase activity. Biosci Biotechnol Biochem 1994; 58(9): 1716-7.
[http://dx.doi.org/10.1271/bbb.58.1716]

[139] Shuto S, Ueda S, Imamura S, Fukukawa K, Matsuda A, Ueda T. A facile one-step synthesis of 5'-phosphatidylnucleosides by an enzymatic two-phase reaction. Tetrahedron Lett 1987; 28(2): 199-202.
[http://dx.doi.org/10.1016/S0040-4039(00)95685-5]

[140] Shuto S, Itoh H, Ueda S, *et al.* A facile enzymatic synthesis of 5'-(3-sn-phosphatidyl)nucleosides and their antileukemic activities. Chem Pharm Bull (Tokyo) 1988; 36(1): 209-17.
[http://dx.doi.org/10.1248/cpb.36.209] [PMID: 3378284]

[141] Koketsu M, Nitoda T, Sugino H, *et al.* Synthesis of a novel sialic acid derivative (sialylphospholipid) as an antirotaviral agent. J Med Chem 1997; 40(21): 3332-5.
[http://dx.doi.org/10.1021/jm9701280] [PMID: 9341907]

[142] Takami M, Suzuki Y. Synthesis of novel phosphatidyldihydroxyacetone via transphosphatidylation reaction by phospholipase D. Biosci Biotechnol Biochem 1994; 58(12): 2136-9.
[http://dx.doi.org/10.1271/bbb.58.2136]

[143] Yamamoto Y, Hosokawa M, Kurihara H, Maoka T, Miyashita K. Synthesis of phosphatidylated-monoterpene alcohols catalyzed by phospholipase D and their antiproliferative effects on human cancer cells. Bioorg Med Chem Lett 2008; 18(14): 4044-6.
[http://dx.doi.org/10.1016/j.bmcl.2008.05.113] [PMID: 18556205]

[144] Hidaka N, Takami M, Suzuki Y. Enzymatic phosphatidylation of thiamin, pantothenic acid, and their derivatives. J Nutr Sci Vitaminol (Tokyo) 2008; 54(3): 255-61.
[http://dx.doi.org/10.3177/jnsv.54.255] [PMID: 18635914]

[145] Takami M, Suzuki Y. Enzymatic synthesis of novel phosphatidylgenipin, and its enhanced cytotoxicity. Biosci Biotechnol Biochem 1994; 58(10): 1897-8.
[http://dx.doi.org/10.1271/bbb.58.1897]

[146] Hossen M, Hernandez E. Phospholipase D-catalyzed synthesis of novel phospholipid-phytosterol conjugates. Lipids 2004; 39(8): 777-82.
[http://dx.doi.org/10.1007/s11745-004-1295-2] [PMID: 15638246]

[147] Ponting CP, Kerr ID. A novel family of phospholipase D homologues that includes phospholipid synthases and putative endonucleases: identification of duplicated repeats and potential active site residues. Protein Sci 1996; 5(5): 914-22.
[http://dx.doi.org/10.1002/pro.5560050513] [PMID: 8732763]

[148] Koonin EV. A duplicated catalytic motif in a new superfamily of phosphohydrolases and phospholipid synthases that includes poxvirus envelope proteins. Trends Biochem Sci 1996; 21(7): 242-3.
[http://dx.doi.org/10.1016/S0968-0004(96)30024-8] [PMID: 8755242]

[149] Sung TC, Roper RL, Zhang Y, *et al.* Mutagenesis of phospholipase D defines a superfamily including a trans-Golgi viral protein required for poxvirus pathogenicity. EMBO J 1997; 16(15): 4519-30.
[http://dx.doi.org/10.1093/emboj/16.15.4519] [PMID: 9303296]

[150] Gottlin EB, Rudolph AE, Zhao Y, Matthews HR, Dixon JE. Catalytic mechanism of the phospholipase D superfamily proceeds via a covalent phosphohistidine intermediate. Proc Natl Acad Sci USA 1998; 95(16): 9202-7.
[http://dx.doi.org/10.1073/pnas.95.16.9202] [PMID: 9689058]

[151] Leiros I, McSweeney S, Hough E. The reaction mechanism of phospholipase D from *Streptomyces* sp. strain PMF. Snapshots along the reaction pathway reveal a pentacoordinate reaction intermediate and an unexpected final product. J Mol Biol 2004; 339(4): 805-20.
[http://dx.doi.org/10.1016/j.jmb.2004.04.003] [PMID: 15165852]

[152] Leiros I, Secundo F, Zambonelli C, Servi S, Hough E. The first crystal structure of a phospholipase D. Structure 2000; 8(6): 655-67.
[http://dx.doi.org/10.1016/S0969-2126(00)00150-7] [PMID: 10873862]

[153] Stuckey JA, Dixon JE. Crystal structure of a phospholipase D family member. Nat Struct Biol 1999; 6(3): 278-84.

[http://dx.doi.org/10.1038/6716] [PMID: 10074947]

[154] Peng X, Frohman MA. Mammalian phospholipase D physiological and pathological roles. Acta Physiol (Oxf) 2012; 204(2): 219-26.
[http://dx.doi.org/10.1111/j.1748-1716.2011.02298.x] [PMID: 21447092]

[155] Li M, Hong Y, Wang X. Phospholipase D- and phosphatidic acid-mediated signaling in plants. Biochim Biophys Acta 2009; 1791(9): 927-35.
[http://dx.doi.org/10.1016/j.bbalip.2009.02.017] [PMID: 19289179]

[156] Mansfeld J, Ulbrich-Hofmann R. Modulation of phospholipase D activity *in vitro*. Biochim Biophys Acta 2009; 1791(9): 913-26.
[http://dx.doi.org/10.1016/j.bbalip.2009.03.003] [PMID: 19286472]

[157] Nishimasu H, Ishizu H, Saito K, *et al.* Structure and function of Zucchini endoribonuclease in piRNA biogenesis. Nature 2012; 491(7423): 284-7.
[http://dx.doi.org/10.1038/nature11509] [PMID: 23064230]

[158] Grazulis S, Manakova E, Roessle M, *et al.* Structure of the metal-independent restriction enzyme *Bfi*I reveals fusion of a specific DNA-binding domain with a nonspecific nuclease. Proc Natl Acad Sci USA 2005; 102(44): 15797-802.
[http://dx.doi.org/10.1073/pnas.0507949102] [PMID: 16247004]

[159] Davies DR, Interthal H, Champoux JJ, Hol WGJ. The crystal structure of human tyrosyl-DNA phosphodiesterase, Tdp1. Structure 2002; 10(2): 237-48.
[http://dx.doi.org/10.1016/S0969-2126(02)00707-4] [PMID: 11839309]

[160] Zhu Y, Huang W, Lee SSK, Xu W. Crystal structure of a polyphosphate kinase and its implications for polyphosphate synthesis. EMBO Rep 2005; 6(7): 681-7.
[http://dx.doi.org/10.1038/sj.embor.7400448] [PMID: 15947782]

[161] Krug M, Lee S-J, Boos W, Diederichs K, Welte W. The three-dimensional structure of TrmB, a transcriptional regulator of dual function in the hyperthermophilic archaeon *Pyrococcus furiosus* in complex with sucrose. Protein Sci 2013; 22(6): 800-8.
[http://dx.doi.org/10.1002/pro.2263] [PMID: 23576322]

[162] Xie Z, Ho W-T, Exton JH. Association of N- and C-terminal domains of phospholipase D is required for catalytic activity. J Biol Chem 1998; 273(52): 34679-82.
[http://dx.doi.org/10.1074/jbc.273.52.34679] [PMID: 9856987]

[163] Iwasaki Y, Horiike S, Matsushima K, Yamane T. Location of the catalytic nucleophile of phospholipase D of *Streptomyces antibioticus* in the C-terminal half domain. Eur J Biochem 1999; 264(2): 577-81.
[http://dx.doi.org/10.1046/j.1432-1327.1999.00669.x] [PMID: 10491106]

[164] Yang SF, Freer S, Benson AA. Transphosphatidylation by phospholipase D. J Biol Chem 1967; 242(3): 477-84.
[PMID: 6022844]

[165] Stanacev NZ, Stuhne-Sekalec L. On the mechanism of enzymatic phosphatidylation. Biosynthesis of cardiolipin catalyzed by phospholipase D. Biochim Biophys Acta 1970; 210(2): 350-2.
[http://dx.doi.org/10.1016/0005-2760(70)90183-9] [PMID: 5476267]

[166] Bruzik K, Tsai MD. Phospholipids chiral at phosphorus. Synthesis of chiral phosphatidylcholine and stereochemistry of phospholipase D. Biochemistry 1984; 23(8): 1656-61.
[http://dx.doi.org/10.1021/bi00303a012] [PMID: 6722117]

[167] Bruzik K, Tsai MD. Phospholipids chiral at phosphorus. 1. Stereochemistry of transphosphatidylation catalyzed by phospholipase D. J Am Chem Soc 1982; 104(3): 863-5.
[http://dx.doi.org/10.1021/ja00367a044]

[168] Yang H, Roberts MF. Phosphohydrolase and transphosphatidylation reactions of two *Streptomyces* phospholipase D enzymes: covalent versus noncovalent catalysis. Protein Sci 2003; 12(9): 2087-98.

[http://dx.doi.org/10.1110/ps.03192503] [PMID: 12931007]

[169] Orth ES, Brandão TAS, Souza BS, *et al.* Intramolecular catalysis of phosphodiester hydrolysis by two imidazoles. J Am Chem Soc 2010; 132(24): 8513-23.
[http://dx.doi.org/10.1021/ja1034733] [PMID: 20509675]

[170] Uesugi Y, Arima J, Iwabuchi M, Hatanaka T. Sensor of phospholipids in *Streptomyces* phospholipase D. FEBS J 2007; 274(10): 2672-81.
[http://dx.doi.org/10.1111/j.1742-4658.2007.05802.x] [PMID: 17459102]

[171] Damnjanović J, Takahashi R, Suzuki A, Nakano H, Iwasaki Y. Improving thermostability of phosphatidylinositol-synthesizing Streptomyces phospholipase D. Protein Eng Des Sel 2012; 25(8): 415-24.
[http://dx.doi.org/10.1093/protein/gzs038] [PMID: 22718790]

[172] Schäffner I, Rücknagel K-P, Mansfeld J, Ulbrich-Hofmann R. Genomic structure, cloning and expression of two phospholipase D isoenzymes from white cabbage. Eur J Lipid Sci Technol 2002; 104(2): 79-87.
[http://dx.doi.org/10.1002/1438-9312(200202)104:2<79::AID-EJLT79>3.0.CO;2-C]

[173] Lerchner A, Mansfeld J, Kuppe K, Ulbrich-Hofmann R. Probing conserved amino acids in phospholipase D (*Brassica oleracea* var. *capitata*) for their importance in hydrolysis and transphosphatidylation activity. Protein Eng Des Sel 2006; 19(10): 443-52.
[http://dx.doi.org/10.1093/protein/gzl028] [PMID: 16845127]

[174] Iwasaki Y, Mishima N, Mizumoto K, Nakano H, Yamane T. Extracellular production of phospholipase D of *Streptomyces antibioticus* using recombinant *Escherichia coli.* J Ferment Bioeng 1995; 79(5): 417-21.
[http://dx.doi.org/10.1016/0922-338X(95)91254-3]

[175] Mishima N, Mizumoto K, Iwasaki Y, Nakano H, Yamane T. Insertion of stabilizing *loci* in vectors of T7 RNA polymerase-mediated *Escherichia coli* expression systems: a case study on the plasmids involving foreign phospholipase D gene. Biotechnol Prog 1997; 13(6): 864-8.
[http://dx.doi.org/10.1021/bp970084o] [PMID: 9413145]

[176] Ogino C, Daido H, Ohmura Y, *et al.* Remarkable enhancement in PLD activity from *Streptoverticillium cinnamoneum* by substituting serine residue into the GG/GS motif. Biochim Biophys Acta 2007; 1774(6): 671-8.
[http://dx.doi.org/10.1016/j.bbapap.2007.04.004] [PMID: 17499030]

[177] Hatanaka T, Negishi T, Mori K. A mutant phospholipase D with enhanced thermostability from *Streptomyces* sp. Biochim Biophys Acta 2004; 1696(1): 75-82.
[http://dx.doi.org/10.1016/j.bbapap.2003.09.013] [PMID: 14726207]

[178] Negishi T, Mukaihara T, Mori K, *et al.* Identification of a key amino acid residue of *Streptomyces* phospholipase D for thermostability by *in vivo* DNA shuffling. Biochim Biophys Acta 2005; 1722(3): 331-42.
[http://dx.doi.org/10.1016/j.bbagen.2005.01.009] [PMID: 15777623]

[179] Iwasaki Y, Masayama A, Mori A, Ikeda C, Nakano H. Composition analysis of positional isomers of phosphatidylinositol by high-performance liquid chromatography. J Chromatogr A 2009; 1216(32): 6077-80.
[http://dx.doi.org/10.1016/j.chroma.2009.06.064] [PMID: 19580974]

[180] Ozaki A, Masayama A, Nakano H, Iwasaki Y. Synthesis of phosphatidylinositols having various inositol stereoisomers by engineered phospholipase D. J Biosci Bioeng 2010; 109(4): 337-40.
[http://dx.doi.org/10.1016/j.jbiosc.2009.09.045] [PMID: 20226373]

[181] Damnjanović J, Nakano H, Iwasaki Y. Deletion of a dynamic surface loop improves stability and changes kinetic behavior of phosphatidylinositol-synthesizing Streptomyces phospholipase D. Biotechnol Bioeng 2014; 111(4): 674-82.
[http://dx.doi.org/10.1002/bit.25149] [PMID: 24222582]

[182] Damnjanović J, Iwasaki Y. Phospholipase D as a catalyst: application in phospholipid synthesis, molecular structure and protein engineering. J Biosci Bioeng 2013; 116(3): 271-80.
[http://dx.doi.org/10.1016/j.jbiosc.2013.03.008] [PMID: 23639419]

[183] Zambonelli C, Casali M, Roberts MF. Mutagenesis of putative catalytic and regulatory residues of *Streptomyces chromofuscus* phospholipase D differentially modifies phosphatase and phosphodiesterase activities. J Biol Chem 2003; 278(52): 52282-9.
[http://dx.doi.org/10.1074/jbc.M310252200] [PMID: 14557260]

[184] Geng D, Baker DP, Foley SF, Zhou C, Stieglitz K, Roberts MFA. A 20-kDa domain is required for phosphatidic acid-induced allosteric activation of phospholipase D from Streptomyces chromofuscus. Biochim Biophys Acta 1999; 1430(2): 234-44.
[http://dx.doi.org/10.1016/S0167-4838(99)00005-9] [PMID: 10082951]

[185] Lajoie DM, Roberts SA, Zobel-Thropp PA, *et al.* Variable substrate preference among phospholipase D toxins from sicariid spiders. J Biol Chem 2015; 290(17): 10994-1007.
[http://dx.doi.org/10.1074/jbc.M115.636951] [PMID: 25752604]

[186] Tokumura A, Majima E, Kariya Y, *et al.* Identification of human plasma lysophospholipase D, a lysophosphatidic acid-producing enzyme, as autotaxin, a multifunctional phosphodiesterase. J Biol Chem 2002; 277(42): 39436-42.
[http://dx.doi.org/10.1074/jbc.M205623200] [PMID: 12176993]

[187] van Meeteren LA, Ruurs P, Christodoulou E, *et al.* Inhibition of autotaxin by lysophosphatidic acid and sphingosine 1-phosphate. J Biol Chem 2005; 280(22): 21155-61.
[http://dx.doi.org/10.1074/jbc.M413183200] [PMID: 15769751]

[188] Verger R, Mieras MCE, de Haas GH. Action of phospholipase A at interfaces. J Biol Chem 1973; 248(11): 4023-34.
[PMID: 4736081]

[189] Deems RA, Eaton BR, Dennis EA. Kinetic analysis of phospholipase A_2 activity toward mixed micelles and its implications for the study of lipolytic enzymes. J Biol Chem 1975; 250(23): 9013-20.
[PMID: 1194274]

[190] Dennis EA. Phospholipase A_2 activity towards phosphatidylcholine in mixed micelles: surface dilution kinetics and the effect of thermotropic phase transitions. Arch Biochem Biophys 1973; 158(2): 485-93.
[http://dx.doi.org/10.1016/0003-9861(73)90540-7] [PMID: 4798723]

[191] Hendrickson HS, Dennis EA. Kinetic analysis of the dual phospholipid model for phospholipase A_2 action. J Biol Chem 1984; 259(9): 5734-9.
[PMID: 6715370]

[192] Roberts MF, Deems RA, Dennis EA. Dual role of interfacial phospholipid in phospholipase A_2 catalysis. Proc Natl Acad Sci USA 1977; 74(5): 1950-4.
[http://dx.doi.org/10.1073/pnas.74.5.1950] [PMID: 266715]

[193] Berg OG, Gelb MH, Tsai M-D, Jain MK. Interfacial enzymology: the secreted phospholipase A($_2$)-paradigm. Chem Rev 2001; 101(9): 2613-54.
[http://dx.doi.org/10.1021/cr990139w] [PMID: 11749391]

[194] Jain MK, Berg OG. The kinetics of interfacial catalysis by phospholipase A_2 and regulation of interfacial activation: hopping versus scooting. Biochim Biophys Acta 1989; 1002(2): 127-56.
[http://dx.doi.org/10.1016/0005-2760(89)90281-6] [PMID: 2649150]

[195] Brown HA, Henage LG, Preininger AM, Xiang Y, Exton JH. Biochemical analysis of phospholipase D Methods in Enzymology. Academic Press 2007; Vol. 434: pp. 49-87.

[196] Henage LG, Exton JH, Brown HA. Kinetic analysis of a mammalian phospholipase D: allosteric modulation by monomeric GTPases, protein kinase C, and polyphosphoinositides. J Biol Chem 2006; 281(6): 3408-17.
[http://dx.doi.org/10.1074/jbc.M508800200] [PMID: 16339153]

[197]   Chalifa-Caspi V, Eli Y, Liscovitch M. Kinetic analysis in mixed micelles of partially purified rat brain phospholipase D activity and its activation by phosphatidylinositol 4,5-bisphosphate. Neurochem Res 1998; 23(5): 589-99.
[http://dx.doi.org/10.1023/A:1022422418388] [PMID: 9566596]

[198]   Qin C, Wang C, Wang X. Kinetic analysis of Arabidopsis phospholipase Ddelta. Substrate preference and mechanism of activation by Ca2+ and phosphatidylinositol 4,5-biphosphate. J Biol Chem 2002; 277(51): 49685-90.
[http://dx.doi.org/10.1074/jbc.M209598200] [PMID: 12397060]

[199]   Abousalham A, Nari J, Teissère M, Ferté N, Noat G, Verger R. Study of fatty acid specificity of sunflower phospholipase D using detergent/phospholipid micelles. Eur J Biochem 1997; 248(2): 374-9.
[http://dx.doi.org/10.1111/j.1432-1033.1997.00374.x] [PMID: 9346291]

[200]   Wissing JB, Grabo P, Kornak B. Purification and characterization of multiple forms of phosphatidylinositol-specific phospholipases D from suspension cultured *Catharanthus roseus* cells. Plant Sci 1996; 117(1-2): 17-31.
[http://dx.doi.org/10.1016/0168-9452(96)04409-3]

[201]   Matsumoto Y, Sugimori D. Substrate recognition mechanism of *Streptomyces* phospholipase D and enzymatic measurement of plasmalogen. J Biosci Bioeng 2015; 120(4): 372-9.
[http://dx.doi.org/10.1016/j.jbiosc.2015.02.020] [PMID: 25900053]

[202]   Majd S, Yusko EC, Yang J, Sept D, Mayer M. A model for the interfacial kinetics of phospholipase D activity on long-chain lipids. Biophys J 2013; 105(1): 146-53.
[http://dx.doi.org/10.1016/j.bpj.2013.05.018] [PMID: 23823233]

[203]   D'Arrigo P, Piergianni V, Scarcelli D, Servi S. A spectrophotometric assay for phospholipase D. Anal Chim Acta 1995; 304(2): 249-54.
[http://dx.doi.org/10.1016/0003-2670(94)00613-Q]

[204]   Iwasaki Y, Nishikawa S, Tsuneda M, Takahashi T, Yamane T. Detection of phospholipase D on solid materials. Anal Biochem 2004; 329(1): 157-9.
[http://dx.doi.org/10.1016/j.ab.2004.02.040] [PMID: 15136182]

[205]   Liu S-J, Wen Q, Tang L-J, Jiang J-H. Phospholipid-graphene nanoassembly as a fluorescence biosensor for sensitive detection of phospholipase D activity. Anal Chem 2012; 84(14): 5944-50.
[http://dx.doi.org/10.1021/ac300539s] [PMID: 22746286]

[206]   Imamura S, Horiuti Y. Purification of *Streptomyces chromofuscus* phospholipase D by hydrophobic affinity chromatography on palmitoyl cellulose. J Biochem 1979; 85(1): 79-95.
[http://dx.doi.org/10.1093/oxfordjournals.jbchem.a132334] [PMID: 762053]

[207]   Zhou M, Diwu Z, Panchuk-Voloshina N, Haugland RP. A stable nonfluorescent derivative of resorufin for the fluorometric determination of trace hydrogen peroxide: applications in detecting the activity of phagocyte NADPH oxidase and other oxidases. Anal Biochem 1997; 253(2): 162-8.
[http://dx.doi.org/10.1006/abio.1997.2391] [PMID: 9367498]

[208]   Dippe M, Ulbrich-Hofmann R. Spectrophotometric determination of phosphatidic acid via iron(III) complexation for assaying phospholipase D activity. Anal Biochem 2009; 392(2): 169-73.
[http://dx.doi.org/10.1016/j.ab.2009.05.048] [PMID: 19497293]

[209]   Aurich I, Hirche F, Ulbrich-Hofmann R. The determination of phospholipase D activity in emulsion systems. Anal Biochem 1999; 268(2): 337-42.

CHAPTER 4

Food Microbial Cultures

Sander Sieuwerts[1,*] and **Gurjot Deepika**[2]

[1] *Arla Innovation Centre, Agro Food Park 19, 8200 Aarhus, Denmark*

[2] *Porton Biopharma Ltd., Manor Farm Road, Porton Down, Salisbury SP4 0JG, UK*

Abstract: Many food products are fermented and this can be for several reasons including taste, shelf-life extension, and improvement of the nutritional value. During the fermentation, the present carbon source is typically converted into organic acid or alcohol. The main fermenting microorganisms are lactic acid bacteria, yeasts and filamentous fungi. Different categories of food products and the benefits that the fermenting microbes bring to these products are discussed. Moreover, there is an emphasis on health-promoting microorganisms, on the mutual effects that microorganisms have on each other, and the resulting effects on the food matrices. Besides, the current and future status of culture research, culture production, and their impact on food fermentation are discussed.

Keywords: Adhesion, Anti-Carcinogens, Anti-Mutagens, Bacteriocin, Bacteriophage, Cheese, Cultures, Complex Consortia, Exopolysaccharides, Food Fermentation, Functional Starter Culture, Functional Foods, Immunomodulation, Lactic Acid Bacteria, Mixed Cultures, Microbial Interactions, Mold, Probiotics, Shelf-Life, Texture, Taste, Yeast, Yoghurt.

INTRODUCTION TO CULTURES

Fermentation is known as the energy-yielding anaerobic metabolic breakdown of a carbon source without net oxidation. It is a natural process that occurs ubiquitously, for example during ripening of fruit [1], during breakdown of organic matter such as foliage [2], and continuously in the human intestine [3]. This fermentation by microorganisms generally yields organic acids, in particular lactic acid, acetic acid, and propionic acid, as well as ethanol, carbon dioxide, or other dissimilation products. Many of today's food products are fermented or contain a component of fermentation in the production process. The process of fermentation as food processing method dates back to the early days of human civilization with reports of fermented products as early as 8,000 years ago [4].

---

[*] **Corresponding author Sander Sieuwerts:** Arla Innovation Centre, Agro Food Park 19, 8200 Aarhus, Denmark; Tel: +4587466703; Fax: +4586281691; E-mail: sander.sieuwerts@arlafoods.com

Ali Osman (Ed.)

Basically, fermentation of foods started with the agricultural revolution, when humankind first started to cultivate crops instead of having a Paleolithic diet. It was then that human food sources became constant and reproducible, which led to some extent of specialization to these food sources by some groups of microorganisms that are still found in our fermented foods today [5]. Here after, microorganisms, first as indigenous biota of food fermentations and later as actual cultures, continued to develop together with their environments, with human technology, and with their applications (Fig. **4.1**). Fermented foods are found in all categories, ranging from plant-based products, such as sauerkraut [6] and sourdough [7] to meat [8, 9], fish [10], and dairy-based products, such as yoghurt [11] and cheese [12]. More examples of fermented foods can be found in Table **4.1**.

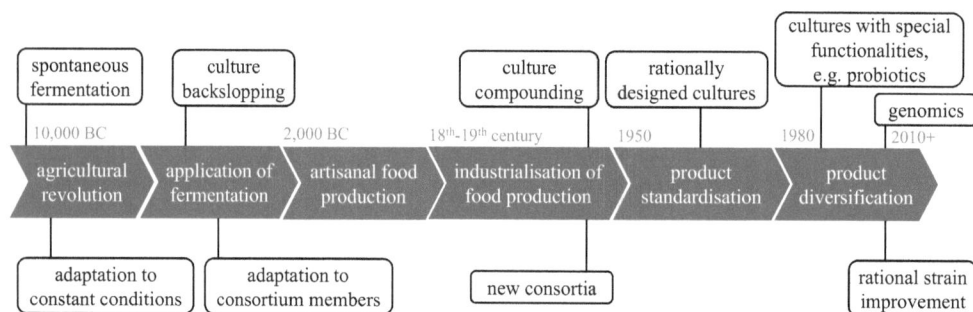

Fig. (4.1). Schematic representation of what happened with food production and its impact on microorganisms and on their use since the beginning of human civilization.

Table 4.1. Examples of fermented foods and the (groups of) microorganisms involved. LAB, lactic acid bacteria. Rhee *et al.* [21] give a good overview of Asian fermented foods in specific.

Product	Origin	Description	Microorganisms Involved	References
Plant based				
Sauerkraut	East Europe and Germany	Sour cabbage	*Lactobacillus, Enterococcus,*	[6, 14]
Sourdough	Europe and Middle East	Dough for bread	*Saccharomyces, Candida, Lactobacillus sanfranciscensis*	[7, 13]
Kimchi	Korea	Fermented vegetables	*Lactobacillus, Leuconostoc, Pediococcus, Weissella*	[22]
Olives	Mediterranean	Olives	*Lactobacillus plantarum, Lactobacillus pentosus*, other *Lactobacillus* spp., yeasts	[23]

(Table 4.1) cont.....

Product	Origin	Description	Microorganisms Involved	References
Wine	Georgia	Fermented grape drink	*Saccharomyces, Hanseniaspora, Pichia, Candida, Metschnikowia, Kluyveromyces, Zygosaccharomyces, Torulaspora, Dekkera, Schizosaccharomyces, Oenococcus oeni*	[24, 25]
Beer	Iraq and Egypt	Fermented grain drink	*Saccharomyces cerevisiae,* other *Saccharomyces* spp., *Kluyveromyces, Rhodotorula, Brettanomyces bruxellensis, Klebsiella, Enterobacter, Escherichia, Citrobacter, Serratia, Pectobacterium* sp., *Pediococcus* sp., *Lactobacillus,* other LAB	[26]
Idli	Southern India	Rice bread	*Leuconostoc*	[21]
Sake	China	Rice wine	*Bacillus, Staphylococcus, Aspergillus flavus* var. *oryzae, Lactobacillus, Saccharomyces cerevisiae*	[27, 28]
Soy sauce	East and South Asia	Mashed soy beans and grains	*Aspergillus oryzae, Aspergillus sojae, Saccharomyces cerevisiae,* LAB, other bacteria	[29, 30]
Meat based				
Salami	Southern Europe	Fermented sausage	*Staphylococcus succinus, Staphylococcus xylosus, Lactobacillus sakei, Leuconostoc carnosum*	[31]
sausage	Southern Europe	Fermented dry sausage	*Lactobacillus sakei, Lactobacillus curvatus,* other LAB, *Staphylococcus xylosus, Penicillium chrysogenum, Penicillium nalgiovense*	[32, 33]
Fish based				
Hákarl	Iceland	Fermented Greenland shark	*Moraxella, Acinetobacter, Lactobacillus*	[10]
Skate	Northern Europe	Fermented ray	*Oceanisphaera, Pseudoalteromonas, Photobacterium, Aliivibrio, Pseudomonas*	[10]
Rakfisk	Norway	Fermented salmon	*Lactobacillus sakei,* other *Lactobacillus* spp., *Psychrobacter*	[10]
Surströmming	Sweden	Fermented herring	*Haloanaerobium*	[10]

(Table 4.1) cont.....

Product	Origin	Description	Microorganisms Involved	References
Plaa-som	Thailand	Fermented freshwater fish	*Pediococcus pentosaceus, Lactobacillus alimentarius, Lactocailuus farciminis, Weisella confusa, Lactobacillus plantarum, Lactococcus garviae, Zygosaccharomyces rouxii*	[34]
Dairy based				
Yoghurt	Iraq and Syria	Acidified milk	*Streptococcus thermophilus, Lactobacillus delbrueckii* subsp. *bulgaricus*	[11]
Gouda-type cheese	Netherlands	Hard cheese	*Lactococcus lactis* subsp. *cremoris* and subsp. *lactis* biovar. *diacelylactis, Leuconostoc mesenteroides*	[12]
Surface-ripened cheese	Europe	Semi-hard/hard Cheese	LAB, *Debaryomyces hansenii, Geotrichum candidum, Arthrobacter, Brevibacterium linens, Corynebacterium ammoniagenes, Staphyolococcus*	[16]
Camembert	France	Mold cheese	*Penicillium camemberti, Geotrichum candidum, Debaryomyces hansenii, Kluyveromyces lactis, Lactobacillus acidophilus, Lactobacillus delbrueckii, Lactobacillus fermentum, Lactobacillus helveticus, Lactobacillus paracasei, Lactobacillus plantarum, Lactobacillus rhamnosus, Lactococcus lactis, Leuconostoc mesenteroides*	[35 - 37]
Kefir	Eastern Europe and Central Asia	Acidified milk	*Lactobacillus kefiranofaciens, Saccharomyces cerevisiae*	[15]
Other basis				
Honey	Unknown	Concentrated sugar from nectar	*Lactobacillus melliventris, Lactobacillus helsingborgensis, Lactobacillus kinbladii, Lactobacillus kullabergensis, Lactobacillus apis, Lactobacillus mellifer, Lactobacillus melis, Lactobacillus kunkeii, Lactobacillus apinorium, Bifidobacterium asteroides, Bifidobacterium melis, Bifidobacterium melliventris, Bifidobacterium corineforme*	[38]

(Table 4.1) cont.....

Product	Origin	Description	Microorganisms Involved	References
Sikhae	Korea	Fermented fish and grains	*Leuconostoc mesenteroides, Lactobacillus, Pediococcus*	[21]

Raw food materials can be fermented for many reasons, for example for shelf-life extension, improvement of organoleptic properties, increasing the availability of nutrients by production of nutrients by the microorganisms or by improved digestibility, and for improved microbial stability, thus improved safety [17]. The metabolic potential of the fermenting microbial cultures will eventually have a huge impact on the exact characteristics of the product. With their fermentative metabolism, levels of organic acids or alcohols in the final product are often sufficient to limit growth of other microorganisms, leading to longer product shelf-lives and reduced chances of the presence of pathogens. This is discussed more in the section on classifying microbial interactions. Additional effects of acids and alcohols are the changes in taste and texture. For example, in milk the lactic acid produced by lactic cultures causes coagulation of caseins, which make up about 80% of the protein content of cow milk. The coagulated protein is referred to as curd and the remaining liquid, containing whey proteins, is called whey. Coagulation by lactic cultures, enzymatic treatment with the enzyme rennet or a combination thereof is the first step in cheese production. In the case of casein precipitation by alcohol, the protein forms complexes with the calcium ions present in milk instead of calcium being released as is the case with acid coagulation [18]. Other ways in which cultures can contribute to texture are for example the production of exopolysaccharides (EPS) that interact with other macromolecules in the food matrix to form a network that can trap liquid [19], and the production of gasses that provide air cavities in the food matrix [20]. With respect to flavor, not only the main fermentation end products contribute to the taste and aroma of food, but also the metabolic intermediates and breakdown products of food matrix macromolecules. While there are only five basic tastes perceived by the tongue, *i.e.* sweet, sour, bitter, salty, and umami, the aroma components perceived by the olfactory system in the nose are in principle unlimited. Their interplay, together with the textural and other organoleptic properties, such as sound and sight, determine how the product is perceived by the consumer.

Traditionally, cultures used for food fermentation are propagated by back-slopping and this is still common practice for artisanal products and part of the ones produced at industrial scale. The often-long history of repeated propagation under more or less stable conditions mainly produced cultures of a mixture of many species and strains of which the composition always stays within certain limits [17]. This stability is considered a great advantage for producing fermented foods with a consistent quality. On the other hand, the use of direct vat inoculation (DVI) cultures is steadily increasing. These cultures generally have a limited complexity, allowing the culture suppliers to have better control over the characteristics of their vendibles. With limited complexity, it will for instance be easier to add one strain or substitute one strain for another with different properties, paving the way for product diversification. Another advantage of the DVI cultures is that single strains can be produced and applied to food products, for example strains with proven specific probiotic properties allowing a probiotic product claim, and strains with specific antimicrobial properties extending the shelf-life by counteracting the development of spoilage microbes. The disadvantage of cultures with limited complexity, however, is that they are more prone to fermentation failure due to sensitivity to small changes in environmental factors or to phage predation. The standard method to deal with phage risk is to apply culture rotation. Here, cultures with similar functionalities but different phage sensitivities are alternately applied, giving phage no chance to multiply to levels high enough to cause problems. More methods to counteract phage are discussed below.

LACTIC ACID BACTERIA CULTURES

The lactic acid bacteria (LAB), or lactobacillales, are a clade of rod or coccus shaped bacteria that have lactic acid as their main fermentation end product. This group consists of the thirteen genera *Abiotrophia, Aerococcus, Carnobacterium, Enterococcus, Lactobacillus, Lactococcus, Leuconostoc, Oenococcus, Pediococcus, Streptococcus, Tetragenococcus, Vagococcus* and *Weissella*, of which *Lactobacillus, Lactococcus, Leuconostoc* and *Streptococcus* are the most widely spread among food fermentations. Although bifidobacteria belong to another phylum, they are also considered lactic acid bacteria. The most important reason to use LAB is their fermentative metabolism. High levels of weak acids, such as lactic acid are effective growth inhibitors of most microorganisms, thereby limiting spoilage and outgrowth of pathogenic microbes [39]. LAB can be divided into three groups based on their metabolic end products: obligate homofermentatives, facultative heterofermentatives, and obligate heterofer-mentatives. While homofermentative metabolism typically results in lactic acid as the main carbohydrate dissimilation product, heterofermentative metabolism generally results in equimolar quantities of lactic acid and ethanol, but acetic acid formation is also widely spread. This difference between homofermantative and heterofermentative metabolism is caused by the presence of enzymes in the Embden-Meyerhof pathway (aldolase, homofermentative) and the pentose phosphate pathway (phosphoketolase, heterofermentative). The presence of a functional pentose phosphate pathway is important but not essential for biosynthesis of nucleic acids, fatty acids, and aromatic amino acids as well as for generating NADPH that can prevent against oxidative stress. It must be noted that even LAB with homofermentative metabolism do not solely produce lactic acid as the fermentation end product. It is common that pathways using pyruvate lead to other organic acids, such as succinic acid and formic acid as shown in *Streptococcus thermophilus* [40]. In addition, the primary metabolism may also produce typical aroma components, such as acetaldehyde and diacetyl [41], which are the main flavor components in yoghurt and butter, respectively. Some of the products from primary metabolism may act as intermediates, *e.g.* in cross-feeding interactions between different species of microorganisms in mixed cultures.

Although LAB can only generate energy by converting sugars, indicating that they typically require an environment rich in carbohydrates, they are considered versatile microorganisms, but the exact extent of versatility depends very much on their habitat and evolutionary background. For example, the genome of *Lactobacillus plantarum*, found on many different plants and in the human oral cavity, encodes many sugar transport systems [42] and some strains even express amylases in order to degrade plant polysaccharides [43]. This is an example indicating that LAB versatility is mainly related to effective use of environmental resources rather than producing biomolecules *de novo*. This not only counts for carbohydrate metabolism, but also for nucleobase, fatty acid, and amino acid metabolism; most LAB species lack genes in pathways for biosynthesis of many of these components, rendering the pathway non-functional except if an intermediate is supplied by an external source.

The versatility of LAB is also reflected in their wide range of habitats. Table **4.1** is dominated by LAB species and indicates that they are present in food products from plant, meat, fish, and dairy origin. Additional to their natural appearance in decaying plant material, many species live as commensals on or in the bodies of animals, such as in the gastrointestinal tract. Their prevalence in dairy is hypothesized to be a result of plant resident species that have adapted to the rich and relatively constant dairy environment [44], in some cases leading to severe reductions in gene content and metabolic potential [45]. There are not many examples of (opportunistic) pathogens among LAB, and many LAB have received GRAS (generally regarded as safe) status for application in food. Still, some of them, in particular streptococci, are close relatives of pathogenic bacteria. This close family relation, however, may exactly be the reason why some LAB are so effective in colonizing the human intestine, thereby reducing the chance of infection by pathogens. LAB with this special probiotic property have to be used with consideration though, as there are also reports of patients that developed sepsis and bacteraemia following probiotic administration [46]. Probiotic bacteria are further discussed in a separate section. There are many LAB species that produce exopolysaccharides. In many fermented foods, in particular those that have a dairy basis, these EPS play important roles in texturizing the food matrix, such as in yoghurt [47], or increasing the water holding capacity thereby providing softness or stability, as is the case with cheese [48]. EPS, which basically are large molecules made up of sugar monomers, can occur in two different types, capsular (forming a capsule around the bacterium, also referred to as capsular polysaccharides or CPS) and secreted into the environment. Moreover, EPS can be branched or unbranched, and they may contain only one type of monosaccharide (homo-polysaccharides) or repeated units of different monosaccharides (hetero-polysaccharides), mostly glucose, galactose, and fructose [49]. Although mainly a polysaccharide, an EPS molecule can also contain non-carbohydrate parts, such as pyruvate, succinate, acetate, and acyl groups from other oxoacids. The polysaccharide normally has electrostatic interactions with DNA molecules, proteins, and other (macro-) molecules in the environment and thereby aids in building and maintaining structure in (semi-) liquid environments. In the case of fermented dairy, such as yoghurt, it is particularly the interaction between EPS and some of the milk proteins that determine key product characteristics, such as viscosity and mouthfeel [50]. With such an enormous variety in the types of EPS and the possible interactions with other molecules, it is not surprising that the exact type and amount of EPS produced by a microbial culture largely determine these product characteristics. For example, the high acylated form of gellan, a type of EPS isolated from the bacterium *Sphingomonas elodea*, was shown to interact with the proteins from a whey protein fraction but not casein, while the effect of low acyl gellan was opposite [51]. Similar observations were made on EPS produced by different commercial yoghurt cultures [50]. It is therefore always important for food companies to evaluate the performance of a culture in the given food matrix rather than assuming that high EPS production naturally leads to high viscosity. Similarly, the large variety of EPS allows exact modulation of the structure of yoghurt by picking the right combination of culture, ratio between whey protein and casein, and the optimal fermentation conditions.

An important functionality of LAB in food fermentations is the ability to produce flavor components. These are often by- or end-products of primary metabolism, amino acid conversion, or protein breakdown [52]. In cheese, it is often the intracellular enzymes that are released upon (auto)lysis that largely determine the flavor development during ripening [41].

LAB are also applied for their potential to produce healthy metabolites, such as vitamins. For example, yoghurt has an elevated folic acid content compared to plain milk as a result of production by the LAB *Streptococcus thermophilus*. In addition, if the bacteria that ferment a certain product do not produce a certain component, but this is wanted, it is relatively simple to re-route their metabolism towards becoming a producer. This is particularly facilitated by the increasing availability of genome sequences and genome modification tools, which include both GMO (genetic modified organism) and non-GMO approaches. An example here comes from *Lactobacillus gasseri* that changed from a folate consumer to a producer by transformation with a plasmid containing five genes from the *Lactococcus lactis* folate biosynthesis pathway [53].

YEAST AND MOLD CULTURES

Yeasts and filamentous fungi (molds) are two important groups in industrial biotechnology; both belong to the kingdom of fungi. In particular, molds are used widely for the breakdown of complex substrates, due to their often-versatile metabolic network and their resilience, and for the production of complex molecules and heterologous enzymes that find many applications. Therewith, they often pose viable alternatives for chemical production of, for example, active components used in the pharmaceutical industry. Moreover, yeasts and molds are common residents of raw substrates used in food production and they are often applied in fermented foods, either as fermenting microorganism or as supplementary culture providing important product characteristics, such as taste and shelf-life.

Within the yeast family *Saccharomycetaceae*, the genera *Saccharomyces* and *Kluyveromyces* are the most common in fermented foods, and in particular *Saccharomyces cerevisiae* is well studied and widely applied. Examples of foods in which this species is used are bread, beer, and wine [7, 25, 26]. In bread, it is applied to provide taste and structure. *S. cerevisiae* is capable of producing significant amounts of carbon dioxide, which forms air bubbles in the unbaked dough and this leads to a less firm structure when the dough is baked. In addition, the produced levels of ethanol and organic acids, notably acetic acid and succinic acid, impact the agglomeration of gluten thereby influencing the dough extensibility [54]. Moreover, it was shown that the volatile profile, and therewith taste of bread depends very much on the exact strain used for leavening and its interaction with flour and other raw materials used in bread making [55]. This indicates the importance of picking the right species or strain for acquiring desired product characteristics. In beer and wine fermentation, it is the alcohol produced during fermentation that is considered the main functionality of yeast, but also the choice of culture has a large influence on the final taste [56]. Moreover, in beer fermentation it is of key importance that the used microorganism – *Saccharomyces* species, other yeasts or bacteria – are tolerant to the antimicrobial components that are present in hop [57].

The characteristics that differentiate molds from yeasts and bacteria are (i) their multicellular filamentous growth allows building of structures (*e.g.* in tempeh or at the surface of white mold cheese), (ii) their relatively high proteolytic and lipolytic activities, (iii) their ability to grow in dry environments (water activity as low as 0.62), and (iv) the fact that they are mostly obligate aerobes [58]. Moreover, many molds have the ability to use organic acids as carbon source, making them possible contaminants of acid fermented foods or consortium members together with LAB in products with sequential proliferation of different microorganisms. This means that molds give functionalities to fermentations and fermented products that are not possible with yeasts and bacteria. Moreover, it is possible to ferment under harsher conditions. It also means that care must be taken to prevent contamination with molds in many cases. In terms of food fermentation, the most industrially relevant filamentous fungi are the genera *Aspergillus, Geotrichum,* and especially *Penicillium*. Furthermore, *Mucor* and *Bothrytis* species are also sometimes used, for example in various oriental fermented foods and wine. Except for some species being pathogenic and many species occurring as spoilage microorganisms, the versatile – often complementary to that of lactic acid bacteria – metabolism and possible production of antimicrobial components establish *Aspergillus, Geotrichum,* and *Penicillium* species as solid members of food-resident microbial communities. In mold cheeses, such as camembert, brie (*G. candidum* and *P. camemberti*), and roquefort (*P. roqueforti*), the fungi have a ripening function. During this ripening, the mold grows mostly on the surface or in pinched holes, because they require oxygen for optimal proliferation, and produce metabolites and enzymes that determine the characteristic flavor and texture of the cheese [59]. The latter is mainly a result of proteolysis. The initial firmness of the cheese is mainly due to the protein network, which is broken down by proteolytic enzymes secreted by the molds, thereby releasing peptides and amino acids. These, together with broken down fatty acids by the mold's lipolytic activity, are the basis of aroma metabolites and provide both the bitter taste and ammonia smell that characterizes ripe mold cheese [59, 60]. *Penicillium* species, for example *P. chrysogenum* and *P. nalgiovense*, also play a role in ripening sausages, where their metabolic activities, among which especially proteolysis and lipolysis, lead to aroma formation and texture softening. Moreover, they contribute to shelf-life by producing antimicrobials [33]. In fermented rice and sake (fermented rice wine), both are very known products in East Asia, the molds belonging to the *Monascus* species group [61], and *A. oryzae* are most commonly applied [27, 28]. Their primary action is to degrade polysaccharides, mainly starch, into small monomers and oligosaccharides and subsequent fermentation of these sugars to primarily ethanol. In contrast to mold cheese, the fungi are the drivers behind the fermentation of rice. However, the ethanol production and flavor development occur often in combination with *S. cerevisiae* and LAB (Table **4.1**). Another example of co-fermentation of yeasts, especially *S. cerevisiae* and *Aspergillus* molds, notably *A. oryzae* and *A. sojae*, is soy sauce, although bacterial species may also be part of the community [29, 30]. Here, the function of the yeast is mainly to produce ethanol, which can be converted to numerous flavor components. The molds are responsible for the main functionality of the sauce; their high amino acid biosynthetic capabilities and proteolytic activity release vast amounts of glutamic acid and peptides from soy and wheat or other grain proteins giving the typical umami taste [62, 63]. Finally, a remark should be made on the selection or development of the right strains of yeast and mold, because they might produce potentially unhealthy compounds [27, 64].

PROBIOTIC CULTURES

Probiotics are live microbial food supplements, which benefit the health of consumers by maintaining or improving their intestinal microbial balance [65]. Due to their perceived health benefits, probiotic bacteria have been increasingly included in yoghurts and fermented milks. Most commonly, they are LAB, such as *Lactobacillus acidophilus* and *bifidobacteria,* often referred to as 'bifidus'. There is growing scientific evidence to support the concept that the maintenance of healthy gut microbiota may provide protection against gastrointestinal disorders including gastrointestinal infections, inflammatory bowel diseases, and even cancer [66 - 68]. The use of probiotic bacterial cultures stimulates the growth of preferred microorganisms, crowds out potentially harmful bacteria, and reinforces the body's natural defense mechanisms. Today, plenty of evidence exists on the positive effects of probiotics on human health. Table **4.2** summarizes a few of such beneficial effects.

Table 4.2. Clinical effects of some probiotic strains and yoghurt strains [124].

Strain	Clinical Effects in Humans	References
Lactobacillus rhamnosus GG (ATCC 53103)	Reduction of antibiotic-associated diarrhoea in children, treatment and prevention of rotavirus and acute diarrhoea in children, treatment of relapsing *Clostridium difficile* diarrhoea, immune response modulation, alleviation of atopic dermatitis symptoms in children	[69 - 72]
Lactobacillus johnsonii (*acidophilus*) LJ-1 (La1)	Modulation of intestinal flora, immune enhancement, adjuvant in *Helicobacter pylori* treatment	[73 - 75]
Bifidobacterium lactis BB-12	Prevention of traveller's diarrhoea, treatment of rotavirus diarrhoea, modulation of intestinal flora, improvement of constipation, modulation of immune response, alleviation of atopic dermatitis symptoms	[73, 76]
Lactobacillus reuteri (BioGaia Biologics)	Shortening of rotavirus diarrhoea in children, treatment of acute diarrhoea in children	[77 - 79]

(Table 4.2) cont.....

Strain	Clinical Effects in Humans	References
Lactobacillus casei Shirota	Modulation of intestinal flora, lowering faecal enzyme activities, positive effects on superficial bladder cancer and cervical cancer, no influence on the immune system of healthy subjects	[80, 81]
Lactobacillus plantarum DSM9843 (299v)	Modulation of intestinal flora, increase in faecal short-chain fatty acid content	[82, 83]
Yoghurt strains (*Streptococcus thermophilus* and/or *L. delbrueckii* subsp. *bulgaricus*)	No effect on rotavirus diarrhoea, no immune enhancing effect during rotavirus diarrhoea, no effect on faecal enzymes	[75, 84]

Selection Criteria of Probiotics

The selection criteria for a strain to be used as 'probiotic' include the following abilities to: (i) exert a beneficial effect on the host; (ii) survive in foodstuff at high cell counts and remain viable throughout the shelf-life of the product; (iii) survive passage through the upper gastrointestinal (GI) tract and arrive alive at its site of action and function in the gut environment; (iv) adhere to the intestinal epithelium cell lining and colonize; (v) produce antimicrobial substances towards pathogens; and (vi) stabilize the intestinal microflora and be associated with health benefits.

Probiotics must have a good shelf-life in food or preparations containing a high number of viable cells at the time of consumption and should not be pathogenic or toxic to the consumer. The most extensively studied and widely used probiotics are the LAB, particularly the *Lactobacillus* and *Bifidobacterium* species as detailed in Table **4.3**.

Table 4.3. The most commonly used species of lactic acid bacteria in probiotic preparations.

Lactobacillus sp.	*Bifidobacterium* sp.	*Enterococcus* sp.	*Streptococcus* sp.
L. acidophilus	*B. bifidum*	*Ent. Faecalis*	*S. cremoris*
L. casei	*B. adolescentis*	*Ent. Faecium*	*S. salivarious*
L. delbrueckii ssp. (bulgaricus)	*B. animalis*		*S. diacetylactis*
L. cellobiosus	*B. infantis*		*S. intermedius*
L. curvatus	*B. thermophilum*		
L. fermentum	*B. longum*		
L. ruteri			
L. brevis			

Functional Properties of Probiotics

While selecting a preferable probiotic strain, several aspects of functionality have to be considered, such as adhesion ability to the host intestinal tract, immunoregulatory, antagonistic, anti-mutagenic, and anti-carcinogenic properties. These properties are discussed in detail below.

Adhesion of probiotic strains to the intestinal surface and subsequent colonization of the human GI-tract have been suggested as an important pre-requisite for probiotic action. Adherent strains of probiotic bacteria are likely to persist longer in the intestinal tract and thus have better possibilities of showing metabolic and immunomodulatory effects than non-adhering strains. Adhesion provides an interaction with the mucosal surface facilitating the contact with gut associated lymphoid tissue and mediating local and systemic immune effects. Thus, only adherent probiotic strains have been thought to effectively induce immune effects and stabilise the intestinal mucosal barrier [85]. Adhesion may also provide means of competitive exclusion of pathogenic bacteria from the intestinal epithelium; exclusion of pathogens by lactic acid bacteria and bifidobacteria has been shown *in vitro* using Caco-2 and HT-29-MTX cell lines [86 - 88]. In the inhibition of pathogen, adhesion *in vitro* of both living and heat-killed *L. acidophilus* cells has been effective [89]. Effective adhesion is mediated by many different cell surface resident proteins that recognise the specific surfaces of human intestinal epithelial cells, which fit their niche. In particular, fibronectin, a multidomain glycoprotein that among others plays a role in cell adhesion in the human body, is a common target for a range of specific fibronectin-binding proteins present on the bacterial surface, also referred to as adhesins [90]. Many of these adhesins thereby also play significant roles in the modification of signalling pathways inside the host cells. These specific interactions between adhesins and other bacterial cell surface proteins, on one side, and receptor proteins on the intestinal epithelial cells is very important for a better understanding of the mechanism through which probiotic microorganisms exert their health-promoting functions [91]. Moreover, the exact nature and effect of these interactions may aid in the design of the most appropriate and effective products containing probiotics.

Additional to intestinal epithelial cells, gut associated lymphoid tissue may have contact with adhesive probiotic strains and their components and in this way adhesion provokes immune effects. Human studies have shown that probiotic bacteria can have positive effects on the immune system of their host. However, differences between probiotic bacteria in respect to their immunomodulatory effects have been observed [92].

To have an impact on the colonic microbiota, it is important for probiotic strains to show antagonism against pathogenic bacteria *via* antimicrobial substance production or competitive exclusion [93 - 95]. *in vitro Lactobacillus rhamnosus* GG produces some low molecular weight antimicrobial products, *i.e.* short chain fatty acids with inhibitory activity against *Clostridium, Bacteroides, Enterobacteriaceae, Pseudomonas, Staphylococcus,* and *Streptococcus*, but not against other lactobacilli [96].

Anti-mutagenic and anti-carcinogenic properties of probiotic bacteria have been widely studied for a number of years. Bacterial anti-carcinogenic properties are considered to represent one or more of the following types: binding and degradation of (pro-) carcinogens, production of anti-mutagenic compounds, modulation of pro-carcinogenic enzymes in the gut, and suppression of tumours by an immune response mechanism [97 - 100].

Probiotic Bacteria as Functional Starter Culture

The market for food products with health-promoting properties, *i.e.* functional foods, has shown a remarkable growth over the last few years; it increased by ca. 25% in 2017 as compared to data from 2013 [101]. Besides, the use of food additives is regarded as unnatural and unsafe [102]. Yet, additives are needed to preserve food products from spoilage and improve the organoleptic properties. The demand for a reduced use of additives and processing seems contradictory with the market preference for products that are fresh, safe, tasty, low in sugar, fat, and salt, and easy to prepare. In cheese-making, for instance, the use of raw milk permits the manufacture of high-value traditional artisan varieties but brings about safety risks, *e.g.* the development of *Listeria monocytogenes*. On the other hand, pasteurization of the milk results in loss of flavor.

In food fermentation, one of the key points for intervention seems to be on the level of the starter culture. The increased understanding of the genomics and metabolomics of food microbes opens perspectives for starter improvement. Through molecular biology, it is now possible to express desirable properties of starter cultures and suppress their undesirable characteristics [103].

Functional starter cultures are starters that possess at least one inherent functional property. The latter can contribute to food safety and/or offer one or more organoleptic, technological, nutritional, or health advantages. The implementation of carefully selected strains as starter cultures in fermentation processes can help to achieve *in situ* expression of the desired property, maintaining a perfectly natural and healthy product. Examples are probiotic LAB strains that are able to produce antimicrobial substances, sugar polymers, sweeteners, aromatic compounds, useful enzymes, or nutraceuticals with health-promoting properties.

This represents a way of replacing chemical additives by natural compounds [104 - 107]. To fully exploit the functional properties of probiotic bacteria, the processes used to manufacture dairy products must be modified to meet the requirements of the probiotics. When this is not possible, other probiotic strains must be tested or new products must be developed.

Main Factors Influencing the Application of Probiotics in Dairy Products

Fermented milk and cheese are the most common foods that contain probiotics. For probiotic strains to be used in fermented dairy products, they must possess good technological properties so that they can be incorporated into food products without losing viability and functionality or creating unpleasant flavors and textures.

In probiotic fermented dairy products mainly *Lactobacillus* and *Lactococcus* species, but also yeast and propionic acid bacteria are used [108]. Active microorganisms interact intensively with their environment by exchanging components of the medium for metabolic products. Thus, the chemical composition of the dairy product is of paramount importance for the metabolic activities of the microorganisms. Essential factors include the type and amount of carbohydrates available, the degree of hydrolysis of milk proteins (which determines the availability of essential amino acids), and the composition and degree of hydrolysis of milk lipids (which determines the availability of short-chain fatty acids in particular) [109]. On the other hand, the proteolytic [110] and lipolytic properties of probiotics may be important for further degradation of proteins and lipids. These two properties may have considerable effects on the taste and flavor of dairy products [109].

The main commercial probiotic culture preparations are supplied in highly concentrated forms, and most of them are constructed for DVI application [111]. Use of these highly concentrated DVI cultures is common due to the difficulties involved in propagating probiotic microorganisms at the production site. The DVI cultures are supplied either as highly concentrated frozen cultures or as freeze-dried cultures. The cultures are filled in gas and light proof containers to protect the cultures against light and humidity. Most often alu-foil coated cartons or pouches are used. Usually deep-frozen cultures contain more than 10^{10} cfu g^{-1}, whereas freeze-dried cultures typically contain more than 10^{11} cfu g^{-1} [112]. The cell concentration per gram of product varies with the culture and the type of organism used.

The physiological state of the probiotics is of special importance which is determined by how fermentation is terminated. Several investigations showed that bacteria from the logarithmic phase are much more susceptible to environmental stresses than bacteria from the stationary phase. The environmental factors that signal the transition from the logarithmic phase to the stationary phase may have a considerable effect on survival rates during the stationary phase [88, 113 - 116]. Thus, a starvation signal, triggered by depletion of carbon sources, appears to be much more favorable for survival than a low pH in the presence of sufficient carbon sources.

Probiotic Interaction with Starter Culture Strains

Commercially available probiotic cultures may consist of a single strain or a mixture of several strains, mainly lactic acid bacteria and *bifidobacteria* [117]. In most cases, the probiotic properties are affected by the way the culture has been produced [118]. Therefore, specific information on strain-specific properties becomes very important for process optimization.

In fermented probiotic products, it is important that the probiotic culture used contributes to good sensory properties. Therefore, it is quite common to use probiotic bacteria together with other types of bacteria (starters) for the fermentation of the specific product. The interactions between probiotic and starter strain generally have an impact on the product quality. It has been shown that it is possible to produce fermented dairy products with excellent sensory properties and good survival of the bacteria by using starter and probiotic organisms together. Suitable starters might be *Streptococcus thermophilus*, yoghurt cultures, and mesophilic starters with different combinations of *Lactococcus* strains. The most suitable combination of starter and a specific probiotic bacterium has to be determined using a screening process evaluating the impact of different starters on the sensory properties and on the survival of the probiotic strain [119].

Although little is known about this interaction, both synergistic and antagonistic effects between different starter organisms are well established. For example, the below discussed classic yogurt culture is characterized by a symbiosis between *Streptococcus thermophilus* and *Lactobacillus delbrueckii* subsp. *bulgaricus*. This synergism, seen as an accelerated and efficient acidification of the milk and multiplication of the culture organisms and based on cross-feeding of both organisms, is not a property of the 2 species but of specific strains of these species [120 - 122]. Antagonism, on the other hand, is often based on the production of substances that inhibit or inactivate specifically other related starter organisms or even unrelated bacteria. Most importantly, antagonism can be caused by bacteriocins, which are peptides or proteins exhibiting antibiotic properties [123, 124]. The ability to produce bacteriocins is often discussed as a desirable property of probiotics [125]; however, antagonism to starter cultures and vice versa may be a limiting factor for combinations of starters and probiotics [126]. Further antagonistic activities produced by lactic acid bacteria have been described and the substances involved are hydrogen peroxide, benzoic acid, biogenic amines (formed by decarboxylation of amino acids), and lactic acid [127 - 129]. An example of a positive interaction between a starter culture, in this case *Saccharomyces cerevisiae*, and a probiotic bacterium, in this case *Lactobacillus rhamnosus*, led to enhanced viability of the latter in dairy products with high acid levels and ambient storage [130, 131]. This was shown to be attributed to direct cell-cell contact and co-aggregation with the yeast cells and/or cell wall components or metabolites produced by the yeast.

An overview of the probiotic starter bacteria most commonly applied in dairy fermentations and some of their relevant physiological properties is given in Table **4.4** and more details on microbial interactions are described in a separate section below.

The intensity of the interactions between probiotics and both the food matrix and the starter organisms largely depends on the time that probiotics are added to the product, *i.e.*, whether they are present during fermentation or are added afterwards. In the latter case, interactions may be minimal because addition may occur immediately before or even after cooling below 8°C. The metabolic activity of starters and probiotics is drastically reduced at these temperatures. However, with extended storage, even small interactions may yield measurable effects [132].

Table 4.4. The most commonly used species of probiotics and their fermentation characteristics.

| Species* | Growth Temperature (°C) | | | Lactic Acid Fermentation | | | |
	Min.	Optimal	Max	Homo-Fermentative	Hetero-Fermentative	Lactic Acid (%)	pH
Lb. delbrueckii subsp. *bulgaricus*	22	45	52	+		1.5–1.8	3.8
Lb. delbrueckii subsp. *lactis*	18	40	50	+		1.5–1.8	3.8
Lb. helveticus	22	42	54	+		1.5–2.2	3.8
Lb. acidophilus	27	37	48	+		0.3–1.9	4.2
Lb. kefir	8	32	43		+	1.2–1.5	—
Lb. brevis	8	30	42		+	1.2–1.5	—
Lb. casei subsp. *casei*		30		+		1.2–1.5	—

(Table 4.4) cont.....

Species*	Growth Temperature (°C)			Lactic Acid Fermentation			
	Min.	Optimal	Max	Homo-Fermentative	Hetero-Fermentative	Lactic Acid (%)	pH
S. thermophilus	22	40	52	+		0.6–0.8	4.5
Lc. lactis subsp. *lactis*	8	30	40	+		0.5–0.7	4.6
Lc. lactis subsp. *cremoris*	8	22	37	+		0.5–0.7	4.6
Lc. lactis subsp. *lactis* biovar.*diacetylactis*	8	22–28	40	+		0.5–0.7	4.6
Ln. mesenteroides subsp. *cremoris*	4	20–28	37		+	0.1–0.2	5.6
Ln. mesenteroides subsp. *dextranicum*	4	20–28	37		+	0.1–0.2	5.6
Bifidobacterium (*bifidum, infantis* etc)	22	37	48			0.1–1.4	4.5

*Lb: *Lactobacillus*; S: *Streptococcus*; Lc: *Lactococcus*; Ln: *Leuconostoc*.

In selecting starter microorganisms, reliable acid-forming ability is one of the most important characteristics. As the environment within the GI-tract and within the food might be quite different, the probiotic strains are often not suitable as a starter organism [112, 118]. The growth rate might be too slow and they might give off-flavors [107].

For milk-based products, the probiotic strains are often mixed with *S. thermophilus* and *L. bulgaricus* to achieve the desired flavor and texture. In many cases the consumers find products fermented with *L. bulgaricus* too acidic and with too heavy acetaldehyde flavor (yoghurt flavor). Therefore, mixed probiotic cultures have been developed to bring out the preferred flavors in the products in which they are used. Examples of such cultures are the so called ABT cultures (ABT standing for *L. acidophilus, Bifidobacterium* and *S. thermophilus*) [133].

Another possibility is to improve the suitability of the food as a substrate for the probiotic by adding energy sources (*e.g.* glucose), growth factors (*e.g.* yeast extract and protein hydrolysates) or suitable antioxidants, minerals, or vitamins [134, 135].

Efficacy and Safety of Probiotics

The US Food and Drug Administration (FDA) considers probiotic organisms added to food as GRAS. In spite of inherent difficulties, establishing good measures of probiotic efficacy [136], studies on lactose intolerance, diarrhea, and colon cancer show that a daily dose of lactic acid bacteria is needed for any measurable effect [137]. Unfortunately, the concentration of probiotics in food products varies tremendously and there are currently no set standards to identify the levels of live probiotic bacteria required in yogurt or other fermented food products. Epidemiological data on the safety of dairy products [136, 138, 139] and a thorough review of the safety data on probiotics [140] suggest no evidence of probiotics being involved in human infections. The food industry needs to carefully assess the safety and efficacy of all new species and strains of probiotics before incorporating them into food products.

MIXED CULTURES

By far most of the commercial food cultures and especially the artisanal ones, maintained by back-slopping, are mixed cultures. These can be mixtures of different species and different strains of the same species. Typically, microbial consortia are more stable than mono cultures, just like any other ecosystem [141]. This is a result of (i) communication between the consortium members, (ii) division of the labor, and (iii) an improved resistance against disease and other disturbing factors [17]. For example, large areas of mono cultures of crops are more prone to disease then when the farmland is filled with a variety of crops, both in space and in time, simply because the disease can spread easier [142]. In the case of fermentation with a bacterial culture, bacteriophage attack would disrupt the fermentation completely if all cells are susceptible to the given phage. It is well recognized that food fermentations, especially those repeatedly carried out in the same equipment, are highly vulnerable to phage predation. This predation may suddenly inactivate important strains in a fermenting culture, leading to failure and product losses [143]. In this case, a collection of closely related strains, each having the same functionality and different bacteriophage sensitivity, will still results in an effective fermentation. If the phage predates one of the strains, the other strains still function. This is also the basis behind many complex starter cultures, where density dependent phage predation ensures that none of the strains can become dominant [144]. Phages only thrive if they find hosts to multiply in. In a complex consortium, it is either unlikely that one strain is present in such numbers that the phage wipes out a significant fraction of the whole culture, or the phage attacks and reduces the cell counts of the dominant strain, allowing the outgrowth of other strains with similar functionalities but different phage sensitivities. Since the separate strains will either increase or decrease in relation to other strains in the mixed consortium as a result of both phage and many other disturbing environmental factors, long-term propagation of this culture will not lead to any significant strain loss. The increased stability of a culture when it is made up of many species or strains does, thus, not only count in the case of phage predation, but also in the case of variations in environmental factors, such as pH, water activity, temperature, oxygen pressure, and nutrient availability. Some strains may be more susceptible to extremes of these factors or have specific nutritional requirements, while other strains thrive well and perform worse under other conditions. This increased stability of mixed cultures compared to mono cultures is called robustness.

Classifying Microbial Interactions that Structure Consortia

Additional to phage attack and abiotic factors, there is another factor structuring microbial communities and that is how the microbes interact with their environment, more precisely with other microbes. This includes both communication and interactions related to labor division. These interactions exist at multiple levels, both directly (involving physical contact or growth factor exchange) and indirectly (through metabolites that can act as signals to other microorganisms or as effects induced by changed physico-chemical properties of the environment) [145, 146]. Moreover, interactions can be directed both towards the own species or strain and towards others, and can have a positive, neutral, or negative influence on both the giver and the receiver (Table **4.5**). In this way, it is possible to classify interactions based on their mutually beneficial and detrimental effects on fitness, *i.e..* their competitive advantage [147, 148]. The interactions can be divided into six main classes: neutralism, amensalism, competition, commensalism, parasitism and mutualism. An overview of microbial interactions, divided over these classes, from several food systems is found in Table **4.6**. Here, we mainly focus on the interactions that involve multiple species with some examples from the food industry.

Table 4.5. Overview modes of microbial interaction and their effects on the receiver. +, positive effect;-, negative effect; N, neutral or possible positive and possible negative effect.

Targets	Intentional	Unintentional
Own species or strain	Quorum sensing (+)	Acid or alcohol production (-) Competing for (micro)nutrients (-)
Other species or strains	Cross-feeding (+) Bacteriocin, antibiotic (-) Reactive oxygen species (-) Quorum sensing (N)	Acid or alcohol production (-) Cross-feeding (+) Competing for (micro)nutrients (-)

Neutralism is defined as the complete absence of interactions and could therefore also not be considered as a class of interactions. In practice, neutralism will not take place within microbial communities, because the changing physico-chemical properties of the environment as a result of one microorganism's metabolism will always affect the metabolism of all the consortium members. It is, however, not always easy to identify these effects if they are not directly related to growth or survival. In that sense, if interactions are classified based on their effects on fitness, neutralism can also be considered a type of interaction without any direct linkage to fitness.

Table 4.6. Selection of microbial (interspecies) interactions observed in various food products, taking place either during fermentations or in the fermented products, and their effects on the products. (Adapted from Sieuwerts *et al.* [147]). LAB, lactic acid bacteria. (x/y) indicates effect on effector (x) and receiver or target (y) with 0 being neutral, + being positive and - being negative.

Products	Effectors	Targets	Effects on Targets and Products	References
Neutralism (0/0)				
No obvious examples				
Amensalism (0/-)				
Yoghurt	Yoghurt starter	*Listeria monocytogenes, Staphylococcus aureus*	No presence of pathogens due to bacteriocin	[162]
Yoghurt	*Streptococcus thermophilus, Lactobacillus bulgaricus*		Inhibition of each other by bacteriocin production	[163 - 165]
Cheese	*Lactobacillus plantarum*	*Staphylococcus aureus, Salmonella typhimurium*	No spoilage due to lactic acid	[166]
Surface-ripened cheese	*L. plantarum*	*Listeria monocytogenes*	No presence of pathogens due to bacteriocin	[151]
Sausage	*Lactobacillus curvatus, Lactococcus lactis*	*Listeria monocytogenes*	No presence of pathogens due to bacteriocin	[167]
Wine	*Lactobacillus hilgardii*	*Pediococcus pentosaceus*	No spoilage due to lactic acid and hydrogen peroxide	[168]
Lettuce	*LAB*	*Listeria monocytogenes*	No presence of pathogens due to lactic acid and bacteriocin	[151]
Various	*Lactococcus lactis*	*Aspergillus parasiticus*	No presence of toxin due to lactic acid	[169]
Various	LAB	*Listeria monocytogenes*	No presence of pathogens due to bacteriocin	[170, 171]

(Table 4.5) cont.....

Products	Effectors	Targets	Effects on Targets and Products	References
Various	LAB	*Salmonella enteritidis*	No presence of pathogens due to lactic acid	[172]
Competition (-/-)				
Yoghurt	*Streptoccus thermophilus, Lactobacillus bulgaricus*		Suboptimal performance due to competition for nutrients	[173]
Fermented milk	LAB, Yeasts		Suboptimal performance due to competition for nutrients	[174, 175]
Commensalism (0/+)				
Cheese	LAB	Propionibacteria	Consumption of lactic acid, produced by LAB, by propionibacteria	[176]
Surface-ripened cheese	LAB, *Debaryomyces hansenii, Geotrichum candidum*	*Debaryomyces hansenii, Geotrichum candidum, Arthrobacter* sp., *Brevibacterium linens, Corynebacterium ammoniagenes,* staphyolococci	Consumption of lactic acid, produced by LAB, by *D. hansenii* and *G. candidum* allow the other consortium members to develop and ripen the cheese, giving it its characteristic flavour	[16]
Parasitism (+/-)				
Milk	Phage	LAB	Phage predate the bacteria and that may lead to failure of fermentation and loss of product.	[143, 177]
Mutualism (+/+)				
Wine	*Saccharomyces cerevisiae, Oenococcus oeni*		More flavour due to mutual stimulation	[24]
Yoghurt	*Streptoccus thermophilus, Lactobacillus bulgaricus*		More flavour and better texture due to mutual stimulation by exchanging growth factors	[11, 45, 178, 181]
Kefir	*Lactobacillus kefiranofaciens, Saccharomyces cerevisiae*		More flavour and better texture due to mutual stimulation by exchanging growth factors	[15]
Fermented milk	LAB, Yeasts		More flavour and better texture due to mutual stimulation	[174, 175]

(Table 4.5) cont.....

Products	Effectors	Targets	Effects on Targets and Products	References
Sourdough	*Saccharomyces exiguous, Candida humilis, Lactobacillus sanfranciscensis*		More flavour and better texture due to mutual stimulation by exchanging carbon sources	[7, 13, 160]

Amensalism is a mode of interaction in which one organism adversely affects another without being affected itself. It frequently occurs in food fermentations. The major end products of primary metabolism, such as carboxylic acids and alcohols are effective growth inhibitors of other microbes, among which food spoilage organisms. This preservative effect of fermentation is one of the reasons why fermented food became such an important part of the human diet [39, 149]. In fact, the metabolism of LAB is optimized for fast acid production, thereby preventing growth of other microorganisms, instead of efficient use of the available resources [150]. This feature, a fermentative metabolism with acid and/or alcohol as the main end products, is shared among all fermenting microorganisms and is a result of growth rate optimization more than the aim of making the environment unfavorable for other microorganisms. This is because the fermentative metabolism provides the microorganism with energy faster than when it would respire. Another example of amensalism, one that is specifically aimed at reduction of potential competitors, is the production of antimicrobial compounds, such as penicillin produced by *Penicillium* fungal species and bacteriocins that are produced by many bacterial species. These compounds can play important roles in the population dynamics of many mixed cultures. Typically, antibiotic and bacteriocin producing strains contain a dedicated immunity system that protects themselves from detrimental effects, while the compound has an inhibitive effect on other microorganisms in the environment. Therefore, many bacteriocin-producing LAB strains or their produced compounds, also referred to as lantibiotics, have found their way into the human food chain as natural means for inhibiting the outgrowth of pathogens and spoilage microbes [151, 152].

Probably the most widespread class of interactions is competition. In nearly every environment, microorganisms compete for energy sources and nutrients. In many food substrates, carbon and nitrogen sources are present in high concentrations. Competition in these environments, therefore, typically relates to rapid uptake of these nutrients and conversion into biomass rather than employing systems to scavenge for these resources, as often is the case for micronutrients, such as iron [153]. In dairy, there is an excess of carbon source (lactose), but most of the nitrogen is bound in protein complexes. Therefore, organisms initially compete for the small amount of free amino acids and small peptides, where in the later stages of fermentation they compete for the peptides released by the action of proteolytic enzymes, if expressed. Many dairy-resident bacteria have therefore relatively few genes for carbohydrate uptake and lack those for the breakdown of polysaccharides compared to their plant-resident relatives, but they may produce extracellular proteases, peptide and amino acid (AA) transport systems and peptidases for the exploitation of milk proteins. Growth rate and population dynamics in mixed culture fermentations in dairy are, therefore, largely determined by the ability to utilize amino acids efficiently [154].

Commensalism is defined as a mode of interaction in which one organism benefits from the interaction while the other organism is not affected. Although in some cases it may be hard to prove that not both organisms are affected by the interaction, a clear case of commensalism would be where organisms grow sequentially and the second benefits from the first (a type of trophic interaction). This is not uncommon in fermented food. As an example, in surface-ripened cheese, there are multiple trophic interactions. Yeasts and filamentous fungi consume lactic acid that is produced by the starter culture, leading to de-acidification of the cheese surface that enables the outgrowth of aerobic bacteria [16].

Parasitism is the mode of interaction in which one species benefits at the expense of the other. The aforementioned bacteriophage case could be considered a parasite as a bacteriophage utilizes the host cell to multiply itself, eventually resulting in the death of the host. In recent years, the understanding of phage biology and the interactions with their hosts has increased significantly, notably for industrially important LAB, such as *Lactococcus lactis* and *Streptococcus thermophilus* [143]. Suppliers of industrial cultures try to employ the genetic potential of the microorganism's own phage resistance systems, among which CRISPR-Cas, restriction/modification, and abortive infection systems [155], to make their strains more robust [156, 157]. This is particularly important now as producers of fermented foods more often choose DVI cultures with limited complexity instead of traditional complex bulk starters. A second case of parasitism occurs when the bacterium *Bdellovibrio bacteriovorus* invades a host cell and uses the available nutrients to multiply, but this is not common in food fermentations as predatory bacteria typically predate Gram-negative bacteria [158, 159], and LAB are Gram-positives.

With mutualism, also referred to as synergism, and in the case of nutrient exchange as protocooperation, both participating microorganisms benefit. Probably the best-known example of mutualism in the field of food fermentation is the one taking place within the yoghurt consortium. Mutualism is an important mode of interaction for many cultures consisting of mixtures of yeasts, LAB, and fungi. Therewith this mode of interaction is of key importance for the exact properties of a broad range of fermented foods. For instance, in kefir granules, normally containing *Saccharomyces cerevisiae* and *Lactobacillus kefiranofaciens*, the yeast utilizes the lactic acid produced by the LAB as carbon source, which enables more growth of the LAB that would otherwise be limited by the low pH [15]. In the fermentation of sourdough, there is a more complex synergistic interaction between yeasts and LAB, in particular the species *Lactobacillus sanfranciscensis*. Amylolytic activity of the yeast releases maltose from starch, which is then fermented by the bacterium. Part of the glucose released upon maltose splitting is excreted by the bacterium and can be used by maltose-negative yeast strains. Moreover, the yeasts increase the availability of amino acids and peptides, either through proteolysis or as a result of autolysis, which benefits *L. sanfranciscensis* [160]. Similar cross-feeding interactions occur between yeasts and LAB in wine [24].

Quorum sensing is not a class of interaction, but a communication method that is mostly targeted at the own species, although it has also been reported that one species produces a compound that acts as quorum sensing signal in another species [161]. Generally, quorum sensing is applied as a process to execute behavior that is only favorable when high cell densities are reached, and that gives benefit to the effector, but the effect on the target can vary.

The Yoghurt Consortium as the Paradigm of Microbial Interactions in Mixed Cultures

Yoghurt is normally made of bovine milk and fermented by the two LAB species *Streptococcus thermophilus* and *Lactobacillus delbrueckii* subsp. *bulgaricus*. These two species stimulate each other's growth, both by the direct exchange of metabolites and as a result of their behavior. The fact that this consortium consists of only two species, and often only one strain of each, makes studying the interactions between these two species relatively simple compared to more complex ecosystems. Previous studies [178, 179] have found multiple interactions (Fig. **4.2**), but their presence and extent may depend on the exact combination of strains. Typically, the yoghurt fermentation is started by growth of *S. thermophilus*, which is more tolerant to neutral pH and more effective in taking up amino acids and trace elements than *L. bulgaricus*. During this growth, *S. thermophilus* produces formic acid and folic acid, which can help purine biosynthesis in *L. bulgaricus* as precursor and as co-factor, respectively. *L. bulgaricus* is missing genes in the folic acid biosynthesis pathway and is therewith impaired in effective purine biosynthesis. In addition, *S. thermophilus* consumes oxygen and produces carbon dioxide, ensuring the necessary anaerobic environment for *L. bulgaricus* growth. As most strains do not possess an exo-protease, the levels of free AA, notably sulfur AA and branched-chain amino acids (BCAA), in milk limit further growth of *S. thermophilus*. The subsequent growth of *L. bulgaricus* and its expression of the protease gene *prt*B increase the levels of oligopeptides, which allows a second exponential growth phase of *S. thermophilus* until both strains get inhibited by the rising level of lactic acid [178]. The activity of the cell-wall resident PrtB is sufficient to provide both species with their amino acid needs, except for sulfur AA and BCAA, which are only small fractions of casein, and therefore genes involved for *de novo* biosynthesis of these AA are the only AA production pathway genes that are upregulated during this second exponential growth phase. Other genes upregulated in this phase are those involved in producing long-chain fatty acids (LCFA) in *S. thermophilus* of which *L. bulgaricus*, having an incomplete pathway for *de novo* biosynthesis of these compounds, may benefit. Moreover, genes involved in the production of EPS are also switched on and this is confirmed by the amount of EPS that is present in the system. It is hypothesized that the EPS play a role in ensuring close proximities between the two species, thereby facilitating the exchange of metabolites [11]. A possible exchange of ornithine and putrescine has also been reported, but the function of this exchange is not yet certain. Possible functions include the production of carbon dioxide and the use of putrescine as co-factor in cell division or as metal ion chelation agent [11]. Hereby, it is noteworthy that metal ions, in particular iron and manganese, are quite scarce in milk and efficient systems to take up these ions may certainly benefit growth of the microorganisms [179]. Finally, two recent reports indicated that urease activity and glutathione production by *S. thermophilus* benefits *L. bulgaricus* by locally increasing the intracellular pH [180] and relieving acid stress [181], respectively. The long history of co-cultivation of the two yoghurt bacteria [182] in dairy made the two species adapt well to their relatively rich environment and to each other's metabolism by allowing the development of the above-mentioned mutualistic interactions. This is in particular exemplified by the reduced genome of *L. bulgaricus* [45]. In fact, the relative stability and predictability of agricultural food sources has led to domestication of many microorganisms by genomic specialization through mechanisms, such as pseudogenization and genome decay [5]. Many genes in *L. bulgaricus* are missing or non-functional, rendering their pathways incomplete. Some of these pathways are complemented by the more versatile *S. thermophilus*. Some strains of *S. thermophilus* express the exo-protease PrtS, which makes them independent of *L. bulgaricus* for their AA supply. In this case, the otherwise mutualistic relationship would turn into a commensalistic one in which only *L. bulgaricus* benefits from the formic acid, folic acid, LCFA, and carbon dioxide produced by *S. thermophilus*. The growth rate of the latter, however, is normally so much higher that *L. bulgaricus* will never reach the same cell densities in combination with a proteolytic *S. thermophilus* as it does with a non-proteolytic one. Finally, cases of amensalism, in which one of the species is affected by the bacteriocin produced by the other, are also reported, but this is not very common in the yoghurt consortium [163 - 165].

As the exact extent and nature of the mutual interactions between the yoghurt bacteria will largely determine the performances of both species, the interactions will also have a large effect on product quality (Fig. **4.2**) [148].

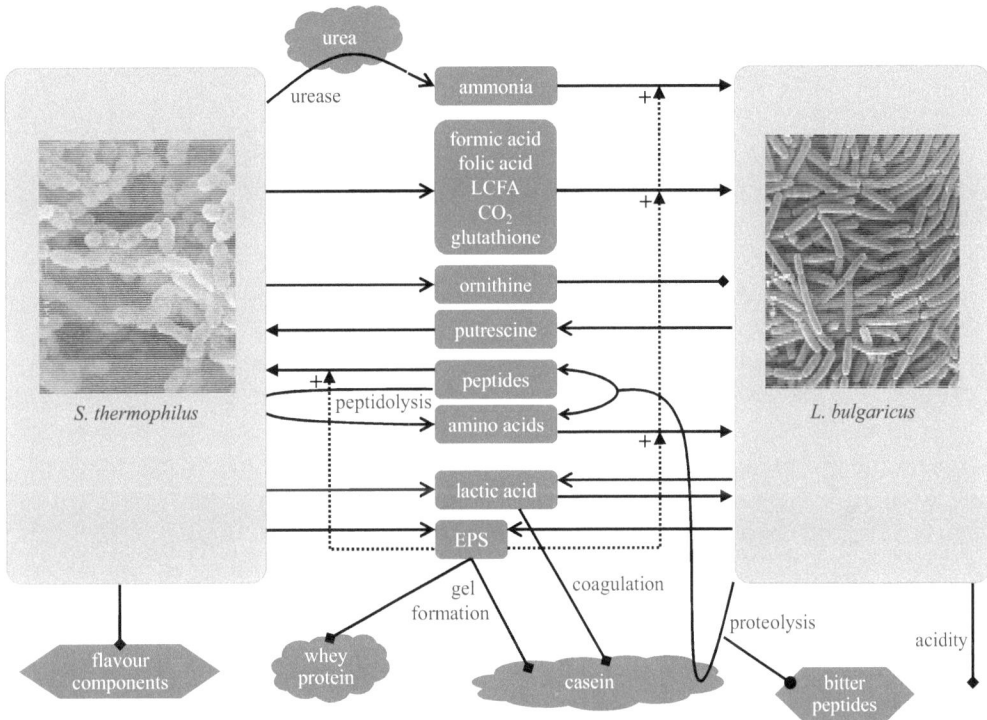

Fig. (4.2). Schematic representation of the possible interactions between *S. thermophilus* and *L. bulgaricus* in yoghurt (adapted from Sieuwerts [148]). Open arrows indicate production; closed arrows a positive effect of the component; lines ending in a circle indicate a negative effect of the component; lines ending in a lozenge indicate a neutral or yet to be confirmed effect of the component; the dotted line indicates that EPS is hypothesized to facilitate the exchange of metabolites by ensuring close proximities between the two species. EPS, exopolysaccharides; LCFA, long-chain fatty acids.

For example, the amount of lactic acid will determine the acidity and generally higher levels of *L. bulgaricus* give higher acidity. Moreover, the amount and the nature of the EPS, which differ between strains and also between the two species, will largely determine organoleptic properties, such as viscosity/mouthfeel and smoothness. It is mostly *S. thermophilus* that is the largest contributor of EPS. In addition, *S. thermophilus* is the main producer of aroma compounds, notably acetoin and diacetyl that together with acetaldehyde make the typical yoghurt flavor, while the PrtB of *L. bulgaricus* may give rise to accumulation of bitter tasting peptides. With the current trend in the dairy industry towards milder yoghurts with a good mouthfeel (high viscosity), it is not unsurprising that culture producers try to engineer combinations of *S. thermophilus* and *L. bulgaricus* strains that result in high counts of the former and low counts of the latter. It is, however, not possible to eliminate *L. bulgaricus* completely from the consortium because this would have a negative influence on the cell counts of the non-proteolytic *S. thermophilus* and because in many countries it is obligatory to have both species, or at least both cocci and bacilli, present and alive in order to be able to call the product yoghurt [182].

The existing interactions and basically the full metabolism of each consortium member should be considered when including other species, such as probiotics. For example, one of the yoghurt-based products in the market contains the bacteria *Lactobacillus acidophilus* LA-5, *Lactobacillus paracasei* subsp. *paracasei* F19, and *Bifidobacterium animalis* subsp. *lactis* BB-12. This results in a significant increase in possible interactions (Table **4.7**). The genomes of both *L. acidophilus* and *L. paracasei* encode an exo-protease, of which the consortium members, in particular the non-proteolytic *S. thermophilus* and *B. lactis*, may benefit [183]. In addition, *L. paracasei* is reported to be a formic acid producer [184], which could stimulate *L. bulgaricus*. In turn, all three added bacteria have special needs for optimal growth that could be provided by the yoghurt bacteria or the other consortium members. *L. acidophilus* grows best at a pH below 5 and will therefore benefit from the initial acidification of the medium by the growth of the others. Moreover, it is incapable of synthesizing many cofactors and vitamins, such as riboflavin, pyridoxine, nicotinate, nicotinamide, biotin, and folic acid and it cannot produce pyrimidines *de novo* [185]. Here, the bacterium may benefit from the folic acid that is produced by *S. thermophilus*, but it is not known whether one of the other components, or precursors thereof, are secreted in the medium by one of the consortium members. *L. paracasei* is impaired in (tetrahydro)folate cycling [186] and thus purine biosynthesis. It may, therefore, also benefit from the folic acid, or purine precursors, provided by *S. thermophilus* or *B. lactis*. This last species, like *L. acidophilus*, cannot synthesize many cofactors and vitamins including riboflavin, pantothenate, lipoate, and biotin. Moreover, its growth is reported to be promoted by amino acids and peptides, the LCFA oleic acid, ascorbic acid, and cobalamin [187, 188]. As *B. lactis* is poorly oxygen tolerant, it will benefit from the anaerobic environment provided by *S. thermophilus*. Another stimulatory effect may come from the provided LCFA, similar to *L. bulgaricus*. One of the primary difficulties that *B. lactis* will experience during growth in milk, also during co-culture with any of the other species, is a low availability of sulfur amino acids, particularly cysteine, as these are poorly abundant in milk as free amino acids and in caseins, and they are rapidly consumed by the other microorganisms. Since none of the consortium members is a hydrogen sulfide producer, *B. lactis* has no means to synthesize cysteine by itself. In conclusion, introducing new bacteria to an existing consortium may provide benefits to both the existing consortium members and support growth of new members, which would not perform well in the given environment on their own, by means of cross-feeding interactions that can have both a mutualistic and a commensalistic character. In order to make these new consortia stable upon propagation (back-slopping), the criterion that has to be met, at least, is that all consortium members meet their nutritional requirements, either through availability of the nutrients in the medium or by provision by one of their consortium members.

Table 4.7. Overview of interactions between members in a consortium consisting of the yoghurt bacteria *Streptococcus thermophilus* and *Lactobacillus bulgaricus* supplemented with the bacteria *Lactobacillus acidophilus*, *Lactobacillus paracasei* and *Bifidobacterium lactis*. Interactions can be strain specific and have a positive effect on the receiver except when (-) is added. AA, amino acids; EPS, exopolysaccharides; FOS, fructooligosaccharides; LCFA, long-chain fatty acids; [a], as result of activity of an exo-protease; [b], confirmed interactions while others are predicted based on literature and genome information. The vitamin and cofactor pool includes: thiamine (B1)[2], riboflavin (B2)[1,3], pantothenate (B5)[3], pyridoxine (B6)[1,3], biotin (B7)[1,3], cobalamin (B12)[3], nicotinate[1], nicotinamide[1], lipoate[3]. [1], needed by *L. acidophilus*; [2], needed by *L. paracasei*. [3], needed by *B. lactis*.

Effector\Receiver	*S. thermophilus*	*L. bulgaricus*	*L. acidophilus*	*L. paracasei*	*B. lactis*
S. thermophilus		See Fig. **4.1**	Formic acid Folic acid Pyrimidine (precursors) CO_2 O_2 consumption Galactose	Purine (precursors) O_2 consumption Galactose EPS	O_2 consumption LCFA Sulfur AA
L. bulgaricus	See Fig. **4.1**		Galactose Bacteriocin(-)	Galactose Bacteriocin (-) EPS	AA[a] Sulfur AA Bacteriocin (-)
L. acidophilus	AA[ab] Bacteriocin (-)	Bacteriocin (-)		Bacteriocin (-) EPS	AA[a] Sulfur AA Bacteriocin (-)
L. paracasei	AA[ab]	Formic acid	Formic acid		AA[a] Sulfur AA
B. lactis	-	-	FOS	FOS EPS	
General	-	Low pH[b]	Low pH[b] Vitamin and cofactor pool	Low pH (-)[b] Vitamin and cofactor pool	Low pH (-)[b] Vitamin and cofactor pool

Towards Interaction Studies in More Complex Consortia

With the importance of microbial interactions for determining the compositions of microbial consortia in food fermentations and therewith influencing the final product properties, it is essential to understand these interactions in order to rationally design the optimal mixtures of species and/or strains for desired product characteristics. The yoghurt consortium is relatively simple with only two species and still the interactions could only be unraveled in the genomics era. As shown in the previous section, the situation already becomes more complex when a few other species are added. How can interactions be studied in more complex consortia? Instead of looking at each separate consortium member and its position in the ecosystem, it is possible to use a metagenomics approach and look at functionalities, such as the presence of specific pathways to produce flavors or vitamins. Even though it can give clues about which functionalities can be included by adding other strains, it may not provide many leads for how successful tinkering with a consortium will be. Another method, first identifying the consortium members and their metabolic potentials and then tracking their performances and behaviors through time, is exemplified in a recent study on community dynamics of starter cultures for Gouda-type cheeses [12]. Here, amplified fragment length polymorphism (AFLP) typing combined with genome sequencing and 16S rDNA typing showed that the traditional cheese starter UR, maintained by back-slopping, consisted of many strains in eight genetic lineages divided over the two species *Lactococcus lactis* subsp. *cremoris* and subsp. *lactis* biovar. *diacetylactis* and *Leuconostoc mesentoides*. These genetic lineages were all collections of strains with variable plasmid contents and phage sensitivities. It was confirmed that density dependent phage predation was one of the driving factors behind maintaining the diversity in closely related strains in each genetic lineage [144]. Moreover, the type strain representing one of the genetic lineages was stimulated greatly by addition of arginine [12]. Carbon dioxide and glutamine are the main substrates for arginine and pyrimidine biosynthesis and folic acid is an essential co-factor in both the biosynthesis of purines. The strain is impaired in both biosynthetic pathways. The type strain representing another genetic lineage is an efficient producer of arginine and therefore it is possible that the impaired strain benefits from a cross-feeding interaction. In addition, a cross-feeding interaction was found in which glutamate was converted into γ- amino-butyric acid (GABA) which was subsequently secreted by *L. lactis* and imported by and converted to succinic acid by *L. mesenteroides*. Similar occasions may be the supply of folic acid by one or more of the strains in another genetic lineage, aiding the *de novo* biosynthesis of purines, and the protease-positive strains that provide AA to protease-negative ones. By applying different back-slopping regimes or modulating the propagation temperature, it is possible to change the ratios between the genetic lineages and thereby the metabolic potential and functionalities of the culture as a whole, for example leading to different flavor profiles [12]. This indicates that, although the interactions between the microbes largely determine which strains are the stable parts of the consortium, environmental factors also play a large role in the exact makeup and potential of a culture. Other culture-independent techniques, such as meta-transcriptomics, have the potential to supplement genome and metagenome sequencing by shedding light on what a culture is actually doing instead of barely identifying which microorganisms are present in which levels, *i.e.*, it is possible to look at their activities. Such culture-independent techniques are particularly useful in very complex consortia with many so far uncultured members [189]. This is exemplified in a study in which the researchers characterized the microbial community structure and gene expression during ripening of traditional Italian Caciocavallo Silano cheese, where after the initial acidification by thermophilic LAB non-starter mesophilic LAB become dominant, expressing typical ripening-associated pathways in the amino acid and lipid metabolism [190]. Moreover, they were able to show that activity of ripening bacteria was higher in the core of the cheese than in the crust and which pathways were elevated by ripening the cheese at higher temperature. These conclusions derived from meta-transcriptome data were in line with a measured increase in related flavor compounds and other metabolites. The availability of genome sequences may become a good aid in rationally designing mixed cultures with desired properties or optimizing fermentation and storage conditions to exactly acquire the desired product characteristics. Not only does the genome sequence provide information on the metabolism of a strain, but also building stoichiometry-based genome scale metabolic models allow researchers to pinpoint exactly which pathways of multiple strains can complement each other [191 - 193]. A recent study even applies general tools to visualize networks of food bacterial communities from different sources in a standardized way [194]. As genome sequencing and annotation can largely occur automated, reduce the cost, and increase both the efficiency and throughput of sequencing projects, it is anticipated that culture produces will apply metabolic models for rational design more often in the future rather than the trial and error combination of available strains that is currently common practice.

FUTURE PROSPECTS FOR FOOD CULTURES

There are several trends in the current use of food cultures. This chapter already touched upon the increase in use of DVI cultures with relatively low microbial complexity instead of the traditional complex bulk starter cultures. It is anticipated that this trend will continue. This also implies that industrially produced cultures will be used more often at the expense of traditionally propagated artisanal cultures. Diversification of product portfolio is, however, also a trend and this requires the use of many different cultures. With the mixing of human cultures and the increasing international trade of fermented food products, it is expected that product categories will move more across borders and that new microbial cultures will be applied to these product categories in order to match the local taste. Therefore, it is important that food companies and in particular culture suppliers look out in the world which microorganisms are available to provide interesting functionalities for product portfolio accretion. In this respect, rational optimization or design of new consortia suited for specific food matrices will become more important. Here, the application of genomics related technologies, such as genome sequencing, genome-scale metabolic modeling, and analysis of microbe-microbe and microbe-matrix interactions will greatly advance the knowledge and possibilities that developers of cultures and fermented products have. In addition, as GMO's are still unwanted in the food chain in large parts of the world and non-GMO techniques often give unfortunate random mutations, these genomics related technologies will aid in avoiding the need to perform any kind of genetic modification by allowing rational combining of strains that together perform the needed functionality. Moreover, knowing the genetic and metabolic potential of a strain or a mixed culture allows prediction of the outcome of an adaptive evolution exercise to acquire the certain functionality [11, 195]. Finally, we see completely new fermented food types emerging, such as fermented macroalgae like seaweed, and microalgae [196]. After enzymatic breakdown of cellulose by cellulases produced by *e.g.* fungal species, lactic fermentation by *Lactobacillus* species, particularly *Lactobacillus brevis*, *Lactobacillus casei,* and *Lactobacillus plantarum*, or ethanol fermentation are good possibilities to turn one of the most abundant groups of plants into healthy and tasty products. With more information becoming available on the health-promoting aspects of certain microorganisms, the number of approved probiotic claims will rise. This is the step-up towards specialized fermented products intended for subjects with diseases or syndromes, such as diabetes and obesity. With respect to food shelf-life and safety, the hunt for microorganisms with specific antimicrobial activities, *e.g.* through bacteriocin production will continue. In addition, since many spoilage and pathogenic bacteria are Gram-negatives, there is a potential application of specific phages and microorganisms like *Bdellovibrio bacteriovorus* for their control. *B. bacteriovorus* has been shown to effectively reduce numbers of *Salmonella* and *Escherichia coli* on beef [197] and control *Pseudomonas tolaasii* on packed mushrooms [198]. The application of both phages and predatory bacteria to combat unwanted microbes is still largely unexplored, but more research in this area may advance the fight against spoilage and thereby allow longer shelf-lives of fermented products, facilitating overseas transport.

CONSENT FOR PUBLICATION

Not applicable.

CONFLICT OF INTEREST

The authors declare no conflict of interest, financial or otherwise.

ACKNOWLEDGEMENTS

Declared none.

ABBREVIATIONS

AA   amino acid;

ABT   *Lactobacillus acidophilus*, *Bifidobacterium*, *Streptococcus thermophilus*;

AFLP amplified fragment length polymorphism;

BCAA branched-chain amino acid;

DVI   direct vat innoculation;

EPS   exopolysaccharides;

FDA   food and drug administration;

FOS   fructooligosaccharides;

GABA gamma-aminobutyric acid;

GI   gastrointestinal;

GRAS generally regarded as safe;

LAB   lactic acid bacteria;

LCFA long-chain fatty acids

REFERENCES

[1]   Ho QT, Verboven P, Ambaw A, Verlinden BE, Nicolaï BM. Transport properties of fermentation metabolites inside 'Conference' pear fruit. Postharvest Biol Technol 2016; 117: 38-48. [http://dx.doi.org/10.1016/j.postharvbio.2016.01.008]

[2]   Berhe AA. Decomposition of organic substrates at eroding vs. depositional landform positions. Plant Soil 2012; 350(1-2): 261-80. [http://dx.doi.org/10.1007/s11104-011-0902-z]

[3]   Ou JZ, Yao CK, Rotbart A, Muir JG, Gibson PR, Kalantar-zadeh K. Human intestinal gas measurement systems: *in vitro* fermentation and gas capsules. Trends Biotechnol 2015; 33(4): 208-13. [http://dx.doi.org/10.1016/j.tibtech.2015.02.002] [PMID: 25772639]

[4]   Fox PF, Law J, McSweeney PLH, Wallace J. Biochemistry of cheese ripening.In: fCheese: Chemistry, Physics and Microbiology. 2nd ed. London: Chapman & Hall Ltd 1993; Vol. 1: pp. 389-438. [http://dx.doi.org/10.1007/978-1-4615-2650-6_10]

[5]   Gibbons JG, Rinker DC. The genomics of microbial domestication in the fermented food environment. Curr Opin Genet Dev 2015; 35: 1-8.

[http://dx.doi.org/10.1016/j.gde.2015.07.003] [PMID: 26338497]

[6]   Chen W, Xu W, Zheng X. A *Lactobacillus plantarum* strain newly isolated from Chinese sauerkraut with high γ-aminobutyric acid productivity and its culture conditions optimization. Metal Mining Industry 2015; 7(9): 388-93.

[7]   De Vuyst L, Neysens P. The sourdough microflora: biodiversity and metabolic interactions. Trends Food Sci Technol 2005; 16: 43-56.
        [http://dx.doi.org/10.1016/j.tifs.2004.02.012]

[8]   Bassi D, Puglisi E, Cocconcelli PS. Comparing natural and selected starter cultures in meat and cheese fermentations. Curr Opin Food Sci 2015; 2: 118-22.
        [http://dx.doi.org/10.1016/j.cofs.2015.03.002]

[9]   Leroy F, Geyzen A, Janssens M, De Vuyst L, Scholliers P. Meat fermentation at the crossroads of innovation and tradition: A historical outlook. Trends Food Sci Technol 2013; 31(2): 130-7.
        [http://dx.doi.org/10.1016/j.tifs.2013.03.008]

[10]   Skåra T, Axelsson L, Stefánsson G, Ekstrand B, Hagen H. Fermented and ripened fish products in the northern European countries. J Ethnic Foods 2015; 2(1): 18-24.
        [http://dx.doi.org/10.1016/j.jef.2015.02.004]

[11]   Sieuwerts S. Analysis of molecular interactions between yoghurt bacteria by an integrated genomics approach. PhD dissertation: Wageningen University. 2009.

[12]   Erkus O. Community dynamics of complex starter cultures for Gouda-type cheeses and its functional consequences. PhD dissertation: Wageningen University. 2014.

[13]   Gobbetti M, Corsetti A, Rossi J. The sourdough microflora. Interactions between lactic acid bacteria and yeasts: metabolism of carbohydrates. Appl Microbiol Biotechnol 1994; 41: 456-60.
        [http://dx.doi.org/10.1007/BF00939035]

[14]   M'hir S, Minervini F, Di Cagno R, Chammem N, Hamdi M. Technological, functional and safety aspects of enterococci in fermented vegetable products: A mini-review. Ann Microbiol 2012; 62(2): 469-81.
        [http://dx.doi.org/10.1007/s13213-011-0363-x]

[15]   Cheirsilp B, Shoji H, Shimizu H, Shioya S. Interactions between *Lactobacillus kefiranofaciens* and *Saccharomyces cerevisiae* in mixed culture for kefiran production. J Biosci Bioeng 2003; 96(3): 279-84.
        [http://dx.doi.org/10.1016/S1389-1723(03)80194-9] [PMID: 16233522]

[16]   Mounier J, Gelsomino R, Goerges S, *et al.* Surface microflora of four smear-ripened cheeses. Appl Environ Microbiol 2005; 71(11): 6489-500.
        [http://dx.doi.org/10.1128/AEM.71.11.6489-6500.2005] [PMID: 16269673]

[17]   Smid EJ, Lacroix C. Microbe-microbe interactions in mixed culture food fermentations. Curr Opin Biotechnol 2013; 24(2): 148-54.
        [http://dx.doi.org/10.1016/j.copbio.2012.11.007] [PMID: 23228389]

[18]   Lowe B. Experimental Cookery From The Chemical And Physical Standpoint. 3rd ed., London: Chapman & Hall Ltd 1943.

[19]   Buldo P, Benfeldt C, Bibiloni R, *et al.* Exopolysaccharide-producing cultures and milk protein ingredients: their effect on microstructure, textural and sensorial properties of stirred yoghurts. Proceedings of the 7th international symposium on food rheology and structure. 2015 June 7-11; Zurich, Switzerland.

[20]   Pedersen TB, Vogensen FK, Ardö Y. Effect of heterofermentative lactic acid bacteria of DL-starters in initial ripening of semi-hard cheese. Int Dairy J 2016; 57: 72-9.
        [http://dx.doi.org/10.1016/j.idairyj.2016.02.041]

[21]   Rhee SJ, Lee J-E, Lee C-H. Importance of lactic acid bacteria in Asian fermented foods. Microb Cell

Fact 2011; 10(S1) (Suppl. 1): S5.
[http://dx.doi.org/10.1186/1475-2859-10-S1-S5] [PMID: 21995342]

[22] Choi H-J, Lee N-K, Paik H-D. Health benefits of lactic acid bacteria isolated from kimchi, with respect to immunomodulatory effects. Food Sci Biotechnol 2015; 24(3): 783-9.
[http://dx.doi.org/10.1007/s10068-015-0102-3]

[23] Hurtado A, Reguant C, Bordons A, Rozès N. Lactic acid bacteria from fermented table olives. Food Microbiol 2012; 31(1): 1-8.
[http://dx.doi.org/10.1016/j.fm.2012.01.006] [PMID: 22475936]

[24] Alexandre H, Costello PJ, Remize F, Guzzo J, Guilloux-Benatier M. *Saccharomyces cerevisiae-Oenococcus oeni* interactions in wine: current knowledge and perspectives. Int J Food Microbiol 2004; 93(2): 141-54.
[http://dx.doi.org/10.1016/j.ijfoodmicro.2003.10.013] [PMID: 15135953]

[25] Capozzi V, Garofalo C, Chiriatti MA, Grieco F, Spano G. Microbial terroir and food innovation: The case of yeast biodiversity in wine. Microbiol Res 2015; 181(1): 75-83.
[http://dx.doi.org/10.1016/j.micres.2015.10.005] [PMID: 26521127]

[26] Bokulich NA, Bamforth CW. The microbiology of malting and brewing. Microbiol Mol Biol Rev 2013; 77(2): 157-72.
[http://dx.doi.org/10.1128/MMBR.00060-12] [PMID: 23699253]

[27] Kitagaki H, Kitamoto K. Breeding research on sake yeasts in Japan: history, recent technological advances, and future perspectives. Annu Rev Food Sci Technol 2013; 4(1): 215-35.
[http://dx.doi.org/10.1146/annurev-food-030212-182545] [PMID: 23464572]

[28] Bokulich NA, Ohta M, Lee M, Mills DA. Indigenous bacteria and fungi drive traditional kimoto sake fermentations. Appl Environ Microbiol 2014; 80(17): 5522-9.
[http://dx.doi.org/10.1128/AEM.00663-14] [PMID: 24973064]

[29] Wei Q, Wang H, Chen Z, Lv Z, Xie Y, Lu F. Profiling of dynamic changes in the microbial community during the soy sauce fermentation process. Appl Microbiol Biotechnol 2013; 97(20): 9111-9.
[http://dx.doi.org/10.1007/s00253-013-5146-9] [PMID: 24037306]

[30] Tanaka Y, Watanabe J, Mogi Y. Monitoring of the microbial communities involved in the soy sauce manufacturing process by PCR-denaturing gradient gel electrophoresis. Food Microbiol 2012; 31(1): 100-6.
[http://dx.doi.org/10.1016/j.fm.2012.02.005] [PMID: 22475947]

[31] Greppi A, Ferrocino I, La Storia A, Rantsiou K, Ercolini D, Cocolin L. Monitoring of the microbiota of fermented sausages by culture independent rRNA-based approaches. Int J Food Microbiol 2015; 212: 67-75.
[http://dx.doi.org/10.1016/j.ijfoodmicro.2015.01.016] [PMID: 25724303]

[32] Aquilanti L, Garofalo C, Osimani A, Clementi F. Ecology of lactic acid bacteria and coagulase negative cocci in fermented dry sausages manufactured in Italy and other Mediterranean countries: An overview. Int Food Res J 2016; 23(2): 429-45.

[33] Laich F, Fierro F, Cardoza RE, Martin JF. Organization of the gene cluster for biosynthesis of penicillin in *Penicillium nalgiovense* and antibiotic production in cured dry sausages. Appl Environ Microbiol 1999; 65(3): 1236-40.
[PMID: 10049889]

[34] Paludan-Müller C, Madsen M, Sophanodora P, Gram L, Møller PL. Fermentation and microflora of plaa-som, a thai fermented fish product prepared with different salt concentrations. Int J Food Microbiol 2002; 73(1): 61-70.
[http://dx.doi.org/10.1016/S0168-1605(01)00688-2] [PMID: 11883675]

[35] Lessard M-H, Bélanger G, St-Gelais D, Labrie S. The composition of Camembert cheese-ripening

cultures modulates both mycelial growth and appearance. Appl Environ Microbiol 2012; 78(6): 1813-9.
[http://dx.doi.org/10.1128/AEM.06645-11] [PMID: 22247164]

[36]   Firmesse O, Alvaro E, Mogenet A, *et al.* Fate and effects of Camembert cheese micro-organisms in the human colonic microbiota of healthy volunteers after regular Camembert consumption. Int J Food Microbiol 2008; 125(2): 176-81.
[http://dx.doi.org/10.1016/j.ijfoodmicro.2008.03.044] [PMID: 18554738]

[37]   Henri-Dubernet S, Desmasures N, Guéguen M. Diversity and dynamics of lactobacilli populations during ripening of RDO Camembert cheese. Can J Microbiol 2008; 54(3): 218-28.
[http://dx.doi.org/10.1139/W07-137] [PMID: 18388993]

[38]   Vásquez A, Forsgren E, Fries I, *et al.* Symbionts as major modulators of insect health: lactic acid bacteria and honeybees. PLoS One 2012; 7(3): e33188.
[http://dx.doi.org/10.1371/journal.pone.0033188] [PMID: 22427985]

[39]   Caplice E, Fitzgerald GF. Food fermentations: role of microorganisms in food production and preservation. Int J Food Microbiol 1999; 50(1-2): 131-49.
[http://dx.doi.org/10.1016/S0168-1605(99)00082-3] [PMID: 10488849]

[40]   Pastink MI, Teusink B, Hols P, Visser S, de Vos WM, Hugenholtz J. Genome-scale model of *Streptococcus thermophilus* LMG18311 for metabolic comparison of lactic acid bacteria. Appl Environ Microbiol 2009; 75(11): 3627-33.
[http://dx.doi.org/10.1128/AEM.00138-09] [PMID: 19346354]

[41]   Pastink MI, Sieuwerts S, de Bok FAM, *et al.* Genomics and high-throughput screening approaches for optimal flavour production in dairy fermentation. Int Dairy J 2008; 18(8): 781-9.
[http://dx.doi.org/10.1016/j.idairyj.2007.07.006]

[42]   Kleerebezem M, Boekhorst J, van Kranenburg R, *et al.* Complete genome sequence of *Lactobacillus plantarum* WCFS1. Proc Natl Acad Sci USA 2003; 100(4): 1990-5.
[http://dx.doi.org/10.1073/pnas.0337704100] [PMID: 12566566]

[43]   Sanni AI, Morlon-Guyot J, Guyot JP. New efficient amylase-producing strains of *Lactobacillus plantarum* and *L. fermentum* isolated from different Nigerian traditional fermented foods. Int J Food Microbiol 2002; 72(1-2): 53-62.
[http://dx.doi.org/10.1016/S0168-1605(01)00607-9] [PMID: 11843413]

[44]   Bachmann H, Starrenburg MJC, Molenaar D, Kleerebezem M, van Hylckama Vlieg JET. Microbial domestication signatures of *Lactococcus lactis* can be reproduced by experimental evolution. Genome Res 2012; 22(1): 115-24.
[http://dx.doi.org/10.1101/gr.121285.111] [PMID: 22080491]

[45]   van de Guchte M, Penaud S, Grimaldi C, *et al.* The complete genome sequence of *Lactobacillus bulgaricus* reveals extensive and ongoing reductive evolution. Proc Natl Acad Sci USA 2006; 103(24): 9274-9.
[http://dx.doi.org/10.1073/pnas.0603024103] [PMID: 16754859]

[46]   Di Cerbo A, Palmieri B, Aponte M, Morales-Medina JC, Iannitti T. Mechanisms and therapeutic effectiveness of lactobacilli. J Clin Pathol 2016; 69(3): 187-203.
[http://dx.doi.org/10.1136/jclinpath-2015-202976] [PMID: 26578541]

[47]   Zhang L, Folkenberg DM, Amigo JM, Ipsen R. Effect of exopolysaccharide-producing starter cultures and post-fermentation mechanical treatment on textural properties and microstructure of low fat yoghurt. Int Dairy J 2016; 53: 10-9.
[http://dx.doi.org/10.1016/j.idairyj.2015.09.008]

[48]   Hassan AN, Biswas AC. Selection criteria of exopolysaccharide-producing cultures for reduced fat cheese with improved texture. Milchwissenschaft 2012; 67(4): 447-50.

[49]   Bajpai VK, Rather IA, Majumder R, *et al.* Exopolysaccharide and lactic acid bacteria: Perception,

functionality and prospects. Bangladesh J Pharmacol 2016; 11(1): 1-23.
[http://dx.doi.org/10.3329/bjp.v11i1.23819]

[50]   Buldo P, Benfeldt C, Folkenberg DM, *et al.* Applied aspects of exopolysaccharide-producing cultures and whey-based protein ingredients. Lebensm Wiss Technol 2016; 72: 189-98.
[http://dx.doi.org/10.1016/j.lwt.2016.04.050]

[51]   Buldo P, Benfeldt C, Carey JP, *et al.* Interactions of milk proteins with low and high acyl gellan: effect on microstructure and textural properties of acidified milk. Food Hydrocoll 2016; 60: 225-31.
[http://dx.doi.org/10.1016/j.foodhyd.2016.03.041]

[52]   Yanachkina P, McCarthy C, Guinee T, Wilkinson M. Effect of varying the salt and fat content in Cheddar cheese on aspects of the performance of a commercial starter culture preparation during ripening. Int J Food Microbiol 2016; 224: 7-15.
[http://dx.doi.org/10.1016/j.ijfoodmicro.2016.02.006] [PMID: 26905194]

[53]   Wegkamp A, Starrenburg M, de Vos WM, Hugenholtz J, Sybesma W. Transformation of folate-consuming *Lactobacillus gasseri* into a folate producer. Appl Environ Microbiol 2004; 70(5): 3146-8.
[http://dx.doi.org/10.1128/AEM.70.5.3146-3148.2004] [PMID: 15128580]

[54]   Rezaei MN, Jayaram VB, Verstrepen KJ, Courtin CM. The impact of yeast fermentation on dough matrix properties. J Sci Food Agric 2016; 96(11): 3741-8.
[http://dx.doi.org/10.1002/jsfa.7562] [PMID: 26676687]

[55]   Makhoul S, Romano A, Capozzi V, *et al.* Volatile Compound Production During the Bread-Making Process: Effect of Flour, Yeast and Their Interaction. Food Bioprocess Technol 2015; 8(9): 1925-37.
[http://dx.doi.org/10.1007/s11947-015-1549-1]

[56]   Marconi O, Rossi S, Galgano F, Sileoni V, Perretti G. Influence of yeast strain, priming solution and temperature on beer bottle conditioning. J Sci Food Agric 2016; 96(12): 4106-15.
[http://dx.doi.org/10.1002/jsfa.7611] [PMID: 26748817]

[57]   Hazelwood LA, Walsh MC, Pronk JT, Daran J-M. Involvement of vacuolar sequestration and active transport in tolerance of Saccharomyces cerevisiae to hop iso-α-acids. Appl Environ Microbiol 2010; 76(1): 318-28.
[http://dx.doi.org/10.1128/AEM.01457-09] [PMID: 19915041]

[58]   Lásztity R. Food Quality and Standards. EOLSS Publishers Company Ltd 2009.

[59]   Lucey JA, Johnson ME, Horne DS. Invited review: perspectives on the basis of the rheology and texture properties of cheese. J Dairy Sci 2003; 86(9): 2725-43.
[http://dx.doi.org/10.3168/jds.S0022-0302(03)73869-7] [PMID: 14507008]

[60]   Salvatore E, Addis M, Pes M, Fiori M, Pirisi A. Evaluation of lipolysis and volatile compounds produced by three *Penicillium roqueforti* commercial cultures in a blue-type cheese made from ovine milk. Ital J Food Sci 2015; 27(4): 437-42.

[61]   Chen W, He Y, Zhou Y, *et al.* Edible Filamentous Fungi from the Species *Monascus*: Early Traditional Fermentations, Modern Molecular Biology, and Future Genomics. Compr Rev Food Sci Food Saf 2015; 14(5): 555-67.
[http://dx.doi.org/10.1111/1541-4337.12145]

[62]   Zhuang M, Lin L, Zhao M, *et al.* Sequence, taste and umami-enhancing effect of the peptides separated from soy sauce. Food Chem 2016; 206: 174-81.
[http://dx.doi.org/10.1016/j.foodchem.2016.03.058] [PMID: 27041313]

[63]   Zhao G, Yao Y, Wang C, *et al.* Transcriptome and proteome expression analysis of the metabolism of amino acids by the fungus *Aspergillus oryzae* in fermented soy sauce. BioMed Res Int 2015; 2015: 456802.
[PMID: 25945335]

[64]   Iacumin L, Chiesa L, Boscolo D, *et al.* Moulds and ochratoxin A on surfaces of artisanal and industrial dry sausages. Food Microbiol 2009; 26(1): 65-70.

[http://dx.doi.org/10.1016/j.fm.2008.07.006] [PMID: 19028307]

[65]   Fuller R. Probiotics in man and animals. J Appl Bacteriol 1989; 66(5): 365-78.
       [http://dx.doi.org/10.1111/j.1365-2672.1989.tb05105.x] [PMID: 2666378]

[66]   Haenel H, Bendig J. Intestinal flora in health and disease. Prog Food Nutr Sci 1975; 1(1): 21-64.
       [PMID: 1223980]

[67]   Mitsuoka T. Recent trends in research on intestinal flora. Bifido Microflo 1982; 1: 3-24.
       [http://dx.doi.org/10.12938/bifidus1982.1.1_3]

[68]   Salminen S, von Wright A, Morelli L, *et al.* Demonstration of safety of probiotics -- a review. Int J
       Food Microbiol 1998; 44(1-2): 93-106.
       [http://dx.doi.org/10.1016/S0168-1605(98)00128-7] [PMID: 9849787]

[69]   Pelto L, Isolauri E, Lilius EM, Nuutila J, Salminen S. Probiotic bacteria down-regulate the milk-
       induced inflammatory response in milk-hypersensitive subjects but have an immunostimulatory effect
       in healthy subjects. Clin Exp Allergy 1998; 28(12): 1474-9.
       [http://dx.doi.org/10.1046/j.1365-2222.1998.00449.x] [PMID: 10024217]

[70]   Rautanen T, Isolauri E, Salo E, Vesikari T. Management of acute diarrhoea with low osmolarity oral
       rehydration solutions and *Lactobacillus* strain GG. Arch Dis Child 1998; 79(2): 157-60.
       [http://dx.doi.org/10.1136/adc.79.2.157] [PMID: 9797599]

[71]   Arvola T, Laiho K, Torkkeli S, *et al.* Prophylactic *Lactobacillus* GG reduces antibiotic-associated
       diarrhea in children with respiratory infections: a randomized study. Pediatrics 1999; 104(5): e64.
       [http://dx.doi.org/10.1542/peds.104.5.e64] [PMID: 10545590]

[72]   Alander M, Satokari R, Korpela R, *et al.* Persistence of colonization of human colonic mucosa by a
       probiotic strain, *Lactobacillus rhamnosus* GG, after oral consumption. Appl Environ Microbiol 1999;
       65(1): 351-4.
       [PMID: 9872808]

[73]   Schiffrin EJ, Rochat F, Link-Amster H, Aeschlimann JM, Donnet-Hughes A. Immunomodulation of
       human blood cells following the ingestion of lactic acid bacteria. J Dairy Sci 1995; 78(3): 491-7.
       [http://dx.doi.org/10.3168/jds.S0022-0302(95)76659-0] [PMID: 7782506]

[74]   Michetti P, Dorta G, Wiesel PH, *et al.* Effect of whey-based culture supernatant of *Lactobacillus
       acidophilus* (johnsonii) La1 on *Helicobacter pylori* infection in humans. Digestion 1999; 60(3): 203-9.
       [http://dx.doi.org/10.1159/000007660] [PMID: 10343133]

[75]   Donnet-Hughes A, Rochat F, Serrant P, Aeschlimann JM, Schiffrin EJ. Modulation of nonspecific
       mechanisms of defense by lactic acid bacteria: effective dose. J Dairy Sci 1999; 82(5): 863-9.
       [http://dx.doi.org/10.3168/jds.S0022-0302(99)75304-X] [PMID: 10342225]

[76]   Fukushima Y, Li S, Hara H, Terada A, Mitsuoka T. Effect of follow-up formula containing
       Bifidobacteria (NAN BF) on fecal flora and fecal metabolites in healthy children. Biosci Micro 1997;
       16(2): 65-72.

[77]   Wolf BW, Wheeler KB, Ataya DG, Garleb KA. Safety and tolerance of *Lactobacillus reuteri*
       supplementation to a population infected with the human immunodeficiency virus. Food Chem
       Toxicol 1998; 36(12): 1085-94.
       [http://dx.doi.org/10.1016/S0278-6915(98)00090-8] [PMID: 9862651]

[78]   Shornikova AV, Casas IA, Isolauri E, Mykkänen H, Vesikari T. *Lactobacillus reuteri* as a therapeutic
       agent in acute diarrhea in young children. J Pediatr Gastroenterol Nutr 1997; 24(4): 399-404.
       [http://dx.doi.org/10.1097/00005176-199704000-00008] [PMID: 9144122]

[79]   Shornikova AV, Casas IA, Mykkänen H, Salo E, Vesikari T. Bacteriotherapy with *Lactobacillus
       reuteri* in rotavirus gastroenteritis. Pediatr Infect Dis J 1997; 16(12): 1103-7.
       [http://dx.doi.org/10.1097/00006454-199712000-00002] [PMID: 9427453]

[80]   Aso Y, Akaza H, Kotake T, Tsukamoto T, Imai K, Naito S. Preventive effect of a *Lactobacillus casei*

preparation on the recurrence of superficial bladder cancer in a double-blind trial. Eur Urol 1995; 27(2): 104-9.
[http://dx.doi.org/10.1159/000475138] [PMID: 7744150]

[81]   Spanhaak S, Havenaar R, Schaafsma G. The effect of consumption of milk fermented by *Lactobacillus casei* strain *Shirota* on the intestinal microflora and immune parameters in humans. Eur J Clin Nutr 1998; 52(12): 899-907.
[http://dx.doi.org/10.1038/sj.ejcn.1600663] [PMID: 9881885]

[82]   Johansson ML, Molin G, Jeppsson B, Nobaek S, Ahrné S, Bengmark S. Administration of different *Lactobacillus* strains in fermented oatmeal soup: *in vivo* colonization of human intestinal mucosa and effect on the indigenous flora. Appl Environ Microbiol 1993; 59(1): 15-20.
[PMID: 8439146]

[83]   Johansson ML, Nobaek S, Berggren A, *et al.* Survival of *Lactobacillus plantarum* DSM 9843 (299v), and effect on the short-chain fatty acid content of faeces after ingestion of a rose-hip drink with fermented oats. Int J Food Microbiol 1998; 42(1-2): 29-38.
[http://dx.doi.org/10.1016/S0168-1605(98)00055-5] [PMID: 9706795]

[84]   Majamaa H, Isolauri E, Saxelin M, Vesikari T. Lactic acid bacteria in the treatment of acute rotavirus gastroenteritis. J Pediatr Gastroenterol Nutr 1995; 20(3): 333-8.
[http://dx.doi.org/10.1097/00005176-199504000-00012] [PMID: 7608829]

[85]   Salminen S, Isolauri E, Salminen E. Probiotics and stabilisation of the gut mucosal barrier. Asia Pac J Clin Nutr 1996; 5(1): 53-6.
[PMID: 24394468]

[86]   Bernet MF, Brassart D, Neeser JR, Servin AL. *Lactobacillus acidophilus* LA 1 binds to cultured human intestinal cell lines and inhibits cell attachment and cell invasion by enterovirulent bacteria. Gut 1994; 35(4): 483-9.
[http://dx.doi.org/10.1136/gut.35.4.483] [PMID: 8174985]

[87]   Deepika G, Charalampopoulos D. Surface and adhesion properties of lactobacilli. Adv Appl Microbiol 2010; 70: 127-52.
[http://dx.doi.org/10.1016/S0065-2164(10)70004-6] [PMID: 20359456]

[88]   Deepika G, Green RJ, Frazier RA, Charalampopoulos D. Effect of growth time on the surface and adhesion properties of *Lactobacillus rhamnosus* GG. J Appl Microbiol 2009; 107(4): 1230-40.
[http://dx.doi.org/10.1111/j.1365-2672.2009.04306.x] [PMID: 19486400]

[89]   Coconnier MH, Bernet MF, Chauvière G, Servin AL. Adhering heat-killed human *Lactobacillus acidophilus*, strain LB, inhibits the process of pathogenicity of diarrhoeagenic bacteria in cultured human intestinal cells. J Diarrhoeal Dis Res 1993; 11(4): 235-42.
[PMID: 8188996]

[90]   Hymes JP, Klaenhammer TR. Stuck in the middle: Fibronectin-binding proteins in gram-positive bacteria. Front Microbiol 2016; 7: 1504.
[http://dx.doi.org/10.3389/fmicb.2016.01504] [PMID: 27713740]

[91]   Yadav AK, Tyagi A, Kumar A, *et al.* Adhesion of Lactobacilli and their anti-infectivity potential. Crit Rev Food Sci Nutr 2017; 57(10): 2042-56.
[http://dx.doi.org/10.1080/10408398.2014.918533] [PMID: 25879917]

[92]   Isolauri E, Salminen S, Mattila-Sandholm T. New functional foods in the treatment of food allergy. Ann Med 1999; 31(4): 299-302.
[http://dx.doi.org/10.3109/07853899908995894] [PMID: 10480762]

[93]   Niku-Paavola ML, Laitila A, Mattila-Sandholm T, Haikara A. New types of antimicrobial compounds produced by *Lactobacillus plantarum*. J Appl Microbiol 1999; 86(1): 29-35.
[http://dx.doi.org/10.1046/j.1365-2672.1999.00632.x] [PMID: 10200070]

[94]   Holzapfel WH, Geisen R, Schillinger U. Biological preservation of foods with reference to protective

cultures, bacteriocins and food-grade enzymes. Int J Food Microbiol 1995; 24(3): 343-62.
[http://dx.doi.org/10.1016/0168-1605(94)00036-6] [PMID: 7710912]

[95]   Helander I, von Wright A, Mattila-Sandholm T. Potential of lactic acid bacteria and novel
        antimicrobials against gram-negative bacteria. Trends Food Sci Technol 1997; 8: 146-50.
        [http://dx.doi.org/10.1016/S0924-2244(97)01030-3]

[96]   Silva M, Jacobus NV, Deneke C, Gorbach SL. Antimicrobial substance from a human *Lactobacillus*
        strain. Antimicrob Agents Chemother 1987; 31(8): 1231-3.
        [http://dx.doi.org/10.1128/AAC.31.8.1231] [PMID: 3307619]

[97]   Zhang XB, Ohta Y, Hosono A. Antimutagenicity and binding of lactic acid bacteria from a Chinese
        cheese to mutagenic pyrolyzates. J Dairy Sci 1990; 73(10): 2702-10.
        [http://dx.doi.org/10.3168/jds.S0022-0302(90)78955-2] [PMID: 1980923]

[98]   Thyagaraja N, Hosono A. Binding properties of lactic acid bacteria from 'Idly' towards food-borne
        mutagens. Food Chem Toxicol 1994; 32(9): 805-9.
        [http://dx.doi.org/10.1016/0278-6915(94)90156-2] [PMID: 7927077]

[99]   El-Nezami H, Kankaanpää P, Salminen S, Ahokas J. Ability of dairy strains of lactic acid bacteria to
        bind a common food carcinogen, aflatoxin B1. Food Chem Toxicol 1998; 36(4): 321-6.
        [http://dx.doi.org/10.1016/S0278-6915(97)00160-9] [PMID: 9651049]

[100]   el-Nezami H, Kankaanpää P, Salminen S, Ahokas J. Physicochemical alterations enhance the ability of
        dairy strains of lactic acid bacteria to remove aflatoxin from contaminated media. J Food Prot 1998;
        61(4): 466-8.
        [http://dx.doi.org/10.4315/0362-028X-61.4.466] [PMID: 9709211]

[101]   Leatherhead Research Publications 24 Nov 2014. https://www.leatherheadfood. com/functional-food-
        -market-increases-in-size

[102]   Ray B, Daeschel M. Food bio-preservatives of microbial origin. Trove Australia 1992.

[103]   Law BA. Controlled and accelerated cheese ripening: the research base for new technology. Int Dairy
        J 2001; 11(4–7): 15.

[104]   Erkkilä S, Petäjä E, Eerola S, Lilleberg L, Mattila-Sandholm T, Suihko ML. Flavour profiles of dry
        sausages fermented by selected novel meat starter cultures. Meat Sci 2001; 58(2): 111-6.
        [http://dx.doi.org/10.1016/S0309-1740(00)00135-2] [PMID: 22062105]

[105]   Chandan RC. Enhancing market value of milk by adding cultures. J Dairy Sci 1999; 82(10): 2245-56.
        [http://dx.doi.org/10.3168/jds.S0022-0302(99)75472-X] [PMID: 10531614]

[106]   Ross RP, Stantona C, Hillb C, Fitzgerald GF, Coffeya A. Novel cultures for cheese improvement.
        Trends Food Sci Technol 2000; 11(3): 14.
        [http://dx.doi.org/10.1016/S0924-2244(00)00057-1]

[107]   Svensson U. Industrial perspectives.Probiotics: a critical review. Horizon Scientific Press 1999.

[108]   Driessen FM, Loones A. Developments in the fermentation process (liquid, stirred and set fermented
        milks). Federation Internationale de Laiterie; Int Dairy Fed 1993; 277: 12.

[109]   Fox PF, Wallace JM, Morgan S, Lynch CM, Niland EJ, Tobin J. Acceleration of cheese ripening.
        Antonie van Leeuwenhoek 1996; 70(2-4): 271-97.
        [http://dx.doi.org/10.1007/BF00395937] [PMID: 8879411]

[110]   Kunji ER, Mierau I, Hagting A, Poolman B, Konings WN. The proteolytic systems of lactic acid
        bacteria. Antonie van Leeuwenhoek 1996; 70(2-4): 187-221.
        [http://dx.doi.org/10.1007/BF00395933] [PMID: 8879407]

[111]   Saarela M, Mogensen G, Fondén R, Mättö J, Mattila-Sandholm T. Probiotic bacteria: safety,
        functional and technological properties. J Biotechnol 2000; 84(3): 197-215.
        [http://dx.doi.org/10.1016/S0168-1656(00)00375-8] [PMID: 11164262]

[112] Oberman H, Libudzisz Z. Fermented milks. Microbiology of Fermented Foods. Springer, US 1997; pp. 308-50.

[113] Deepika G, Karunakaran E, Hurley CR, Biggs CA, Charalampopoulos D. Influence of fermentation conditions on the surface properties and adhesion of *Lactobacillus rhamnosus* GG. Microb Cell Fact 2012; 11: 116.
[http://dx.doi.org/10.1186/1475-2859-11-116] [PMID: 22931558]

[114] Kolter R, Siegele DA, Tormo A. The stationary phase of the bacterial life cycle. Annu Rev Microbiol 1993; 47: 855-74.
[http://dx.doi.org/10.1146/annurev.mi.47.100193.004231] [PMID: 8257118]

[115] Hartke A, Bouche S, Gansel X, Boutibonnes P, Auffray Y. Starvation-Induced Stress Resistance in *Lactococcus lactis* subsp. *lactis* IL1403. Appl Environ Microbiol 1994; 60(9): 3474-8.
[PMID: 16349399]

[116] Rallu F, Gruss A, Maguin E. *Lactococcus lactis* and stress. Antonie van Leeuwenhoek 1996; 70(2-4): 243-51.
[http://dx.doi.org/10.1007/BF00395935] [PMID: 8879409]

[117] Spork AC. New trends for probiotic cultures. Eur Dairy Mag (Germany) 1995; 4(12): 2.

[118] German B, Schiffrin EJ, Reniero R, Mollet B, Pfeifer A, Neeser J-R. The development of functional foods: lessons from the gut. Trends Biotechnol 1999; 17(12): 492-9.
[http://dx.doi.org/10.1016/S0167-7799(99)01380-3] [PMID: 10557163]

[119] Ishibashi N. Bifidobacteria: research and development in Japan. Food Technol 1993; 47(6): 6.

[120] Driessen FM, Kingma F, Stadhouders J. Evidence that *Lactobacillus bulgaricus* in yogurt is stimulated by carbon dioxide produced by *Streptococcus thermophilus.* Netherlands Milk Dairy J (Netherlands) 1983; 36(2): 10.

[121] Radke-Mitchell LC, Sandine WE. Influence of temperature on associative growth of *Streptococcus thermophilus* and *Lactobacillus bulgaricus.* J Dairy Sci 1986; 69(10): 2558-68.
[http://dx.doi.org/10.3168/jds.S0022-0302(86)80701-9] [PMID: 3805441]

[122] Perez PF, de Antoni GL, Anon MC. Formate production by *Streptococcus thermophilus* cultures. J Dairy Sci 1991; 74(9): 5.
[http://dx.doi.org/10.3168/jds.S0022-0302(91)78465-8]

[123] de Vuyst L, Vandamme EJ. Antimicrobial Potential of Lactic Acid Bacteria.Bacteriocins of Lactic Acid Bacteria. 1 ed. Springer US 1994; pp. 91-142.
[http://dx.doi.org/10.1007/978-1-4615-2668-1_3]

[124] Dodd HM, Gasson MJ. Bacteriocins of lactic acid bacteria.Genetics and Biotechnology of Lactic Acid Bacteria. Springer 1994; pp. 211-51.
[http://dx.doi.org/10.1007/978-94-011-1340-3_5]

[125] Salminen S, von Wright A, Morelli L, *et al.* Demonstration of safety of probiotics -- a review. Int J Food Microbiol 1998; 44(1-2): 93-106.
[http://dx.doi.org/10.1016/S0168-1605(98)00128-7] [PMID: 9849787]

[126] Joseph PJ, Dave RI, Shah NP. Antagonism between yoghurt bacteria and probiotic bacteria isolated from commercial starter cultures, commercial yoghurts and a probiotic capsule. Food Aust 2013; 50(1): 4.

[127] Lankaputhra WEV, Shah NP, Britz ML. Survival of bifidobacteria during refrigerated storage in the presence of acid and hydrogen peroxide. Milchwissenschaft 1996; 51(2): 6.

[128] Leuschner RG, Heidel M, Hammes WP. Histamine and tyramine degradation by food fermenting microorganisms. Int J Food Microbiol 1998; 39(1-2): 1-10.
[http://dx.doi.org/10.1016/S0168-1605(97)00109-8] [PMID: 9562873]

[129] Sieber R, Bütikofer U, Bosset J. Benzoic acid as a natural compound in cultured dairy products and cheese. Int Dairy J 1995; 5(3): 19.
[http://dx.doi.org/10.1016/0958-6946(94)00005-A]

[130] Lim PL, Toh M, Liu SQ. *Saccharomyces cerevisiae* EC-1118 enhances the survivability of probiotic *Lactobacillus rhamnosus* HN001 in an acidic environment. Appl Microbiol Biotechnol 2015; 99(16): 6803-11.
[http://dx.doi.org/10.1007/s00253-015-6560-y] [PMID: 25846337]

[131] Suharja AAS, Henriksson A, Liu SQ. Impact of *saccharomyces cerevisiae* on viability of probiotic *Lactobacillus rhamnosus* in fermented milk under ambient conditions. J Food Process Preserv 2014; 38(1): 326-37.
[http://dx.doi.org/10.1111/j.1745-4549.2012.00780.x]

[132] Deepika G, Rastall RA, Charalampopoulos D. Effect of food models and low-temperature storage on the adhesion of *Lactobacillus rhamnosus* GG to Caco-2 cells. J Agric Food Chem 2011; 59(16): 8661-6.
[http://dx.doi.org/10.1021/jf2018287] [PMID: 21756003]

[133] Tamime AY, Marshall VM, Robinson RK. Microbiological and technological aspects of milks fermented by bifidobacteria. J Dairy Res 1995; 62(1): 151-87.
[http://dx.doi.org/10.1017/S002202990003377X] [PMID: 7738242]

[134] Dave RI, Shah NP. Viability of yoghurt and probiotic bacteria in yoghurts made from commercial starter cultures. Int Dairy J 1997; 7(1): 11.
[http://dx.doi.org/10.1016/S0958-6946(96)00046-5]

[135] Gomes AM, Malcata FX, Klaver FA. Growth enhancement of *Bifidobacterium lactis* Bo and *Lactobacillus acidophilus* Ki by milk hydrolyzates. J Dairy Sci 1998; 81(11): 2817-25.
[http://dx.doi.org/10.3168/jds.S0022-0302(98)75840-0] [PMID: 9839223]

[136] Rolfe RD. The role of probiotic cultures in the control of gastrointestinal health. J Nutr 2000; 130 (2S Suppl): 396s-402s.

[137] Rembacken BJ, Snelling AM, Hawkey PM, Chalmers DM, Axon AT. Non-pathogenic *Escherichia coli* versus mesalazine for the treatment of ulcerative colitis: a randomised trial. Lancet 1999; 354(9179): 635-9.
[http://dx.doi.org/10.1016/S0140-6736(98)06343-0] [PMID: 10466665]

[138] Reddy GV, Shahani KM, Banerjee MR. Inhibitory effect of yogurt on Ehrlich Ascites tumor-cell proliferation. J Natl Cancer Inst 1973; 50(3): 815-7.
[http://dx.doi.org/10.1093/jnci/50.3.815] [PMID: 4708161]

[139] Saavedra JM, Bauman NA, Oung I, Perman JA, Yolken RH. Feeding of *Bifidobacterium bifidum* and *Streptococcus thermophilus* to infants in hospital for prevention of diarrhoea and shedding of rotavirus. Lancet 1994; 344(8929): 1046-9.
[http://dx.doi.org/10.1016/S0140-6736(94)91708-6] [PMID: 7934445]

[140] Ouwehand AC, Salminen S, Isolauri E. Probiotics: an overview of beneficial effects. Antonie van Leeuwenhoek 2002; 82(1-4): 279-89.
[http://dx.doi.org/10.1023/A:1020620607611] [PMID: 12369194]

[141] Ives AR, Carpenter SR. Stability and diversity of ecosystems 2007.
[http://dx.doi.org/10.1126/science.1133258]

[142] Cook RJ, Weller DM. In Defense of Crop Monoculture 2004. www.cropscience.org.au

[143] Sturino JM, Klaenhammer TR. Bacteriophage defense systems and strategies for lactic acid bacteria. Adv Appl Microbiol 2004; 56: 331-78.
[http://dx.doi.org/10.1016/S0065-2164(04)56011-2] [PMID: 15566985]

[144] Erkus O, de Jager VC, Spus M, *et al.* Multifactorial diversity sustains microbial community stability.

ISME J 2013; 7(11): 2126-36.
[http://dx.doi.org/10.1038/ismej.2013.108] [PMID: 23823494]

[145] Bull AT, Slater JH. Microbial Interactions and Communities. London: Academic Press 1982.

[146] Taga ME, Bassler BL. Chemical communication among bacteria. Proc Natl Acad Sci USA 2003; 100(S2) (Suppl. 2): 14549-54.
[http://dx.doi.org/10.1073/pnas.1934514100] [PMID: 12949263]

[147] Sieuwerts S, de Bok FA, Hugenholtz J, van Hylckama Vlieg JE. Unraveling microbial interactions in food fermentations: from classical to genomics approaches. Appl Environ Microbiol 2008; 74(16): 4997-5007.
[http://dx.doi.org/10.1128/AEM.00113-08] [PMID: 18567682]

[148] Sieuwerts S. Microbial Interactions in the Yoghurt Consortium: Current Status and Product Implications. SOJ Microbiol Infect Dis 2016; 4(2): 1-5.
[http://dx.doi.org/10.15226/sojmid/4/2/00150]

[149] Lindgren SE, Dobrogosz WJ. Antagonistic activities of lactic acid bacteria in food and feed fermentations. FEMS Microbiol Rev 1990; 7(1-2): 149-63.
[http://dx.doi.org/10.1111/j.1574-6968.1990.tb04885.x] [PMID: 2125429]

[150] Teusink B, Wiersma A, Molenaar D, *et al.* Analysis of growth of *Lactobacillus plantarum* WCFS1 on a complex medium using a genome-scale metabolic model. J Biol Chem 2006; 281(52): 40041-8.
[http://dx.doi.org/10.1074/jbc.M606263200] [PMID: 17062565]

[151] Allende A, Martínez B, Selma V, Gil MI, Suárez JE, Rodríguez A. Growth and bacteriocin production by lactic acid bacteria in vegetable broth and their effectiveness at reducing *Listeria monocytogenes in vitro* and in fresh-cut lettuce. Food Microbiol 2007; 24(7-8): 759-66.
[http://dx.doi.org/10.1016/j.fm.2007.03.002] [PMID: 17613374]

[152] Loessner M, Guenther S, Steffan S, Scherer S. A pediocin-producing *Lactobacillus plantarum* strain inhibits *Listeria monocytogenes* in a multispecies cheese surface microbial ripening consortium. Appl Environ Microbiol 2003; 69(3): 1854-7.
[http://dx.doi.org/10.1128/AEM.69.3.1854-1857.2003] [PMID: 12620882]

[153] Noordman WH, Reissbrodt R, Bongers RS, Rademaker JL, Bockelmann W, Smit G. Growth stimulation of *Brevibacterium* sp. by siderophores. J Appl Microbiol 2006; 101(3): 637-46.
[http://dx.doi.org/10.1111/j.1365-2672.2006.02928.x] [PMID: 16907814]

[154] Juillard V, Foucaud C, Desmazeaud M, Richard J. Utilization of nitrogen sources during growth of *Lactococcus lactis* in milk. Lait 1996; 76: 13-24.
[http://dx.doi.org/10.1051/lait:19961-22]

[155] Samson JE, Moineau S. Bacteriophages in food fermentations: new frontiers in a continuous arms race. Annu Rev Food Sci Technol 2013; 4: 347-68.
[http://dx.doi.org/10.1146/annurev-food-030212-182541] [PMID: 23244395]

[156] Horvath P, Fremaux C, Fourcassie P. *Streptococcus thermophilus* strains. EU Patent WO2015007791 A1 2015.

[157] Kouwen RHM, Hanemaaijer LL, Van Sinderen D, McDonnel B, Mahoy J. Phage resistant lactic acid bacteria. EU patent WO2016012552 2016.

[158] Dwidar M, Monnappa AK, Mitchell RJ. The dual probiotic and antibiotic nature of *Bdellovibrio bacteriovorus*. BMB Rep 2012; 45(2): 71-8.
[http://dx.doi.org/10.5483/BMBRep.2012.45.2.71] [PMID: 22360883]

[159] Sockett RE. Predatory lifestyle of *Bdellovibrio bacteriovorus*. Annu Rev Microbiol 2009; 63: 523-39.
[http://dx.doi.org/10.1146/annurev.micro.091208.073346] [PMID: 19575566]

[160] Gobbetti M, Corsetti A. *Lactobacillus sanfrancisco* a key sourdough lactic acid bacterium: a review. Food Microbiol 1997; 14: 175-87.

[http://dx.doi.org/10.1006/fmic.1996.0083]

[161] Wynendaele E, Gevaert B, Stalmans S, Verbeke F, De Spiegeleer B. Exploring the chemical space of quorum sensing peptides. Biopolymers 2015; 104(5): 544-51.
       [http://dx.doi.org/10.1002/bip.22649] [PMID: 25846138]

[162] Benkerroum N, Oubel H, Mimoun LB. Behavior of *Listeria monocytogenes* and *Staphylococcus aureus* in yogurt fermented with a bacteriocin-producing thermophilic starter. J Food Prot 2002; 65(5): 799-805.
       [http://dx.doi.org/10.4315/0362-028X-65.5.799] [PMID: 12030291]

[163] Ivanova I, Miteva V, Stefanova Ts, *et al.* Characterization of a bacteriocin produced by *Streptococcus thermophilus* 81. Int J Food Microbiol 1998; 42(3): 147-58.
       [http://dx.doi.org/10.1016/S0168-1605(98)00067-1] [PMID: 9728685]

[164] Peirera Martins JF, Luchese RH. The assessment of growth compatibility between strains of *Lactobacillus bulgaricus* and *Streptococcus thermophilus.* Rev Inst Lactic Cândido Tostes (Brasil) 1988; 43: 11-3.

[165] Reddy GV, Shahani KM. Shahani K.M. Isolation of an antibiotic from *Lactobacillus bulgaricus.* J dairy sc. 1971; 54: p. 748.

[166] Stecchini ML, Sarais I, de Bertoldi M. The influence of *Lactobacillus plantarum* culture inoculation on the fate of *Staphylococcus aureus* and *Salmonella typhimurium* in Montasio cheese. Int J Food Microbiol 1991; 14(2): 99-109.
       [http://dx.doi.org/10.1016/0168-1605(91)90096-8] [PMID: 1777389]

[167] Benkerroum N, Daoudi A, Hamraoui T, *et al.* Lyophilized preparations of bacteriocinogenic *Lactobacillus curvatus* and *Lactococcus lactis* subsp. *lactis* as potential protective adjuncts to control *Listeria monocytogenes* in dry-fermented sausages. J Appl Microbiol 2005; 98(1): 56-63.
       [http://dx.doi.org/10.1111/j.1365-2672.2004.02419.x] [PMID: 15610417]

[168] Rodriguez AV, Manca de Nadra CM. Effect of pH and hydrogen peroxide produced by *Lactobacillus hilgardii* on *Pediococcus pentosaceus* growth. FEMS Microbiol Lett 1995; 128: 59-62.
       [http://dx.doi.org/10.1111/j.1574-6968.1995.tb07500.x]

[169] Luchese RH, Harrigan WF. Growth of, and aflatoxin production by *Aspergillus parasiticus* when in the presence of either *Lactococcus lactis* or lactic acid and at different initial pH values. J Appl Bacteriol 1990; 69(4): 512-9.
       [http://dx.doi.org/10.1111/j.1365-2672.1990.tb01543.x] [PMID: 2127265]

[170] Harris LJ, Fleming HP, Klaenhammer TR. Sensitivity and resistance of *Listeria monocytogenes* ATCC 19115, Scott A, and UAL500 to nisin. J Food Prot 1991; 54: 836-40.
       [http://dx.doi.org/10.4315/0362-028X-54.11.836]

[171] Thomas LV, Wimpenny JW, Barker GC. Spatial interactions between subsurface bacterial colonies in a model system: a territory model describing the inhibition of *Listeria monocytogenes* by a nisin-producing lactic acid bacterium. Microbiology 1997; 143(Pt 8): 2575-82.
       [http://dx.doi.org/10.1099/00221287-143-8-2575] [PMID: 9274011]

[172] Park JH, Seok SH, Cho SA, *et al.* Antimicrobial effect of lactic acid producing bacteria culture condensate mixture (LCCM) against *Salmonella enteritidis.* Int J Food Microbiol 2005; 101(1): 111-7.
       [http://dx.doi.org/10.1016/j.ijfoodmicro.2004.11.005] [PMID: 15878412]

[173] Moon NJ, Reinbold GW. Commensalism and competition in mixed cultures of *Lactobacillus bulgaricus* and *Streptococcus thermophilus.* J Milk Food Technol 1976; 39: 337-41.
       [http://dx.doi.org/10.4315/0022-2747-39.5.337]

[174] Gadaga TH, Mutukumira AN, Narvhus JA. The growth and interaction of yeasts and lactic acid bacteria isolated from Zimbabwean naturally fermented milk in UHT milk. Int J Food Microbiol 2001; 68(1-2): 21-32.
       [http://dx.doi.org/10.1016/S0168-1605(01)00466-4] [PMID: 11545217]

[175] Narvhus JA, Gadaga TH. The role of interaction between yeasts and lactic acid bacteria in African fermented milks: a review. Int J Food Microbiol 2003; 86(1-2): 51-60.
[http://dx.doi.org/10.1016/S0168-1605(03)00247-2] [PMID: 12892921]

[176] Codon S, Cogan TM, Piveteau P, O'Callaghan J, Lyons B, Eds. Stimulation of propionic acid bacteria by lactic acid bacteria in cheese manufacture. Cork, Ireland: Irish Agriculture and Food Development Authority 2001.

[177] Brussow H. Phages of dairy bacteria. Annu Rev Microbiol 2001; 55: 283-303.
[http://dx.doi.org/10.1146/annurev.micro.55.1.283] [PMID: 11544357]

[178] Sieuwerts S, Molenaar D, van Hijum SA, *et al.* Mixed-culture transcriptome analysis reveals the molecular basis of mixed-culture growth in *Streptococcus thermophilus* and *Lactobacillus bulgaricus.* Appl Environ Microbiol 2010; 76(23): 7775-84.
[http://dx.doi.org/10.1128/AEM.01122-10] [PMID: 20889781]

[179] Herve-Jimenez L, Guillouard I, Guedon E, *et al.* Postgenomic analysis of *streptococcus thermophilus* cocultivated in milk *with Lactobacillus delbrueckii* subsp. *bulgaricus*: involvement of nitrogen, purine, and iron metabolism. Appl Environ Microbiol 2009; 75(7): 2062-73.
[http://dx.doi.org/10.1128/AEM.01984-08] [PMID: 19114510]

[180] Arioli S, Della Scala G, Remagni MC, *et al. Streptococcus thermophilus* urease activity boosts *Lactobacillus delbrueckii* subsp. *bulgaricus* homolactic fermentation Int J Food Microbiol 2016; pii: S0168-1605(16): 30007-1.
[http://dx.doi.org/10.1016/j.ijfoodmicro.2016.01.006]

[181] Wang T, Xu Z, Lu S, Xin M, Kong J. Effects of glutathione on acid stress resistance and symbiosis between *Streptococcus thermophilus* and *Lactobacillus delbrueckii* subsp. *bulgaricus.* Int Dairy J 2016; 61: 22-8.
[http://dx.doi.org/10.1016/j.idairyj.2016.03.012]

[182] Sieuwerts S. Genome Sequences and Co-Cultures are Momentous for Present-Day Natural Optimization of Food Fermenting Microorganisms. EC Microbiol 2016; ECO.01: 1-2.

[183] Tamime AY. Fermented milks: a historical food with modern applications--a review. Eur J Clin Nutr 2002; 56(S4) (Suppl. 4): S2-S15.
[http://dx.doi.org/10.1038/sj.ejcn.1601657] [PMID: 12556941]

[184] Makras L, Van Acker G, De Vuyst L. *Lactobacillus paracasei* subsp. *paracasei* 8700:2 degrades inulin-type fructans exhibiting different degrees of polymerization. Appl Environ Microbiol 2005; 71(11): 6531-7.
[http://dx.doi.org/10.1128/AEM.71.11.6531-6537.2005] [PMID: 16269678]

[185] Hitt E. Biography of Todd R. Klaenhammer. Proc Natl Acad Sci USA 2005; 102(11): 3903-5.
[http://dx.doi.org/10.1073/pnas.0500351102] [PMID: 15753282]

[186] ERGO [homepage on the internet]. Genome sequence/annotation Lactobacillus paracasei F19 https://ergo.integratedgenomics.com/home

[187] Lee JH, O'Sullivan DJ. Genomic insights into bifidobacteria. Microbiol Mol Biol Rev 2010; 74(3): 378-416.
[http://dx.doi.org/10.1128/MMBR.00004-10] [PMID: 20805404]

[188] Lahtinen S, Ouwehand AC, Salminen S, Von Wright A, Eds. Lactic acid bacteria: microbiological and functional aspects. 4th ed., CRC Press 2012.

[189] Tamang JP, Watanabe K, Holzapfel WH. Review: Diversity of microorganisms in global fermented foods and beverages. Front Microbiol 2016; 7(3): 377.
[PMID: 27047484]

[190] De Filippis F, Genovese A, Ferranti P, Gilbert JA, Ercolini D. Metatranscriptomics reveals temperature-driven functional changes in microbiome impacting cheese maturation rate. Sci Rep 2016;

6: 21871.
[http://dx.doi.org/10.1038/srep21871] [PMID: 26911915]

[191] Smid EJ, Erkus O, Spus M, Wolkers-Rooijackers JC, Alexeeva S, Kleerebezem M. Functional implications of the microbial community structure of undefined mesophilic starter cultures. Microb Cell Fact 2014; 13 (Suppl. 1): S2.
[http://dx.doi.org/10.1186/1475-2859-13-S1-S2] [PMID: 25185941]

[192] Teusink B, Smid EJ. Modelling strategies for the industrial exploitation of lactic acid bacteria. Nat Rev Microbiol 2006; 4(1): 46-56.
[http://dx.doi.org/10.1038/nrmicro1319] [PMID: 16357860]

[193] Smid EJ, Molenaar D, Hugenholtz J, de Vos WM, Teusink B. Functional ingredient production: application of global metabolic models. Curr Opin Biotechnol 2005; 16(2): 190-7.
[http://dx.doi.org/10.1016/j.copbio.2005.03.001] [PMID: 15831386]

[194] Parente E, Cocolin L, De Filippis F, *et al*. FoodMicrobionet: A database for the visualisation and exploration of food bacterial communities based on network analysis. Int J Food Microbiol 2016; 219: 28-37.
[http://dx.doi.org/10.1016/j.ijfoodmicro.2015.12.001] [PMID: 26704067]

[195] Sieuwerts S, Teusink B, de Vos WM, van Hylckama Vlieg JET. Co-evolution of *S. thermophilus* and *L. bulgaricus* reveals the genetic and physiological bases behind their collaborations in the yoghurt consortium. Proceedings of the Annual conference 2015 of the Association for General and Applied Microbiology (VAAM). 2015, March 1-4; Marburg, Germany.

[196] Uchida M, Miyoshi T. Algal fermentation - The seed for a new fermentation industry of foods and related products. Jpn Agric Res Q 2013; 47(1): 53-6.
[http://dx.doi.org/10.6090/jarq.47.53]

[197] Page JA, Lubbers B, Maher J, Ritsch L, Gragg SE. Investigation into the efficacy of *Bdellovibrio bacteriovorus* as a novel preharvest intervention to control *Escherichia coli* O157:H7 and *Salmonella* in cattle using an *in vitro* model. J Food Prot 2015; 78(9): 1745-9.
[http://dx.doi.org/10.4315/0362-028X.JFP-15-016] [PMID: 26319730]

[198] Saxon EB, Jackson RW, Bhumbra S, Smith T, Sockett RE. *Bdellovibrio bacteriovorus* HD100 guards against *Pseudomonas tolaasii* brown-blotch lesions on the surface of post-harvest *Agaricus bisporus* supermarket mushrooms. BMC Microbiol 2014; 14(1): 163.
[http://dx.doi.org/10.1186/1471-2180-14-163] [PMID: 24946855]

CHAPTER 5

Biotechnological Innovations in Pro-, Pre-, and Synbiotics

Andrea Monteagudo-Mera[*] and **Dimitrios Charalampopoulos**

Department of Food and Nutritional Sciences, University of Reading, Whiteknights, P.O. Box 226, Reading RG6 6AP, United Kingdom

Abstract: Evidence has shown that human gut microbiota has an important effect on many aspects of human physiology including metabolism, nutrient absorption and immune function. Perturbation of the intestinal microbiota could lead to chronic diseases, such as inflammatory bowel diseases, obesity, diabetes, and colon cancer. Modulation of the microbiota by dietary interventions, especially by the use of probiotics, prebiotics and synbiotics has shown its potential for the treatment and prevention of diseases. Hence, the number of studies aimed to research the therapeutic effect of probiotic strains and prebiotics as well as the molecular mechanisms involved in the modulation of microbial populations and their environment by dietary intervention had increased considerably in the last 10 years. On the other hand, the current omics technologies are providing the tools needed to examine the community structure and function of the gut microbiota and therefore, understand its role in health and disease. The aim of this chapter is to provide a comprehensive overview of the research carried out on probiotics, prebiotics and synbiotics in the last 10 years as well as present the new biotechnologies that are contributing to the understanding of the host-microbiota interactions and the mechanisms of actions of pro-, pre- and synbiotics.

Keywords: Dysbiosis, Functional Properties, Gut Microbiota, Health Benefits, Host-Microbiota Interactions, Omics Technologies, Probiotics, Prebiotics, Synbiotics, Therapeutic Modulation.

INTRODUCTION

Nowadays, strong scientific evidences show how a healthy diet can provide multiple benefits to an individual's health. Actually, this concept was introduced by Hippocrates 2,400 years ago: "Let food be thy medicine and medicine be thy food". However, it was not until the last three decades when scientists, motivated by the rising population of health-conscious consumers as well as numerous

[*] **Corresponding author Andrea Monteagudo-Mera:** Department of Food and Nutritional Sciences. University of Reading, Whiteknights, P.O. Box 226, Reading RG6 6AP, United Kingdom; Tel: +44 (0) 118 378 7713; Fax: +44 (0) 118 931 0080; E-mail: a.monteagudo@reading.ac.uk

technological advances, began to focus on the investigation and identification of active components of foods with potential to reduce the risk of some chronic diseases. Here starts the era of Functional Foods where the development and sale of numerous health promoting food products are incremented. During the last 20 years, and due to the better knowledge about the role that the gut microbiome plays in the host health and due to the evidences showing how unbalances of the microbiota are involved in intestinal diseases, such as chronic inflammatory bowel diseases and colonic cancer, the concept of functional foods is moving towards the development of dietary supplementation aimed to modulate the composition and/or activity of intestinal microbiota. The most well-known strategies to shape the gastrointestinal tract ecosystem to date are the dietary use of probiotics, prebiotics or their combination, and synbiotics. Currently, some emerging of new technologies, such as Next Generation Sequencing methods (NGS), which allow us to have a better vision of the entire bacterial community in the gut, have triggered an expansion in the investigation of the impact of food products on the intestinal microbiota; this has resulted in an important biotechnological field with remarkable potential for innovation. In this chapter, we introduce the new technologies that have allowed us to study the human microbiota more deeply in the last years as well as the impact that consumption of probiotics, prebiotics and synbiotics can exert on the microbiome. We also present the most recent investigations on new probiotic strains aimed to be used in medical therapy, and novel technologies and economical sources for prebiotic production.

PROBIOTICS

The word probiotic is composed of the Latin preposition *pro* ("for") and the Greek adjective *Biotikos* derived from the noun *Bios* ("life"). This term, referring to microbes, was firstly used by Lilly and Stilwell [1] to define substances produced by microorganisms that stimulate the growth of other microorganisms. However, over a century ago, the Nobel laureate Elie Metchnikoff already theorized about the possibility to modify the gut flora with host-friendly bacteria from fermented products in detriment of proteolytic bacteria. Metchnikoff based his hypothesis on the fact that Bulgarian and Russian populations with a diet rich in dairy products showed a long-life expectancy with a high number of centenarians. Currently, the Food and Agriculture Organization of the United Nations and World Health Organization (FAO/WHO) define probiotic as "live microorganisms which when administered in adequate amounts, confer a health benefit on the host" [2].

Probiotics commercialized for human consumption include both bacteria and yeast and must be generally recognized as safe (GRAS). Most of the probiotics commercially available for human consumption belong to *Lactobacillus* and *Bifidobacterium* genera although species of other bacterial genera such as *Enterococcus, Streptococcus, Lactococcus, Bacillus, Escherichia* as well as the yeast genus *Saccharomyces* are also utilized. The beneficial properties of probiotic bacteria are strain-specific, since even related strains belonging to the same species have unique genetic and physiological features. For this reason, it is essential to perform a complete characterization of the microorganism. The selection criteria of probiotic microorganisms can be classified in three categories: Safety, functional and technological aspects. The final objective is to select potential probiotic strains that fulfil the safety requirements for human consumption (*e.g.*, non-pathogenicity and lack of antibiotic resistance) and possess intrinsic characteristic that allow them to resist the harsh conditions in the gastrointestinal tract (*e.g.*, low pH, bile salts, enzymes activities) and to reach the intestine where they will exert their intended beneficial effects. In addition, the probiotic strain must accomplish some technological aspects to survive during the product processing.

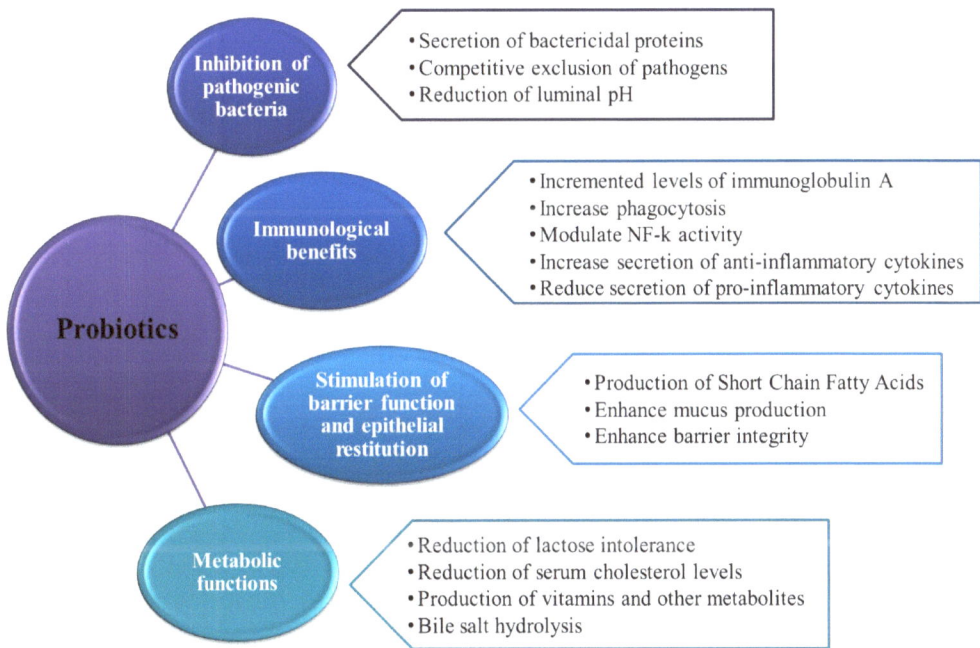

Fig. (5.1). Main mechanisms of probiotic action on human health.

There are several mechanisms of action through which probiotics can exert their beneficial properties on the health including: competitive exclusion of pathogenic microorganisms, enhancement of epithelial barrier, modulation of the immune system and production of beneficial metabolites, such as short chain fatty acids (SCFAs), bacteriocins and/or vitamins (Fig. **5.1**). In the past, several bacteria were isolated, characterized and intended to be used as probiotics on the basis of ecology and phenotypic characteristics. Nowadays, the new technologies have optimized the genomic characterization of novel potential strains reducing labour and cost and easing the study of the probiotic effect on the host microbiota. These new promising advances are contributing towards a deeper knowledge on the interaction between probiotic strains and host and/or gut microbiota, which will help to elucidate the molecular mechanism of probiotic activities.

Use of OMIC Technologies for the Investigation of Probiotic Functionality

The development of new omics-based technologies is contributing to further the understanding of probiotic functionality as well as their interaction with the host and gut microbiota (Fig. **5.2**). Comparative genomic hybridization (CGH) is the first valuable approach to perform genomic comparisons between new isolates and reference genomes. This methodology has been used for bifidobacteria [3] and some species of *Lactobacillus* [4] in order to analyse their genomic diversity; however, a limitation of this technology is its incapability to identify new genes since CGH analysis can only detect genes that are present in the reference genomes. Since the first bacterial genome was sequenced in 1995 [5], the new next generation sequencing tools enabled a rapid and cost effective whole genome sequencing (WGS). This has facilitated the genome sequencing of several lactic acid bacteria strains including probiotic bacteria, resulting in a new discipline called "probiogenomics" [6]. Recent genome comparison and pan-genomic studies (analysis of the genome sequences of a number of members of the same species) of probiotic species, such as *L. rhamnosus* [7], *B. breve* [8], *L. paracasei* [9] and *L. casei* [10], have provided insights into evolution, diversity and adaptability of probiotic bacteria to diverse ecological niches as well as the identification of candidate genes involved in health-promoting activities. For instance, a recent comparative pan-genomic study of *L. rhamnosus* strains demonstrated that the gene operon for Spa CBA pilus-related proteins which enhances cellular adhesion and gut colonization is located outside the core genome of this bacterial species, and therefore, it is not shared by all strains. Only 4 strains (human gut origin) of the 13 genomes analysed displayed the Spa CBA operon. This finding reflects the acquisition of this operon by lateral gene transfer, probably from *L. casei*, offering a functional advantage for the niche adaptation to the intestinal environment of some *L. rhamnosus* strains [7]. These genomic analyses can be a valuable and fast tool for the prediction of phenotypic and functional features of bacteria providing useful information for the selection of potential probiotic strains for further *in vitro* and *in vivo* studies.

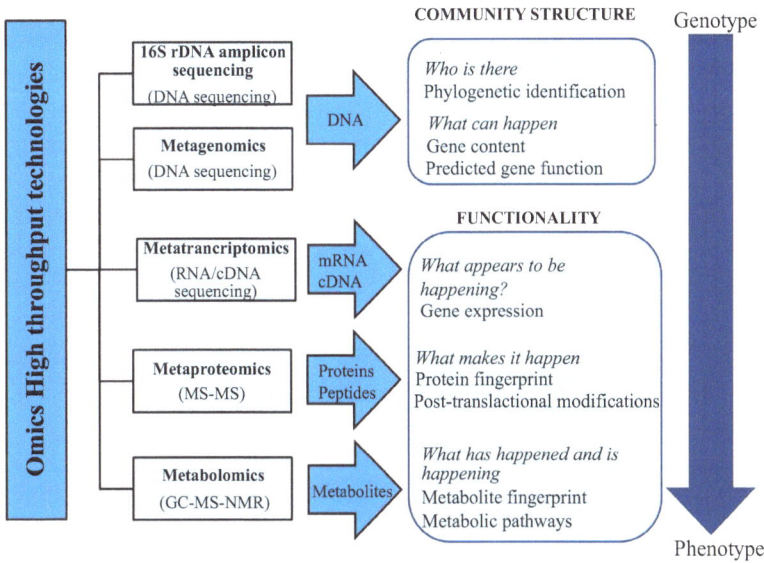

Fig. (5.2). Schematic representation of the different omics technologies used for the analysis of the composition and function of the intestinal microbiota.

The combination of these genomic studies with other "omics" methodologies like transcriptomics and proteomics has contributed to the understanding of the interaction between probiotic bacteria and host and/or microbial communities through the examination of gene expression and/or protein production within their natural ecological niche (Fig. **5.2**). In this regard, numerous *in vitro* studies have been conducted in the last few years in order to elucidate the response of probiotic strains to gastrointestinal conditions such as bile or acid stress [11 - 13], and the effects that food matrix and fermentation have on the viability and functionality of probiotic bacteria [12, 14, 15].

Recently, the introduction of cheap and fast high throughput DNA sequencing methods has also allowed the study of phylogenetic diversity in microbial communities by 16sRNA gene amplicon sequencing and/or the sequencing of the complete gene repertoire of microbial communities directly in their natural environment (Metagenomics) with a growing interest in the human microbiome. During the past few years, different platforms and analytical pipelines have been developed and optimized to generate more and longer reads per run; these include second generation sequencing platforms such as 454 Life Sciences Roche (*e.g.,* GS FLX Titanium), Illumina (*e.g.,* GA II, MiSeq, and HiSeq), Ion torrent Personal Genome machines (PGM) of Life Biosciences and the recently developed Single-Molecule Real-time (SMRT) technique introduced by Pacific Bioscience. During the last decade, there has been an increased awareness of the large number of microorganisms associated with human body (the human microbiota outnumbers the host cells by approximately ten times) [16]. The introduction of these new technologies has eased the study of the gut microbiota composition and provided a better understanding on how it is established during the life of the host. For example, to date it was thought that infants are born with a sterile gut and that the individual microbiome starts to be formed immediately during birth through their mother's vagina in a natural delivery or through the environment in deliveries by C-section. However, recent studies have shown that intestines are not always sterile since bacteria have been found in the meconium (first stool within hours of the birth) [17, 18] and also in the placenta and amniotic fluid during the pregnancy [19, 20]. During the first year of life, the gut microbiome changes dynamically in response to different environmental factors, such as delivery type, infant feeding practices, lifestyle and diet [21 - 24]. After the introduction of solid foods, the microbiota begins to shift toward an adult-type microbiota which possesses a higher bacterial diversity and a more complex, functional ecosystem. During adult life, the gut microbiome is dominated by Firmicutes and Bacteroidetes [25]. An interesting metagenomic study of human faecal microbiome revealed that human faecal samples can be separated into three robust clusters of microorganisms defined as "enterotypes" [26]. Each enterotype is characterized by variations in the abundance of one of the main bacterial genera in the gut: *Bacteroides* (enterotype 1), *Prevotella* (enterotype 2) and *Ruminococcus* (enterotype 3). The study showed no correlation of these three enterotypes with host properties, such as age, nationality, gender or body mass index (BMI); however, Wu and colleagues [27] showed that long-term diets were strongly associated with these three categories. Conversely, this classification is still controversial. Different factors in the methodology used, such as the variable region of the 16s RNA analysed or the sample processing method, could affect the correct categorization of enterotype. Moreover, recent works using next generation sequencing (NGS) have suggested that samples fall into continuous gradients [28, 29]. Therefore, grouping the microbiota of individual subjects into enterotypes may be oversimplifying a complex situation where the microbial composition and abundance varies along the gastrointestinal tract.

Moreover, several studies have focused in the last years on describing the healthy gut microbiota and aberrancies related to disease using NGS with the intention to understand the role of gut microbes in disease conditions. A first and common method is the use of gnotobiotic mice [30], which are delivered in aseptic conditions and raised as germfree in sterile isolators to be later colonized with a defined microbiota. The gnotobiotic mice can be used to replicate an individual's gut microbiome community under highly controlled environmental conditions or can be also colonized with just specific bacterial species in a reductionist approach to study the individual contribution of specific species or groups of bacteria. Gnotobiotic mouse models provide a valuable tool for studying the human microbiota and its modulation by environmental factors, such as diet. For instance, gnotobiotic mice were used in a recent metagenomic study aimed to investigate the role of the gut microbiota in severe acute malnutrition [31]. In this work, germ free mice received microbiota of twins from Malawi with or without severe malnutrition. The results showed that the combination of Malawian diet and microbiome from the individual with severe malnutrition produced a noticeable weight loss in mice as well as altered metabolic functions. Ridaura and colleagues [32] recently conducted a study with gnotobiotic mice fed on a high-fat diet and colonized with microbiota from monozygotic and dizygotic human twin pairs discordant for obesity. In this study, an increment was reported for both fat mass and fat free mass in animals receiving microbes from the obese twins, despite no significant differences between groups in daily food intake. In addition, diet-induced obesity was associated with higher proportions of the Firmicutes class, Erysipelotrichi, and lower abundance of Bacteroidetes phylum. In the same study, it was shown that co-housing mice containing the obese twin's microbiota (Ob) with mice harbouring lean co-twin's microbiota (Ln) prevented the increment in body mass and obese phenotype in Ob mice by the transmission of microbiota from Ln mice to Ob mice by the coprophagic habits of these animals.

However, the knowledge of the microbiota composition does not necessarily lead to an understanding of its function. Different microbial communities in terms of composition could have similar functional profiles. For instance, a study performed in 242 individuals reflected the high inter-individual variation in the microbial composition of 18 different body habitats using 16S rRNA sequencing. However, the metabolic functional profiles obtained by metagenomic studies from the same samples were surprisingly similar showing that different taxa could occupy similar functional niches [33]. Metagenomics has been widely used to characterize the gene repertoire of the human gut microbiota but this approach can just predict functional capabilities. The gap between the presence of genes and their expression has led to the development of other omic technologies like meta-transcriptomics and meta-proteomics where –meta- refers to the totality of the microbiota (Fig. **5.2**). Due to the fact that the place of action of probiotic bacteria is usually a complex microbial ecosystem, these technologies are useful methodologies to assess the impact of probiotic consumption on the gut microbiota providing a better overview of the functionality of probiotic bacteria in their habitat. For example, the impact of the probiotic strain *L. rhamnosus* LGG, in the gut microbiota of healthy elderly people, was studied [34]. This work revealed that although no changes in the composition of the microbiota were observed during the probiotic feeding trial, the transcriptional response of the gut microbiota was modulated by probiotic treatment. In this regard, it was found that LGG consumption increased the expression of genes involved in the flagellar motility, chemotaxis and adhesion of *Bifidobacterium* and the butyrate producers

Roseburia and *Eubacterium.* This finding suggests that LGG may promote interactions between microbiota and host epithelium. Another work carried out by Mac Nulty *et al.* [35] showed that administration of fermented milk strains to adult female monozygotic twin pairs for 4 weeks did not resulted in significant changes in the composition of bacterial species after consumption. However, metagenomics analysis detected significant changes in expression of microbiome-encoded enzymes involved in metabolic pathways related to carbohydrate metabolism.

Although the potential of these new technologies is enormous, the development of improved integrated analytical tools is essential in order to analyse the large meta-data obtained from these works, which can expand our knowledge on functional gut microbiomics. The number of microbiome studies is expected to increase substantially in the near future, allowing the exploration of novel routes for identification and screening of probiotic strains as well as new and effective strategies to define new targets for probiotic intervention. For this purpose, it will be necessary to conduct future clinical trials that provide greater insights to elucidate the therapeutic potential of probiotics.

Probiotics for Therapeutic Modulation

The use of probiotics in food products, especially dairy-based fermented products, has long been an extended practice due to the health benefits associated with their consumption. For example, the consumption of natural non-flavoured yoghurt improves lactose intolerance by the presence of bacterial lactase activity that enhances the hydrolysis of lactose to glucose and galactose in the small intestine, where they are rapidly absorbed or fermented [36]. However, interest in probiotics as a form of medical therapy, mainly in the prevention or treatment of gastroenterological diseases, has increased in recent years as result of the new information about the gut microbiota structure. Although the effectiveness of probiotics in some intestinal diseases such as Crohn's disease is still controversial and needs more clinical research [37], there are numerous human trials that show evidences supporting the efficacy of some probiotic strains against particular intestinal diseases despite the lack of understanding of the mechanism of action. For instance, the use of the probiotic mix preparation VSL#3 in conventional therapy for ulcerative colitis treatment has been described to be more effective in terms of remission rates than conventional therapy alone [38]. Moreover, the use of probiotics has been proven successful for the prevention of necrotizing enterocolitis (NEC) in preterm neonates [39], reducing the severity and duration of acute infectious diarrhoea [40, 41], and for the alleviation of constipation [42]. The Goldenberg Cochrane Database System Review indicates that probiotics are also helpful in the prevention of *C. difficile*-associated diarrhoea in both adults and children. In this regard, faecal microbial transplantations (FMT) have been shown to be a promising option to treat recurrent *C. difficile* infection [43, 44]. However, some safety issues related to the use of this treatment have led to the exploration of other options. Recently, a synthetic human stool mixture, composed of 33 different bacterial strains, has been developed and used to successfully treat two patients with severe *C. difficile* infection [45].

Although there has been a lot of interest on the effect of probiotics in the prevention or treatment of intestinal disorders, it has been suggested that probiotics could be effective not only in intestinal diseases but also in reducing the risk of immunologically mediated diseases like eczema. Allergic disorders are associated with a shift in the Th1/Th2 cytokine balance leading to activation of Th2 cytokines and the release of interleukin-4 (IL-4), IL-5 and IL-13 as well as IgE. Although the mechanism is poorly understood, probiotic intake seems to induce the activation of dendritic cells and TH1 response [46]. Although contradictory results have been observed in the efficacy of probiotics for allergic diseases, such as respiratory allergic symptoms and food allergy [47, 48], some promising studies have shown the efficacy of probiotic administration in the prevention of eczema [49, 50]. Marlow and co-workers [50] conducted a study with 331 children who possessed a genetic variation in the Toll-like receptors that confer a high risk of developing eczema. This work showed that the pre- and post-natal supplementation with the probiotic *Lactobacillus rhamnosus* HN001 significantly reduced the risk of eczema or eczema severity. Based on this point, we should also mention the topical probiotics as an emerging approach, which is still in the phase of early development, to treat atopic dermatitis and barrier repair by showing activity against pathogenic biofilms. A recent *in vitro* study has demonstrated that the topical application of the recognized probiotic strain LGG increases the epidermal keratinocyte survival during exposure to *S. aureus* [51]. Besides, topical probiotics could be of special interest in the treatment of severe acne. Some studies have reported that probiotic species like *B. longum* and *L. paracasei* [52, 53] could attenuate skin inflammation and have inhibitory activity against *Propionibacterium acnes*. A hypothesis on the topical use of probiotics to promote the healing of diabetic foot ulcers (DFU) and prevent infection has also been proposed lately, due to the adverse effects obtained with antibiotic therapies [54]. Probiotic bacteria have been described to possess antimicrobial activity by competitive exclusion of pathogens and/or production of antimicrobial metabolites. In addition, new data based on human fibroblast cultures and DNA microarrays suggest that probiotics possess the ability to naturally stimulate the skin's immune response, which would support the hypothesis of using topical formulations of probiotics for the treatment of DFU. However, the use of topical probiotic is not aimed only to skin disorders. For example, the toothpaste PerioBiotic™, that contains a probiotic strain of *L. paracasei* LP-33, was launched in the market by Designs for Health™ in 2009. A previous study reported that this probiotic strain is able to co-aggregate with *Streptococcus mutans*, which presents a potential to alter the plaque biofilm and therefore reduce the dental caries [55]. However, more *in vitro* and *in vivo* studies are needed to be conducted in order to support the preliminary results of topical use of probiotics and also evaluate the long-term effects of using these products.

On the other hand, specific strains of probiotic bacteria have also shown a potential role as adjuvants for improving the vaccination response, and thus reduce the risk of infection. An adjuvant needs to fulfil certain requirements, such as targeting specific immune cell types, inducing sIgA responses as well as being safe for consumption [56, 57]. In this regard, probiotics appear to meet all these requirements, since they are GRAS microorganisms and can modulate DCs, T and B cell populations as well as cytokine and antibody responses [57, 58]. Although most of the studies published so far have been conducted in animals, some promising results have been obtained against the human rotavirus (HRV) using *L. rhamnosus* (LGG) and *B.lactis* (Bb12) [59, 60]. On the other hand, some works have also shown the possible efficacy of using probiotic bacteria as adjuvants in influenza vaccines. However, most of the studies at this respect are small with a low statistical power [61]. Moreover, the different responses obtained in the different works reflect the need for optimizing the design of human trials to understand the possible role of probiotic bacteria as adjuvants.

Future Studies and Next Generation of Probiotics

Most of the current strains of probiotics belong to the genera of *Lactobacillus* and *Bifidobacterium* and their utilization as treatment or prevention of gastrointestinal diseases is still limited. However, the use of NGS technologies has recently opened the possibility to define the composition of the gut microbiota in healthy and disease status providing new information that could be used for the development of novel probiotic strains, that have therapeutic effects in specific intestinal diseases. For instance, the depletion of *Faecalibacterium prausnitzii*, a butyrate producer, in intestinal bowel diseases (IBD) has been extensively studied [62]. Several studies carried out in a mouse colitis model showed that the metabolic activity of this bacterial species, belonging to the *Clostridium* cluster IV, is linked to anti-inflammatory effects in the colon [63, 64]. Another example is the propionate-producing *Akkermansia muciniphila*, which is suggested for improving metabolic syndrome. A recent study conducted by Shin *et al.* [65] demonstrated how the increase of this species in obese mice contributed to the improvement in energy and glucose metabolism. *Bacteroides uniformis* [66] and species of Clostridia clusters IV, XIVa and XVIII [67] have been also proposed as potential probiotic strains.

The progress in knowledge of the microbiota function and their composition has also led to the exploration of the effect of the intestinal microbiome not only on the gastrointestinal tract but also on the general host physiology. In this regard, in the past few years, numerous studies using animal models have found an interesting association between gut microbiota and stress-related behaviours. This gut–brain axis has become a novel research topic with a huge potential in the neuroscience field. Several studies using rodents have shown the importance of the microbiota in the development of brain systems related with emotional and stress behaviours [68 - 70]. Other works have also revealed how germ free (GF) mice show reduced stress-related behaviours [71, 72]. Microbiota activities, such as the production of bioactive compounds, nutrient absorption at the mucosal surface or regulation of the barrier integrity can impact the neurological function through the vague nerve or in some cases by systemic circulation. Therefore, a potential strategy to modify the brain function and behaviour could be through dietary intervention aimed to modify the composition of the microbiota and their metabolic activity. For instance, high fat diets have been related with an increase in the anxiety of mice; however, the administration of *L. helveticus* to wild type (wt) mice fed on high fat diet during 21 days prevented anxiety-like behaviours. Interestingly, this prevention was not effective in mice knockout for IL-10 revealing the importance of immune signalling in the modulation of behaviour [73]. The strong evidence obtained with animal experiments has led to the investigation of the gut-brain axis in humans. A recent study using magnetic resonance imaging showed that the intake of probiotic yoghurt by healthy women for 4 weeks affected the activity of brain regions controlling central processing of emotion and sensation [74]. In the last years, a special interest has been given to autism spectrum disorder (ASD). As people diagnosed with ASD frequently suffer from gastrointestinal disorders, some of the current works are specifically aimed to improve certain ASD symptoms through dietary intervention, including probiotic therapies. A double-blind, placebo-controlled study conducted by Parracho *et al.* [75] showed that probiotic supplementation with the strain *L. plantarum* for 3 months resulted in reduced GI problems and an increment in the abundance of *Lactobacillus* species. Another recent trial tested the effect of *L. acidophilus* supplementation for 2 months in autistic children and showed an improvement in the concentration ability and a reduction of D-arabinitol, a metabolite of *Candida* species, whose excretion is elevated in ASD patients [76]. Despite this evidence, more clinical and mechanistic studies are necessary to understand the effect of microbiota on gut-brain axis and other physiological functions in order to translate this knowledge into clinical therapies by developing effective probiotic interventions, that allow us to improve and/or prevent symptoms of some neurological disorders.

Probiotic Products

Dairy products are the main fermented products containing probiotic bacteria available in the market nowadays. The increase in popularity of vegetarian diets, the high cholesterol content of dairy products together with lactose intolerance, suffered by some parts of the population, have led to the exploration of other alternative food matrices in which probiotics are used, such as vegetables or cereals. In this regard, research works aimed to develop novel probiotic cereal foods, such as beverages [77], sorghum flour based yoghurt [78] or bread [79], can be found in the recent literature in the last years. Indeed, non-dairy probiotic products using cereals have a great potential since there are several cereal based-foods processed by fermentation. The research of novel probiotic/starter strains to develop these products is a research topic of growing interest, especially in some African countries where the access to dairy products is low [80] and the fermentation of cereals by lactic acid bacteria (LAB) cultures is the main processing method for the production of beverages, gruels and porridge [81]. Due to the fact that traditional fermented products enriched with probiotics may be more easily accepted than dairy products by the population of Africa, we can find works focused on the selection of potential probiotic microbial strains isolated from traditional fermented cereal foods of Africa that could be exploited as starter cultures to produce cereal-based functional products [78, 82], and hence could contribute to the improvement of the health status of children in these countries.

Fruit juices are also promising option as probiotic carriers and a challenge for the probiotic industry. These beverages have a well-established market sector due to their high content of essential nutrients and the fact that their taste profiles appeal to all age groups. However, one of the main inconveniences is the acidic conditions of these beverages that could affect to the viability of probiotic strains. Some fruit juices can reach pH values of around 3.5, while dairy fermented drinks have pH values between 4 and 5. In this regard, innovation on appropriate technologies to improve the viability of probiotics in fruits juices as well as selection of more suitable strains for this purpose is the main challenge. For instance, Bhat and colleagues [83] investigated a biodegradable polymer -poly-γ-glutamic acid (γ-PGA)-, produced from a non-toxic microbial species of *Bacillus*, as a new cryoprotectant that can be used to improve the survival of *Bifidobacterium* strains both during storage in acidic orange and pomegranate juices and during the gastrointestinal passage. Other approaches investigated are the fortification of the probiotic juices with prebiotics, and the adaptation and induction of resistance by exposure of probiotics to a sub-lethal stress [84, 85]. Other challenges in the production of these beverages, fortified with probiotics, is to maintain the sensory properties of the fruit juice, since although the probiotic bacteria can survive in the product, occasionally the sensory quality of the beverage is not acceptable. In these cases, researchers have focused their efforts on improving the sensory attributes of these products by adding pleasant aroma [86]. Other works though have shown that the addition of probiotic bacteria such as *L. reuteri* and *L. casei* in specific fruit juices did not affect the overall acceptance [87, 88]. The research in non-dairy food carriers has led to the search and selection of new potential probiotic strains from non-dairy sources as well as the investigation of different methods to enhance the survival of probiotic bacteria in order to ensure their functional properties, by improving the adaptability and application of these strains in the food product.

Although dairy products and probiotic-fortified foods are the most common forms of probiotics, other systems, such as capsules, tablets or powder have been also developed to deliver probiotic bacteria. These probiotic formulations have advantages compared to other probiotic foods, such as accurate dosages, ease of administration, good patient acceptability and suitability for large-scale production. On the other hand, the production of these commercial supplements requires the production of probiotics at large scale. Most of probiotic bacteria currently marketed belong to the LAB group. This group of bacteria are fastidious microorganisms, that need a complex medium to grow such as MRS, M17, and skim milk. The increasing demand of probiotics in the last years has led to the search of new low-cost growing media to reduce the cost of probiotic biomass production at large scale. A cheap alternative is the use of agro-industrial wastes as substrates. Usually, residues that are not going to be used for other purposes are eliminated, but firstly, they should be pre-treated to reduce the polluting load generating a cost to the industries. In this regard, the potential of wastes obtained from some fruits [89], dairy liquid [90] and cereals [91] starts to be investigated as growing media for LAB and *Bifidobacterium*. The utilization of these wastes would lead to significant savings for the industry and increase the economic value of the wastes as well as contribute to sustainable development.

New Trends in Microencapsulation

The incorporation of probiotics in foods is one of the main research challenges, as the decrease in cell viability in the food carriers during storage and in the gastrointestinal tract can compromise the probiotic efficacy. One of the techniques that has been developed during years to improve the bacterial survival in the food products is microencapsulation. Microencapsulation can be defined as the process in which cells are retained within an encapsulating membrane to reduce cell injury or cell death and potentially control the release of the probiotic in the gut [92]. During the last decade, different techniques of microencapsulation, such as emulsion, extrusion or spray drying have been developed in order to improve the protection of probiotic bacteria. The choice of the most suitable technology depends on different factors, such as the conditions of food processing, storage conditions or cost constraints. Different materials, such as polysaccharides (alginate, starch, gellan), proteins (milk, gelatin) and lipids are used to encapsulate probiotic bacteria. A current trend in the microencapsulation process is the use of mixtures of different types of biopolymers to improve cell viability by altering the release rates of microencapsulation systems. For instance, in the work performed by Sandoval-Castilla *et al.* [93], beads made with sodium alginate and amidated low methoxyl pectin blends in 1:4 and 1:6 ratios provided a much better protection to the entrapped probiotic than using alginate or pectin alone under simulated gastrointestinal conditions. Another example is the work conducted by Cabuk *et al.* [94], where blend microspheres from pullulan and whey proteins enhanced the survival of probiotic *L. acidophilus* NRRL-B 4495 during spray drying. The addition of some protectants, such as prebiotics is also of interest, since production of synbiotic microcapsules may have considerable potential for the food industry. Co-encapsulation of some probiotic strains like *E. faecium* with chicory in alginate beads [95] or *Lactobacillus acidophilus* and *L.casei* with galactooligosaccharides [96] have also been performed to evaluate the survival of probiotics in these capsules *in vitro*. Future *in vivo* studies using animal models are needed to evaluate the efficacy of the co-encapsulation of bacteria with prebiotics and to determine the synbiotic effects when they are released in the gastrointestinal tract.

Another important factor that should be considered is the place of action of the probiotic bacteria, in order to choose the most appropriate technique and material of encapsulation; the aim here is to release the probiotics in the area where they should exert the beneficial effect. For instance, probiotics administrated as a therapy for ulcerative colitis (UC) must be viable in the small intestine and reach the large colon where they should be released. To achieve the release of encapsulated probiotics in the target region, there are some approaches like mechanical breaking by peristaltic movements in the colon, pH gradient, time dependent systems or utilization of material carriers, such as pectin that only are degraded by enzymatic activity of colon microbiota [97]. This last approach is maybe more interesting, taking into account the large number of bacterial population and their enzymatic activities, and the fact that the capsule degradation in this case would be an exclusive event of the colon region independently of the GI transit time and pH. So, although microencapsulation techniques started to be investigated decades ago, the synthesis and characterization of new capsules for target delivery of probiotics, aimed to treat intestinal diseases, continue nowadays. In this regard, a recent *in vitro* and *in vivo* study investigated the co-encapsulation in pectin based beads of mesalazine (a drug to treat UC) with the probiotic *L. acidophilus* which helps to restore the microbiota. Results from this work showed that both mesalazine and the encapsulated probiotic strain were protected in gastric environment and reached the colonic region; the *in vivo* studies performed in mice showed a reduction of ulcer scores in induced colitis [98]. In addition, other microencapsulation delivery systems are being investigated in order to improve the release of probiotic bacteria in colon. For instance, Chen *et al.* [99] showed that sub-100uM Ca^{2+} alginate microcapsules enhanced bacterial survival in the gastrointestinal tract; however, calcium ions reduced the muco-adhesive properties of alginate hydrogel. In a following study by the same research group [100], chitosan coated Ca^{2+} alginate microcapsules with improved muco-adhesive ability to colonic mucosa tissue at slight acid neutral pH were developed in order to prolong the time of these capsules in human colon, where probiotics should exert its action.

Although many *in vitro* studies have been reported to evaluate bacterial survival in microcapsules, few *in vivo* studies have been performed so far. A recent study conducted by Wurth and colleagues [101] showed that milk-protein-based microcapsules increased survival of *Lactobacillus* strains in simulated gastric juice, however a lack of protection was observed in a murine gastrointestinal system 3h or 24h after oral administration. Although, this study cannot be extrapolated and does not prove that microcapsules do not protect probiotic bacteria in the gastrointestinal tract, it demonstrates the need of developing new *in vivo* or more complex *in vitro* systems to validate the potential protective effect of microencapsulation.

PREBIOTICS

The concept of prebiotic was introduced for the first time by Gibson and Roberfroid in 1995 [102] as a 'non-digestible food ingredient that beneficially affects the host by selectively stimulating the growth and/or activity of one or a limited number of bacteria in the colon, and thus improves host health'. Since the introduction of this concept, the interest in prebiotics has quickly incremented resulting in the investigation of different non-digestible oligosaccharides (NDO) and their potential effects on the microbiota modulation. The concept of prebiotic has been discussed and refined several times since the first time it was introduced. The last definition of prebiotic was proposed recently by Gibson *et al.* [81] and describes prebiotics as a "selectively fermented ingredient that results in specific changes in the composition and/or activity of the gastrointestinal microbiota, thus conferring benefit(s) upon host health'. It is important to mention that although the definition of prebiotic and dietary fibre overlaps causing confusion, the main difference between these two terms is the selectivity of microbial fermentation. While most of prebiotics are dietary fibres, most of dietary fibres do not induce a specific change in the composition of the intestinal microbiome. In this regard, species belonging to *Bifidobacterium* and *Lactobacillus* genera are the preferred target microorganisms for prebiotics. Especially, a specific bifidogenic response is frequently desired when the prebiotic activity of novel compounds is tested. However, the introduction of NGS tools has revealed that the consumption of prebiotics can also enrich other members of the intestinal community related to gut health status, such as *Faecalibacterium prausnitzii* and *Akkermansia muciniphila* [103]. Moreover, it has been even observed that bacterial species belonging to groups of bacteria, such as Clostridia or *Bacteroides*, that were considered detrimental for years, could also show beneficial effects [66, 67, 104]. To this end, it has been recently published that some strains of Clostridia showed to have anti-inflammatory effects in a colitis murine model [67]. These new findings suggest that stimulation of *Bifidobacterium* should not necessarily be the only aim of prebiotics, as other beneficial groups of bacteria could appear as novel prebiotic targets in next years.

A food ingredient with prebiotic activity must not be hydrolysed or absorbed in the stomach or small intestine in order to reach the colon area, where it will exert its beneficial effects. The majority of the beneficial health effects associated with prebiotics are related to the production of short chain fatty acids (SCFAs), as result of the prebiotic fermentation in the colon, or to the immunological activities resulting from microbiota modulation (Fig. **5.3**). SCFAs have multiple beneficial effects on health. For instance, SCFAs contribute to the acidification of the luminal pH, which influences the composition of the colonic microbiota and increases mineral absorption. Acetate, propionate and butyrate are the main products generated by prebiotic fermentation (usually in a molar rate of approximately 60: 20: 20 [105, 106]). Butyrate is of particular interest because it is an important source of energy for colonic cells (it provides around 50% of the daily energy requirements of colonic mucosa) and it has tumor-suppressive properties in mammalian intestinal mucosa through the inhibition of histone deacetylase expression (an important epigenetic determinant in cellular differentiation and potential anti-cancer mediator) [107]. In addition, butyrate also shows a strong anti-inflammatory effect by inhibiting the expression of lipopolysaccharide (LPS)-induced cytokines, such as IL-6 and IL-12p40 [108].

Galacto-oligosaccharides (GOS) and inulin type fructans such as inulin and oligofructose (FOS) are the most extensively studied prebiotics; however, other candidate prebiotics such as xylo-oligosaccharides (XOS) are being investigated recently, due to their beneficial effects and low cost of production, since they can be obtained from agricultural wastes.

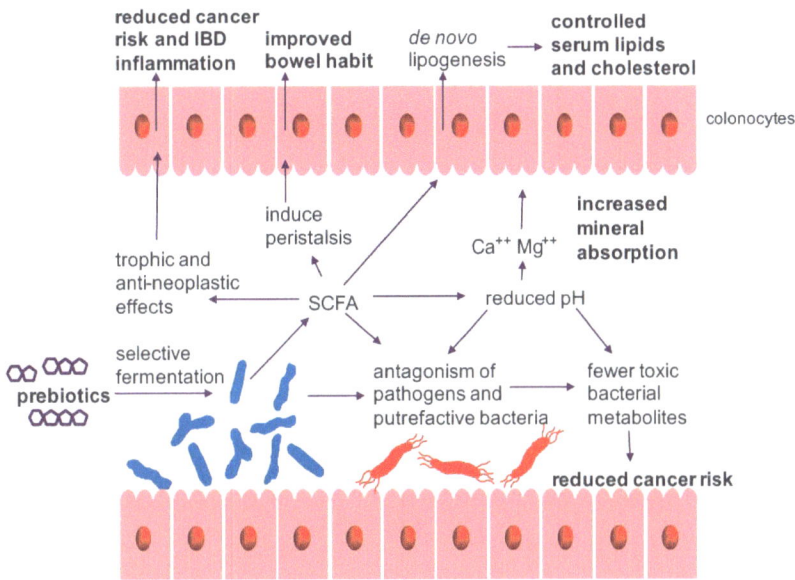

Fig. (5.3). Diagram illustrating the mechanisms of prebiotic action through stimulation of beneficial microbial activities. Figure taken by Crittenden [191].

Most Common Prebiotics

Inulin-Type Fructans

Fructans are carbohydrates with one or more fructosyl-fructose links, constituting the majority of glycosidic bonds. These oligosaccharides can be inulin type, that consists of short to medium length chains of β-D fructans, in which the fructosyl units are bound by a β-(2-1) glycosidic linkage, or levan type fructans with β-(6,2)-D-fructofuranosyl units.

Inulin type fructans are the best known and studied fructans. They can be obtained from sucrose by synthesis or hydrolysis of inulin; inulin occurs naturally in chicory, garlic, onion, asparagus, banana, leek, artichoke and other vegetables. Fructans represent one of the best studied prebiotics and the most commercialized products in the market. Inulin type fructans includes inulin (DP= 2-60, DP av= 12), inulin HP (DP=10-60, DPav= 25) and FOS or oligofructose (DP=2-8, DPav= 4) [109].

Inulin is moderately soluble in water (up to 10% in room temperature) while fructooligosaccharides (FOS) show higher solubility and sweetness than inulin. Both are stable in a pH range of 4.0-7.0 and are stable to heat at pH > 5 [110]. The initial source for the extraction of inulin-type fructans is mainly chicory root, and usually they are extracted by hot water to be later processed by enzymatic hydrolysis. FOS can be produced on industrial scale by two different approaches: a) by the hydrolysis of inulin through the action of endoinulinases or b) by the transfructosylation of sucrose using fructosyl transferase (FTase) enzymes, that are usually found in fungi such as *Aspergillus* sp., *Aureobasidium* sp. or *Penicillium* [111]. FOS obtained from sucrose usually possesses a DP between 2 and 4 and are also named short chain FOSs. Usually, the short chain FOS, synthesised from sucrose, are used as prebiotic ingredients while longer chain oligosaccharides obtained from inulin are more commonly used as fat replacers. The purity of oligofructoses in the market ranges between 55 and 99% [112]. One of the limitations in the production of short chain FOS from sucrose is the low yield obtained, due to the high amount of glucose co-produced during the enzymatic synthesis, which inhibits the transfructosylating reactions [113]. In this regard, different biotechnological approaches have been developed in the last years in order to improve the production efficiency. One of these methods is the use of mixed enzyme systems or microorganisms containing β-fructofuranosidase and glucose oxidase, which convert the glucose generated into gluconic acid, which is removed later by adsorption using ion exchange resins or precipitation techniques [114, 115]. In addition, several ion-exchange resins are also being evaluated for use in chromatography systems to improve the purification of FOS from mixtures of sugars at large scale [116].

Galacto-oligosaccharides

The term galacto-oligosaccharides is used for non-digestible oligosaccharides derived from lactose with a degree of polymerization (DP) between 2 and 10, comprising of galactose units with a terminal glucose and/or galactose. GOS can have α- or β- configurations depending on the nature of the glycosidic bonds but β-(1-4) GOS are the most commonly used. The most reported effect of GOS is the selective growth stimulation of bifidobacteria. This strong bifidogenic effect makes GOS the main prebiotics used in infant formulas in order to mimic the beneficial effects of the oligosaccharides present in human breast milk [34], although the structures are different. The different GOS composition and structures between GOS products can influence their effect on the intestinal microbiota [117]. The composition of GOS in the commercial products varies depending on the enzyme and production method used. Traditionally, GOS have been produced using β-galactosidases which require high concentrations of lactose for the synthesis to occur. Some commercially available GOS preparations such us Oligomate 55 (Yakult Pharma, Ind. Co.), Vivinal GOS (Friesland Campina) or Bimuno (Clasado Inc) contain a GOS purity between 52 and 72% [118]. In recent years, multiple β-galactosidases and reactor designs have been assessed in order to improve the production efficiency of GOS and achieve purer formulations [26, 27]. β-Hexosyl-transferases have also received attention recently, since different studies have shown that these enzymes have transgalactosylation activity even at low lactose concentrations and the conversion into GOS is close to 75%. For instance, a recent work conducted by Dagher and colleagues [119] showed a soluble recombinant β-hexosyl transferase from *Sporobolomyces singularis*, that permitted a more controllable process to reach a purity of 95% GOS with only a residual small proportion of lactose.

Novel Oligosaccharides: Xylo-oligosaccharides

XOS are sugar oligomers made up of xylose units that can be found naturally in plant cell walls and some algae [120]. Until few years ago, the only prebiotics available in the market were mainly obtained from lactose, sucrose or inulin. The lignocellulosic biomass is an emerging resource that presents high potential as feedstock and as source for prebiotics. In this regard, XOS are the only oligosaccharides currently marketed obtained from lignocellulosic materials although their market is still small in Europe and US, with Asia being the main producer of commercial XOS. Production of XOS is mainly performed by autohydrolysis of xylan, which consists of a backbone chain of xylose units linked by β-(1,4)-glycosidic bonds and branched by α-(1,2)-glycosidic bonds with 4--methylglucuronic acid groups. In addition, hydrolysis of xylan with additional branches of arabinofuranose units linked to the backbone results in the generation of arabino-xylooligosaccharides (AXOS) [121].

XOS/AXOS possess some advantages over other prebiotics already established in the market. For instance, XOS exhibit higher resistance to pH and high temperatures than FOS [122]. These properties could make them suitable for use in specific products such as low pH beverages or in high temperatures food processing. Moreover, studies have shown that XOS might provide biological effects at lower doses than other oligosaccharides. Daily doses between 5 and 20 g depending on the type of prebiotic are usually needed to exert beneficial effects on the consumer. However, XOS have shown to be effective with a smaller daily intake [123]. In this regard, a recent human study conducted by Finegold and colleagues [124] demonstrated that doses as low as 1.4 g/day were enough to detect an increase in bifidobacteria after 8 weeks of treatment. Recently XOS have also displayed some promising benefits regarding immunomodulatory effects. An *in vitro* study in macrophages [125] reported the induction of pro-inflammatory TNF-α, IL-1B, and IL-6 indicating the ability of stimulation of macrophages. However, since a big stimulation can cause collateral damages to cells, the anti-inflammatory activity was also tested when macrophages were stimulated with LPS in the presence of XOS. In this case, XOS were able to inhibit TNF-a, IL1B and IL6 production in LPS stimulated cells. Furthermore, a human trial [126] showed reduced plasma LPS concentration and immunomodulatory effects, when XOS in combination with inulin were administrated to healthy individuals for 4 weeks. The beneficial properties of XOS/AXOS and the fact that they can be produced from different agricultural residues, such as oil palm empty fruit bunch [93] or brewer's spent grain [94], makes them promising prebiotic candidates.

Other Novel Sources of Prebiotics

Research on the prebiotic potential of polysaccharides, extracted from seaweed, is increasing in recent years due to the increased consumption of seaweeds and their significant potential as functional food components or nutraceuticals [127]. The composition of polysaccharides, which are mainly located in the seaweeds cell walls, depends on several factors, such as age, species and habitat of algae. For instance, red algae are rich in floridean starch and sulphated galactans while alginates are predominant in brown algae [128]. A recent study evaluated the potential prebiotic activity of polysaccharides extracted from brown algae, alginate and laminaran on the intestinal microbiome of rats using pyrosequencing. Data obtained from this work concluded that these compounds induced a higher production of SCFA as well as promoted specific changes in the composition of the microbiota, leading to an increase in unspecific groups of bacteria, such as Lachnospiraceae family, *Allobacum* genus and *Clostridium ramosum*.

Yacon root (*Smallanthus sonchifolius*) is a traditional food among the Andean population. However, in Europe it is considered as a novel food product and must therefore fulfil the safety requirements before being introduced in the market. Yacon tubers are rich in fructans, especially in inulin-type compounds with potential prebiotic activity [129, 130]. Additionally, this product lacks starch, which makes it also suitable for diabetics' diet.

Non-digestible long chain β-glucans extracted mainly from edible mushrooms, oat or barley have been also proposed and recently investigated as novel prebiotics with promising results [131 - 134].

Health Benefits Associated with Prebiotics

The main effect of prebiotics is to selectively modify the gut microbiome by stimulating beneficial groups of bacteria, such as *Bifidobacterium* and *Lactobacillus*. The growth promotion of bifidobacteria or also called "bifidogenic effect" is usually the main consequence after prebiotic administration. The fermentation of prebiotics in the colon especially by *Bifidobacterium* induces positive physiological effects in the gut environment that generally result in a decreased risk of diseases, mainly the intestinal ones. Additionally, recent investigations have shown that these oligosaccharides *per se* can also show a direct effect on the host's immunomodulation, enhancing the immune response.

Colorectal Cancer (CRC)

A high-fat diet which is low in dietary fibre has been associated with increased risk of CRC, which is the most common cancer in Western populations (35 cases per 100,000 hab) [135]. In the last years, a number of research studies have shown the potential of prebiotics in reducing the risk of CRC as well as other intestinal diseases [136].

Animal model studies using different types of prebiotics have shown a positive role in the prevention and treatment of CRC in the last decade [137 - 140]. One of the mechanisms proposed for CRC prevention is the increase of butyrate as a result of the prebiotic fermentation. Butyrate is the main energy source of healthy colonocytes; however, under colorectal conditions lower doses of butyrate are metabolized due to the Warburg effect [141]. This effect is associated with the utilization of glucose by cancerous colonocytes as main energy source resulting in the accumulation of butyrate as a tumor-suppressive metabolite in the nucleus where it functions as a histone deacetylase (HDAC) inhibitor to stimulate histone acetylation, induce apoptosis and inhibit cell proliferation. A recent study [142] performed in gnotobiotic mice with induced CRC strongly supported this mechanism. Mice, colonized with four commensal bacteria plus or minus the butyrate producer *Butyrivibrio fibrisolvens*, were fed on diets supplemented with 6% FOS/Inulin. After the treatment, an increased butyrate concentration was observed as well as a reduction in the number of tumors when the mice were inoculated with *B. fribrisolvens* and received the prebiotic diet; in contrast, neither *B. fibrisolvens* nor the high fibre diet had a protective effect on their own. Also, the consumption of AXOS enriched diet by rats has been reported to reduce preneoplastic lesions along the colon; this effect is more pronounced in the distal colon [143]. However, more human studies with better experimental designs are needed to support the results of the studies with animal models. Moreover, it should be taken into account that SCFAs production rates and the ratios of the different fatty acids can vary depending on the type of prebiotic administrated and the type and abundance of bacterial species present in the host.

Mineral Absorption

A well-recognized health effect of prebiotics consumption is the promotion of the colonic absorption of minerals, such as iron, calcium, zinc, magnesium and copper. As a result of the fermentation of prebiotics by colonic bacteria, SCFAs are generated and released in the lumen causing a reduction in the pH, which helps dissolving minerals. In addition, some SCFAs such as lactate and butyrate promote the proliferation of colonocytes and thus enlarging the absorption surface. Studies in animal models have shown that prebiotics, such as FOS and GOS increase mineral absorption in the colon, which has been associated with improved bone mineral density [144 - 146]. However, human trials are scarce so far. Whisner and colleagues [147] administrated GOS to young girls aged 10-13 years old for 3 weeks. The prebiotic treatment increased the calcium absorption as well as bifidobacteria in the intestine. In a posterior study [148], the same group also showed that administration of inulin, FOS or soluble maize fibre with potential prebiotic activity to healthy adolescents was associated with increases in bone mineral density (BMD) and calcium absorption. However, in a recent human study with postmenopausal women who consumed calcium and FOS, no effects in bone mineral density were observed although the bone turnover increased in the women with FOS and calcium supplementation [149]. Although the effects of prebiotics in mineral absorption seem promising, there is a need for more double-blind placebo-controlled human studies in specific human populations to draw general conclusions, as depending on the prebiotic used and the population groups studied, the results could provide different insights.

Inmunomodulation

Prebiotics possess immunomodulatory effects that can be exerted in an indirect manner through the modification of the intestinal microbiota and/or by direct interaction with immune cells. Although limited information is available about the direct immunomodulatory effects exerted by oligosaccharides, an *in vitro* study conducted with HMOs showed that the neonatal Th-2type T-cell phenotype, which is related to the development of atopic diseases, shifted toward a more balanced Th1/Th2 profile with a higher production of IGN- and IL-10, indicating allergy-preventive immunologic effects [150]. In addition, in the same study, evidence was provided for the *in vitro* transfer of oligosaccharides, specifically HMOs, through the epithelial monolayer. A more recent study aiming to characterize the direct effect of carbohydrate-based low-digestible fibres on immunity revealed that not all fibres have the same immunomodulatory effect [151]. Initially, an *in vitro* study showed that only one type of fibre of all tested, a glucose-based fibre, called by the authors F1, displayed immunosuppressive effects inhibiting the production of anti-and proinflammatory cytokines induced by bacteria in peripheral blood mononuclear cells (PBM). Subsequently, this fibre (F1) and glucose as control were administrated in mice models of colitis for short and long periods. Both treatments showed a similar impact on the modulation of intestinal microbiota and production of SCFAs. However, the results showed that F1 significantly reduced the levels of inflammatory markers, such as IL-6 and serum amyloid A (SAA) over both treatment periods, whereas no immuno-modulatory effect was observed with glucose. Another recent *in vitro* study with human peripheral blood mononuclear (HPBM) cells, showed that lower DP inulin-type fructans via TLR2 increased the ratio of IL-10 to IL-12, inducing a more anti-inflammatory balance [152]. Despite the evidence showing the direct effect of prebiotics in the immune system, further investigations are needed to determine the mode of action of different oligosaccharides that could be influenced by different factors including DP, cell type and/or disease of the study subjects.

Other Health Benefits

There is strong evidence to suggest that consumption of prebiotics can influence the bowel habit. All indigestible carbohydrates increase the bacterial mass and osmotic water binding capacity during colon transit which contributes to increase faecal bulk and improves stool consistency [153]. In addition, the production of some SCFAs by prebiotic fermentation in the colon, such as acetate propionate or butyrate can induce peristaltic activity [154]. Numerous studies have also reported the ability of prebiotics to relieve constipation and improve abdominal pain, bloating and/or flatulence [155 - 157].

On the other hand, the administration of prebiotics could also regulate glucose and insulin levels by different mechanisms, such as reducing the level of glucose into the bloodstream by shortening intestinal transit time, or modulating glycemia and insulinemia by SCFAs generated by prebiotic fermentation. In this regard, recent clinical studies with women with type 2 diabetes who received a daily supplement of prebiotics (resistant dextrin and a combination of inulin and oligofructose) for 8 weeks resulted in reduced metabolic syndrome markers that could improve the insulin resistance in these women [158, 159].

Prebiotics in Infant Formula

Numerous studies with breast fed infants have shown that the intestinal microbiota is dominated by *Bifidobacterium* (up to 90%), while formula fed infants possess a microbiota which is more similar to adults [124 - 126]. A microbiota enriched with this genus due to maternal milk feeding generally results in a lower incidence of bacterial infections, compared to formula fed infants. Human milk contains between 8 and 13 g/L of a complex mixture of least 115 different human milk oligosaccharides (HMOs) [127]. Prebiotics, particularly GOS and FOS, can mimic the beneficial effects of HMOs. Since breastfeeding is not always possible, improved formulas supplemented with prebiotics are emerging in the market supported by evidence showing that infant fed formulae supplemented with prebiotics result in a higher abundance of bifidobacteria compared to the placebo group [128 - 130]. Although GOS and FOS seem to be an effective strategy for mimicking the effects of HMOs, they are very different structurally. GOS and FOS are linear chains, whereas HMOs are more complex and possess branches with residues of fucose, sialic acid and N-acetylglucosamine, which have been suggested to play an important role in the immunomodulation function. For instance, an *in vitro* study conducted by Eiwegger and colleagues [131] showed that HMOs, unlike GOS and FOS, induced the production of IFN-γ and iL-10 in cord blood-derived monocytes. Based on the above, a challenge in the last years is to develop and produce milk derived oligosaccharides (MOs) from alternative sources with prebiotic activity and similar health benefits to those provided by HMOs. In this regard, bovine milk oligosaccharides (BMOs) obtained mainly from bovine milk and dairy streams, are the main alternatives. However, BMOs are found in low amounts compared with HMOs in human milk [132] so a major challenge is the development of scalable industrial methods for the production of novel MOs. Moreover, an important aspect of this is to optimize the purification and concentration processes in order to increase the amounts of specific types of oligosaccharides that are also found in HMOs, and minimise other residuals such as lactose, minerals and salts.

SYNBIOTICS

Synbiotics ("syn" –together and "bios" –life) were firstly defined as "mixtures of probiotics and prebiotics that beneficially affect the host by improving the survival and implantation of live microbial dietary supplements in the gastrointestinal tract, by selectively stimulating the growth and/or by activating the metabolism of one or a limited number of health-promoting bacteria" [102]. Recently, two different synbiotic approaches have been proposed by Kolida and Gibson [160]:

• Synergistic concept: The prebiotic is chosen to support the growth and colonization of the selected probiotic strain as well as to enhance its beneficial activity. In this case, the prebiotic not only stimulates the probiotic but also promote the growth of beneficial members of the microbiota.
• Complementary concept: Probiotic and prebiotic are independently chosen. The probiotic strain is selected based on its beneficial activity on the host while the prebiotic is chosen to selectively promote beneficial members of the microbiota. However, the prebiotic could also indirectly stimulate the growth and activity of the probiotic strain.

Although both concepts are valid according to the definition of synbiotic, the synergistic approach is the most widely accepted and desired for the development of new synbiotic formulations.

Nowadays, one of the potential applications of synbiotic formulations is their use in early life nutrition. During the last decade, positive clinical evidence has led to the development of infant formula supplemented with pro- and prebiotics aiming to mimic the bifidogenic activity of breast milk. Recently, new infant formula milks containing synbiotics have been proposed and the data available to date suggest that these formulations may reduce the risk of infectious disease in healthy infants [161, 162]. Moreover, studies investigating the effect of synbiotic consumption in different pathologies, such as IBD, non-alcoholic steatohepatitis, obesity and/or diabetes have also been reported in the last few years [163 - 166] (Table **5.1**). Preliminary synbiotic interventions in rodent models have demonstrated significant protection against the development of CRC or colitis, and have been shown to be potentially more effective as preventative therapies compared to the use of probiotic or prebiotic alone [167 - 169]. However, research in inflammatory bowel diseases using synbiotic treatments in humans is still scarce and not well documented. Few double-blind placebo-controlled human trials have been conducted investigating the immunomodulatory effect of synbiotic formulations in IBD. Recently, Steed and colleagues [163] conducted a human trial where the synbiotic combination of *Bifidobacterium longum* and Sinergy 1 (Inulin + FOS) seemed to be effective in the attenuation of symptoms of active Crohn's disease. This treatment significantly reduced the activity indices of the disease, histological scores and TNF-α expression in patients who consumed the synbiotic (Table **5.1**). The same formulation was also tested in a group of elderly people to determine if synbiotic consumption exerted positive effects on the immune system and improved the intestinal microbiota structure. In this case, significant reductions in the cytokines IL-6, IL-8 and TNF-α were observed. This synbiotic also displayed a significant impact on the intestinal bifidobacterial population [170]. Also, synbiotic interventions in diabetic patients have shown promising results, reducing serum insulin levels [171, 172] (Table **5.1**).

Table 5.1. **Emerging clinical trials investigating the use of synbiotic agents in the prevention and/or treatment of different disease conditions.**

Study-year	Target Condition and Population	Sample Size (n)	Study Design	Treatment Duration	Synbiotic	Study Outcome
Steed *et al.* [163]	Crohn's disease; Adults 18-79 years	35	Randomized, double-blind, placebo-controlled trial	6 months	*B. longum* plus Sinergy 1	A significant reduction in both Crohn's disease activity indices and histological scores. Mucosal bifidobacteria proliferated in synbiotic patients.
Van Der Aa *et al.* [173]	Atopic dermatitis (AD); Infants < 7 months	90	Randomized, double-blind, placebo-controlled trial	12 weeks	*B. breve* M16V plus GOS/FOS mixture	No effect on AD.

(Table 5.1) cont.....

Study-year	Target Condition and Population	Sample Size (n)	Study Design	Treatment Duration	Synbiotic	Study Outcome
Picaud *et al.* [161]	infectious diseases in infants; Infants 4-6 months	771	Randomized, double-blind, placebo-controlled trial	3 months	*B. longum* and *S. thermophilus* plus FOS	A significant reduction in the incidence of infectious disease. A higher weight gain was observed in infants fed synbiotic.
Van Der Aa *et al.* [174]	Asthma-like symptoms in infants with atopic dermatitis; Infants < 7 months	90	Randomized, double-blind, placebo-controlled trial	12 weeks	*B. breve* M16V plus galacto-fructooligosaccharide mixture	Prevent asthma-like symptoms in infants with AD.
Vandenplas *et al.* [175]	Acute gastroenteritis; Infants 40 months	111	Randomized, double-blind, placebo-controlled trial	7 days	*S. thermophilus, L. rhamnosus, L. acidophilus, B. lactis, B. infantis* plus FOS	Shorten the media duration by 1 day of diarrhoea in infants treated with the synbiotic.
Ishikawa *et al.* [176]	Ulcerative colitis; Adults	41	Randomized, double-blind, placebo-controlled trial	1 year	*B. breve* (Yakult) plus GOS	A significant improvement in endoscopic grading (Matts classification) in the synbiotic group.
Eguchi *et al.* [177]	Elective living-donor liver transplantation (LDLT); Adults 25-68 years	50	Randomized, double-blind, placebo-controlled trial	2 weeks	*L. casei* (Shirota), *B. breve* (Yakult) plus GOS oligomate	A significant reduction in infectious complications after LDLT.
Van de Pol *et al.* [178]	Allergic disease; Adults	29	Randomized, double-blind, placebo-controlled trial	4 weeks	*B. breve* M16V plus galacto (90%)-fructooligosaccharide (10%) mixture	No effect on bronchial inflammation. A significant reduction in Th2 cytokines after allergen challenge and improved peak expiratory flow.
Usami *et al.* [179]	Hepatic surgery; Adults 18-90 years	61	Randomized, double-blind, placebo-controlled trial	25 days	*L. casei* (Shirota), *B. breve* (Yakult) plus GOS oligomate	Synbiotic treatment attenuated the decrease in intestinal integrity and reduced the rate of infectious complications.
Passariello *et al.* [180]	Acute diarrhoea Children 3-36 months	107	Randomized, double-blind, placebo-controlled trial	5 days	*L. paracasei* B21060 plus arabinogalactan and XOS	A significant reduction in the duration and severity of diarrhoea.

(Table 5.1) cont.....

Study-year	Target Condition and Population	Sample Size (n)	Study Design	Treatment Duration	Synbiotic	Study Outcome
Malaguarnera *et al.* [165]	Non-alcoholic steatohepatitis; Adults	66	Randomized, double-blind, placebo-controlled trial	24 weeks	*B. longum* plus FOS	Synbiotic treatment with lifestyle modification significantly reduced TNF-α, CRP, serum AST levels, HOMA-IR, serum endotoxin, steatosis, and the NASH activity index.
Moroti *et al.* [164]	Elderly people with type2 diabetes mellitus Adults 50-60 years	20	Randomized, double-blind, placebo-controlled trial	30 days	*L. acidophilus, B. bifidum* plus FOS	Non-significant reduction in total cholesterol and triglycerides. Significant increase in HDL and significant reduction in fasting glycemia.
Waitzberg et al. [181]	Constipation; Adult women 18-75 years	100	Randomized, double-blind, placebo-controlled trial	30 days	*L. paracasei* (Lpc-37), *L. rhamnosus* (HN001, *L. acidophilus* (NCFM). *B. lactis* (HN019) plus FOS	Increased frequency of evacuation and stool consistency after 4 weeks of treatment. Significant reduction. in constipation intensity (Agachan score).
Ostadrahimi *et al.* [182]	Lactating mothers	80	Randomized, double-blind, placebo-controlled trial	90 days	*L. casei* PXN 37, *L. rhamnosus* PXN 54, *S. thermophilus* PXN 66, *B. breve* PXN 25, *L.acidophilus* plus FOS	Synbiotics prevented weight loss in lactating mothers that resulted in a significant infant's weight gain.
Rogha *et al.* [183]	Irritable Bowel Syndrome Adults	85	Randomized, double-blind, placebo-controlled trial	12 weeks	*Bacillus coagulans* plus FOS	A significant reduction in abdominal pain frequency. Significant decrease in the diarrhoea frequency.

(Table 5.1) cont.....

Study-year	Target Condition and Population	Sample Size (n)	Study Design	Treatment Duration	Synbiotic	Study Outcome
Asemi *et al.* [171]	Excess weight with type 2 diabetes; Adults	62	doubled blind cross-over controlled trial	6 weeks	*L. sporogenes* plus inulin HPX	A significant decrease in serum insulin levels and hs-CRP levels as well as a significant increase in uric acid and plasma total GSH levels.
Eslamparast *et al.* [184]	Nonalcoholic fatty liver disease (NAFLD); Adults 18 years	52	Randomized, double-blind, placebo-controlled trial	28 weeks	*L. casei,L. rhamnosus, S. thermophilus,B.breve, L. acidophilus*, *B. longum* and *L. bulgaricus* plus FOS	A significant attenuation of inflammatory markers in the body from week 14.
Childs *et al.* [185]	Healthy adults; (25-65 years)	41	doubled blind cross-over controlled trial	21 days	*B. animalis* subsp. *lactis plus XOS*	Synbiotic treatment resulted in immunostimulatory effects, promoting Th1 responses and lowering Th2 activity.
Bazzocchi *et al.* [186]	Constipation; Adults	29	Randomized, double-blind, placebo-controlled trial	8 weeks	*L. paracasei, B. animalis* subsp. *lactis, L. acidophilus* plus Psyllium	A significant reduction in intestinal transit time was found in individuals treated with the synbiotic.
Islek *et al.* [187]	Acute infectious diarrhoea in children	156	Randomized, double-blind, placebo-controlled trial	5 days	*B. lactis* B94 plus Inulin	The duration of diarrhoea was significantly reduced in the synbiotic group vs the placebo group.
Famouri *et al.* [187]	Weight gain in children with failure to thrive	84	Randomized, double-blind, placebo-controlled trial	6 months	*B. coagulans* plus FOS	The increase in weight was significantly higher in synbiotics group than in controls.
Nascimento *et al.* [188]	Radiation-induced acute proctitis in patients with prostate cancer Adults	20	Randomized, double-blind, placebo-controlled trial	5 weeks	*L. reuteri* plus *inulin (Nestle)*	Proctitis symptoms were highest scored in the placebo group in both the second and third weeks than in the synbiotic group.

(Table 5.1) cont.....

Study-year	Target Condition and Population	Sample Size (n)	Study Design	Treatment Duration	Synbiotic	Study Outcome
Dilli *et al.* [189]	Necrotizing enterocolitis in very low birth weight in infants	400	Randomized, double-blind, placebo-controlled trial	8 weeks	*B. lactis* plus inulin	The rate of NEC was lower in probiotic (2.0%) and synbiotic (4.0%) groups vs prebiotic (12.0%) and placebo (18.0%) groups (P < .001).
Beserra *et al.* [166]	Lipid profile and glucose homeostasis in overweight or obese adults (18-65 years)	513	Meta-analysis Study. Four synbiotic randomized controlled trials	30-196 days	*Lactobacillus* combined with inulin or *Lactobacillus* and *Bifidobacterium* combined with FOS	A significant reduction in the triglycerides and fasting insulin
Rammohan *et al.* [190]	Surgery for chronic pancreatitis Adults	75	Randomized, single-blind, placebo-controlled trial	15 days	*S. faecalis, C. butyricum, B. mesentericus, L. sporogenes* plus FOS	The incidence of postoperative infectious complications, duration of antibiotic therapy and length of hospital stay were significantly lower in the synbiotic group.

However, despite the positive clinical evidence for the use of synbiotics as a treatment of different diseases, human studies are scarce, therefore further investigations are needed. Proper controls (probiotic and prebiotic alone) must be included in the human trials in order to evaluate the additive effect caused by the synbiotic. Additionally, more preliminary studies should be conducted to analyse different mixtures of prebiotics and probiotics and select the most suitable combinations that have a high synergistic effect.

CONCLUSIONS

In the last years, there have been significant knowledge advances on the composition and role of the human microbiota, triggered mainly by the transition from conventional culture techniques to culture independent genomic-based techniques, such as next generation sequencing. The introduction of these new techniques has allowed a more accurate characterization of the gut microbiota structure revealing the full diversity of the gut ecosystem as well as its metabolic activity. The human gut microbiota is unique to each individual. Despite the fact that most of critical metabolic functions are conserved across the population, the composition of the microbiota can vary considerably from one individual to another. Some diseases have been correlated with dysbiosis; however, it still remains unclear if dysbiosis is a cause or a consequence. On the other hand, the development of new functional genomic biotechnologies has allowed us to more deeply study the impact of new prebiotics and synbiotics formulations on the microbiota function. Therefore, a crucial challenge will be to elucidate the molecular mechanisms underlying the symbiotic host-bacterial relationships that will allow us to define the functional characteristics of a healthy microbiome. Moreover, in-depth understanding of molecular mechanisms of action of pro-, pre- and synbiotics, could potentially lead to the development of personalised effective strategies aimed to treat specific diseases and improve the health status of people.

CONSENT FOR PUBLICATION

Not applicable.

CONFLICT OF INTEREST

The authors declare no conflict of interest, financial or otherwise.

ACKNOWLEDGEMENTS

Declared none.

REFERENCES

[1]   Lilly DM, Stillwell RH. Probiotics: Growth-promoting factors produced by microorganisms. Science 1965; 147(3659): 747-8.
[http://dx.doi.org/10.1126/science.147.3659.747] [PMID: 14242024]

[2]   Hill C, Guarner F, Reid G, *et al.* Expert consensus document. The International Scientific Association for Probiotics and Prebiotics consensus statement on the scope and appropriate use of the term probiotic. Nat Rev Gastroenterol Hepatol 2014; 11(8): 506-14.
[http://dx.doi.org/10.1038/nrgastro.2014.66] [PMID: 24912386]

[3]   O'Connell Motherway M, Zomer A, Leahy SC, *et al.* Functional genome analysis of *Bifidobacterium breve* UCC2003 reveals type IVb tight adherence (Tad) pili as an essential and conserved host-colonization factor. Proc Natl Acad Sci USA 2011; 108(27): 11217-22.
[http://dx.doi.org/10.1073/pnas.1105380108] [PMID: 21690406]

[4]   Sarmiento-Rubiano LA, Berger B, Moine D, Zúñiga M, Pérez-Martínez G, Yebra MJ. Characterization of a novel *Lactobacillus* species closely related to *Lactobacillus johnsonii* using a combination of molecular and comparative genomics methods. BMC Genomics 2010; 11: 504.
[http://dx.doi.org/10.1186/1471-2164-11-504] [PMID: 20849602]

[5]   Fleischmann RD, Adams MD, White O, *et al.* Whole-genome random sequencing and assembly of *Haemophilus influenzae* Rd. Science 1995; 269(5223): 496-512.
[http://dx.doi.org/10.1126/science.7542800] [PMID: 7542800]

[6]   Ventura M, O'Flaherty S, Claesson MJ, *et al.* Genome-scale analyses of health-promoting bacteria: probiogenomics. Nat Rev Microbiol 2009; 7(1): 61-71.
[http://dx.doi.org/10.1038/nrmicro2047] [PMID: 19029955]

[7]   Kant R, Rintahaka J, Yu X, *et al.* A comparative pan-genome perspective of niche-adaptable cell-

surface protein phenotypes in *Lactobacillus rhamnosus*. PLoS One 2014; 9(7): e102762.
[http://dx.doi.org/10.1371/journal.pone.0102762] [PMID: 25032833]

[8]   Bottacini F, O'Connell Motherway M, Kuczynski J, *et al.* Comparative genomics of the *Bifidobacterium breve* taxon. BMC Genomics 2014; 15: 170.
[http://dx.doi.org/10.1186/1471-2164-15-170] [PMID: 24581150]

[9]   Smokvina T, Wels M, Polka J, *et al. Lactobacillus paracasei* comparative genomics: towards species pan-genome definition and exploitation of diversity. PLoS One 2013; 8(7): e68731.
[http://dx.doi.org/10.1371/journal.pone.0068731] [PMID: 23894338]

[10]   Broadbent JR, Neeno-Eckwall EC, Stahl B, *et al.* Analysis of the *Lactobacillus casei* supragenome and its influence in species evolution and lifestyle adaptation. BMC Genomics 2012; 13: 533.
[http://dx.doi.org/10.1186/1471-2164-13-533] [PMID: 23035691]

[11]   Alcántara C, Zúñiga M. Proteomic and transcriptomic analysis of the response to bile stress of Lactobacillus casei BL23. Microbiology 2012; 158(Pt 5): 1206-18.
[http://dx.doi.org/10.1099/mic.0.055657-0] [PMID: 22322960]

[12]   An H, Douillard FP, Wang G, *et al.* Integrated transcriptomic and proteomic analysis of the bile stress response in a centenarian-originated probiotic *Bifidobacterium longum* BBMN68. Mol Cell Proteomics 2014; 13(10): 2558-72.
[http://dx.doi.org/10.1074/mcp.M114.039156] [PMID: 24965555]

[13]   Jin J, Zhang B, Guo H, *et al.* Mechanism analysis of acid tolerance response of *Bifidobacterium longum* subsp. *longum* BBMN 68 by gene expression profile using RNA-sequencing. PLoS One 2012; 7(12): e50777.
[http://dx.doi.org/10.1371/journal.pone.0050777] [PMID: 23236393]

[14]   Laakso K, Koskenniemi K, Koponen J, *et al.* Growth phase-associated changes in the proteome and transcriptome of *Lactobacillus rhamnosus* GG in industrial-type whey medium. Microb Biotechnol 2011; 4(6): 746-66.
[http://dx.doi.org/10.1111/j.1751-7915.2011.00275.x] [PMID: 21883975]

[15]   Siragusa S, De Angelis M, Calasso M, *et al.* Fermentation and proteome profiles of *Lactobacillus plantarum* strains during growth under food-like conditions. J Proteomics 2014; 96: 366-80.
[http://dx.doi.org/10.1016/j.jprot.2013.11.003] [PMID: 24231110]

[16]   Mitsuoka T. The human gastrointestinal tract.The lactic acid bacteria. New York: Elsevier Applied Science 1992; pp. 69-114.

[17]   Ardissone AN, de la Cruz DM, Davis-Richardson AG, *et al.* Meconium microbiome analysis identifies bacteria correlated with premature birth. PLoS One 2014; 9(3): e90784.
[http://dx.doi.org/10.1371/journal.pone.0090784] [PMID: 24614698]

[18]   Moles L, Gómez M, Heilig H, *et al.* Bacterial diversity in meconium of preterm neonates and evolution of their fecal microbiota during the first month of life. PLoS One 2013; 8(6): e66986.
[http://dx.doi.org/10.1371/journal.pone.0066986] [PMID: 23840569]

[19]   Satokari R, Grönroos T, Laitinen K, Salminen S, Isolauri E. *Bifidobacterium* and *Lactobacillus* DNA in the human placenta. Lett Appl Microbiol 2009; 48(1): 8-12.
[http://dx.doi.org/10.1111/j.1472-765X.2008.02475.x] [PMID: 19018955]

[20]   DiGiulio DB. Diversity of microbes in amniotic fluid. Semin Fetal Neonatal Med 2012; 17(1): 2-11.
[http://dx.doi.org/10.1016/j.siny.2011.10.001] [PMID: 22137615]

[21]   Thompson AL, Monteagudo-Mere A, Cadenae MB, Lampi ML, Azcarate-Peril MA. Milk- and solid-feeding practices and daycare attendance are associated with differences in bacterial diversity, predominant communities, and metabolic and immune function of the infant gut microbiome. Front Cell Infect Mi 2015; p. 5.

[22]   Dominguez-Bello MG, Costello EK, Contreras M, *et al.* Delivery mode shapes the acquisition and structure of the initial microbiota across multiple body habitats in newborns. Proc Natl Acad Sci USA

2010; 107(26): 11971-5.
[http://dx.doi.org/10.1073/pnas.1002601107] [PMID: 20566857]

[23]   Fallani M, Young D, Scott J, *et al.* Intestinal microbiota of 6-week-old infants across Europe: geographic influence beyond delivery mode, breast-feeding, and antibiotics. J Pediatr Gastroenterol Nutr 2010; 51(1): 77-84.
[http://dx.doi.org/10.1097/MPG.0b013e3181d1b11e] [PMID: 20479681]

[24]   Azad MB, Konya T, Maughan H, *et al.* Infant gut microbiota and the hygiene hypothesis of allergic disease: impact of household pets and siblings on microbiota composition and diversity. Allergy Asthma Cl Im 2013; p. 9.

[25]   Eckburg PB, Bik EM, Bernstein CN, *et al.* Diversity of the human intestinal microbial flora. Science 2005; 308(5728): 1635-8.
[http://dx.doi.org/10.1126/science.1110591] [PMID: 15831718]

[26]   Arumugam M, Raes J, Pelletier E, *et al.* Enterotypes of the human gut microbiome. Nature 2011; 473(7346): 174-80.
[http://dx.doi.org/10.1038/nature09944] [PMID: 21508958]

[27]   Wu GD, Chen J, Hoffmann C, *et al.* Linking long-term dietary patterns with gut microbial enterotypes. Science 2011; 334(6052): 105-8.
[http://dx.doi.org/10.1126/science.1208344] [PMID: 21885731]

[28]   Knights D, Ward TL, McKinlay CE, *et al.* Rethinking "enterotypes". Cell Host Microbe 2014; 16(4): 433-7.
[http://dx.doi.org/10.1016/j.chom.2014.09.013] [PMID: 25299329]

[29]   Koren O, Knights D, Gonzalez A, *et al.* A guide to enterotypes across the human body: meta-analysis of microbial community structures in human microbiome datasets. PLOS Comput Biol 2013; 9(1): e1002863.
[http://dx.doi.org/10.1371/journal.pcbi.1002863] [PMID: 23326225]

[30]   Smith K, McCoy KD, Macpherson AJ. Use of axenic animals in studying the adaptation of mammals to their commensal intestinal microbiota. Semin Immunol 2007; 19(2): 59-69.
[http://dx.doi.org/10.1016/j.smim.2006.10.002] [PMID: 17118672]

[31]   Smith MI, Yatsunenko T, Manary MJ, *et al.* Gut microbiomes of Malawian twin pairs discordant for kwashiorkor. Science 2013; 339(6119): 548-54.
[http://dx.doi.org/10.1126/science.1229000] [PMID: 23363771]

[32]   Ridaura VK, Faith JJ, Rey FE, *et al.* Gut microbiota from twins discordant for obesity modulate metabolism in mice. Science 2013; 341(6150): 1241214.
[http://dx.doi.org/10.1126/science.1241214] [PMID: 24009397]

[33]   Lozupone CA, Stombaugh JI, Gordon JI, Jansson JK, Knight R. Diversity, stability and resilience of the human gut microbiota. Nature 2012; 489(7415): 220-30.
[http://dx.doi.org/10.1038/nature11550] [PMID: 22972295]

[34]   Eloe-Fadrosh EA, Brady A, Crabtree J, *et al.* Functional dynamics of the gut microbiome in elderly people during probiotic consumption. MBio 2015; 6(2): e00231-15.
[http://dx.doi.org/10.1128/mBio.00231-15] [PMID: 25873374]

[35]   McNulty NP, Yatsunenko T, Hsiao A, *et al.* The impact of a consortium of fermented milk strains on the gut microbiome of gnotobiotic mice and monozygotic twins. Sci Transl Med 2011; 3(106): 106ra106.
[http://dx.doi.org/10.1126/scitranslmed.3002701] [PMID: 22030749]

[36]   de Vrese M, Stegelmann A, Richter B, Fenselau S, Laue C, Schrezenmeir J. Probiotics--compensation for lactase insufficiency. Am J Clin Nutr 2001; 73(2) (Suppl.): 421S-9S.
[http://dx.doi.org/10.1093/ajcn/73.2.421s] [PMID: 11157352]

[37]   Ghouri YA, Richards DM, Rahimi EF, Krill JT, Jelinek KA, DuPont AW. Systematic review of

randomized controlled trials of probiotics, prebiotics, and synbiotics in inflammatory bowel disease. Clin Exp Gastroenterol 2014; 7: 473-87.
[PMID: 25525379]

[38]   Mardini HE, Grigorian AY. Probiotic mix VSL#3 is effective adjunctive therapy for mild to moderately active ulcerative colitis: a meta-analysis. Inflamm Bowel Dis 2014; 20(9): 1562-7.
[http://dx.doi.org/10.1097/MIB.0000000000000084] [PMID: 24918321]

[39]   Yang Y, Guo Y, Kan Q, Zhou XG, Zhou XY, Li Y. A meta-analysis of probiotics for preventing necrotizing enterocolitis in preterm neonates 2014.
[http://dx.doi.org/10.1590/1414-431X20143857]

[40]   Allen SJ, Martinez EG, Gregorio GV, Dans LF. Probiotics for treating acute infectious diarrhoea. Cochrane Database Syst Rev 2010; (11): CD003048.
[PMID: 21069673]

[41]   Salari P, Nikfar S, Abdollahi M. A meta-analysis and systematic review on the effect of probiotics in acute diarrhea. Inflamm Allergy Drug Targets 2012; 11(1): 3-14.
[http://dx.doi.org/10.2174/187152812798889394] [PMID: 22309079]

[42]   Choi CH, Chang SK. Alteration of gut microbiota and efficacy of probiotics in functional constipation. J Neurogastroenterol Motil 2015; 21(1): 4-7.
[http://dx.doi.org/10.5056/jnm14142] [PMID: 25611063]

[43]   Kassam Z, Lee CH, Yuan Y, Hunt RH. Fecal microbiota transplantation for *Clostridium difficile* infection: systematic review and meta-analysis. Am J Gastroenterol 2013; 108(4): 500-8.
[http://dx.doi.org/10.1038/ajg.2013.59] [PMID: 23511459]

[44]   Gough E, Shaikh H, Manges AR. Systematic review of intestinal microbiota transplantation (fecal bacteriotherapy) for recurrent *Clostridium difficile* infection. Clin Infect Dis 2011; 53(10): 994-1002.
[http://dx.doi.org/10.1093/cid/cir632] [PMID: 22002980]

[45]   Petrof EO, Gloor GB, Vanner SJ, *et al.* Stool substitute transplant therapy for the eradication of *Clostridium difficile* infection: 'RePOOPulating' the gut. Microbiome 2013; 1(1): 3.
[http://dx.doi.org/10.1186/2049-2618-1-3] [PMID: 24467987]

[46]   Winkler P, Ghadimi D, Schrezenmeir J, Kraehenbuhl JP. Molecular and cellular basis of microflora-host interactions. J Nutr 2007; 137(3) (Suppl. 2): 756S-72S.
[http://dx.doi.org/10.1093/jn/137.3.756S] [PMID: 17311973]

[47]   Cuello-Garcia CA, Brożek JL, Fiocchi A, *et al.* Probiotics for the prevention of allergy: A systematic review and meta-analysis of randomized controlled trials. J Allergy Clin Immunol 2015; 136(4): 952-61.
[http://dx.doi.org/10.1016/j.jaci.2015.04.031] [PMID: 26044853]

[48]   Elazab N, Mendy A, Gasana J, Vieira ER, Quizon A, Forno E. Probiotic administration in early life, atopy, and asthma: a meta-analysis of clinical trials. Pediatrics 2013; 132(3): e666-76.
[http://dx.doi.org/10.1542/peds.2013-0246] [PMID: 23958764]

[49]   Panduru M, Panduru NM, Sălăvăstru CM, Tiplica GS. Probiotics and primary prevention of atopic dermatitis: a meta-analysis of randomized controlled studies. J Eur Acad Dermatol Venereol 2015; 29(2): 232-42.
[http://dx.doi.org/10.1111/jdv.12496] [PMID: 24698503]

[50]   Marlow G, Han DY, Wickens K, *et al.* Differential effects of two probiotics on the risks of eczema and atopy associated with single nucleotide polymorphisms to Toll-like receptors 2015.
[http://dx.doi.org/10.1111/pai.12371]

[51]   Mohammedsaeed W, McBain AJ, Cruickshank SM, O'Neill CA. *Lactobacillus rhamnosus* GG inhibits the toxic effects of *Staphylococcus aureus* on epidermal keratinocytes. Appl Environ Microbiol 2014; 80(18): 5773-81.
[http://dx.doi.org/10.1128/AEM.00861-14] [PMID: 25015889]

[52]   Gueniche A, Benyacoub J, Philippe D, *et al. Lactobacillus paracasei* CNCM I-2116 (ST11) inhibits substance P-induced skin inflammation and accelerates skin barrier function recovery in vitro. Eur J Dermatol 2010; 20(6): 731-7.
[PMID: 20965806]

[53]   Guéniche A, Bastien P, Ovigne JM, *et al. Bifidobacterium longum* lysate, a new ingredient for reactive skin. Exp Dermatol 2010; 19(8): e1-8.
[http://dx.doi.org/10.1111/j.1600-0625.2009.00932.x] [PMID: 19624730]

[54]   Sonal Sekhar M, Unnikrishnan MK, Vijayanarayana K, Rodrigues GS, Mukhopadhyay C. Topical application/formulation of probiotics: will it be a novel treatment approach for diabetic foot ulcer? Med Hypotheses 2014; 82(1): 86-8.
[http://dx.doi.org/10.1016/j.mehy.2013.11.013] [PMID: 24296233]

[55]   Cannon ML. A Review of Probiotic Therapy in Preventive Dental Practice. Probiotics Antimicrob Proteins 2011; 3(2): 63-7.
[http://dx.doi.org/10.1007/s12602-011-9072-9] [PMID: 26781571]

[56]   Licciardi PV, Tang ML. Vaccine adjuvant properties of probiotic bacteria. Discov Med 2011; 12(67): 525-33.
[PMID: 22204769]

[57]   Konieczna P, Groeger D, Ziegler M, *et al. Bifidobacterium infantis* 35624 administration induces Foxp3 T regulatory cells in human peripheral blood: potential role for myeloid and plasmacytoid dendritic cells. Gut 2012; 61(3): 354-66.
[http://dx.doi.org/10.1136/gutjnl-2011-300936] [PMID: 22052061]

[58]   Evrard B, Coudeyras S, Dosgilbert A, *et al.* Dose-dependent immunomodulation of human dendritic cells by the probiotic *Lactobacillus rhamnosus* Lcr35. PLoS One 2011; 6(4): e18735.
[http://dx.doi.org/10.1371/journal.pone.0018735] [PMID: 21533162]

[59]   Vlasova AN, Chattha KS, Kandasamy S, *et al.* Lactobacilli and bifidobacteria promote immune homeostasis by modulating innate immune responses to human rotavirus in neonatal gnotobiotic pigs. PLoS One 2013; 8(10): e76962.
[http://dx.doi.org/10.1371/journal.pone.0076962] [PMID: 24098572]

[60]   Liu F, Wen K, Li G, *et al.* Dual functions of *Lactobacillus acidophilus* NCFM as protection against rotavirus diarrhea. J Pediatr Gastroenterol Nutr 2014; 58(2): 169-76.
[http://dx.doi.org/10.1097/MPG.0000000000000197] [PMID: 24126832]

[61]   Yaqoob P. Ageing, immunity and influenza: a role for probiotics? Proc Nutr Soc 2014; 73(2): 309-17.
[http://dx.doi.org/10.1017/S0029665113003777] [PMID: 24300282]

[62]   Cao Y, Shen J, Ran ZH. Association between Faecalibacterium prausnitzii Reduction and Inflammatory Bowel Disease: A Meta-Analysis and Systematic Review of the Literature. Gastroent Res Pract 2014.

[63]   Sokol H, Pigneur B, Watterlot L, *et al. Faecalibacterium prausnitzii* is an anti-inflammatory commensal bacterium identified by gut microbiota analysis of Crohn disease patients. Proc Natl Acad Sci USA 2008; 105(43): 16731-6.
[http://dx.doi.org/10.1073/pnas.0804812105] [PMID: 18936492]

[64]   Qiu X, Zhang M, Yang X, Hong N, Yu C. *Faecalibacterium prausnitzii* upregulates regulatory T cells and anti-inflammatory cytokines in treating TNBS-induced colitis. J Crohn's Colitis 2013; 7(11): e558-68.
[http://dx.doi.org/10.1016/j.crohns.2013.04.002] [PMID: 23643066]

[65]   Shin NR, Lee JC, Lee HY, *et al.* An increase in the *Akkermansia* spp. population induced by metformin treatment improves glucose homeostasis in diet-induced obese mice. Gut 2014; 63(5): 727-35.
[http://dx.doi.org/10.1136/gutjnl-2012-303839] [PMID: 23804561]

[66]   Gauffin Cano P, Santacruz A, Moya Á, Sanz Y. *Bacteroides uniformis* CECT 7771 ameliorates metabolic and immunological dysfunction in mice with high-fat-diet induced obesity. PLoS One 2012; 7(7): e41079.
[http://dx.doi.org/10.1371/journal.pone.0041079] [PMID: 22844426]

[67]   Atarashi K, Tanoue T, Oshima K, *et al.* Treg induction by a rationally selected mixture of Clostridia strains from the human microbiota. Nature 2013; 500(7461): 232-6.
[http://dx.doi.org/10.1038/nature12331] [PMID: 23842501]

[68]   Cryan JF, Dinan TG. Mind-altering microorganisms: the impact of the gut microbiota on brain and behaviour. Nat Rev Neurosci 2012; 13(10): 701-12.
[http://dx.doi.org/10.1038/nrn3346] [PMID: 22968153]

[69]   Crumeyrolle-Arias M, Jaglin M, Bruneau A, *et al.* Absence of the gut microbiota enhances anxiety-like behavior and neuroendocrine response to acute stress in rats. Psychoneuroendocrinology 2014; 42: 207-17.
[http://dx.doi.org/10.1016/j.psyneuen.2014.01.014] [PMID: 24636517]

[70]   De Palma G, Blennerhassett P, Lu J, *et al.* Microbiota and host determinants of behavioural phenotype in maternally separated mice. Nat Commun 2015; 6: 7735.
[http://dx.doi.org/10.1038/ncomms8735] [PMID: 26218677]

[71]   Neufeld KM, Kang N, Bienenstock J, Foster JA. Reduced anxiety-like behavior and central neurochemical change in germ-free mice 2011.
[http://dx.doi.org/10.1111/j.1365-2982.2010.01620.x]

[72]   Diaz Heijtz R, Wang S, Anuar F, *et al.* Normal gut microbiota modulates brain development and behavior. Proc Natl Acad Sci USA 2011; 108(7): 3047-52.
[http://dx.doi.org/10.1073/pnas.1010529108] [PMID: 21282636]

[73]   Ohland CL, Kish L, Bell H, *et al.* Effects of *Lactobacillus helveticus* on murine behavior are dependent on diet and genotype and correlate with alterations in the gut microbiome. Psychoneuroendocrinology 2013; 38(9): 1738-47.
[http://dx.doi.org/10.1016/j.psyneuen.2013.02.008] [PMID: 23566632]

[74]   Tillisch K, Labus J, Kilpatrick L, *et al.* Consumption of fermented milk product with probiotic modulates brain activity 2013.
[http://dx.doi.org/10.1053/j.gastro.2013.02.043]

[75]   Parracho HMRT, Gibson GR, Knott F, Bosscher D, Kleerebezem M, McCartney AL. A double-blind, placebo-controlled, crossover-dessigned probiotic feeding study in childrem diagnosed with autistic spectrum disorders. Int J Probiotics Prebiotics 2010; 5(2): 69-74.

[76]   Kałużna-Czaplińska J, Błaszczyk S. The level of arabinitol in autistic children after probiotic therapy. Nutrition 2012; 28(2): 124-6.
[http://dx.doi.org/10.1016/j.nut.2011.08.002] [PMID: 22079796]

[77]   Rathore S, Salmerón I, Pandiella SS. Production of potentially probiotic beverages using single and mixed cereal substrates fermented with lactic acid bacteria cultures. Food Microbiol 2012; 30(1): 239-44.
[http://dx.doi.org/10.1016/j.fm.2011.09.001] [PMID: 22265307]

[78]   Sanni A, Franz C, Schillinger U, Huch M, Guigas C, Holzapfel W. Characterization and Technological Properties of Lactic Acid Bacteria in the Production of "Sorghurt," a Cereal-Based Product. Food Biotechnol 2013; 27(2): 178-98.
[http://dx.doi.org/10.1080/08905436.2013.781949]

[79]   Soukoulis C, Yonekura L, Gan HH, Behboudi-Jobbehdar S, Parmenter C, Fisk I. Probiotic edible films as a new strategy for developing functional bakery products: The case of pan bread. Food Hydrocoll 2014; 39(100): 231-42.
[http://dx.doi.org/10.1016/j.foodhyd.2014.01.023] [PMID: 25089068]

[80]   Franz CM, Huch M, Mathara JM, *et al.* African fermented foods and probiotics. Int J Food Microbiol 2014; 190: 84-96.
[http://dx.doi.org/10.1016/j.ijfoodmicro.2014.08.033] [PMID: 25203619]

[81]   Vijaya Kumar B, Vijayendra SVN, Reddy OVS. Trends in dairy and non-dairy probiotic products - a review. J Food Sci Technol 2015; 52(10): 6112-24.
[http://dx.doi.org/10.1007/s13197-015-1795-2] [PMID: 26396359]

[82]   Ogunremi OR, Sanni AI, Agrawal R. Probiotic potentials of yeasts isolated from some cereal-based Nigerian traditional fermented food products. J Appl Microbiol 2015; 119(3): 797-808.
[http://dx.doi.org/10.1111/jam.12875] [PMID: 26095794]

[83]   Bhat AR, Irorere VU, Bartlett T, *et al.* Bacillus subtilis natto: a non-toxic source of poly-γ-glutamic acid that could be used as a cryoprotectant for probiotic bacteria. AMB Express 2013; 3(1): 36.
[http://dx.doi.org/10.1186/2191-0855-3-36] [PMID: 23829836]

[84]   Gobbetti M, Cagno RD, De Angelis M. Functional microorganisms for functional food quality. Crit Rev Food Sci Nutr 2010; 50(8): 716-27.
[http://dx.doi.org/10.1080/10408398.2010.499770] [PMID: 20830633]

[85]   Saarela M, Alakomi HL, Matto J, Ahonen AM, Tynkkynen S. Acid tolerant mutants of Bifidobacterium animalis subsp lactis with improved stability in fruit juice. Lebensm Wiss Technol 2011; 44(4): 1012-8.
[http://dx.doi.org/10.1016/j.lwt.2010.11.004]

[86]   Luckow T, Sheehan V, Fitzgerald G, Delahunty C. Exposure, health information and flavour-masking strategies for improving the sensory quality of probiotic juice. Appetite 2006; 47(3): 315-23.
[http://dx.doi.org/10.1016/j.appet.2006.04.006] [PMID: 16857295]

[87]   Ellendersen LDN, Granato D, Guergoletto KB, Wosiacki G. Development and sensory profile of a probiotic beverage from apple fermented with Lactobacillus casei. Eng Life Sci 2012; 12(4): 475-85.
[http://dx.doi.org/10.1002/elsc.201100136]

[88]   Perricone M, Corbo MR, Sinigaglia M, Speranza B, Bevilacqua A. Viability of *Lactobacillus reuteri* in fruit juices. J Funct Foods 2014; 10: 421-6.
[http://dx.doi.org/10.1016/j.jff.2014.07.020]

[89]   Pyar H, Liong MT, Peh KK. Potentials of pinneapple waste as growth medium for lactobacillus species. Int J Pharm Pharm Sci 2014; 6(1): 142-5.

[90]   Alfano A, Donnarumma G, Cimini D, *et al. Lactobacillus plantarum*: microfiltration experiments for the production of probiotic biomass to be used in food and nutraceutical preparations. Biotechnol Prog 2015; 31(2): 325-33.
[http://dx.doi.org/10.1002/btpr.2037] [PMID: 25582766]

[91]   Panesar PS, Kaur S. Bioutilisation of agro-industrial waste for lactic acid production. Int J Food Sci Technol 2015; 50(10): 2143-51.
[http://dx.doi.org/10.1111/ijfs.12886]

[92]   Sultana K, Godward G, Reynolds N, Arumugaswamy R, Peiris P, Kailasapathy K. Encapsulation of probiotic bacteria with alginate-starch and evaluation of survival in simulated gastrointestinal conditions and in yoghurt. Int J Food Microbiol 2000; 62(1-2): 47-55.
[http://dx.doi.org/10.1016/S0168-1605(00)00380-9] [PMID: 11139021]

[93]   Sandoval-Castilla O, Lobato-Calleros C, Garcia-Galindo HS, Alvarez-Ramirez J, Vernon-Carter EJ. Textural properties of alginate-pectin beads and survivability of entrapped *Lb. casei* in simulated gastrointestinal conditions and in yoghurt. Food Res Int 2010; 43(1): 111-7.
[http://dx.doi.org/10.1016/j.foodres.2009.09.010]

[94]   Cabuk B, Harsa S. Whey Protein-Pullulan (WP/Pullulan) polymer blend for preservation of viability of *Lactobacillus acidophilus*. Dry Technol 2015; 33(10): 1223-33.
[http://dx.doi.org/10.1080/07373937.2015.1021008]

[95] Sathyabama S, Kumar MR, Devi PB, Vijayabharathi R, Priyadharisini VB. Co-encapsulation of probiotics with prebiotics on alginate matrix and its effect on viability in simulated gastric environment. Lebensm Wiss Technol 2014; 57(1): 419-25.
[http://dx.doi.org/10.1016/j.lwt.2013.12.024]

[96] Krasaekoopt W, Watcharapoka S. Effect of addition of inulin and galactooligosaccharide on the survival of microencapsulated probiotics in alginate beads coated with chitosan in simulated digestive system, yogurt and fruit juice. Lebensm Wiss Technol 2014; 57(2): 761-6.
[http://dx.doi.org/10.1016/j.lwt.2014.01.037]

[97] Chen J, Wang Q, Liu CM, Gong J. Issues deserve attention in encapsulating probiotics: critical review of existing literature. Crit Rev Food Sci Nutr 2017; 57(6): 1228-38.

[98] Vishwakarma N, Ganeshpurkar A, Pandey V, Dubey N, Bansal D. Mesalazine-probiotics beads for acetic acid experimental colitis: formulation and characterization of a promising new therapeutic strategy for ulcerative colitis. Drug Deliv 2015; 22(1): 94-9.
[http://dx.doi.org/10.3109/10717544.2013.872711] [PMID: 24491122]

[99] Chen S, Zhao Q, Ferguson LR, Shu Q, Weir I, Garg S. Development of a novel probiotic delivery system based on microencapsulation with protectants. Appl Microbiol Biotechnol 2012; 93(4): 1447-57.
[http://dx.doi.org/10.1007/s00253-011-3609-4] [PMID: 21975694]

[100] Chen S, Cao Y, Ferguson LR, Shu Q, Garg S. Evaluation of mucoadhesive coatings of chitosan and thiolated chitosan for the colonic delivery of microencapsulated probiotic bacteria. J Microencapsul 2013; 30(2): 103-15.
[http://dx.doi.org/10.3109/02652048.2012.700959] [PMID: 22746548]

[101] Wurth R, Hormannsperger G, Wilke J, Foerst P, Haller D, Kulozik U. Protective effect of milk protein based microencapsulation on bacterial survival in simulated gastric juice versus the murine gastrointestinal system. J Funct Foods 2015; 15: 116-25.
[http://dx.doi.org/10.1016/j.jff.2015.02.046]

[102] Gibson GR, Roberfroid MB. Dietary modulation of the human colonic microbiota: introducing the concept of prebiotics. J Nutr 1995; 125(6): 1401-12.
[PMID: 7782892]

[103] Everard A, Lazarevic V, Gaïa N, *et al.* Microbiome of prebiotic-treated mice reveals novel targets involved in host response during obesity. ISME J 2014; 8(10): 2116-30.
[http://dx.doi.org/10.1038/ismej.2014.45] [PMID: 24694712]

[104] Ulsemer P, Toutounian K, Kressel G, *et al.* Safety and tolerance of *Bacteroides xylanisolvens* DSM 23964 in healthy adults. Benef Microbes 2012; 3(2): 99-111.
[http://dx.doi.org/10.3920/BM2011.0051] [PMID: 22417778]

[105] Cummings JH. Short chain fatty acids in the human colon. Gut 1981; 22(9): 763-79.
[http://dx.doi.org/10.1136/gut.22.9.763] [PMID: 7028579]

[106] Wong JMW, de Souza R, Kendall CWC, Emam A, Jenkins DJA. Colonic health: fermentation and short chain fatty acids. J Clin Gastroenterol 2006; 40(3): 235-43.
[http://dx.doi.org/10.1097/00004836-200603000-00015] [PMID: 16633129]

[107] Chang PV, Hao L, Offermanns S, Medzhitov R. The microbial metabolite butyrate regulates intestinal macrophage function via histone deacetylase inhibition. Proc Natl Acad Sci USA 2014; 111(6): 2247-52.
[http://dx.doi.org/10.1073/pnas.1322269111] [PMID: 24390544]

[108] Nastasi C, Candela M, Bonefeld CM, *et al.* The effect of short-chain fatty acids on human monocyte-derived dendritic cells. Sci Rep-Uk 2015; p. 5.

[109] Roberfroid MB. Inulin-type fructans: functional food ingredients. J Nutr 2007; 137(11) (Suppl.): 2493S-502S.

[http://dx.doi.org/10.1093/jn/137.11.2493S] [PMID: 17951492]

[110] Schaafsma G, Slavin JL. Significance of Inulin Fructans in the Human Diet. Compr Rev Food Sci Food Saf 2015; 14(1): 37-47.
[http://dx.doi.org/10.1111/1541-4337.12119]

[111] Singh RS, Singh RP. Production of Fructooligosaccharides from Inulin by Endoinulinases and Their Prebiotic Potential. Food Technol Biotechnol 2010; 48(4): 435-50.

[112] Nobre C, Teixeira JA, Rodrigues LR. New Trends and Technological Challenges in the Industrial Production and Purification of Fructo-oligosaccharides. Crit Rev Food Sci Nutr 2015; 55(10): 1444-55.
[http://dx.doi.org/10.1080/10408398.2012.697082] [PMID: 24915327]

[113] Yun JW, Song SK. The Production of High-Content Fructo-Oligosaccharides from Sucrose by the Mixed-Enzyme System of Fructosyltransferase and Glucose-Oxidase. Biotechnol Lett 1993; 15(6): 573-6.
[http://dx.doi.org/10.1007/BF00138542]

[114] Sheu DC, Chang JY, Wang CY, Wu CT, Huang CJ. Continuous production of high-purity fructooligosaccharides and ethanol by immobilized Aspergillus japonicus and Pichia heimii. Bioprocess Biosyst Eng 2013; 36(11): 1745-51.
[http://dx.doi.org/10.1007/s00449-013-0949-8] [PMID: 23568753]

[115] Nobre C, Castro CC, Hantson AL, Teixeira JA, De Weireld G, Rodrigues LR. Strategies for the production of high-content fructo-oligosaccharides through the removal of small saccharides by co-culture or successive fermentation with yeast. Carbohydr Polym 2016; 136: 274-81.
[http://dx.doi.org/10.1016/j.carbpol.2015.08.088] [PMID: 26572356]

[116] Kuhn RC, Mazutti MA, Maugeri Filho F. Separation and purification of fructooligosaccharides on a zeolite fixed-bed column. J Sep Sci 2014; 37(8): 927-33.
[http://dx.doi.org/10.1002/jssc.201300979] [PMID: 24510747]

[117] Cardelle-Cobas A, Corzo N, Olano A, Peláez C, Requena T, Ávila M. Galactooligosaccharides derived from lactose and lactulose: influence of structure on *Lactobacillus*, *Streptococcus* and *Bifidobacterium* growth. Int J Food Microbiol 2011; 149(1): 81-7.
[http://dx.doi.org/10.1016/j.ijfoodmicro.2011.05.026] [PMID: 21700354]

[118] Guerrero C, Vera C, Novoa C, Dumont J, Acevedo F, Illanes A. Purification of highly concentrated galacto-oligosaccharide preparations by selective fermentation with yeasts. Int Dairy J 2014; 39(1): 78-88.
[http://dx.doi.org/10.1016/j.idairyj.2014.05.011]

[119] Dagher SF, Azcarate-Peril MA, Bruno-Bárcena JM. Heterologous expression of a bioactive β-hexosyltransferase, an enzyme producer of prebiotics, from *Sporobolomyces singularis*. Appl Environ Microbiol 2013; 79(4): 1241-9.
[http://dx.doi.org/10.1128/AEM.03491-12] [PMID: 23241974]

[120] Vazquez MJ, Alonso JL, Dominguez H, Parajo JC. Xylooligosaccharides: manufacture and applications. Trends Food Sci Technol 2000; 11(11): 387-93.
[http://dx.doi.org/10.1016/S0924-2244(01)00031-0]

[121] Demirbas A. Pyrolysis of biomass for fuels and chemicals. Energ Source Part A 2009; 31(12): 1028-37.
[http://dx.doi.org/10.1080/15567030801909383]

[122] Wang J, Sun BG, Cao YP, Tian Y. Enzymatic preparation of wheat bran xylooligosaccharides and their stability during pasteurization and autoclave sterilization at low pH. Carbohydr Polym 2009; 77(4): 816-21.
[http://dx.doi.org/10.1016/j.carbpol.2009.03.005]

[123] Singh RD, Banerjee J, Arora A. Prebiotic potential of oligosaccharides: A focus on xylan derived

oligosaccharides. Bioactive Carbohydrates and Dietary Fibre 2015; 5(1): 19-30.
[http://dx.doi.org/10.1016/j.bcdf.2014.11.003]

[124] Finegold SM, Li Z, Summanen PH, *et al.* Xylooligosaccharide increases bifidobacteria but not lactobacilli in human gut microbiota. Food Funct 2014; 5(3): 436-45.
[http://dx.doi.org/10.1039/c3fo60348b] [PMID: 24513849]

[125] Chen HH, Chen YK, Chang HC, Lin SY. Immunomodulatory Effects of Xylooligosaccharides. Food Sci Technol Res 2012; 18(2): 195-9.
[http://dx.doi.org/10.3136/fstr.18.195]

[126] Lecerf JM, Dépeint F, Clerc E, *et al.* Xylo-oligosaccharide (XOS) in combination with inulin modulates both the intestinal environment and immune status in healthy subjects, while XOS alone only shows prebiotic properties. Br J Nutr 2012; 108(10): 1847-58.
[http://dx.doi.org/10.1017/S0007114511007252] [PMID: 22264499]

[127] Gupta S, Abu-Ghannam N. Bioactive potential and possible health effects of edible brown seaweeds. Trends Food Sci Technol 2011; 22(6): 315-26.
[http://dx.doi.org/10.1016/j.tifs.2011.03.011]

[128] Zaporozhets TS, Besednova NN, Kuznetsova TA, *et al.* The prebiotic potential of polysaccharides and extracts of seaweeds. Russ J Mar Biol 2014; 40(1): 1-9.
[http://dx.doi.org/10.1134/S1063074014010106]

[129] Campos D, Betalleluz-Pallardel I, Chirinos R, Aguilar-Galvez A, Noratto G, Pedreschi R. Prebiotic effects of yacon (Smallanthus sonchifolius Poepp. & Endl), a source of fructooligosaccharides and phenolic compounds with antioxidant activity. Food Chem 2012; 135(3): 1592-9.
[http://dx.doi.org/10.1016/j.foodchem.2012.05.088] [PMID: 22953898]

[130] Velez E, Castillo N, Mesón O, Grau A, Bibas Bonet ME, Perdigón G. Study of the effect exerted by fructo-oligosaccharides from yacon (Smallanthus sonchifolius) root flour in an intestinal infection model with Salmonella Typhimurium. Br J Nutr 2013; 109(11): 1971-9.
[http://dx.doi.org/10.1017/S0007114512004230] [PMID: 23137694]

[131] Zhao J, Cheung PCK. Comparative proteome analysis of *Bifidobacterium longum* subsp. *infantis* grown on β-glucans from different sources and a model for their utilization. J Agric Food Chem 2013; 61(18): 4360-70.
[http://dx.doi.org/10.1021/jf400792j] [PMID: 23577653]

[132] Murphya P, Dal Bello F, O'Doherty J, Arendt EK, Sweeney T, Coffey A. The effects of liquid versus spray-dried Laminaria digitata extract on selected bacterial groups in the piglet gastrointestinal tract (GIT) microbiota. Anaerobe 2013; 21: 1-8.
[http://dx.doi.org/10.1016/j.anaerobe.2013.03.002] [PMID: 23542115]

[133] Murphy P, Bello FD, O'Doherty JV, Arendt EK, Sweeney T, Coffey A. Effects of cereal β-glucans and enzyme inclusion on the porcine gastrointestinal tract microbiota. Anaerobe 2012; 18(6): 557-65.
[http://dx.doi.org/10.1016/j.anaerobe.2012.09.005] [PMID: 23022204]

[134] Shen RL, Dang XY, Dong JL, Hu XZ. Effects of oat β-glucan and barley β-glucan on fecal characteristics, intestinal microflora, and intestinal bacterial metabolites in rats. J Agric Food Chem 2012; 60(45): 11301-8.
[http://dx.doi.org/10.1021/jf302824h] [PMID: 23113683]

[135] Jemal A, Bray F, Center MM, Ferlay J, Ward E, Forman D. Global cancer statistics. CA Cancer J Clin 2011; 61(2): 69-90.
[http://dx.doi.org/10.3322/caac.20107] [PMID: 21296855]

[136] Randi G, Edefonti V, Ferraroni M, La Vecchia C, Decarli A. Dietary patterns and the risk of colorectal cancer and adenomas. Nutr Rev 2010; 68(7): 389-408.
[http://dx.doi.org/10.1111/j.1753-4887.2010.00299.x] [PMID: 20591107]

[137] Verma A, Shukla G. Administration of prebiotic inulin suppresses 1,2 dimethylhydrazine

dihydrochloride induced procarcinogenic biomarkers fecal enzymes and preneoplastic lesions in early colon carcinogenesis in Sprague Dawley rats. J Funct Foods 2013; 5(2): 991-6.
[http://dx.doi.org/10.1016/j.jff.2013.02.006]

[138]   Komiyama Y, Mitsuyama K, Masuda J, *et al.* Prebiotic treatment in experimental colitis reduces the risk of colitic cancer. J Gastroenterol Hepatol 2011; 26(8): 1298-308.
[http://dx.doi.org/10.1111/j.1440-1746.2011.06690.x] [PMID: 21303406]

[139]   Le Leu RK, Brown IL, Hu Y, Esterman A, Young GP. Suppression of azoxymethane-induced colon cancer development in rats by dietary resistant starch. Cancer Biol Ther 2007; 6(10): 1621-6.
[http://dx.doi.org/10.4161/cbt.6.10.4764] [PMID: 17932462]

[140]   Verghese M, Walker LT, Shackelford L, Chawan CB. Inhibitory effects of nondigestible carbohydrates of different chain lengths on azoxymethane-induced aberrant crypt foci in Fisher 344 rats. Nutr Res 2005; 25(9): 859-68.
[http://dx.doi.org/10.1016/j.nutres.2005.09.007]

[141]   Donohoe DR, Collins LB, Wali A, Bigler R, Sun W, Bultman SJ. The Warburg effect dictates the mechanism of butyrate-mediated histone acetylation and cell proliferation. Mol Cell 2012; 48(4): 612-26.
[http://dx.doi.org/10.1016/j.molcel.2012.08.033] [PMID: 23063526]

[142]   Donohoe DR, Holley D, Collins LB, *et al.* A gnotobiotic mouse model demonstrates that dietary fiber protects against colorectal tumorigenesis in a microbiota- and butyrate-dependent manner. Cancer Discov 2014; 4(12): 1387-97.
[http://dx.doi.org/10.1158/2159-8290.CD-14-0501] [PMID: 25266735]

[143]   Femia AP, Salvadori M, Broekaert WF, *et al.* Arabinoxylan-oligosaccharides (AXOS) reduce preneoplastic lesions in the colon of rats treated with 1,2-dimethylhydrazine (DMH). Eur J Nutr 2010; 49(2): 127-32.
[http://dx.doi.org/10.1007/s00394-009-0050-x] [PMID: 19711111]

[144]   Bryk G, Coronel MZ, Pellegrini G, *et al.* Effect of a combination GOS/FOS® prebiotic mixture and interaction with calcium intake on mineral absorption and bone parameters in growing rats. Eur J Nutr 2015; 54(6): 913-23.
[http://dx.doi.org/10.1007/s00394-014-0768-y] [PMID: 25241022]

[145]   Lobo AR, Gaievski EHS, De Carli E, Alvares EP, Colli C. Fructo-oligosaccharides and iron bioavailability in anaemic rats: the effects on iron species distribution, ferroportin-1 expression, crypt bifurcation and crypt cell proliferation in the caecum. Br J Nutr 2014; 112(8): 1286-95.
[http://dx.doi.org/10.1017/S0007114514002165] [PMID: 25192308]

[146]   Weaver CM, Martin BR, Nakatsu CH, *et al.* Galactooligosaccharides improve mineral absorption and bone properties in growing rats through gut fermentation. J Agric Food Chem 2011; 59(12): 6501-10.
[http://dx.doi.org/10.1021/jf2009777] [PMID: 21553845]

[147]   Whisner CM, Martin BR, Schoterman MHC, *et al.* Galacto-oligosaccharides increase calcium absorption and gut bifidobacteria in young girls: a double-blind cross-over trial. Br J Nutr 2013; 110(7): 1292-303.
[http://dx.doi.org/10.1017/S000711451300055X] [PMID: 23507173]

[148]   Whisner CM, Martin BR, Nakatsu CH, *et al.* Soluble maize fibre affects short-term calcium absorption in adolescent boys and girls: a randomised controlled trial using dual stable isotopic tracers. Br J Nutr 2014; 112(3): 446-56.
[http://dx.doi.org/10.1017/S0007114514000981] [PMID: 24848974]

[149]   Slevin MM, Allsopp PJ, Magee PJ, *et al.* Supplementation with calcium and short-chain fructo-oligosaccharides affects markers of bone turnover but not bone mineral density in postmenopausal women (vol 144, pg 297, 2014). J Nutr 2014; 144(7): 1125.

[150]   Eiwegger T, Stahl B, Haidl P, *et al.* Prebiotic oligosaccharides: in vitro evidence for gastrointestinal epithelial transfer and immunomodulatory properties. Pediatr Allergy Immunol 2010; 21(8): 1179-88.

[http://dx.doi.org/10.1111/j.1399-3038.2010.01062.x] [PMID: 20444147]

[151] Breton J, Plé C, Guerin-Deremaux L, *et al.* Intrinsic immunomodulatory effects of low-digestible carbohydrates selectively extend their anti-inflammatory prebiotic potentials. BioMed Res Int 2015; 2015: 162398.
[http://dx.doi.org/10.1155/2015/162398] [PMID: 25977916]

[152] Vogt L, Ramasamy U, Meyer D, *et al.* Immune modulation by different types of β2→1-fructans is toll-like receptor dependent. PLoS One 2013; 8(7): e68367.
[http://dx.doi.org/10.1371/journal.pone.0068367] [PMID: 23861894]

[153] Rosendale D, Vetharaniam I, Cookson AL, Roy N. Modelling the effect of undigested dietary carbohydrate on the health and function of the large bowel. Bioactive Carbohydrates and Dietary Fibre 2015; 5(1): 86-98.
[http://dx.doi.org/10.1016/j.bcdf.2014.12.007]

[154] Soret R, Chevalier J, De Coppet P, *et al.* Short-chain fatty acids regulate the enteric neurons and control gastrointestinal motility in rats. Gastroenterology 2010; 138(5): 1772-82.
[http://dx.doi.org/10.1053/j.gastro.2010.01.053] [PMID: 20152836]

[155] Drakoularakou A, Tzortzis G, Rastall RA, Gibson GR. A double-blind, placebo-controlled, randomized human study assessing the capacity of a novel galacto-oligosaccharide mixture in reducing travellers' diarrhoea. Eur J Clin Nutr 2010; 64(2): 146-52.
[http://dx.doi.org/10.1038/ejcn.2009.120] [PMID: 19756029]

[156] Silk DBA, Davis A, Vulevic J, Tzortzis G, Gibson GR. Clinical trial: the effects of a trans-galactooligosaccharide prebiotic on faecal microbiota and symptoms in irritable bowel syndrome. Aliment Pharmacol Ther 2009; 29(5): 508-18.
[http://dx.doi.org/10.1111/j.1365-2036.2008.03911.x] [PMID: 19053980]

[157] Yen CH, Tseng YH, Kuo YW, Lee MC, Chen HL. Long-term supplementation of isomalto-oligosaccharides improved colonic microflora profile, bowel function, and blood cholesterol levels in constipated elderly people--a placebo-controlled, diet-controlled trial. Nutrition 2011; 27(4): 445-50.
[http://dx.doi.org/10.1016/j.nut.2010.05.012] [PMID: 20624673]

[158] Aliasgharzadeh A, Dehghan P, Gargari BP, Asghari-Jafarabadi M. Resistant dextrin, as a prebiotic, improves insulin resistance and inflammation in women with type 2 diabetes: a randomised controlled clinical trial. Br J Nutr 2015; 113(2): 321-30.
[http://dx.doi.org/10.1017/S0007114514003675] [PMID: 27028002]

[159] Aliasgharzadeh A, Khalili M, Mirtaheri E, *et al.* A combination of prebiotic inulin and oligofructose improve some of cardiovascular disease risk factors in women with type 2 diabetes: A randomized controlled clinical trial. Adv Pharm Bull 2015; 5(4): 507-14.
[http://dx.doi.org/10.15171/apb.2015.069] [PMID: 26819923]

[160] Kolida S, Gibson GR. Synbiotics in health and disease. Annu Rev Food Sci Technol 2011; 2: 373-93.
[http://dx.doi.org/10.1146/annurev-food-022510-133739] [PMID: 22129388]

[161] Picaud JC, Chapalain V, Paineau D, Zourabichvili O, Bornet FRJ, Duhamel JF. Incidence of infectious diseases in infants fed follow-on formula containing synbiotics: an observational study. Acta Paediatr 2010; 99(11): 1695-700.
[http://dx.doi.org/10.1111/j.1651-2227.2010.01896.x] [PMID: 20560895]

[162] Burks AW, Harthoorn LF, Langford JE, *et al.* Functional effects of an amino-acid based formula with synbiotics in cow's milk allergic infants. Allergy 2013; 68: 703.

[163] Steed H, Macfarlane GT, Blackett KL, *et al.* Clinical trial: the microbiological and immunological effects of synbiotic consumption - a randomized double-blind placebo-controlled study in active Crohn's disease. Aliment Pharmacol Ther 2010; 32(7): 872-83.
[http://dx.doi.org/10.1111/j.1365-2036.2010.04417.x] [PMID: 20735782]

[164] Moroti C, Souza Magri LF, de Rezende Costa M, Cavallini DCU, Sivieri K. Effect of the consumption

of a new symbiotic shake on glycemia and cholesterol levels in elderly people with type 2 diabetes mellitus. Lipids Health Dis 2012; 11: 29.
[http://dx.doi.org/10.1186/1476-511X-11-29] [PMID: 22356933]

[165] Malaguarnera M, Vacante M, Antic T, *et al. Bifidobacterium longum* with fructo-oligosaccharides in patients with non alcoholic steatohepatitis. Dig Dis Sci 2012; 57(2): 545-53.
[http://dx.doi.org/10.1007/s10620-011-1887-4] [PMID: 21901256]

[166] Beserra BT, Fernandes R, do Rosario VA, Mocellin MC, Kuntz MG, Trindade EB. A systematic review and meta-analysis of the prebiotics and synbiotics effects on glycaemia, insulin concentrations and lipid parameters in adult patients with overweight or obesity. Clin Nutr 2015; 34(5): 845-58.
[http://dx.doi.org/10.1016/j.clnu.2014.10.004] [PMID: 25456608]

[167] Simeoli R, Mattace Raso G, Lama A, *et al.* Preventive and therapeutic effects of *Lactobacillus paracasei* B21060-based synbiotic treatment on gut inflammation and barrier integrity in colitic mice. J Nutr 2015; 145(6): 1202-10.
[http://dx.doi.org/10.3945/jn.114.205989] [PMID: 25926411]

[168] Le Leu RK, Hu Y, Brown IL, Woodman RJ, Young GP. Synbiotic intervention of *Bifidobacterium lactis* and resistant starch protects against colorectal cancer development in rats. Carcinogenesis 2010; 31(2): 246-51.
[http://dx.doi.org/10.1093/carcin/bgp197] [PMID: 19696163]

[169] Verma A, Shukla G. Synbiotic (*Lactobacillus rhamnosus+Lactobacillus acidophilus*+inulin) attenuates oxidative stress and colonic damage in 1,2 dimethylhydrazine dihydrochloride-induced colon carcinogenesis in Sprague-Dawley rats: a long-term study. Eur J Cancer Prev 2014; 23(6): 550-9.
[http://dx.doi.org/10.1097/CEJ.0000000000000054] [PMID: 25025584]

[170] Macfarlane S, Cleary S, Bahrami B, Reynolds N, Macfarlane GT. Synbiotic consumption changes the metabolism and composition of the gut microbiota in older people and modifies inflammatory processes: a randomised, double-blind, placebo-controlled crossover study. Aliment Pharmacol Ther 2013; 38(7): 804-16.
[http://dx.doi.org/10.1111/apt.12453] [PMID: 23957631]

[171] Asemi Z, Khorrami-Rad A, Alizadeh SA, Shakeri H, Esmaillzadeh A. Effects of synbiotic food consumption on metabolic status of diabetic patients: a double-blind randomized cross-over controlled clinical trial. Clin Nutr 2014; 33(2): 198-203.
[http://dx.doi.org/10.1016/j.clnu.2013.05.015] [PMID: 23786900]

[172] Tajadadi-Ebrahimi M, Bahmani F, Shakeri H, *et al.* Effects of daily consumption of synbiotic bread on insulin metabolism and serum high-sensitivity C-reactive protein among diabetic patients: a double-blind, randomized, controlled clinical trial. Ann Nutr Metab 2014; 65(1): 34-41.
[http://dx.doi.org/10.1159/000365153] [PMID: 25196301]

[173] van der Aa LB, Heymans HS, van Aalderen WM, *et al.* Effect of a new synbiotic mixture on atopic dermatitis in infants: a randomized-controlled trial 2010.
[http://dx.doi.org/10.1111/j.1365-2222.2010.03465.x]

[174] van der Aa LB, van Aalderen WMC, Heymans HSA, *et al.* Synbiotics prevent asthma-like symptoms in infants with atopic dermatitis. Allergy 2011; 66(2): 170-7.
[http://dx.doi.org/10.1111/j.1398-9995.2010.02416.x] [PMID: 20560907]

[175] Vandenplas Y, De Hert SG, Grp P-S. Randomised clinical trial: the synbiotic food supplement Probiotical vs. placebo for acute gastroenteritis in children. Aliment Pharmacol Ther 2011; 34(8): 862-7.
[http://dx.doi.org/10.1111/j.1365-2036.2011.04835.x] [PMID: 21899583]

[176] Ishikawa H, Matsumoto S, Ohashi Y, *et al.* Beneficial effects of probiotic *bifidobacterium* and galacto-oligosaccharide in patients with ulcerative colitis: a randomized controlled study. Digestion 2011; 84(2): 128-33.

[http://dx.doi.org/10.1159/000322977] [PMID: 21525768]

[177] Eguchi S, Takatsuki M, Hidaka M, Soyama A, Ichikawa T, Kanematsu T. Perioperative synbiotic treatment to prevent infectious complications in patients after elective living donor liver transplantation: a prospective randomized study. Am J Surg 2011; 201(4): 498-502.
[http://dx.doi.org/10.1016/j.amjsurg.2010.02.013] [PMID: 20619394]

[178] van de Pol MA, Lutter R, Smids BS, Weersink EJ, van der Zee JS. Synbiotics reduce allergen-induced T-helper 2 response and improve peak expiratory flow in allergic asthmatics. Allergy 2011; 66(1): 39-47.
[http://dx.doi.org/10.1111/j.1398-9995.2010.02454.x] [PMID: 20716319]

[179] Usami M, Miyoshi M, Kanbara Y, *et al.* Effects of perioperative synbiotic treatment on infectious complications, intestinal integrity, and fecal flora and organic acids in hepatic surgery with or without cirrhosis. JPEN J Parenter Enteral Nutr 2011; 35(3): 317-28.
[http://dx.doi.org/10.1177/0148607110379813] [PMID: 21527594]

[180] Passariello A, Terrin G, Cecere G, *et al.* Randomised clinical trial: efficacy of a new synbiotic formulation containing *Lactobacillus paracasei* B21060 plus arabinogalactan and xilooligosaccharides in children with acute diarrhoea. Aliment Pharmacol Ther 2012; 35(7): 782-8.
[http://dx.doi.org/10.1111/j.1365-2036.2012.05015.x] [PMID: 22324448]

[181] Waitzberg DL, Logullo LC, Bittencourt AF, *et al.* Effect of synbiotic in constipated adult women - a randomized, double-blind, placebo-controlled study of clinical response. Clin Nutr 2013; 32(1): 27-33.
[http://dx.doi.org/10.1016/j.clnu.2012.08.010] [PMID: 22959620]

[182] Ostadrahimi A, Nikniaz L, Mahdavi R, Hejazi MA, Nikniaz Z. Effects of synbiotic supplementation on lactating mothers' energy intake and BMI, and infants' growth. Int J Food Sci Nutr 2013; 64(6): 711-4.
[http://dx.doi.org/10.3109/09637486.2013.775229] [PMID: 23480276]

[183] Rogha M, Esfahani MZ, Zargarzadeh AH. The efficacy of a synbiotic containing *Bacillus coagulans* in treatment of irritable bowel syndrome: a randomized placebo-controlled trial. Gastroenterol Hepatol Bed Bench 2014; 7(3): 156-63.
[PMID: 25120896]

[184] Eslamparast T, Poustchi H, Zamani F, Sharafkhah M, Malekzadeh R, Hekmatdoost A. Synbiotic supplementation in nonalcoholic fatty liver disease: a randomized, double-blind, placebo-controlled pilot study. Am J Clin Nutr 2014; 99(3): 535-42.
[http://dx.doi.org/10.3945/ajcn.113.068890] [PMID: 24401715]

[185] Childs CE, Röytiö H, Alhoniemi E, *et al.* Xylo-oligosaccharides alone or in synbiotic combination with *Bifidobacterium animalis* subsp. *lactis* induce bifidogenesis and modulate markers of immune function in healthy adults: a double-blind, placebo-controlled, randomised, factorial cross-over study. Br J Nutr 2014; 111(11): 1945-56.
[http://dx.doi.org/10.1017/S0007114513004261] [PMID: 24661576]

[186] Bazzocchi G, Giovannini T, Giussani C, Brigidi P, Turroni S. Effect of a new synbiotic supplement on symptoms, stool consistency, intestinal transit time and gut microbiota in patients with severe functional constipation: a pilot randomized double-blind, controlled trial. Tech Coloproctol 2014; 18(10): 945-53.
[http://dx.doi.org/10.1007/s10151-014-1201-5] [PMID: 25091346]

[187] İşlek A, Sayar E, Yılmaz A, Baysan BO, Mutlu D, Artan R. The role of *Bifidobacterium lactis* B94 plus inulin in the treatment of acute infectious diarrhea in children. Turk J Gastroenterol 2014; 25(6): 628-33.
[http://dx.doi.org/10.5152/tjg.2014.14022] [PMID: 25599772]

[188] Nascimento M, Aguilar-Nascimento JE, Caporossi C, Castro-Barcellos HM, Motta RT. Efficacy of synbiotics to reduce acute radiation proctitis symptoms and improve quality of life: a randomized, double-blind, placebo-controlled pilot trial. Int J Radiat Oncol Biol Phys 2014; 90(2): 289-95.

[http://dx.doi.org/10.1016/j.ijrobp.2014.05.049] [PMID: 25304789]

[189] Dilli D, Aydin B, Fettah ND, *et al.* The propre-save study: effects of probiotics and prebiotics alone or combined on necrotizing enterocolitis in very low birth weight infants 2015.
[http://dx.doi.org/10.1016/j.jpeds.2014.12.004]

[190] Rammohan A, Sathyanesan J, Palaniappan R, Govindan M. Synbiotics in Surgery for Chronic Pancreatitis - A Hobson's Choice? Ann Surg 2015.
[PMID: 28591023]

[191] Gibson GR, Rastall RA, Eds. R C. Emerging prebiotic carbohydrates.Prebiotics: Development and Application. Chichester, UK: John Wiley & sons Ltd 2006; pp. 111-33.

CHAPTER 6

Recent Biotechnological Advances in Food Additives

Valker A. Feitosa[1,2], Daniela A.V. Marques[3], Angela F. Jozala[4], Jorge F. B. Pereira[5] and Valéria C. Santos-Ebinuma[5,*]

[1] *Department of Biochemical and Pharmaceutical Technology, School of Pharmaceutical Sciences, University of São Paulo–USP, São Paulo, SP, Brazil*

[2] *Bionanomanufacturing Center, Institute for Technological Research–IPT, São Paulo, SP, Brazil*

[3] *Serra Talhada Campus, University of Pernambuco–UPE, Serra Talhada, PE, Brazil*

[4] *Department of Technological and Environmental Processes, University of Sorocaba–UNISO, Sorocaba, SP, Brazil*

[5] *Department of Bioprocesses and Biotechnology, School of Pharmaceutical Sciences, Univ. Estadual Paulista – UNESP, Araraquara, SP, Brazil*

Abstract: Food additives are substances incorporated into foods to improve quality, nutritional value, functional properties, and consumer acceptance. There is a tendency to replace synthetic additives by natural ones not only because of toxicity of the first but also because natural additives can enhance the characteristics of foodstuff. In this sense, the use of bioprocess technologies can improve the production of natural substances by using wild or genetically modified microorganisms, metabolic engineering, or simply by developing new food formulations. The major goal of this chapter is to present the main food additives used nowadays, mainly focusing on the ones produced *via* biotechnology. The food additives discussed include amino acids, antimicrobial peptides, colorants, organic acids, vitamins, and sweeteners. Although there are many food additives biotechnologically produced on research level, a lot of effort is required to improve their production yields and reduce their overall cost, enabling their large industrial production and commercialization. Anyway, food additives need to be approved by the regulatory authorities before they become industrially favorable. Despite this, biotechnology seems to be very promising in producing natural food additives. Some of the advances in this area are described in the present chapter.

Keywords: Additives, Amino Acids, Antimicrobials, Biotechnology, Colorants, Natural Compounds, Organic Acids, Sweeteners, Vitamins.

* **Corresponding author Valéria C. Santos-Ebinuma**: Department of Bioprocesses and Biotechnology; Univ. Estadual Paulista - UNESP, Rodovia Araraquara-Jaú, Km 01, 14800-903, Araraquara, SP, Brazil; Tel: +55-16-33-1-4647; E-mail: valeriac@fcfar.unesp.br

INTRODUCTION

Many substances are incorporated into foods for functional purposes, and, in many cases, these ingredients can also be found naturally in food. However, when they are used in processed foods, these substances are known as "food additives" [1]. Food additives play a fundamental role in maintaining food quality and the characteristics that consumers demand, *i.e.*, keeping food safe, wholesome, and appealing from farm-to-fork [2]. A food additive is defined as any substance the intended use of which results or may reasonably be expected to result, directly or indirectly, in it becoming a component or otherwise affecting the characteristic of any food. This includes any substance with a possible use in producing, manufacturing, packing, processing, preparing, treating, packaging, transporting, or holding food; and any source of radiation intended for any such use; if such substance is not *"generally regarded as safe* (GRAS)" or sanctioned prior to 1958 is excluded from the definition of food additives [3].

According to Food and Agriculture Organization of United Nations/ World Health Organization (FAO/WHO), food additives are "non-nutritive substances used in order to improve appearance, flavor, texture, and storage time". FAO states that "in some cases, the chemicals included to enhance the quality of the product may increase its nutritive value" [4].

Each country has a specific regulation about authorized additives in foods. However, most countries follow *"Codex Alimentarius"*, which presents a general standard for approved food additives. It was established by FAO/WHO *Expert Committee on Food Additives* (JECFA) [5]. Other relevant regulatory agencies of food additives are *Food and Drug Administration* (FDA) in the United States of America (USA) and the *European Food Safety Authority* (EFSA) in the European Union (EU). The role of regulatory agencies is to evaluate if a particular substance is suitable to be used as food additive, taking into account that a food additive should maintain consumers' health. The regulatory agencies also define the *"acceptable daily intake* (ADI)" as an estimate of the amount of a food additive, expressed on a body weight basis, that can be ingested daily over a lifetime without appreciable health risk [6]. So, all food additives necessarily undergo a rigorous scientific safety evaluation before being approved for commercial use [7].

There are various advantages justifying the use of food additives; for example, food additives (i) protect raw materials, and thus avoid waste and improve production; (ii) help to keep the organoleptic and sanitary quality of the product; (iii) increase the product shelf life during transport and storage; (iv) enhance the technical properties of the product and improve certain characteristics of the original food; and (v) keep the consumers' interest [7].

The additives can be natural or artificial (synthetic and semi-synthetic). Industrially, artificial food additives are more used than natural ones; however, many reports are showing the risks associated with consuming these substances [8]. Studics have shown that even additives approved by the regulatory agencies present some potential risks towards health. As an example, sodium benzoate, a widely used preservative in many foods and soft drinks, which is listed among the GRAS compounds by the FDA, presented adverse effects on glucose homeostasis at GRAS doses [8]. Therefore, further studies have to be performed to prove that it is not hazardous on long-term exposure [9].

From the above, natural food additives have been gaining more interest from both the public and food industry since they have lower toxic effects, than the synthetic ones, and can improve some food characteristics due to their biological activity and technological functionalities [8]. The main problem regarding the use of natural additives is their high cost compared to the synthetic ones. In this sense, biotechnology has the potential to improve their production yields and reduce their manufacturing costs.

Biotechnology can be defined as the assembly of technical and industrial processes involving the use of living organisms (bacteria, fungi, algae, plant or animal cells, *etc.*) in the production of goods and services. It is being used for many different purposes, such as the genetic improvement of plants and animals to increase their yields or efficiencies, and the production of bulk chemicals, fine chemicals, functional ingredients, additives, and other types of bio-compounds [4].

The main objective of this chapter is to introduce the main substances used as food additives and to show the alternative sources of these compounds through biotechnology. In the present chapter, amino acids, antimicrobial peptides, colorants, organic acids, vitamins, and sweeteners used as food additives will be described, focusing on the ones produced *via* biotechnology.

AMINO ACIDS

Amino acids are the structural blocks of proteins. There are 20 standard amino acids divided into essential and non-essential ones. The human body can only synthesize 11 amino acids; the other 9 must be supplied through diet. According to a survey conducted by "*Market Research*" [10], the market for amino acids was estimated to hit USD 12.8 billion by the end of 2017 and USD 35.4 billion by 2022. According to a new study by "*Grand View Research*" [11], the period 2015-2022 is expected to witness the highest amino acid market growth.

Biotechnological manufacturing methods of amino acids include extraction from the hydrolysates of animal and plant protein, and synthesis using fermentation and enzymatic processes. Nowadays, several amino acids are being produced by enzymatic (L-alanine, L-aspartic acid, L-cysteine, and L-tryptophan) and fermentation processes (L-arginine, L-glutamic acid, L-glutamine, L-histidine, L-isoleucine, L-leucine, L-lysine-HCl, L-phenylalanine, L-proline, L-serine, L-threonine, L-tyrosine, and L-valine) [12]; L-glutamic acid and L-lysine are the most produced ones [13].

L-glutamic acid emerged as the leading amino acid in the market; it accounted for over 40% of total market volume in 2014. L-lysine, L-threonine, L-methionine, and L-tryptophan are mainly consumed as feed additives for livestock production. L-tryptophan is expected to be the fastest growing amino acid with an estimated compound annual growth rate (CAGR) of 18% from 2015 to 2022 [11].

Bioprocesses for amino acid production have been developed since the end of the 1950s. Nowadays, bioreactors are well standardized for several amino acid production, where batch, fed batch, repeated fed batch, and continuous production regimes can be applied [14].

The major boundary for using batch processes is the osmolality of the initial medium and the low productivity due to the long lag phase. Compared with batch processes, the fed batch processing is still the standard process for manufacturing of amino acids. Fed batch processes ensure a better control of the nutrient concentration, reduce substrate inhibition, decrease overflow metabolism, minimize the osmotic stress of the cells, and reduce the lag phase [14 - 16]. In repeated fed batch or continuous production techniques, the above advantages are even more pronounced and may often lead to an increase in the process productivity [14].

A continuous process for L-glutamic acid production with *Brevibacterium lactofermentum* was developed, and the reported volumetric productivity was two times higher than the productivity achieved with the batch process [17]. Hirao *et al.* [18] studied a continuous process for L-lysine production using a strain of *Corynebacterium glutamicum* ATCC13032, which is resistant to S-(2-aminoethyl)-L-cysteine, rifampicin, streptomycin, and 6-azauracil. Under these constraints, the continuous process exhibited a 3-fold increase in productivity than the standard fed batch process.

C. glutamicum is a Gram-positive bacterium, non-pathogenic and GRAS. This bacterium occupies the leading position among the bacterial amino acid producers. It is the primary producer of L-glutamate and L-lysine; however, this bacterium has also been engineered for the production of other amino acids like

L-serine, L-proline [19], L-arginine, L-valine and L-isoleucine [20]. As L-lysine is easily accumulated during the biosynthesis of other amino acids by *C. glutamicum*, because the direct precursor for this amino acid synthesis is essential for cell wall synthesis of *C. glutamicum*, L-lysine is the main by-product of the metabolism of this bacterium [20]. In this sense, there are some research groups working in the reduction of the accumulation of L-lysine by *C. glutamicum*. To exemplify, Dong *et al.* [20] used metabolic engineering to construct a strain, from a wild strain of *C. glutamicum*, which would be able to produce L-threonine and L-isoleucine by fermentation while reducing the accumulation of L-lysine without affecting the cell growth. Although metabolic engineering is a very effective tool to improve the production of amino acids and other bioproducts, the design of the process is a critical step to allow efficient amino acids production.

In a scaling-up to 500 m³ bioreactor, for instance, p_{CO2} is usually increased as a result of lower aeration in large bioreactor. This point might be crucial for scaling-up of bioprocesses with *Corynebacterium* due to the significance of anaplerotic reactions [21]. This means that some enzymes, such as phosphoenolpyruvate carboxylase and pyruvate carboxylase, proved to be essential for L-glutamate production with a *C. glutamicum* temperature-sensitive mutant [22]. Addition of CO_2 to the process also increased the biomass yield and considerably decreased the formation of organic acids in L-lysine production with leucine and homoserine auxotrophic *C. glutamicum* [23].

In terms of oxygen availability, Shu and Liao [24] revealed that an increase of the oxygen transfer rates (OTR) improved L-phenylalanine productivity. Yao *et al.* [25] showed that the dissolved oxygen concentration influenced the process yield, which increased from 0.300 to 0.343 g. g $^{-1}$ for L-lysine production with *Brevibacterium lactofermentum*. Oxygen limited zones in the bioreactor indicate the presence of overflow metabolism or mixed acid fermentation, thereby inducing stress responses for the host cells. This is also described for bioprocesses with coryneform bacteria [26].

Process temperature can also be changed depending on the demands of the organism. Most studies mention 30–34°C as the optimum temperature for cultivating for *coryneform* bacteria. However, high temperatures are used to induce L-glutamic acid production in certain strains of bacteria [15, 22]. These strains are sometimes preferred for production processes because of the limited amount of cooling water available at some production sites. Furthermore, improving the bioprocess, evaluating different bioreactors, nutritional sources, and aeration, among others can improve the amino acid production.

Apart from the adjustment of production parameters, the producing strains used might be changed in order to increase the yield and the productivity of the corresponding amino acids. The use of *Escherichia coli* recombinant strains, for example, has proven to be particularly advantageous in the production of L-threonine [27] and L-tryptophan [28], and to a lesser extent in the manufacture of L-phenylalanine [29] and L-cysteine [30]. The last two were previously produced mainly by enzymatic methods; however, they can nowadays be obtained more efficiently using fermentation processes.

As cited before, recombinant DNA technology has made its way into the amino acid production field. Microbial strains have been constructed with plasmids bearing amino acid biosynthetic operons. However, some problems still exist. *C. glutamicum* strains, used for L-lysine production, which are obtained by classical mutagenesis and screening, are often less stable than their wild type counterparts. Mutant alleles of three gene targets, namely lysC, hom and pyc, encoding aspartokinase, homoserine dehydrogenase, and pyruvate carboxylase, respectively, which were identified in a classically obtained L-lysine producing strain, were introduced into *C. glutamicum* wild type by metabolic engineering. The mutant strain displayed a high L-lysine yield (around 15%) caused by increased flux into lysine biosynthesis, reduced flux towards threonine biosynthesis, and improved precursor supply [31].

Seibold *et al.* [32] introduced the amyE gene from *Streptomyces griseus* into an L-lysine producing strain of *C. glutamicum*; the resulting strain expressed a heterologousely secreted α-amylase using starch as a substrate for lysine production.

While the improvement of lysine-producing *C. glutamicum* strains was and is a continuous iterative process, serine-overproducing *C. glutamicum* strains have been engineered quite quickly from scratch. De-regulation and overexpression of genes for the three steps of serine biosynthesis, starting from the glycolytic intermediate 3-phosphoglycerate, were not sufficient for serine overproduction. However, when serine degradation to pyruvate was avoided by deleting the serine deaminase gene sdaA and when conversion of serine to glycine was lowered by reducing the transcription of glyA, which is the gene for encoding serine hydroxymethyl transferase, high serine yields could be achieved [33, 34]. Advances in fermentation technology and strain improvement of amino acid producing microorganisms have enabled the industrial-scale production of L-lysine as well as glutamate [21, 35, 36].

A cost-efficient downstream processing is crucial to reduce investment and production costs in the amino acid production [14]. Nowadays, many processes, from the simplest to the most sophisticated, are used for separation and purification of amino acids produced by fermentation, taking into account that chromatographic techniques are the most used for amino acid purification [37]. In the integrative downstream process, a continuous chromatography process separates the amino acids from the fermentation broth with high selectivity, increasing thus the product yields.

Some examples of the most highlighted downstream processes include affinity and ion-exchange chromatography for the separation of different types of peptides [38], reversed phase high pressure liquid chromatography (RP-HPLC), and ultra-high pressure liquid chromatography (UHPLC) with increased throughput, resolution, and sensitivity for the separation of complex protein mixtures [39, 40]. For the accurate measurement of the molecular weight of the amino acids, matrix-assisted laser desorption/ionization-time-of-flight mass spectrometry (MALDI-TOF/MS) can be used [41].

Regarding market volume and application of the amino acids, development has been tremendous in the so-called feed amino acids (*e.g.,* L-lysine, DL-methionine, L-threonine, and L-tryptophan). The share of the food sector in the amino acid market is also substantial and is determined mainly by three amino acids: L-glutamic acid in the form of the flavor enhancer monosodium glutamate (MSG), and the amino acids L-aspartic acid and L-phenylalanine; both of which are starting materials for producing the peptide sweetener aspartame. The other remaining amino acids are mainly used in the pharmaceutical, cosmetics, and agricultural industries [42].

L-glutamic acid is almost entirely produced through fermentation. The sodium salt of glutamic acid, monosodium glutamate (MSG), is a food additive authorized by EFSA and considered as GRAS by FDA. MSG is produced by an efficient bioprocess using *Corynebacterium* strains; however, other genera can also be producers of this amino acid, such as *Brevibacterium, Microbacterium, and Arthobacter* [43]. Currently, MSG is not only used as a food additive but also as (i) a therapeutic agent against gastroenterological disorders, (ii) an agent for improving the functions of liver and brain, (iii) an immune-enhancement agent and (iv) against gastric ulcer and alcoholism [44]. Glutamate is a naturally occurring amino acid that is found in nearly all foods, especially high-protein foods. It is also produced by human body and has an essential role in normal body functioning. When MSG is added to food, it provides a flavor similar to the glutamate that occurs naturally in food. It acts as a flavor enhancer and adds the fifth taste, called "*umami*", which is best described as a savory, broth-like, or meaty taste. According to FDA, MSG can be added to savory and processed foods, such as frozen foods, spice mixes, canned and dry soups, salad dressings and meat, or fish-based products [44].

Compared with L-glutamic acid, L-lysine is almost entirely used as a feed additive. Traditional feedstuffs like corn, wheat, or barley are poor in lysine. In order to increase feed efficiency, either L-lysine rich crops like soybean meal should be used, or L-lysine should be added to these raw feed materials. Adding soybean meal increases the protein level and supplies additional non-limiting amino acids. The dosed addition of L-lysine saves raw materials and reduces nitrogen excretion. The addition of 0.5% L-lysine increases feed quality by as much as adding approximately 20% soybean meal [14].

Like L-lysine, the amino acid L-threonine is almost exclusively used as a feed additive. Especially pig and poultry diets have a high demand of L-threonine. While corn germ meal, for example, contains similar amounts of L-threonine and L-lysine, soybean meal contains almost twice as much L-lysine as L-threonine. Methionine is another amino acid used in animal feed [14].

The use of *E. coli* recombinant strains has proven to be particularly advantageous in the production of L-threonine [27] and L-tryptophan [28], and to a lesser extent in the manufacturing of L-phenylalanine [29] and L-cysteine [30]. The last two were previously mainly produced by enzymatic processes, but can nowadays be obtained more cost-effectively through fermentation.

In general, amino acids and their derivatives are used to increase the nutritional value of food. As there is a tendency of the consumer to give priority to a healthier diet, the use of amino acids as a food additive is expected to grow further in the coming years; thus, the use of metabolic engineering and fermentation processes more efficiently can fulfill this expected high market demand.

ANTIMICROBIAL PEPTIDES

In the last two decades, antimicrobial peptides (AMPs) have gained a lot of attention due to the growing consumer demand for natural preservatives to replace the chemical ones. As a result of the emergence of multi-drug resistant bacteria as a serious problem over the past decades, major research efforts have been focused on finding effective alternatives.

AMPs are small cationic peptides that protect their hosts against bacteria, protozoa, viruses, and fungi. These peptides are vital components of the innate immune response by several forms of life, including bacteria, insects, plants, and vertebrates. They have been recognized as ancient evolutionary molecules that are effectively preserved in mammals [45 - 48].

AMPs demonstrate to be effective against a broad range of microorganisms, including Gram-negative and Gram-positive bacteria, fungi, and viruses [45]. In higher organisms, AMPs contribute to innate immunity; unlike the majority of conventional antibiotics, it appears that AMPs may also have the ability to (i) enhance immunity by functioning as immune-modulators, and (ii) act as a first defense line against harmful microorganisms [45, 48]. In bacteria, the production of AMPs provides a competitive advantage for the producer in certain ecological niches, which is a necessary strategy to decrease the number of competitors and to obtain more nutrients [49].

Bacterial AMPs are called bacteriocins, and they are produced by both Gram-positive and Gram-negative bacteria. However, there are some significant differences between eukaryotic AMPs and bacteriocins. Bacteriocins are often very potent, acting at pico- to nano-molar concentrations, whereas micromolar concentrations are required for the activity of eukaryotic AMPs. Most bacteriocins have a very narrow target spectrum, *i.e.*, being active against only a few species/genera closely related to the producer, while eukaryotic AMPs are generally less specific with a broad target spectrum [48].

More than 99% of bacteria can produce at least one bacteriocin, mostly not identified. These substances may be induced or produced spontaneously, being the producers immune to them. Bacteriocins are classified according to the bacterial spectrum, molecular weight, chemical structure, and mode of action [50].

Due to their protein nature, all bacteriocins are inactivated by one or more proteolytic enzymes, including those of pancreatic (α-chymotrypsin and trypsin) and gastric (pepsin) origin; thus, they are generally harmless to the human body and the surrounding environment [49]. This feature is fascinating for their use in food products.

Because of high potency and specificity, bacteriocins are considered as promising antimicrobial agents for different applications, including food preservation and infection treatment [48, 58], particularly, the bacteriocin produced by lactic acid bacteria (LAB) [59]. Pure or mixed cultures of bacteriocin-producing LAB, and their bacteriocins, can be used as a protective system against common food spoilage bacteria and pathogens [59].

Although many other types of bacteriocin, such as subtilin, cerein, thuricin, plantaricin, *etc.*, have been isolated and characterized from different producing strains, the commercially available sources, to date, are only nisin (*Lactococcus lactis*) and pediocin (*Pediococcus acidilactici*). Other bacteriocins are either not approved yet as food additives or are awaiting to be commercialized as food preservatives [60].

Nisin is already approved for food applications, and thus, it has gained widespread use in the food industry. The bacteriocin is approved for use in over 48 countries, including the EU. The FAO/WHO Codex Committee accepted nisin as a food additive in milk and milk products. Nisin is added to preserve and stabilize fermented dairy products, mayonnaise-type spreads, cream, cheese products, and meat or vegetable compositions. This bacteriocin is safe, sensitive to digestive proteases (can be hydrolyzed into amino acids in the intestine by α-chromotrypsin), and it does not cause changes in the organoleptic properties of food [61, 62]. Its amphipathic nature limits the widespread application of nisin in various products, since it interacts with fat and other food components [63]. This is exemplified by the fact that nisin is generally not as active in the preservation of meat as it is in dairy products. When nisin is used as meat preservative, an interference with meat components, such as phospholipids occurs, limiting nisin's activity especially in food with high-fat content [64]. A recent advance in this field is the use of immobilized nisin in the development of food packaging. It can be achieved by adding antimicrobial agents to the packaging system to prevent the growth of spoilage and pathogenic bacteria [65].

Improvement of growth conditions is crucial to enhance bacteriocin production, especially in cheap growth media. Usually, the production requires complex media and well-controlled parameters, such as temperature and pH [51, 52]. Additionally, the recommended medium for LAB culture usually contains a surplus of proteins (tryptone, peptone, meat extract, yeast extract), a substantial proportion of which remains unconsumed, leading to unnecessary cost and hindering bacteriocin purification [53]. The media for bacteriocins production include, for instance, whey, sugar molasses, mussel-processing wastes, and barley extract obtained from a by-product of *shochu* (traditional Japanese alcoholic beverage) called *shochu kasu*, supplemented with glucose [54 - 57].

Nisin and pediocin are the AMPs available to be used commercially which indicates that this field has a vast area to be explored in order to commercialize other AMPs in the market.

COLORANTS

Colorants, which can also be called pigments or dyes, are additives used to provide color to substances [66]. These compounds have been used to color foods since ancient times. Regarding their source, as other additives, colorants can be classified as natural, nature-identical, and artificial/synthetic [67]. The first colorants application was based on natural sources; however, after the synthesis of Mauve by William H. Perkin in 1856, natural colorants were replaced by synthetic ones, which possessed superior properties concerning tinctorial strength, hue, stability, and low cost [68].

Formerly, the colorants had the function of improving the food color, which was usually lost during processing. Since color was a much-appreciated feature for the consumer, the industry has taken actions to color different types of food products in order to improve consumer acceptance, and thus product sales [7]. Moreover, colorants mend the color changes taking place during storage, processing, packaging or distribution, replacing the unique identifying characteristics of the product, and protecting flavor and vitamins from photo-degradation [68]. According to the report of *"Report Buyer"*, the food color market is projected to reach USD 2.5 billion by 2020, growing at a annual rate of 4.5% [69].

Among all kinds of available colorants, the synthetic ones are the most used because of the advantages previously mentioned; however, toxicological studies have shown that they can be toxic after lengthy use, causing health problems, such as indigestion, anemia and allergic reactions, pathological lesions, tumors and cancer, paralysis, mental retardation, abnormalities in offspring, growth retardation, and eye defects resulting in blindness [70]. As a result, many synthetic colorants authorized for use in the past are no longer used in many countries. For example, in the EU, the colorants sunset yellow, quinoline yellow, azorubine, red 40, tartrazine, and ponceau 4R are banned in five countries, while in other European countries a warning saying *"in this product there may be an adverse effect on activity and attention in children"* should be placed on any food or drink that contains any of the six colorants [71]. Therefore, the interest in natural colorants has increased.

Natural colorants can be extracted from plants, insects, or microorganisms [72]. Although extracts from plants are more used compared to the others, microorganisms can be an attractive source of these natural compounds [73]. In the production of natural colorants by plants, biotechnology has been used to understand all the factors that affect the biosynthesis and accumulation of these compounds [74]. This also included genetic engineering of plants, sequencing, and understanding of the entire genomes, followed by the control of gene expression and transformation. However, the introduction of modified genes into the plant genome and the control of these genes are the most limiting factors for scaling-up many plant species as producers of natural colorants [74]. Besides, there is a vast research field to be explored and a challenge to be solved.

Due to the great variety of natural colorants suitable for use as food colorants, only the colorants mostly applied, or with the potential to be used, as food additives will be further discussed. The chemical structures of some of these colorants are depicted in Fig. (**6.1**).

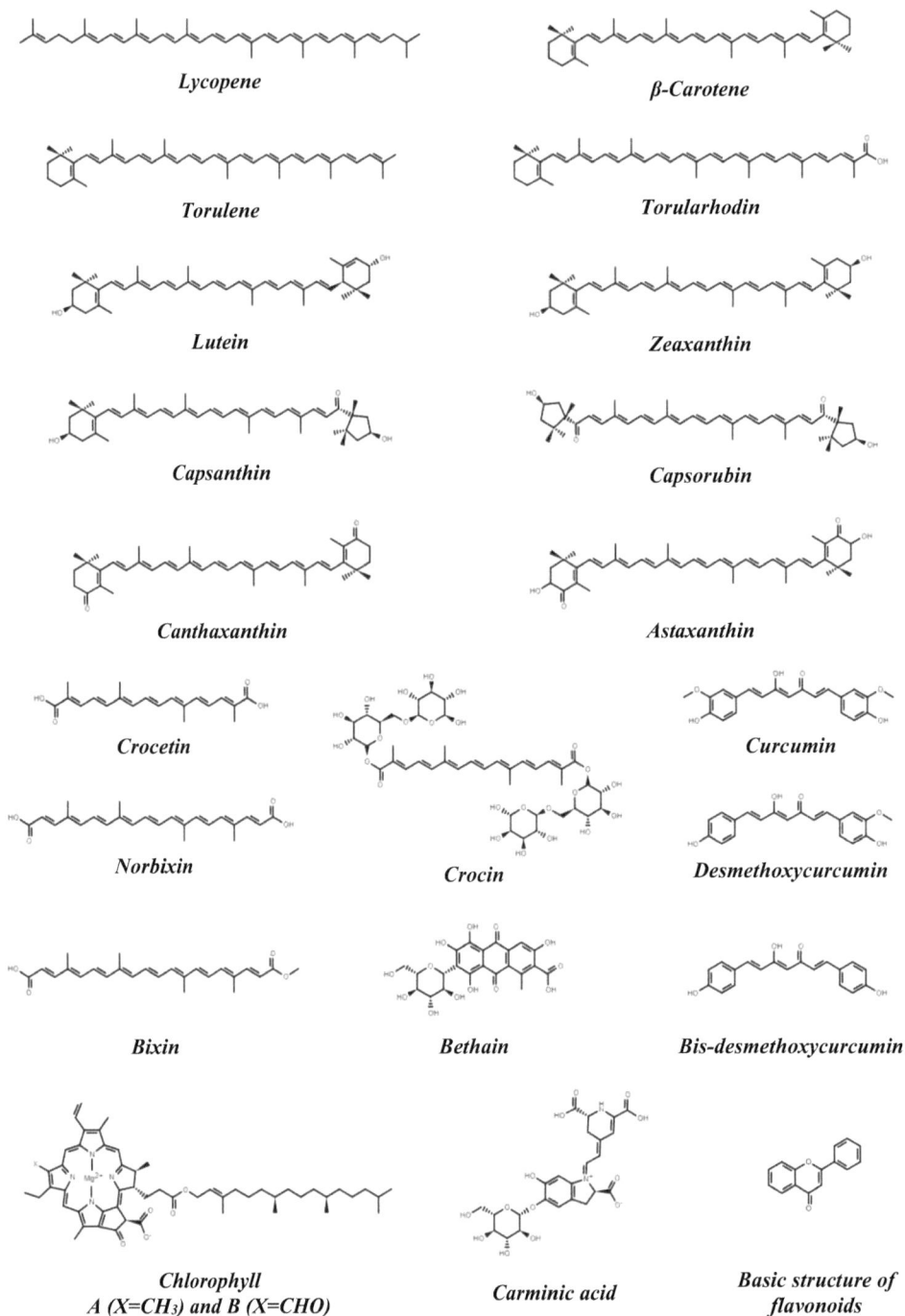

Lycopene

β-Carotene

Torulene

Torularhodin

Lutein

Zeaxanthin

Capsanthin

Capsorubin

Canthaxanthin

Astaxanthin

Crocetin

Curcumin

Norbixin

Crocin

Desmethoxycurcumin

Bixin

Bethain

Bis-desmethoxycurcumin

Chlorophyll
A (X=CH₃) and B (X=CHO)

Carminic acid

Basic structure of
flavonoids

Fig. (6.1). Chemical structure of some colorants used in food.

Carotenoids

Carotenoids are lipid-soluble colorants that present yellow, orange, and red color spectra in nature. More than 700 carotenoids exist, and it is one of the largest and most widespread groups of colorants in nature [75]. Their function is complex; however, when consumed either in foods or as supplements, carotenoids can exert positive effects on health, potentially in the prevention of degenerative or age-related diseases, as precursors of vitamin A, as retinoid-dependent signaling, in cell communication, in the regulation of gene expression, and in filtering of blue light [76]. The carotenoids of major importance are:

i. β-carotene: it is an oil soluble colorant which imparts a yellow color to foods. It can be found in carrot and palm oil seed extracts. This carotenoid can also be produced by microorganisms. In EU, the production of this carotenoid is performed by the fungi *Blakeslea trispora* (approved by EFSA), *Mucor circinelloides*, and *Phycomyces blakesleeanus* [77]. β-carotene finds applications in dairy products, cakes, soup, and confectionery.

ii. Lycopene: it has an intense red color and can be found in tomato (major source), watermelon and red grapefruit. This carotenoid can also be produced by *B. trispora* using a fermentation process, and it is approved as a food additive by EFSA. The potential application of lycopene includes beverages, confectionery, boiled sweets, bread, and cakes [68].

iii. Annatto: it is a permitted natural food colorant derived from *Bixa orellana* [68]. *B. orellana* may have pharmacological effects, namely, antibacterial and antifungal activity, antioxidant and free radical scavenging activity, anti-inflammatory activity, and anticarcinogenic activity, among others [78]. Depending on the extraction method, several carotenoids, including norbixin (yellow colorant) and bixin (red lipid-soluble colorant) [79], can also be extracted from *B. orellana*. These colorants are applied in cakes, biscuits, rice, dairy products, flour, fish, soft drinks, snacks, and meat products [8].

iv. Saffron: it is one of the worldwide earliest food additives. It is a water-soluble extract obtained from the stigma of flowers like *Crocus sativus and Gardenia jasminoides*. In the extract of saffron, crocin, crocetin, zeaxanthin, β-carotene, and certain flavoring compounds can be found. The flavoring compounds impart a distinct spicy flavor, restricting thus the use of saffron extract as just food colorant. It is usually added to foodstuffs, such as curry products, soups, meat, and certain confectionery products, where a spicy flavor is desirable along with the enhanced yellow color of these products [68].

v. Paprika (*Capsicum annum L.*): it is a vegetal, which can produce different compounds. Capsanthin and capsorubin are the main carotenoids in red paprika, while lutein is abundant in yellow and green paprika, and β-carotene is proved to be excessive in orange paprika [80]. Examples of food products colored with paprika are sausage, cheese sauces, gravies, condiments, salad dressings, baked goods, snacks, icings, cereals, soups, and lipsticks.

The advances in biotechnology to produce carotenoids are basically focused on the use of microorganisms. This will be discussed in detail in the section entitled *"Colorants produced by microorganisms"*.

Flavonoids

Flavonoids are a diverse group of polyphenolic compounds widely found in the plant kingdom [81]. They are divided according to their chemical structure into 6 groups: flavonols, flavanones, flavones, isoflavonoids, flavanols, and anthocyanidins [82]. These compounds have biological activities, such as anticarcinogenic, cardiovascular improvement functions, improved metabolic control properties, neuroprotective activity, anti-secretory functions, restoration of intestinal microflora, prevention of metabolic diseases, and intestinal barrier protection [82]. Among the flavonoids, anthocyanins are the most established food colorants and may be found in a wide variety of edible plant materials and flowers [8]. Although these colorants are used as additives in many foods, they have some application restrictions due to their ability to participate in some reactions, causing decolorization. In addition, the color of anthocyanin changes with the pH; at pH 4.0 and below (red color), pH between 4.0 and 6.0 (colorless), and pH between 7.0 and 8.0 (deep blue color). Successful applications of anthocyanins include the coloring of canned fruit, fruit syrups, yogurt, wine, and soft drinks. One of the biotechnological advances associated with the production of flavonoids is their methylation using genetic and metabolic engineering. Methylation of free hydroxyl groups in flavonoids can increase their metabolic stability and enhance their membrane transport, facilitating their absorption and a greater oral bioavailability. As a result, the characteristics of flavonoids will be improved, and these compounds could be sucessfully applied in more products [83].

Betalains

Betalains are natural colorants extracted from different fruits and vegetables. They show antioxidant activity, anti-cancer properties, antilipidemic effects and antimicrobial activity [84]. The only betalain approved for use as food colorant is betanin, known as "beetroot red". It is most stable between pH 4.0 and 6.0, and it exhibits a high sensitive to light. Therefore, betanin can only be used in foodstuffs with a short shelf-life which do not undergo prolonged heat treatment. Betanin is mainly used in food products with high protein content, such as poultry meat sausages, soy protein products, gelatin dessert, and dairy products like yogurt and ice cream. This is because proteins can act as a preservative of betanin color [85].

Besides their application as a food colorant, betalains show some pharmacological activities, such as antioxidant, anticancer, antimicrobial, and antilipidemic activities [84].

Chlorophyll

Chlorophyll is the green pigment found in all green plants as well as in green algae. Among the five different chlorophylls that exist, only two (A and B) are widely used in the food industry as colorants. Their complex structure is difficult to stabilize, which is the main drawback of their industrial application. One mechanism to preserve chlorophylls is a chemical modification by replacing the magnesium center with a copper ion; the copper complex is considered to be safe and it is permitted for use in most countries as a food additive [86]. Commercial chlorophyll is used in the food industry for coloring dairy products, edible oil, soaps, chewing gum, and sugar confectionery [86].

Turmeric

Turmeric is a fluorescent yellow colored extract obtained from the root of the curcuma plant [68]. Traditional applications of turmeric involve grinding the tuber into powder and adding it to the food as a spice rather than as a coloring agent. Three colorants (curcumin, desmethoxycurcumin, and bis-desmethoxycurcumin) are present in the turmeric extract [68]. Curcumin is the most applied colorant industrially. The major disadvantage of using curcumin is that it imparts a characteristic odor and sharp taste to foodstuffs [68]. Curcumin is used to confer an orange color in mustard, ice creams, baked goods, salad dressing and dairy products [8].

Cochineal

Cochineal is an anthraquinone approved for use as a food colorant by FDA and EFSA [87]. The principal cochineal pigment used industrially is carminic acid (E 120), which comes from the female of the genus *Dactylopius coccus* [68]. The color of carminic acid in solution changes with pH [87]; at pH lower than 4.0 it is orange, at pH 4.0 it is red, and at pH 10.0 it is violet. Carminic acid can be complexed with various metals to produce a bright red color. Aluminum is the metal usually applied in the commercial formulation of carminic acid complexes. The resulting product is known as carmine. The intense red color of carmine makes it an attractive coloring agent for jams, syrups, preserves, confectionery, and baked goods. However, studies showed that carmine could cause intense allergy [88] and asthma [89].

Colorants Produced by Microorganism

Like other biomolecules produced by microorganisms, the production of natural colorants depends on several factors. The production can be performed by submerged or by static culture techniques. Most colorants produced by *Monascus* species in Asia are produced under static culture techniques. The production using submerged culture techniques can result in more pronounced yields and more stable products. Regarding the submerged culture processing, many research groups have been largely studying the nutritional requirements and the microorganisms' metabolic routes to produce these colorants. However, scaling up to industrial production is yet a major challenge particularly the effect of process variables (temperature, pH, oxygen concentration, pressure, nutrient concentration, and shear stress) on the cultivation, regardless of the volume of the bioreactor [72]. Table **6.1** shows some examples of natural colorants produced by different classes of microorganisms (filamentous fungi, bacteria, algae, and yeast).

Table 6.1. Examples of natural colorants produced by microorganisms.

Molecule	Class of Microorganisms	Color	Microorganism	Reference*
Ankaflavin	Fungi	Yellow	*Monascus* sp.	[72, 90]
Antraquinone	Fungi	Red	*Penicillium oxalicum*	[72, 90]
			Isaria farinosa (formerly *Paecilomyces farinosus*)	[72]
Astaxanthin	Yeast	Pink-red	*Xantophyllomyces dendrorhous*	[90]
			Phaffia rhodozyma	[72, 90]
	Algae		*Haematococcus pluvialis*	[91]
β-carotene	Algae	Red	*Dunaliella salina*	[91]
	Fungi	Cream	*Blakeslea trispora*	[72, 90]
		Yellow-orange	*Fusarium sporotrichioides*	[72, 90]
			Mucor circinelloides	[72, 90]
			Neurospa crassa	[72, 90]
			Phycomyces blakesleeanus	[72, 90]
Canthaxanthin	Algae	Dark-red	*Bradyrhizobium* sp.	[91]
			Haloferax alexandrinus	[91]
			Hematococcus	[91]
	Fungi	Orange-pink	*Monascus roseus*	[72, 90]

(Table 6.1) cont.....

Molecule	Class of Microorganisms	Color	Microorganism	Reference*
Lycopene	Fungi	Red	*Blakeslea trispora*	[72, 90]
			Fusarium sporotrichioides	[72, 90]
Monascin		Red-yellow	*Monascus purpureus*	[72, 90]
Monascorubramin		Red-orange	*Monascus* sp.	[72, 90]
Naphtoquinone		Deep blood-red	*Cordyceps unilateralis*	[91]
Riboflavin		Yellow	*Asbya gossypi*	[91]
Rubropunctatin		Orange	*Monascus* sp.	[1]
Torularhodin	Yeast	Orange-red	*Rhodotorula* sp.	[91]
Unkown	Fungi	Orange-red	*Aspergillus* sp.	[72]
		Dark-red	*Aspergillus glaucus*	[72]
		Red	*Helminthosporium catenarium*	[72]
			Talaromyces purpurogenus (formerly *Penicillium purpurogenum*)	[72]
			Isaria cicadae (formerly *Paecilomyces sinclairii*)	[72]
		Bronze	*Helminthosporium avenae*	[72]
		Orange	*Penicillium cyclopium*	[72]
		Yellow	*Penicillium nalgiovensis*	[91]
	Yeast	Red	*Cryptococcus* sp.	[91]
		Black	*Saccharomyces neoformans* var. *nigricans*	[91]

*Adapted from cited references.

As can be seen in Table **6.1**, several microbes can be effective producers of natural colorants. Furthermore, genetic engineering appears as an increasingly powerful tool to manipulate microorganisms and consequently improve the production yield. The genetic approach is directed towards the understanding of the biosynthetic machinery [92]. Next, some examples of the biotechnological production of colorants by different microorganisms will be presented.

Filamentous Fungi and Yeast

The colorants from fungi can be broadly classified chemically as carotenoids and polyketides. Among the polyketides, representative classes include anthraquinones, hydroxyanthraquinones, naphthoquinones, and azaphilones; each of these structures exhibits an array of color hues [87]. Furthermore, polyketides may show bioactivities, such as antibiotic-, anticancer-, immuno-suppressor-, antifungal-, and cardiovascular-activities [93].

The red yeast *Sporobolomyces ruberrimus* can produce carotenoids (γ-carotene, β-carotene, torulene, and torularhodin) by submerged culture techniques [94]. Other yeast producers of carotenoids were also described, and they include *Phaffia rhodozyma*, as well the *Sporobolomyces* and *Rhodotorula* genus. The last one can produce β-carotene, torulene, and torularhodin [95]. In Asian countries, colorants produced by *Monascus* strains are popular [96], such as those produced by *M. purpureus* and *M. ruber*. In Japan, the use of colorants produced by *M. purpureus* is authorized as a food additive; however, in USA and EU these colorants are not yet approved. The colorants produced by *Monascus* species can be considered for use in protein-rich foods, such as meat, sausages, processed seafood, milk, and baked goods. The most significant colorants produced by *Monascus* species are azaphilones, such as monascin and ankaflavin (yellow colorants), monascorubrin and rubropunctain (orange colorants), and monascorubramine and rubropunctamine (red colorants) [97]. Colorants produced by *Monascus* exhibit biological activities, such as antimutagenic and anticancer properties, antimicrobial activity, potential anti-obesity activity, anti-inflammatory, antioxidant activities, and as a dietary supplement against hyperlipidemia, ameliorate hypertension, and hypercholesterolemia [72]. The main problem of colorants produced by *Monascus* is the co-production of citrinin, a hepato-nephrotoxic mycotoxin. Meanwhile, several research groups have been working hard to minimize citrinin production or even to develop strains that are incapable of co-producing citrinin [87]. Jia *et al.* [92] evaluated traditional chemical and physical mutagens and a bio-screening method to select mutants of *M. purpureus* SM001. Only one of the *Monascus* mutants resulted in citrinin-free colorant production at yields similar to those obtained by the wild strain.

Nutrients in the cultivation medium, as well as the bioreactors' configuration influence the production of colorants. The review of Torres *et al.* [72] summarizes several reports about the production of natural colorants by filamentous fungi. It is important to clarify that the use of filamentous fungi in bioreactors to produce colorants is a complex process that involves numerous well-controlled physical parameters, such as aeration rate, mass and heat transfer, and pH [98].

Mapari *et al.* [99] evaluated the ability of several strains of *Penicillium* as colorants' producers with a polyketide structure and chromophores similar to the *Monascus* colorants. The strains able to produce colorants with the cited characteristics were *P. purpurogenum, P. pinophilum, P. aculeatum, P. funiculosum,* and *P. minioluteum.* Among the produced metabolites, monascorubrin, xanthomonasin A, and derivatives of rubropunctatin were found in the extract of *P. aculeatum*, while the monascorubrin was identified in the extract of *P. pinophilum*. The main advantage of these strains is that they do not produce citrinin. Fig. (**6.2**) shows the chemical structure of some colorants produced by filamentous fungi as presented by Mapari *et al.* [100].

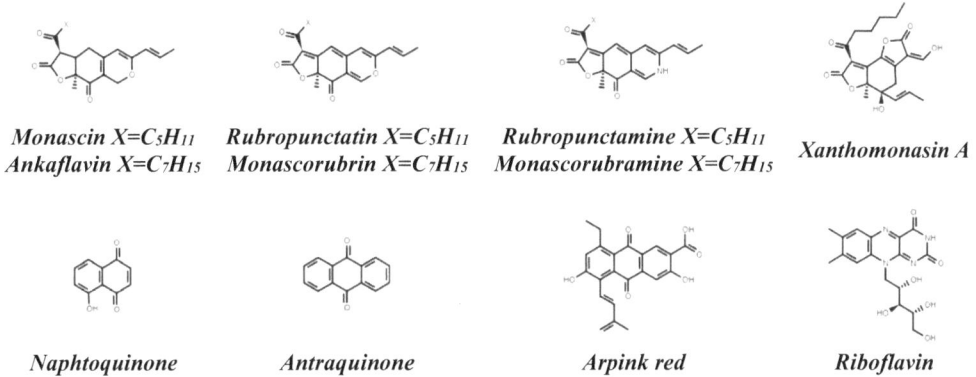

Monascin X=C₅H₁₁ *Rubropunctatin X=C₅H₁₁* *Rubropunctamine X=C₅H₁₁*
Ankaflavin X=C₇H₁₅ *Monascorubrin X=C₇H₁₅* *Monascorubramine X=C₇H₁₅* *Xanthomonasin A*

Naphtoquinone   *Antraquinone*   *Arpink red*   *Riboflavin*

Fig. (6.2). Chemical structure of some colorants produced by filamentous fungi [100].

According to the schematic procedure proposed by Mapari *et al.* [87], the step-by-step guidance for using fungal cell factories for the production of azaphilones includes:

1. pre-select potentially safe colorant producers, which must be chosen against myco-toxigenic and human pathogenic strains, based on their metabolic profile.
2. identify the known or novel colors using high-resolution LC/MS and colorimetric methods for the characterization of colorants.
3. study the enhanced photostability of colorants from selected microorganisms over the commercially available *Monascus* pigments in liquid food model systems.
4. attempt to increase the production capacity of potentially safe strains which produce *Monascus*-like azaphilone colorants.

Cordyceps unilateralis (polyketide naphthoquinone red pigments) [87] and *Isaria farinose* (anthraquinone-related compounds) [73] are also natural colorants' producers. *Penicillium oxalicum* var. *Armeniaca* produces a polyketide anthraquinone-carmine acid analog, called "arpink red". The EU approved this colorant for distribution in the Czech Republic, and it is being manufactured by Ascolor Biotec in that country [87].

Bacteria

Bacillus subtilis can produce riboflavin to be used in foods, vitamin-enriched milk products, and energy drinks. *Flavobacterium* and *Agrobacterium aurantiacum* produce β-carotene with applications as food supplements for humans and as additives for feed. *Flavobacterium* sp. and *Paracoccus xanthinifaciens* produce zeaxanthin, a yellow colorant in the development stage that can be applied as an additive in poultry feed to strengthen the yellow color of the animal skin and to strengthen the color of egg yolk. Other food colorants produced by bacteria are carotenoids from *Streptomyces* sp., canthaxanthin from *Photosynthetic bacterium*, *Bradyrhizobium* sp. and *Halobacterium* sp., and astaxanthin from *Agrobacterium aurantiacum, P. carotinifaciens,* and *H. salinarium* [101].

Algae

Algae can produce a group of colorants known as biliproteins or phycobiliproteins. The algae producers are red algae (*Rhodophyta*), blue-green algae (*Cyanophyta*), and the cryptomonads (*Cryptophyta)*. The biliproteins may further be divided into two groups: the red phycoerythrins and the blue phycocyanins [68]. From the blue-green algae *Spirulina platensis,* a blue colorant can be extracted, where it is already commercialized by the Japanese company Dainippon Ink and Chemicals Inc. The use of this colorant in chewing gum, soft drinks, alcoholic beverages, and fermented milk products such as yogurt, has been patented in Japan. Other applications of the biliprotein include confectioneries, candied ices and sherbets [68]. On the other hand, the microalgae *Phormidium autumnale* can also produce carotenoids, such as canthaxanthin, chlorophyll a, and C-phycocyanin [102].

As can be seen, several microorganisms can produce natural colorants through fermentation. Most of this production is still in the research stage, and before these colors get approved as food additives, some criteria should be improved, such as establishing standard conditions of cultivation, evaluating the stability, toxicity and tinting power of the colorants, and reducing the overall cost of the production process. However, with the current advances in biotechnology, the production of natural colorants from microorganisms or cell cultures can become a reality. These compounds will, therefore, be expected to shape the future of colorants, not only by adding color to food but also by increasing their biological activities.

ORGANIC ACIDS

Organic acids are one of the first biotechnological-based food additives produced and used on a large scale by the food industry, particularly from the earliest applications of citric acid and acetic acid, dated back to 1913 and 1823, respectively [103]. Organic acids, generally defined in the food industry as "food acids", have different roles as additives, and a wide array of these compounds have already been approved and regulated by several international authorities, such as FDA and EFSA. The long history of using organic acids as food additives and preservatives has been consolidated not only by the broad spectrum of applications but also by their resistance in a wide range of food production and processing conditions [104].

Organic acids are organic compounds (carbon-based chemical structures) with acidic properties, which can be produced in synthetic ways, or through biological processes as typical products of microbial metabolism. The benefit of organic acids as additives is that they are natural food ingredients, produced either by microbial processes or available as normal constituents of plants and animal tissues [108]. Various organic acids used industrially are mainly produced through biotechnological approaches, such as citric, lactic, acetic, itaconic, fumaric, propionic and gluconic acids [105]. The main advantages of their production through fermentation processes are the low energy requirements and processing cost-effectiveness, since only simple media formulations (for example, renewable and waste materials) are required [106]. It is important to note that although several organic acids or their derivatives are directly incorporated as additives into human food, they can also be obtained through fermentation during processing or storage, for instance as a result of the activity of starter cultures added to foods [107].

The production of organic acids using biotechnology-based approaches is of interest because of food legislation, and the fact that consumers demand more natural and sustainable products (with low environmental impact). However, the large scale production of organic acids through fermentation has some limitations, mainly related to lack of knowledge of some metabolic routes, the complexity of downstream processing, and the high bioprocesses production costs when compared with the synthetic routes [108].

Commonly, organic acids are classified into different groups according to their chemical structure; each one exhibiting specific properties and applications. The most typical group includes the carboxylic-based acids, which have one or more carboxyl groups (-COOH) in their chemical structure. The widely known organic acids are classified in this group, for example (1) monocarboxylic acids, such as acetic acid, lactic acid and gallic acid, (2) dicarboxylic acids, such as ascorbic acid and malic acid, and (3) tricarboxylic-acid, such as citric acid [108]. Other classes, such as sugar acids, aromatic and phenolic acids are also important for the food industry. On the other hand, organic acids can be categorized according to their natural occurrence, *i.e.*, if they are part of the primary or secondary metabolism of the producing organism. The first ones are the most common in the food industry, acids mostly used as acidulants or acidifiers. From the second type of organic acids, only a few compounds are applied in the production or processing of foodstuffs on industrial scales, such as ferulic acid and lactobionic acid [108].

Although the use of organic acids as food additives has been popular for a long time, their use increased between the late 1990s and the beginning of 2000s when the EU banned the use of antibiotics in food preparations. Thus, the number of acids available increased, and nowadays approximately 300 organic acid-based products are commercialized as single acids or as blends of more than one acid (a combination of more than three acids is not common). In addition, the use of organic acids in conjunction with their salts is an effective approach, since this reduces their odor and makes them easy to handle (reduction of corrosive characteristics) [109]. The international authorities, in particular EFSA and FDA, have legislated a series of food acids or organic acid-based derivatives. Several of these compounds were approved in different safety tests and they are defined as GRAS or incorporated into the "E" classification (reference numbers given to food additives that have passed rigorous testing and are approved for use throughout the EU). An overview of the most common organic acids, corresponding E numbers, and their main functions when applied as food additives are summarized in Fig. (**6.3**).

Citric Acid – E330	Acetic Acid – E260	Ascorbic Acid – E300	Lactic Acid – E270	Fumaric Acid – E297	Others Acids:
Antioxidant Acidulant Flavoring pH stabilizer Preservative	Preservative Flavoring	Antioxidant Preservative Nutrient	Acidulant Flavoring pH stabilizer Preservative	Acidulant Functional ingredient Preservative	*Gallic Acid* (antioxidant; preservative) *Phenyllactic Acid* (preservative) *Carminic Acid* (colorant)
Gluconic Acid – E574	**Malic Acid – E296**	**Propionic Acid – E280**	**Succinic Acid – E363**	**Tartaric Acid – E334**	*Ferulic acid* (Antioxidant) *Carnosic Acid* (antioxidant)
Acidulant Flavoring Preservative	Acidulant Flavoring pH stabilizer Preservative	Flavoring Preservative	pH stabilizer Preservative (sequestrants)	Acidulant	*Rosmarinic acid* (antioxidant) *Lactobionic Acid* (preservative)

Fig. (6.3). Schematic diagram comparing the most common organic acids, E numbers, and their main functions as food additives.

The schematic diagram in Fig. (**6.3**) compiles the most popular organic acids applied as food additives. Their use as food preservatives, flavor enhancers, and acidulants/acidifiers are the main applications. However, the interesting aspect of some organic acids is their multi-function capability, in which, for example, an organic acid can act simultaneously as a pH stabilizer and as flavor enhancer. In addition, weak organic acids have been successfully applied as food preservatives, mainly due to their general antimicrobial activities; nowadays, they are in fact the most useful antimicrobial agents in the food industry [110]. Their effectiveness results not only from their broad spectrum of antimicrobial activity but also from other features, such as their antioxidant, sequestrant and acidulant characteristics. While the pH reduction directly inhibits microbial growth, the antioxidant and sequestrant properties of some organic acids prevent unwanted chemical reactions and consequently avoid food deterioration, which may increase the shelf-life of perishable food ingredients [108]. From the large number of organic acids with high relevance in the food industry (Fig. **6.3**), citric acid and lactic acid are two of the most significant ones, since they can be "naturally" obtained.

Citric Acid

Citric acid is the most widely used food and beverage acidulant, and one of the world's major fermentation products, with a total production which exceeded 1.8 million tons in 2011 [111]. Nowadays, almost 70% of the total citric acid in use by food industry is obtained through fermentation [112, 113]. Citric acid is one of the most common food acids, representing 60% of all food acidulants applied worldwide [114]. EFSA, FDA, and FAO/WHO expert committees on food additives recognize the use of citric acid as "safe" with relevant GRAS status and E number [115]. Beside the use of citric acid as a popular acidulant, it exhibits several other applications, such as pH regulator, flavor enhancer, antioxidant synergist, and the most common use as a preservative to maintain the quality of products and to avoid or delay their deterioration. In bakery, this acid can be also used as a leavening agent.

This organic acid is also important for other industries, such as chemical (electroplating, antifoam, softener, metal processing, *etc.*), petrochemical (removing the iron clogged in the pores of the sand in the oil wells), pharmaceutical, and cosmetics (as effervescent in denture cleansers, anticoagulant, shampoos, *etc.*) industries [115]. Despite the attractiveness of citric acid as an additive in the food and beverage industry, this additive has some disadvantages, such as its sharp taste, hygroscopic nature, uneven particle size, and short-lived tartness flavor [103]. Yet, these are too scarce when compared with the importance of the "natural food additive" characteristics of citric acid. Thus, the use of citric acid as food additive will continue to be important in the food industry worldwide.

Citric acid is a primary metabolite of several microorganisms; it could be produced by many species including filamentous fungi, yeasts, and bacteria. However, their preferential commercial production is by the submerged fermentation of *Aspergillus niger* using low molecular weight carbohydrate solutions as substrates, which allow high citric acid production yields per unit time [112, 114, 115].

Lactic Acid

Like citric acid, lactic acid is also classified as GRAS by the FDA and is approved by EFSA. Lactic acid was first isolated from sour milk and then produced microbiologically (the first organic acid which was produced using a microbial route). Fermentative lactic acid production is also well established in the market with an annual production around 300 to 400,000 tons [102, 116] and an estimated future growth between 5 to 8% per year [117]. Lactic acid is a compound with a wide array of industrial applications, ranging from the conventional applications as chemical feedstock in metal (electroplating bath, plasticizer, and corrosion inhibitor), textile (finishing, antimony lactate) and leather industries (acidulant), to the most recent "bio-applications" in food, pharmaceutical (humectants, lotions, *etc.*) and medical (orthopedic implants, controlled drug release) industries. Despite its widespread applications, lactic acid is mainly used in the food industry [118]. As food additive, lactic acid is added to several foodstuffs (fermented drinks, meats, fish, vegetables, fruit, cake products, *etc.*) to preserve the products, increase their taste and flavor, and to act as an acidulant and a buffering agent (the most common use) [119].

Lactic acid is a food-based product that is a major end product of the fermentation of carbohydrates by two types of lactic acid-producing bacteria, the hetero-fermentative and the homo-fermentative [120]. The first group of bacteria, for example, *Leuconostoc mesenteroides*, produces lactic acid and several byproducts, such as ethanol, diacetyl, acetic acid and carbon dioxide, which makes it unattractive for the commercial production of lactic acid. On the other hand, the second type of lactic acid-producing bacteria is preferred for industrial production because they produce lactic acid as the major end product [116]. The homo-fermentative bacteria comprise a large number of *Lactobacillus* species, facultative anaerobes with a capability to convert the major carbon source into lactic acid (approximately two moles of lactic acid per mole of hexose sugar consumed). However, other Genera, such as *Lactococcus*, *Streptococcus*, *Pediococcus* and *Enterococcus*, include also some lactic acid-producing microorganisms. Some examples of homo-fermentative bacteria are *L. bulgaricus* that consumes lactose as a preferred carbon source, and *L. lactis* which can use glucose, sucrose, and galactose as carbon sources. In addition to the wild-type bacteria, lactic acid can also be produced by engineered producers, such as

Bacillus strains and *E. coli* [116].

Unlike citric acid, lactic acid is not entirely obtained by fermentation, as it can be also obtained through chemical synthesis. However, it is important to note that lactic acid has two optical isomers, D- and L-lactic acid, and when produced chemically the production occurs as a racemic mixture. The L-isomer is the preferred form in food processing, and considering that the optically pure form of lactic acid, *i.e.*, since the L-isomer is preferentially produced by fermentation, biotechnology is therefore the most convenient route to obtain the commercial form of lactic acid-based additives [118].

Like other fermentation processes, the microorganisms, fermentation conditions, and substrates are critical variables to control in the formation of L-lactic acid. Recently, the production of lactic acid using species of the fungus *Rhizopus* showed some advantages over bacterial-based processes, particularly with regards to (i) the amylolytic characteristics of *Rhizopus* that enables the use of various starchy biomasses without previous saccharification [121], (ii) the low nutritional requirements of these fungi [122], and (iii) the valuable fermentation by-products (fungal biomass have a high added-value) [116, 118].

Sodium lactate is an L-lactic acid derivative produced *via* microbial fermentation, which exhibits interesting properties, such as the preservation of food from microbial contamination during cold storage (it exhibits either bactericide or bacteriostatic activity), and the enhancement of flavors in processed meat [123]. Sodium lactate exhibits both bactericidal and bacteriostatic effects, but the second one is more predominant. Despite these interesting anti-bacterial characteristics, its spectrum of activity is only specific to certain type of organisms, for example against the pathogens *Clostridium botulinum* and *L. monocytogenes* [123].

Lactic acid and its derivatives are easily produced, and thus their application in several food-based processes has been increasing, mainly due to their high preservative and pH reduction functions. Lactic acid can be directly used from the bacterial fermentation broth in food preservation, replacing chemical food preservatives [124]. On the other hand, considering that biodegradable and biocompatible polymers can be manufactured with lactic acid, the use of these polymers as edible films or as replacers of conventional food packing materials appear as valuable solutions for food industry, consequently increasing the commercial potential of this acid.

Other Organic Acids

In the last paragraphs, the importance of the two well-known organic acids, citric acid and lactic acid, was highlighted. However, as shown in Fig. (**6.3**), other organic acids are also used in the production and processing of foodstuffs. Thus, some of the new organic acids obtained from microbial sources and recently proposed as interesting food additives are discussed below.

Consumers have been looking for food "free" from additives or at least naturally-based food additives obtained from different natural biological sources, such as microbes, plant or animal tissues [8]. Recently, little-known organic acids have been added to some foodstuffs, showing a set of more interesting properties than synthetic food additives. Some examples of these "new" food acids, highlighted in Fig. (**6.3**), include (i) the acids with antioxidant properties, such as gallic acid, ferulic acid, carnosic acid, and rosmarinic acid; (ii) acids with preservative effects, such as phenyllactic acid and lactobionic acid; (iii) acids with colorant potential, such as carminic acid. The majority of these acids are quite new for food-related processes, and thus, some of them are not yet regulated or accepted by international food authorities [8].

From a biotechnology perspective, ferulic acid appears as the most promising food additive, since it exhibits a wide variety of other biological activities beyond its antioxidant and antimicrobial properties, such as anti-inflammatory, anti-diabetic, anti-allergic, hepatoprotective, anticarcinogenic, antithrombotic, antiviral, modulation of enzyme activity, activation of transcriptional factors, and gene expression activities, among others [125]. Usually, ferulic acid is added to oils, animal fats, sauces, meat, fish, and bakery wares, or it is used as an antioxidant and precursor of other preservatives or flavors (vanillin). Recently, the application of ferulic acid in the formulation of food gels and edible films has also been reported [125]. Ferulic acid is commonly found in plants, vegetables, fruits, grains, *etc.*, but its direct extraction is quite complex, mainly due to its insolubility in water. To overcome this, some biotechnological studies have shown the possibility to produce ferulic acid using a recombinant strain of *Ralstonia euthropha* H16 (eugenol as substrate) [126] or *S. cerevisae* (eugenol and coniferyl as substrate) [127].

Like ferulic acid, carnosic acid is also a main active antioxidant used in foodstuffs. Carnosic acid is generally found in rosemary extracts, and its efficacy as an essential antioxidant was recognized in 2010, when the EU regulated the rosemary extracts (E392) composed of carnosic acid and carnosol as food additives [128]. However, it is important to note that the US, for example, does not consider rosemary extracts as authorized food additives. Therefore, more studies regarding the understanding of the mode of action of this acid are required to allow an extensive use in food-based processes [128].

Another example of an organic acid with high potential for the food industry is lactobionic acid, which is obtained by lactose oxidation and exhibits a strong potential as a bioactive additive [129]. Lactobionic acid has antioxidant, chelating, and humectant properties that are essential for food preservation. Due to its excellent characteristics, particularly the hygroscopicity, lactobionic acid has also been commercialized for pharmaceutical, medical, chemical, and nanotechnology purposes. It has a commercial market with an expected growth around 5% per year [129]. The production of lactobionic acid can be achieved through chemical (not economically feasible), electrochemical, biocatalytic, and heterogeneous oxidation of lactose. The biocatalytic approach uses specific enzymes (glucose-fructose dehydrogenases, cellobiose dehydrogenase, pyranose dehydrogenase) or whole microorganism cells (various species of *Pseudomonas*, *Acetobacter orientalis*, *Paraconiothyrium* sp., *Penicillium chrysogenum, etc.*) as biocatalysts to oxidize the lactose molecule [4]. Taking into account that FDA and EFSA approved lactobionic acid as a food additive, several promising applications have been proposed, for example as a sweet acidulant, as a filler in cheese production, or in the manufacture of meat-based foodstuffs with low water loss upon freezing and subsequent thawing [129].

Herein, we give some examples of "new" organic acids that can be successfully applied as additives in the food industry. However, as final remarks of this section it is important to understand what are the next acid-based food manufacturing challenges, and how biotechnology can overcome these challenges. In our opinion, two important approaches can contribute effectively to the expansion of applications of organic acids in the food industry. The first one is their use in the formulation of antimicrobial packaging, especially in the production of organic acid-based edible films. The direct incorporation of organic acid into films can increase their preservative properties, improve food sensory properties, and enhance the safety and shelf life of foodstuffs. The second challenge is the broad application of the biopreservation concept using organic acid-producing bacteria in foodstuffs. This approach aims at the improvement of food security and quality, and the extension of food shelf life, by combining fermentation and preservation processes. During fermentation, microorganisms produce some metabolites that enhance the flavor and texture, as well as organic acids that inhibit the growth of other microbes [130]. Although the effectiveness of the use microorganisms in food is already demonstrated, some concerns related to the interaction between the microbial-based food additives and consumers should be more investigated. For example, some recent reports have shown that lactic acid producing bacteria can be applied as bioprotectant agents in raw fish materials; however, the research should address some consumer concerns so that this use gains a wide market acceptance. The recent increase in the consumption of probiotics indicates that consumers accept foods containing microorganisms as ingredients. Thus, it seems evident that biotechnology will be an important driver to maintain organic acids as essential ingredients in the manufacturing of foods, making the natural acid-based foods economically attractive, and helping in the discovery of new acids through biotechnological approaches, or simply by developing new food formulations.

SWEETENERS

Sweeteners are food additives used to replace sugar and impart a sweet taste in foodstuffs. As for all food additives, sweeteners are regulated substances subject to safety evaluation before market authorization. Depending on regulations, foods containing intensive sweeteners have to state this fact on the label and include the name of the sweetener in the ingredients list [131]. The sweeteners can be divided into three groups: (i) sugar alcohols–polyols (*e.g.*, sorbitol, mannitol, glycerol, isomalt, maltitol, lactitol, xylitol, erythritol), (ii) artificial sweeteners also referred to as low-calorie sweetener (*e.g.*, acesulfame potassium, aspartame, cyclamate, saccharine, sucralose, alitame, neohesperidin dc, neotame, aspartame-acesulfame salt, advantame), and (iii) natural sweeteners (*e.g.*, thaumatin, glycyrrhizin, steviol glycosides, tagatose, fructo-oligosaccharides). The chemical structures of some sweeteners are depicted in Fig. (**6.4**).

All the above-cited examples are approved for use as food additives by EFSA [132]. Since the 1800's, sweeteners have been one of the most successful achievements of the food industry. In this long historical use, many controversies, conflicting regulations and laws have created a wide debate on the impact of sweeteners on industry, health and human lifestyle [133]. In fact, the global demand for zero-calorie natural sweeteners has increased significantly over the past decade as consumers have become increasingly health conscious [134]. In 2016, the global market for food sweeteners was valued at USD 85 billion. This is estimated to increase at a CAGR of 4.5% to reach nearly USD 112 billion by 2022 [135], which shows the importance of this food additive industrially. Regarding the low-calorie sweeteners, its worldwide market is expected to reach USD 2.2 billion by 2020 and its consumption is growing, especially in Latin America and China [136].

The artificial sweeteners are sweeter than sucrose, and thus, these are added to foods at smaller amounts than sucrose, in order to obtain the same level of sweetness [137]. To exemplify, cyclamate and saccharin, which represent the majority of artificial sweeteners consumed worldwide, are 30 and 300 times sweeter than sucrose by weight, respectively [136]. Although artificial sweeteners present high sweetening-power, some of them exhibit adverse effects. For example, in the study performed by Toigo *et al.* [138], animals exposed to aspartame during the prenatal period showed a high consumption of sweet foods during adulthood. These findings suggest that the use of aspartame during gestation may have deleterious long-term effects and should be consumed with caution. Some of the artificial sweeteners are thought to pose carcinogenicity and toxicity risks on liver, bladder, and fetus malformations [139]. In addition, artificial sweeteners are a newly recognized class of environmental contaminants, due to their extreme persistence and ubiquitous occurrence in various aquatic ecosystems. As they are resistant to wastewater treatment processes, they are continuously introduced into the aquatic environment [140].

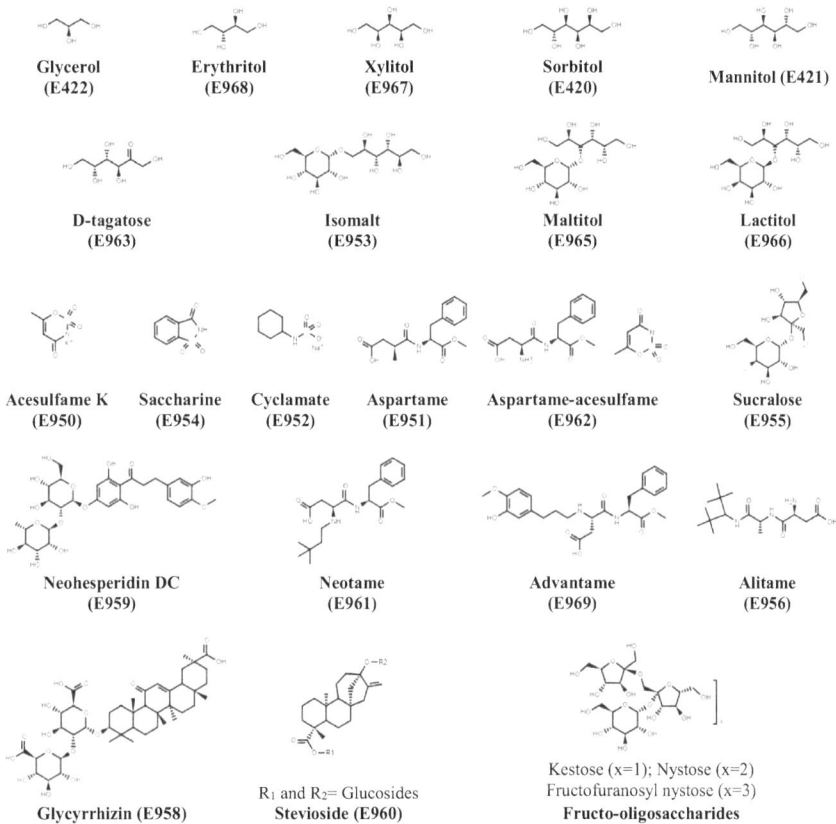

Fig. (6.4). Chemical structures of most representative sweeteners (E numbers).

In this way, the replacement of sucrose by natural sweeteners have been gaining attention by consumers due to (i) the problems of toxicity related to artificial sweetener cited before, (ii) the growing interest in natural products, and (iii) the fact that excessive sugar consumption has increased the incidence of obesity and diabetes [139]. Natural sweeteners can be divided into two groups, bulk sweeteners and high-potency sweeteners [8]. The difference between them is the sweetness potency, which is much more pronounced in the latter. Natural sweeteners need to be safe, have a good taste, exhibit high solubility and stability, and have an acceptable cost-in-use [8].

In the following paragraphs, we will describe some examples of sugar alcohols (polyols) and natural sweeteners, since they are the most interesting choices by consumers and can be produced biotechnologically.

Sugar Alcohols (Polyols)

Polyols are typical sucrose replacers, which have low calories and glycemic indexes; thus, when consumed they may decrease the risk of obesity and diabetes. The application of polyols in bakery products has been studied for decades. Several studies along this line can be found in the literature [141]. Among the polyols available as food additives, erythritol and xylitol have the most considerable market importance.

Erythritol

Erythritol (butane-1,2,3,4-tetraol), a four-carbon sugar alcohol, is a naturally occurring polyol found specially in several seaweeds, mushrooms, algae, fruits, and fermented foods [142]. It has little ability to change insulin levels and, thus, it can be consumed by diabetic's patients. Due to its low energy value, non-insulin stimulant properties, and excellent taste, erythritol is largely used as a pharmaceutical excipient and a non-caloric sweetener [143]. More than 90% of ingested erythritol is not metabolized by the human body, being excreted intact with urine, not changing blood glucose or insulin levels [142]. Furthermore, it has been demonstrated that erythritol is an important inhibitor of the growth and the biofilm formation of the oral *Streptococci* bacteria, which are known to contribute to plaque formation or dental caries [144, 145]. This sweetener is allowed in the USA and the EU although, in the latter, with some restrictions related to its use in beverages [8].

Erythritol occurs naturally in some fruits (*e.g.*, pears, watermelon, and grapes) and fermented foods (*e.g.*, soy sauce, wine, sake), and its application ranges from food to beverage industries, *e.g.*, chewing gum, candies, chocolates, baked foods, and soft drinks [144]. The chemical synthesis of erythritol involves a high-temperature hydrogenation reaction where periodate-oxidized starch or dialdehyde starch is converted to erythritol in the presence of Raney nickel catalyst [143, 146]. However, this method is not environmentally friendly, and due to the low efficiency, this process has not been industrialized [142, 144]. It is important to note that microorganisms can produce erythritol, and many studies have been carried out, not only to obtain organisms with high production yields, but also to improve the fermentation methods used for high-efficiency production. Erythritol has been produced using biotechnological processes by (i) osmophilic yeasts, such as *Pichia sp., Candida magnolia, C. sorbosivorans, Yarrowia lipolytica, Torula sp., Trigonopsis variabilis, Moniliella tomentosa* var. pollinis, and *Trichosporonoides sp.*, (ii) fungi, such as *Aureobasidium sp. and Pseudozyma tsukubaensis*, and (iii) some lactic acid bacteria under anaerobic conditions [142, 144, 145, 147, 148].

The industrial production of erythritol dates back to 1990 when it was produced by the aerobic fermentation of *Trichosporonoides megachiliensis* in Japan. *T. megachiliensis* SN-G42 (previously named *Aureobasidium* sp.) is a high erythritol production mutant obtained from the wild-type *T. megachiliensis* 124A by Ishizuka *et al.* [149]. Saran *et al.* [144] showed that erythritol can be produced by a newly isolated strain of *Candida sorbosivorans* SSE-24. Rywinska *et al.* [142] produced erythritol with a yield of 47 g.L^{-1} and a productivity of 0.85 g.L^{-1}.h^{-1} using 100 g.L^{-1} glycerol as a carbon source for the yeast *Yarrowia lipolytica* Wratislavia K1.

Savergave *et al.* [148] identified that glucose and yeast extract are critical medium components. In this investigation, it was observed that high erythritol production yields were obtained by increasing the yeast extract concentration. The efficient conversion of glucose to erythritol required high level of vital nutrients that are contained in yeast extract.

Xylitol

Xylitol (pentane-1,2,3,4,5-pentol), a five-carbon sugar alcohol ($C_5H_{12}O_5$), is naturally found in certain fruits (*e.g.*, yellow plum, strawberries, raspberries), vegetables (*e.g.*, cauliflower, lamb's lettuce, eggplant, pumpkin, spinach, lettuce, kohlrabi, fennel, onion, carrot) and white mushroom. However, the extraction process from such sources is uneconomical viable, due to the relatively small concentration of xylitol, generally below 1% (w/v), which makes the extraction process not scalable [150 - 152].

In comparison with sucrose, xylitol has almost an equivalent sweetness level but has 33% lower caloric content. Xylitol has been considered as an attractive alternative for non-insulin dependent diabetic patients. It also helps in the treatment of hyperglycemia, since its metabolism is insulin independent [153]. Furthermore, it is considered as a functional food due to its prebiotic nature, and it improves health, prevents infections, inflammatory processes, and osteoporosis, and shows some anticarcinogenic properties [151]. Due to its characteristics, xylitol was chosen as one of the top 12 value-added chemicals from biomass and could be used as a chemical platform for polymer synthesis, such as xylaric acid and glycols, directly polymerized for the production of unsaturated polyester resins [160]. Therefore, the demand of xylitol has increased for food, chemical, and pharmaceuticals purposes [152, 154, 155].

The primary interest in xylitol is centered on its properties and potential uses as an alternative sweetener [152]. Xylitol is added to improve shelf life, color, and taste of food products. It is used exclusively, or in combination with other sugar substitutes, in the production of sugar-free chocolate, chewing gum, hard candies, wafer fillings, chocolate, pastilles, and other candies for patients with diabetes. When xylitol is added to bakery products, it provides a characteristic flavor and color to baked goods. However, some browning can be expected from the reducing sugar present in the flour [151].

Xylitol can be used as an excipient to improve the taste of many pharmaceutical preparations, such as syrups, tonics, vitamin preparations, chewing gums, and tablets. Because xylitol is chemically inert, it does not undergo Maillard reactions or react with other excipients or active pharmaceutical ingredients; therefore, it does not darken or reduce the nutritional value of proteins in food products. It is also applied as a stabilizing agent in protein extractions to prevent denaturation of proteins. In addition to sweetness, non-reactivity, and non-fermentability, it is highly soluble at body temperature, and it offers a pleasing and cooling sensation [151, 152].

Xylitol can be produced either by chemical hydrogenation of xyloses or by biotechnological processes [151]. At industrial scale, xylitol is mostly produced from corn cobs (China) and hardwoods, such as birch (USA) through a chemical processing [150]. The precursor of xylitol is xylose, one of the major sugars in cellulosic biomass. This sugar is converted to xylitol *via* catalytic dehydrogenation in the presence of Raney nickel catalyst and hydrogen gas at over ambient conditions, such as 80–120°C and up to 5–5.5 MPa [151, 154]. However, the hydrogenated solution requires further processing (*e.g.*, chromatographic fractionation, concentration, and crystallization). At the end of the downstream processing, at least about 50–60% of xylose is converted into pure xylitol [151]. This method is complex, expensive, energy-intensive, and not environmentally friendly, due to the use of a toxic nickel catalyst and high-pressure hydrogen gas [152].

Xylose fermenting microorganisms and xylose reductase enzymes have been studied for the bioconversion of xylose to xylitol as an alternative way to chemical processing [153, 156]. Because of the instability and high cost of xylose reductase and the fact that cofactors (NADPH and NADH) are involved in the enzymatic process, the biotechnological production of xylose is preferentially carried out using microbial systems [154].

Indeed, microbial xylitol production has the advantage to be safe and environmentally friendly, since the process does not requires toxic catalysts and can be conducted under mild conditions of pressure and temperature [152]. The bioconversion is highly productive and can use cheap raw material (agricultural, agro-industrial, and forestry residues), such as corn fiber, sugarcane bagasse, spent brewing grain, olive tree pruning, soybean hull, palm oil empty fruit bunch fiber, rice straw, banana peel, mung bean hull, peanut hull, and oat hull [151, 157].

Lignocellulosic biomass is a widely spread, abundant, renewable, and cost-effective source of polysaccharides that can be used for xylitol production. Xylose is produced from agricultural wastes by enzymatic or chemical hydrolysis and can be microbiologically converted to xylitol through an economic process that requires less energy than chemical processing [151]. Microbial xylitol production has been considered the most favorable and sustainable for industrial applications [152] but, despite these advantages, the biotechnological routes of xylitol production represent a small portion of the market share so far [150].

Hydrolysis under acidic conditions is yet the most common pretreatment method of hemicellulosic biomass. Various microbial growth inhibitors (*e.g.*, furfurals, toxic phenolic compounds, aliphatic acids, aldehydes, and heavy metals) are produced during acid hydrolysis; decreasing xylitol yield. There are four different approaches to reduce the inhibitory effects of acid hydrolysate, namely: (i) use of a bioconversion friendly hydrolysis process; (ii) detoxification of hydrolysates before fermentation; (iii) use of microorganisms that are resistant to inhibitors or metabolically engineered microbial strains, which can tolerate inhibitors; (iv) conversion of toxic compounds into non-toxic counterparts [151]. Considering these facts, the detoxification may increase the cost of xylitol production. The most used physical and chemical methods for hydrolysate detoxification are vacuum evaporation, pH adjustment, activated charcoal, neutralization, sulfite treatment, calcium hydroxide, sulfuric acid, adsorption with ion-exchange resins and extraction with organic solvents [151].

Among several microorganisms that can convert xylose into xylitol, *Candida* species (*e.g.*, *C. boidinii, C. guillermondii, C. parapsilosis,* and *C. tropicalis)* are recognized as good xylitol producers [158], being used in industry [151, 152]. The production of xylitol from xylose in aerobic cultures of the yeast *C. mogii* TISTR 5892 was strongly stimulated by supplementing the culture medium with the growth inhibitor sodium benzoate [155]. Other xylitol microbial producers are

Serratia, Cellulomonas, Corynebacterium, and some genetically modified microorganisms, such as *E. coli, B. subtilis, Hypocrea jecorina, T. reesei* [158], *K. marxianus* [153] and *S. cerevisiae* [159]. As example, engineered *S. cerevisiae* expressing xylose reductase cannot metabolize xylitol into the central carbon metabolism. So, it converts xylose to xylitol at almost the theoretical yield (~100%). This is a promising strategy to meet the industrial demands for microbial xylitol production [154].

Natural Sweeteners

Natural sweeteners are derived from natural sources, such as plants and microorganisms. Some polyol sweeteners can also be considered as natural ones; however, this depends on their origin. In the next paragraphs, some characteristics of the most applied natural sweeteners in foodstuff will be discussed.

Although the market of natural sweeteners derived from plants is growing, its large use is limited due to agricultural sustainability, undesirable taste qualities, apparent safety, and commercial viability [134]. The biotechnological production of natural sweeteners can overcome some of the limitations associated with plant-based production processes, such as scaling-up and overall sustainability. In addition, the development of plant or microbial cell-based production platforms can allow a rapid modification of pathway enzymes to generate novel sweeteners with reduced or negligible taste attributes [134].

The biotechnological production of natural sweeteners includes (i) developing of cost-effective plant cell culture methods, or (ii) the metabolic engineering of functional pathways in microbial hosts followed by an optimization of production strains. In both cases, the scale-up of the fermentation process is a key step to further allow an economical production and purification of natural sweeteners. Besides, it is always important to evaluate key criteria, such as taste quality, sweetness intensity, safety for consumption, chemical stability, cost-effectiveness, and patentability, among others [160].

Steviol Glycosides

Steviol glycosides, mainly steviosides and rebaudiosides, are examples of well-known natural sweeteners used worldwide and known just as stevia, stevioside or steviol. Stevia, already approved by FDA and EFSA, has drawn a lot of attention because of its marketable qualities in terms of low caloric high sweetness, non-carcinogenic, and antidiabetic properties [166]. Industrially, it has been produced by extraction from the leaves of the plant *Stevia rebaudiana Bertoni* [8]. Studies using metabolic engineering of *ent*-kaurene, a precursor pool responsible for synthesizing steviol glycosides, in recombinant *E. coli* [160] has shown potential to obtain high-level production of stevia through fermentation processes; this represents a significant advance in the manufacture of this natural sweetener.

Rebaudioside A is mainly derived from plants; however, it can also be obtained from fermentation using a genetically engineered strain of *Yarrowia lipolytica*, which express the steviol glycoside metabolic pathway of the plant *S. rebaudiana*. Purified rebaudioside A (> 95 wt%) is isolated from the fermented broth by crystallization, dried and packaged [161, 162]. This compound is not approved by FDA and EFSA, but some studies have been performed to prove its safety [161], aiming at its entrance into the market.

Glycyrrhizin

Glycyrrhizin, also known as glycyrrhizinic acid or glycyrrhizic acid, is usually extracted from the liquorice plant, *Glycyrrhiza glabra*. This compound exhibits several positive physiological effects, such as anti-ulcer, anti-inflammatory, anti-viral, anti-carcinogenic, and anti-spasmodic. It also has corticoid activity, influencing steroid metabolism to maintain blood pressure and volume, and to regulate glucose/glycogen balance. However, besides these positive attributes, glycyrrhizin shows several side-effects, which typically involve cardiac dysfunction, edema, and hypertension; all these adverse effects were observed in subjects that received high doses of glycyrrhizin-based pharmaceuticals or consumed large amounts of liquorice-containing confectionery over a prolonged period [163]. This compound is legally used in the USA and the EU under the form of mono-ammonium glycyrrhizinate and ammoniated glycyrrhizin. In food processing industry, it is mainly applied in liquorice, baked goods, frozen dairy products, beverages, confectionery, and chewing gum [8].

D-Tagatose

D-Tagatose is a rare natural hexo-ketose and an isomer of D-galactose. It naturally occurs as a component of the gum exudate of *Sterculia setigera*. It is also found in small quantities in dairy products, such as the in-container sterilized cow's milk, various cheeses, and yogurts. This monosaccharide has a low caloric index, a low glycemic index, prevents tooth decay, and promotes the growth of intestinal probiotics. D-Tagatose is certified as GRAS by FDA and authorized by EFSA as a food ingredient. The addition of tagatose to cereals, beverages, yogurts, chewing gum, chocolate, fudge, caramel, fondant, and ice cream is its most common application [8]. Since D-tagatose rarely exists in nature, its massive production is highly dependent of chemical or biochemical synthesis. The isomerization of D-galactose using calcium hydroxide under alkaline conditions into tagatose, followed by strong acid conditions to neutralize the calcium-tagatose complex is the traditional chemical synthetic route. Nevertheless, this is not environmentally friendly, as it promotes the formation of several by-products.

Afterwards, a biotechnological method using D-galactitol was established, which was oxidized into D-tagatose using the bacteria *Arthrobacter globiformis* [164]. However, this process is also not economically attractive for large scale production, due to the high price and commercial unavailability of D-galactitol [171]. To overcome this problem, Xu *et al.* [165], purposed a single-step production of D-tagatose from lactose by a whole recombinant *E. coli* strain, co-expressing the thermophilic enzymes L-arabinose isomerase (from *L. fermentum* CGMCC2921) and β-galactosidase (from *Thermus thermophilus* HB27) into a polycistronic plasmid. This process involves the hydrolysis of lactose by β-galactosidase and the isomerization of released D-galactose into D-tagatose by L-arabinose isomerase [165].

Fructo-oligosaccharides (FOS)

Fructo-oligosaccharides (FOS), new alternative sweeteners, include mainly 1-kestose, nystose, and 1-β-fructofuranosyl nystose, which are produced from sucrose or inulin. FOS have low caloric values, are non-carcinogenic, and have bifidus-stimulating properties. They also help gut absorption of minerals, and decrease the levels of lipids and cholesterol. The industrial production of FOS involves the action of enzymes with transfructosylating activity (β-D-fructofuranosidases and fructosyltransferases) using different biocatalytic methods, such as whole cells, isolated enzymes, and immobilized biocatalysts (either whole cell or enzyme immobilization). The main FOS producing microorganisms are fungi (*e.g.A. japonicus, A. niger, A. sydowi, A. foetidus, A. oryzae, A. pullulans, P. citrinum, P. frequentans, and F. oxysporum, Aureobasidium sp.*), bacteria (*e.g.Arthrobacter sp., Zymomonas mobilis, L. reuteri* and *B. macerans)*, and yeasts (*e.g.Kluyveromyces* and *Candida*). Application of enzymes obtained from plant sources is less common, due to the seasonal variations; therefore, microbial enzymes are still preferred for the production of FOS [166].

Sweet Plant Proteins

Other plant-derived natural products which induce sweet-responses or modulate sweetness include terpenoids, phenylpropanoids, dihydroisocoumarins, flavonoids, steroids, proanthocyanidins, amino acids, and proteins [134].

Sweet plant proteins are safe, natural, and low-caloric sweeteners. They also may be suitable replacements of sugars in the food and beverage industry [167]. Currently, some types of sweet proteins have been found in plants, such as thaumatin, mabinlin, monellin, pentadin, brazzein, curculin (neoculin), and miraculin. Among them, only mabinlin is stored in seeds [134, 167, 168].

Thaumatin is a mixture of five proteins (taumatin I, I, III, A, and B) extracted from the fruit of *Thaumatococcus daniellii* Benth, a plant native to Africa. Due to its liquorice cool aftertaste, thaumatin is not used in high quantities, although when mixed with other sweeteners it confers a good taste and reduces bitterness in foodstuff. Natural proteins are usually added either as sweeteners or flavor enhancers into sauces, soups, fruit juices, poultry, egg products, chewing gum, and processed vegetables among others [8].

Mabinlin is isolated from the seeds of *Capparis masaikai* Levl., which grows in subtropical regions in China. The mabinlin family has four sweet proteins, mabinlins I, II, III, and IV. Among them, mabinlin II possesses the most interesting properties, as it has around 400-fold the sweetness of sucrose on a molar basis, and its activity remains unchanged after 48 h of incubation at boiling temperatures [167].

Since the extraction of sweet proteins from raw plants is complex and expensive, their large use as naturally occurring sweetners in the food industry is limited. Recombinant DNA technology is being proposed as suitable alternative to obtain cheaper and high amounts of these proteins [167]. Thaumatin, for example, has been expressed in *E. coli* [169 - 171], *Pichia pastoris* [172 - 175], *Aspergillus* fungus [176, 177], and transgenic plants [178, 179]. Miraculin has been extensively produced in transgenic plants, especially in tomatoes [180 - 190], citrus cell suspension culture [191], and microorganisms, such as *E. coli* [192] and *A. oryzae* [193]. Brazzein has been expressed in *E. coli* [200], lactic acid bacteria, such as *L. lactis* [201] and *Lactobacillus* spp [202], in yeasts, such as *P. pastoris* [203 - 205] and *K. lactis* [206], as well as in the milk of transgenic mice [207]. Curculin has been expressed in *E. coli* [208, 209] and neoculin has been produced in *A. oryzae* [210, 211]. On the other hand, monellin has been expressed in bacteria, such as *E. coli* [194] and *B. subtilis* [195], in yeasts, such as *C. utilis* [196] and *S. cerevisiae* [197], as well as in transgenic tomatoes [198]. Moreover, recombinant monellin was also purified from the yeast strain ABllO [199]. More recently, mabinlin II was expressed in *E. coli* and in *L. lactis* [167].

As highlighted previously in the organic acids section, the use of recombinant lactic acid bacteria (LAB), *L. lactis* and *Lactobacillus* spp, can represent a new application for these bacteria by creating a system in which they express sweet-tasting proteins *in situ* in dairy products, thereby removing the need for added sugar and contributing to the value of LAB as producers of functional foods [167, 201, 202].

As shown, many "natural or biotechnological" sweeteners are available in the market. However, the most interesting sweeteners from commercialization and consumers' perspective are those derived from natural sources, since they usually exhibit low caloric indexes. To improve the production of these compounds, biotechnology presents various opportunities through metabolic engineering of plants or by improving batch culture systems using wild and genetically modified microorganisms.

VITAMINS

Vitamins are nutrients required for normal physiological functioning. Many of them play vital roles within the body in a variety of processes. While vitamins have been investigated for a wide range of physiological roles, recently, researchers have begun to examine how these are involved in dysfunction of the central nervous system, from chronic diseases to acute insults [212]. Many vitamins are easily obtained through chemical synthesis, but, some of these bring serious disadvantages, such as the toxicity of the reactants. Thus, replacing chemical processes by biotechnology has already gained a lot of interest, in particular due to its low environmental impact. Taking this into consideration, the majority of fat- and water-soluble vitamins can be already produced through microbial fermentation processes.

The main vitamins with antioxidant potential already in use as food additives are vitamin C (ascorbic acid) and vitamin E (tocopherols). Both vitamins can work in synergy due to the regeneration of vitamin E through vitamin C from the tocopheroxyl radical to an intermediate, reinstating once again its antioxidant potential. For this reason, they are usually used together to extend the shelf life of foodstuffs [213, 214].

Vitamin C is an essential vitamin for humans and is only acquired through a nutritional diet [215]. It is important to note that Vitamin C is also named as Ascorbic Acid, and thus, it can be also classified as organic acid or food acid. The most important function of ascorbic acid is to protect tissues from harmful oxidative products, *i.e.*, free radicals as superoxide radical anions, hydrogen peroxide, hydroxyl radicals, singlet oxygen, and reactive nitrogen oxide [213, 215]. Some bacteria, algae, and yeasts are responsible for producing vitamin C. The main process used is the fermentation of D-glucose to 2-keto-L-gulonic acid followed by chemical conversion to L-ascorbic acid. This method can be carried out using the enzyme 2,5-diketo-D-gluconic acid reductase or biologically using microorganisms, such as *Cynobacterium* sp., among others, including recombinant organisms. As observed for other food additives, genetic engineering has been used for strain improvement and for enhancing the production yields [216].

When applied in food industry, ascorbic acid appears as one of the most used antioxidants in meat, beverage, fish, and bakery products, among others. One of the most known applications is in baking, as it strengthens the dough and prevents the collapse of the dough during fermentation and baking. Furthermore, ascorbic acid absorbs oxygen in food, oxidizing itself to dehydroascorbic acid. In this way, the available oxygen is reduced, acting also as a food preservative. This means that ascorbic acid, when added to foodstuffs during processing or before packing, maintains color, aroma, and nutrient contents. In addition to the antioxidant mechanism, ascorbic acid also acts as an anti-browning agent by reconverting quinones back to the phenolic form and avoiding flavor deterioration in beverages [213, 215].

Vitamin E comprises a group of lipid-soluble compounds that have 4 isoforms (α-, β-, γ-, and δ-tocopherol) and 4 tocotrienols (α-, β-, γ-, and δ-tocotrienol) [214]. Among these isoforms, α-tocopherol has the highest antioxidant activity *in vivo* [217]. This vitamin exerts its biological activity, in particular against lipid peroxidation and rancidification, by donating its phenolic hydrogen to the peroxyl radicals forming tocoperoxyl radicals that, despite also being radicals, are unreactive and unable to continue the oxidative chain reaction [218].

Like other vitamins, α-tocopherol is obtained by chemical synthesis, and can also be directly extracted from vegetable oils. Both traditional processes raise some concerns in terms of sustainability, because in the first process non-environmental reactants are used, while the second process is not efficient. In addition, as observed for other vitamins, synthetic α-tocopherol is not identical to its natural counterpart [216]. A good example of the biotechnological production of α-tocopherol is the use of the freshwater microalgae *Euglena gracilis* and Z. (1-BPV) and the marine microalgae *Dunaliella tertiolecta* [220, 221]. The simultaneous production of high amounts of β-carotene, vitamin C, and vitamin E has also been successfully demonstrated by *E. gracilis Z.* using two-step culture in two different media. The two-step culture using high cell densities allowed high productivity of antioxidant vitamins [219, 221].

Riboflavin (vitamin B$_2$) is a powerful antioxidant found in meat and dairy sources. It rapidly reduces the oxidized iron [222]; the high levels of which lead to free radical damage and lipid peroxidation [223, 224]. Vitamin B$_2$ is absorbed and phosphorylated to become flavin mononucleotide and is then converted into flavin adenine dinucleotide [225]. Additionally, riboflavin can be converted to dihydroriboflavin, which reduces the heme protein with high oxidative states of iron, a process that further reduces oxidative damage [222, 223, 226]. The production of Vitamin B$_2$ through fermentative conversion of glucose by *Eremothecium ashbyii, Ashbya gossypii, Bacillus sp.*, and other microorganisms

[216] was already proposed. The highest riboflavin level was obtained using a recombinant *B. subtilis* strain, which efficiently produces riboflavin directly from glucose in fed-batch operation [227]. When applied as a food additive, it can play a role to improve nutrition. On the other hand, it can also be simply used as a yellow colorant [216]. Riboflavin is added to baby foods, breakfast cereals, pasta, and vitamin-enriched meal replacement products. In some countries, such as the USA, the addition of this vitamin is also used to enrich the white flour.

There are other vitamins present in the market; however, due to their low stability, they are not used as food additives.

CONCLUDING REMARKS

The main objective of this work was to describe the food additives produced through biotechnological approaches. For this purpose, amino acids, antimicrobial peptides, colorants, organic acids, sweeteners, and vitamins were presented and discussed as important food additives. Despite the efforts made by several researchers, the use of biotechnology to produce food additives is mainly feasible on a lab scale. Certainly, there are some exceptions like the citric and lactic acids that have been produced industrially by microorganisms since many years ago. We have also shown that new food additives can be produced by biologically using wild microorganisms, genetically modified microorganisms, or they can be extracted from plants sources. It is clear that more studies are still required and should be performed to fill the experimental gaps and the lack of knowledge, regarding the toxicology of these compounds. As discussed, there is an awareness of consumers about the use of synthetic additives and their health implications due to their side-toxicological effects. So, natural additives appear as interesting alternatives, since they exhibit low adverse effects and in some cases, they even improve the characteristics of food due to their biological activities. However, studies to reduce their production costs, improve the scale-up process, and some of their toxicological effects still have to be performed. In this context, biotechnology will be the key for developing new and viable technologies to obtain food additives from microbial and plants sources. To conclude, the use of biotechnology can shape the future of food additives with enhanced characteristics, and some examples of this progress were already highlighted in the current chapter.

CONSENT FOR PUBLICATION

Not applicable.

CONFLICT OF INTEREST

The authors declare no conflict of interest, financial or otherwise.

ACKNOWLEDGEMENTS

The authors are grateful to *São Paulo Research Foundation* (FAPESP) [grant numbers 2014/01580-3, 2014/16424-6, 2015/04751-6], *National Council for Scientific and Technological Development* (CNPq) [grant number 443984/2014-0], *Coordination for Higher Level Graduate Improvements* (CAPES) and *Institute for Technological Research* (IPT).

ABBREVIATIONS

ADI	Acceptable daily intake
AMPs	Antimicrobial peptides
CAGR	Compound annual growth rate
EFSA	European Food Safety Authority
EU	European Union
FAO/WHO	Food and Agriculture Organization of the United Nations/World Health Organization
JECFA	FAO/WHO Expert Committee on Food Additives
FDA	Food and Drug Administration
FOS	Fructo-oligosaccharides
GRAS	Generally regarded as safe
LAB	Lactic acid bacteria
MSG	Monosodium glutamate
OTR	Oxygen transfer rates
USA	United States of America

REFERENCES

[1]   Damodaran S, Parkin KL, R. FO. Fennema's food chemistry. 4th ed., Taylor & Francis 2007.

[2]   Gilbert J, Senyuva H. Bioactive compounds in foods. Blackwell Publishing 2008; p. 409. [http://dx.doi.org/10.1002/9781444302288]

[3]   FDA. 2015. Available from: http://www.fda.gov/

[4]   FDO. Food and Agriculture Organization of the United Nations 2016. Available from: http://www.fao.org/

[5]   FAO/WHO. Codex alimentarius Procedural manual. 23rd ed., U.S. Food and Agriculture Organization/World Health Organization 2015.

[6]   WHO. Principles for the safety assessment of food additives and contaminants in food. World Health Organization 1987.

[7]   Evangelista J. Technology of foods. 2nd ed., Brazil: Atheneu 2001.

[8]   Carocho M, Morales P, Ferreira I. Natural food additives: *Quo vadis*? Trends Food Sci Technol 2015; 45(2): 284-95.
        [http://dx.doi.org/10.1016/j.tifs.2015.06.007]

[9]   Lennerz BS, Vafai SB, Delaney NF, *et al.* Effects of sodium benzoate, a widely used food preservative, on glucose homeostasis and metabolic profiles in humans. Mol Genet Metab 2015; 114(1): 73-9.
        [http://dx.doi.org/10.1016/j.ymgme.2014.11.010] [PMID: 25497115]

[10]   Research BCC. 2015. Market-Research. World Markets for Fermentation Ingredients:

[11]   Grand View Research I. Aminoacids market. 2016. Available from: http://www.grandviewresearch.com/industry-analysis/amino-acids-market

[12]   Demain AL. The business of biotechnology. Ind Biotechnol (New Rochelle NY) 2007; 3(3): 269-83.
        [http://dx.doi.org/10.1089/ind.2007.3.269]

[13]   Encyclopedia of food microbiology 2014.

[14]   Hermann T. Industrial production of amino acids by coryneform bacteria. J Biotechnol 2003; 104(1-3): 155-72.
        [http://dx.doi.org/10.1016/S0168-1656(03)00149-4] [PMID: 12948636]

[15]   Gourdon P, Lindley ND. Metabolic analysis of glutamate production by *Corynebacterium glutamicum*. Metab Eng 1999; 1(3): 224-31.
        [http://dx.doi.org/10.1006/mben.1999.0122] [PMID: 10937937]

[16]   Enfors SO, Jahic M, Rozkov A, *et al.* Physiological responses to mixing in large scale bioreactors. J Biotechnol 2001; 85(2): 175-85.
        [http://dx.doi.org/10.1016/S0168-1656(00)00365-5] [PMID: 11165362]

[17]   Koyoma Y, Ishii T, Kawahara Y, Koyoma Y, Shimizu E, Yoshioka T. Inventors method for producing L-glutamic acid by continuous fermentation, 1998.

[18]   Hirao T, Nakano T, Azuma T, Sugimoto M, Nakanishi T. L-Lysine production in continuous culture of an L-lysine hyperproducing mutant of *Corynebacterium glutamicum*. Appl Microbiol Biotechnol 1989; 32(3): 269-73.
        [http://dx.doi.org/10.1007/BF00184972]

[19]   Peters-Wendisch P, Götker S, Heider SAE, *et al.* Engineering biotin prototrophic *Corynebacterium glutamicum* strains for amino acid, diamine and carotenoid production. J Biotechnol 2014; 192(Pt B): 346-54.
        [http://dx.doi.org/10.1016/j.jbiotec.2014.01.023] [PMID: 24486440]

[20]   Dong X, Zhao Y, Hu J, Li Y, Wang X. Attenuating l-lysine production by deletion of *ddh* and *lysE* and their effect on l-threonine and l-isoleucine production in *Corynebacterium glutamicum*. Enzyme Microb Technol 2016; 93-94: 70-8.
        [http://dx.doi.org/10.1016/j.enzmictec.2016.07.013] [PMID: 27702487]

[21]   de Graaf AA, Eggeling L, Sahm H. Metabolic engineering for L-lysine production by *Corynebacterium glutamicum*. Adv Biochem Eng Biotechnol 2001; 73: 9-29.
        [http://dx.doi.org/10.1007/3-540-45300-8_2] [PMID: 11816814]

[22]   Delaunay S, Lapujade P, Engasser JM, Goergen JL. Flexibility of the metabolism of *Corynebacterium glutamicum* 2262, a glutamic acid-producing bacterium, in response to temperature upshocks. J Ind Microbiol Biotechnol 2002; 28(6): 333-7.
        [http://dx.doi.org/10.1038/sj.jim.7000251] [PMID: 12032806]

[23]   Sassi AH, Deschamps AM, Lebeault JM. Process analysis of L-lysine fermentation with *Corynebacterium glutamicum* under different oxygen and carbon dioxide supplies and redox potentials. Process Biochem 1996; 31(5): 493-7.
        [http://dx.doi.org/10.1016/0032-9592(95)00087-9]

[24]   Shu CH, Liao CC. Optimization of L-phenylalanine production of *Corynebacterium glutamicum* under product feedback inhibition by elevated oxygen transfer rate. Biotechnol Bioeng 2002; 77(2): 131-41.
[http://dx.doi.org/10.1002/bit.10125] [PMID: 11753919]

[25]   Yao HM, Tian YC, Tade MO, Ang HM. Variations and modelling of oxygen demand in amino acid production. Chem Eng Process 2001; 40(4): 401-9.
[http://dx.doi.org/10.1016/S0255-2701(01)00112-X]

[26]   Xu B, Jahic M, Blomsten G, Enfors SO. Glucose overflow metabolism and mixed-acid fermentation in aerobic large-scale fed-batch processes with *Escherichia coli*. Appl Microbiol Biotechnol 1999; 51(5): 564-71.
[http://dx.doi.org/10.1007/s002530051433] [PMID: 10390814]

[27]   Debabov VG. The Threonine Story.Microbial Production of L-Amino Acids. Berlin, Heidelberg: Springer Berlin Heidelberg 2003; pp. 113-36.
[http://dx.doi.org/10.1007/3-540-45989-8_4]

[28]   Ikeda M, Katsumata R. Hyperproduction of tryptophan by *Corynebacterium glutamicum* with the modified pentose phosphate pathway. Appl Environ Microbiol 1999; 65(6): 2497-502.
[PMID: 10347033]

[29]   Cordwell SJ. Microbial genomes and "missing" enzymes: redefining biochemical pathways. Arch Microbiol 1999; 172(5): 269-79.
[http://dx.doi.org/10.1007/s002030050780] [PMID: 10550468]

[30]   Wacker. Cysteine from wacker – fermentation synthesis for the highest demand. 2004.

[31]   Ohnishi J, Katahira R, Mitsuhashi S, Kakita S, Ikeda M. A novel gnd mutation leading to increased L-lysine production in *Corynebacterium glutamicum*. FEMS Microbiol Lett 2005; 242(2): 265-74.
[http://dx.doi.org/10.1016/j.femsle.2004.11.014] [PMID: 15621447]

[32]   Seibold G, Auchter M, Berens S, Kalinowski J, Eikmanns BJ. Utilization of soluble starch by a recombinant *Corynebacterium glutamicum* strain: growth and lysine production. J Biotechnol 2006; 124(2): 381-91.
[http://dx.doi.org/10.1016/j.jbiotec.2005.12.027] [PMID: 16488498]

[33]   Netzer R, Peters-Wendisch P, Eggeling L, Sahm H. Cometabolism of a nongrowth substrate: L-serine utilization by *Corynebacterium glutamicum*. Appl Environ Microbiol 2004; 70(12): 7148-55.
[http://dx.doi.org/10.1128/AEM.70.12.7148-7155.2004] [PMID: 15574911]

[34]   Peters-Wendisch P, Stolz M, Etterich H, Kennerknecht N, Sahm H, Eggeling L. Metabolic engineering of *Corynebacterium glutamicum* for L-serine production. Appl Environ Microbiol 2005; 71(11): 7139-44.
[http://dx.doi.org/10.1128/AEM.71.11.7139-7144.2005] [PMID: 16269752]

[35]   Scheper T, Ed. M I. Amino acid production processes. Biotechnological manufacture of lysine. Advances in Biochemical Engineering. Springer 2003.

[36]   Pfefferle W, Möckel B, Bathe B, Marx A. Biotechnological Manufacture of Lysine.Microbial Production of L-Amino Acids. Berlin, Heidelberg: Springer Berlin Heidelberg 2003; pp. 59-112.
[http://dx.doi.org/10.1007/3-540-45989-8_3]

[37]   de Castro RJS, Sato HH. Biologically active peptides: Processes for their generation, purification and identification and applications as natural additives in the food and pharmaceutical industries. Food Res Int 2015; 74: 185-98.
[http://dx.doi.org/10.1016/j.foodres.2015.05.013] [PMID: 28411983]

[38]   Ortiz-Martinez M, Winkler R, García-Lara S. Preventive and therapeutic potential of peptides from cereals against cancer. J Proteomics 2014; 111: 165-83.
[http://dx.doi.org/10.1016/j.jprot.2014.03.044] [PMID: 24727098]

[39]   Fekete S, Guillarme D. Ultra-high-performance liquid chromatography for the characterization of

therapeutic proteins. Trac-Trends in Analytical Chemistry 2014; 63: 76-84.
[http://dx.doi.org/10.1016/j.trac.2014.05.012]

[40] Yang Y, Boysen RI, Chowdhury J, Alam A, Hearn MTW. Analysis of peptides and protein digests by reversed phase high performance liquid chromatography-electrospray ionisation mass spectrometry using neutral pH elution conditions. Anal Chim Acta 2015; 872: 84-94.
[http://dx.doi.org/10.1016/j.aca.2015.02.055] [PMID: 25892073]

[41] Panchaud A, Affolter M, Kussmann M. Mass spectrometry for nutritional peptidomics: How to analyze food bioactives and their health effects. J Proteomics 2012; 75(12): 3546-59.
[http://dx.doi.org/10.1016/j.jprot.2011.12.022] [PMID: 22227401]

[42] Leuchtenberger W, Huthmacher K, Drauz K. Biotechnological production of amino acids and derivatives: current status and prospects. Appl Microbiol Biotechnol 2005; 69(1): 1-8.
[http://dx.doi.org/10.1007/s00253-005-0155-y] [PMID: 16195792]

[43] Kumar R, Vikramachakravarthi D, Pal P. Production and purification of glutamic acid: A critical review towards process intensification. Chem Eng Process 2014; 81: 59-71.
[http://dx.doi.org/10.1016/j.cep.2014.04.012]

[44] Kusumoto I. Industrial production of L-glutamine. J Nutr 2001; 131(9) (Suppl.): 2552S-5S.
[http://dx.doi.org/10.1093/jn/131.9.2552S] [PMID: 11533312]

[45] Izadpanah A, Gallo RL. Antimicrobial peptides. J Am Acad Dermatol 2005; 52(3 Pt 1): 381-90.
[http://dx.doi.org/10.1016/j.jaad.2004.08.026] [PMID: 15761415]

[46] Guaní-Guerra E, Santos-Mendoza T, Lugo-Reyes SO, Terán LM. Antimicrobial peptides: general overview and clinical implications in human health and disease. Clin Immunol 2010; 135(1): 1-11.
[http://dx.doi.org/10.1016/j.clim.2009.12.004] [PMID: 20116332]

[47] Reddy KVR, Yedery RD, Aranha C. Antimicrobial peptides: premises and promises. Int J Antimicrob Agents 2004; 24(6): 536-47.
[http://dx.doi.org/10.1016/j.ijantimicag.2004.09.005] [PMID: 15555874]

[48] Hassan M, Kjos M, Nes IF, Diep DB, Lotfipour F. Natural antimicrobial peptides from bacteria: characteristics and potential applications to fight against antibiotic resistance. J Appl Microbiol 2012; 113(4): 723-36.
[http://dx.doi.org/10.1111/j.1365-2672.2012.05338.x] [PMID: 22583565]

[49] Yang SC, Lin CH, Sung CT, Fang JY. Antibacterial activities of bacteriocins: application in foods and pharmaceuticals. Front Microbiol 2014; 5.

[50] Todorov SD, Dicks LMT. Effect of modified MRS medium on production and purification of antimicrobial peptide ST4SA produced by *Enterococcus mundtii*. Anaerobe 2009; 15(3): 65-73.
[http://dx.doi.org/10.1016/j.anaerobe.2008.11.002] [PMID: 19100330]

[51] De Vuyst L, Vandamme EJ. Influence of the carbon source on nisin production in *Lactococcus lactis* subsp. lactis batch fermentations. J Gen Microbiol 1992; 138(3): 571-8.
[http://dx.doi.org/10.1099/00221287-138-3-571] [PMID: 1593266]

[52] Kumar M, Jain AK, Ghosh M, Ganguli A. Statistical optimization of physical parameters for enhanced bacteriocin production by L-casei. Biotechnol Bioprocess Eng; BBE 2012; 17(3): 606-16.
[http://dx.doi.org/10.1007/s12257-011-0631-4]

[53] Vázquez JA, González MP, Murado MA. Preliminary tests on nisin and pediocin production using waste protein sources. Factorial and kinetic studies. Bioresour Technol 2006; 97(4): 605-13.
[http://dx.doi.org/10.1016/j.biortech.2005.03.020] [PMID: 15913992]

[54] Guerra NP, Pastrana L. Nisin and pediocin production on mussel-processing waste supplemented with glucose and five nitrogen sources. Lett Appl Microbiol 2002; 34(2): 114-8.
[http://dx.doi.org/10.1046/j.1472-765x.2002.01054.x] [PMID: 11849506]

[55] de Arauz LJ, Jozala AF, Baruque-Ramos J, Mazzola PG, Pessoa A Junior, Vessoni Penna TC. Culture

medium of diluted skimmed milk for the production of nisin in batch cultivations. Ann Microbiol 2012; 62(1): 419-26.
[http://dx.doi.org/10.1007/s13213-011-0278-6]

[56]   Jozala AF, Silva DP, Vicente AA, Teixeira JA, Pessoa A Junior, Penna TCV. Processing of bypro-ducts to improve nisin production by *Lactococcus lactis*. Afr J Biotechnol 2011; 10(66): 14920-5.
[http://dx.doi.org/10.5897/AJB11.979]

[57]   Furuta Y, Maruoka N, Nakamura A, Omori T, Sonomoto K. Utilization of fermented barley extract obtained from a by-product of barley shochu for nisin production. J Biosci Bioeng 2008; 106(4): 393-7.
[http://dx.doi.org/10.1263/jbb.106.393] [PMID: 19000617]

[58]   Zacharof MP, Lovitt RW. Bacteriocins Produced by Lactic Acid Bacteria A Review Article. 3rd International Conference on Biotechnology and Food Science (Icbfs 2012). 50-6.
[http://dx.doi.org/10.1016/j.apcbee.2012.06.010]

[59]   Balciunas EM, Martinez FAC, Todorov SD, Franco B, Converti A, Oliveira RPD. Novel biotechnological applications of bacteriocins: A review. Food Control 2013; 32(1): 134-42.
[http://dx.doi.org/10.1016/j.foodcont.2012.11.025]

[60]   Bali V, Panesar PS, Bera MB, Kennedy JF. Bacteriocins: Recent Trends and Potential Applications. Crit Rev Food Sci Nutr 2014.
[PMID: 25117970]

[61]   Guerra NP, Agrasar AT, Macias CL, Bernardez PF, Castro LP. Dynamic mathematical models to describe the growth and nisin production by *Lactococcus lactis* subsp lactis CECT 539 in both batch and re-alkalized fed-batch cultures. J Food Eng 2007; 82(2): 103-13.
[http://dx.doi.org/10.1016/j.jfoodeng.2006.11.031]

[62]   Pongtharangkul T, Demirci A. Evaluation of agar diffusion bioassay for nisin quantification. Appl Microbiol Biotechnol 2004; 65(3): 268-72.
[http://dx.doi.org/10.1007/s00253-004-1579-5] [PMID: 14963617]

[63]   Taylor TM, Davidson PM, Zhong Q. Extraction of nisin from a 2.5% commercial nisin product using methanol and ethanol solutions. J Food Prot 2007; 70(5): 1272-6.
[http://dx.doi.org/10.4315/0362-028X-70.5.1272] [PMID: 17536693]

[64]   Deegan LH, Cotter PD, Hill C, Ross P. Bacterlocins: Biological tools for bio-preservation and shelf-life extension. Int Dairy J 2006; 16(9): 1058-71.
[http://dx.doi.org/10.1016/j.idairyj.2005.10.026]

[65]   Malhotra B, Keshwani A, Kharkwal H. Antimicrobial food packaging: potential and pitfalls. Front Microbiol 2015; 6: 611.
[http://dx.doi.org/10.3389/fmicb.2015.00611] [PMID: 26136740]

[66]   Saron C, Felisberti MI. Ação de colorantes na degradação e estabilização de polímeros. Quim Nova 2006; 29: 124-8.
[http://dx.doi.org/10.1590/S0100-40422006000100022]

[67]   Santos-Ebinuma VC, Roberto IC, Simas Teixeira MF, Pessoa A Jr. Improving of red colorants production by a new *Penicillium purpurogenum* strain in submerged culture and the effect of different parameters in their stability. Biotechnol Prog 2013; 29(3): 778-85.
[http://dx.doi.org/10.1002/btpr.1720] [PMID: 23554384]

[68]   Lee Y-K, Khng H-P. Natural Color Additives. In: Branen AL, Davidson PM, Salminen S, Thorngate-III JH, editors. Food additives 2001.
[http://dx.doi.org/10.1201/9780824741709.ch17]

[69]   Report-Buyer UK. Food colorants market-growth, trends and forecast. Mordor Intelligence LLP 2016; pp. 2015-20.

[70]   El-Wahab HM, Moram GS. Toxic effects of some synthetic food colorants and/or flavor additives on

male rats. Toxicol Ind Health 2013; 29(2): 224-32.
[http://dx.doi.org/10.1177/0748233711433935] [PMID: 22317828]

[71]   EFSA. European Food Safety Authority 2016. Available from: http://www.efsa.europa.eu/

[72]   Torres FA, Zaccarim BR, de Lencastre Novaes LC, *et al.* Natural colorants from filamentous fungi. Appl Microbiol Biotechnol 2016; 100(6): 2511-21.
[http://dx.doi.org/10.1007/s00253-015-7274-x] [PMID: 26780357]

[73]   Velmurugan P, Lee YH, Nanthakumar K, *et al.* Water-soluble red pigments from *Isaria farinosa* and structural characterization of the main colored component. J Basic Microbiol 2010; 50(6): 581-90.
[http://dx.doi.org/10.1002/jobm.201000097] [PMID: 20806258]

[74]   Matthews PD, Wurtzel ET. Biotechnology of Food Colorant Production.Food Colorants: Chemical and Functional Properties: Taylor & Francis Group. 2007; pp. 347-98.
[http://dx.doi.org/10.1201/9781420009286.sec5b]

[75]   Lerfall J. Carotenoids: occurrence, properties and determination.Encyclopedia of food and health. 2016; pp. 663-9.
[http://dx.doi.org/10.1016/B978-0-12-384947-2.00119-7]

[76]   Sy C, Dangles O, Borel P, Caris-Veyrat C. Stability of bacterial carotenoids in the presence of iron in a model of the gastric compartment - comparison with dietary reference carotenoids. Arch Biochem Biophys 2015; 572: 89-100.
[http://dx.doi.org/10.1016/j.abb.2014.12.030] [PMID: 25595845]

[77]   Kumar A, Vishwakarma HS, Singh J, Dwivedi S, Kumar M. Microbial pigments: production and their applications in various industries. Int J Pharm Chem Biol Sci 2015; 5: 203-12.

[78]   Shahid-ul-Islam. Rather LJ, Mohammad F. Phytochemistry, biological activities and potential of annatto in natural colorant production for industrial applications – A review. J Adv Res 2015.

[79]   Santos LF, Dias VM, Pilla V, Andrade AA, Alves LP, Munin E, *et al.* Spectroscopic and photothermal characterization of annatto: Applications in functional foods. Dyes Pigments 2014; 110: 72-9.
[http://dx.doi.org/10.1016/j.dyepig.2014.05.018]

[80]   Kim JS, An CG, Park JS, Lim YP, Kim S. Carotenoid profiling from 27 types of paprika (*Capsicum annuum* L.) with different colors, shapes, and cultivation methods. Food Chem 2016; 201: 64-71.
[http://dx.doi.org/10.1016/j.foodchem.2016.01.041] [PMID: 26868549]

[81]   Rymbai H, Sharma RR, Srivastav M. Biocolorants and its implications in health and food industry - A review. Int J Pharm Tech Res 2011; 3(4): 2228-44.

[82]   Hoenscha HP, Oertelb R. The value of flavonoids for the human nutrition: Short review and perspectives. Clinical Nutrition Experimental 2015; 3: 8-14.
[http://dx.doi.org/10.1016/j.yclnex.2015.09.001]

[83]   Koirala N, Thuan NH, Ghimire GP, Thang DV, Sohng JK. Methylation of flavonoids: Chemical structures, bioactivities, progress and perspectives for biotechnological production. Enzyme Microb Technol 2016; 86: 103-16.
[http://dx.doi.org/10.1016/j.enzmictec.2016.02.003] [PMID: 26992799]

[84]   Gengatharan A, Dykes GA, Choo WS. Betalains: Natural plant pigments with potential application in functional foods. Lebensm Wiss Technol 2015; 64(2): 645-9.
[http://dx.doi.org/10.1016/j.lwt.2015.06.052]

[85]   Coultate TP. Corantes Alimentos A química dos seus componentes. 3rd ed., Porto Alegre: Artmed 2004.

[86]   Hosikian A, Lim S, Halim R, Danquah MK. Chlorophyll Extraction from *Microalgae*: A Review on the Process Engineering Aspects. Int J Chem Eng 2010; 2010: 11.
[http://dx.doi.org/10.1155/2010/391632]

[87]   Mapari SAS, Thrane U, Meyer AS. Fungal polyketide azaphilone pigments as future natural food

colorants? Trends Biotechnol 2010; 28(6): 300-7.
[http://dx.doi.org/10.1016/j.tibtech.2010.03.004] [PMID: 20452692]

[88] Gallen C, Pla J. Allergy and intolerance to food additives. Rev Fr Allergol 2013; 53: S9-S18.
[http://dx.doi.org/10.1016/S1877-0320(13)70044-7]

[89] Ferrer A, Marco FM, Andreu C, Sempere JM. Occupational asthma to carmine in a butcher. Int Arch Allergy Immunol 2005; 138(3): 243-50.
[http://dx.doi.org/10.1159/000088725] [PMID: 16215325]

[90] Dufosse L. Microbial production of food grade pigments. Food Technol Biotechnol 2006; 44(3): 313-21.

[91] Mali KK, Tokkas J, Goyal S. Microbial Pigments: A review. International Journal of Microbial Resource Technology 2012; 1(4): 2278-3822.

[92] Jia XQ, Xu ZN, Zhou LP, Sung CK. Elimination of the mycotoxin citrinin production in the industrial important strain *Monascus purpureus* SM001. Metab Eng 2010; 12(1): 1-7.
[http://dx.doi.org/10.1016/j.ymben.2009.08.003] [PMID: 19699814]

[93] Mapari SAS, Meyer AS, Thrane U. Photostability of natural orange-red and yellow fungal pigments in liquid food model systems. J Agric Food Chem 2009; 57(14): 6253-61.
[http://dx.doi.org/10.1021/jf900113q] [PMID: 19534525]

[94] Cardoso LAC, Jäckel S, Karp SG, Framboisier X, Chevalot I, Marc I. Improvement of *Sporobolomyces ruberrimus* carotenoids production by the use of raw glycerol. Bioresour Technol 2016; 200: 374-9.
[http://dx.doi.org/10.1016/j.biortech.2015.09.108] [PMID: 26512861]

[95] Cheng Y-T, Yang C-F. Using strain *Rhodotorula mucilaginosa* to produce carotenoids using food wastes. Journal of the Taiwan Institute of Chemical Engineers 2016; 1-6.
[http://dx.doi.org/10.1016/j.jtice.2015.12.027]

[96] Lin YL, Wang TH, Lee MH, Su NW. Biologically active components and nutraceuticals in the *Monascus*-fermented rice: a review. Appl Microbiol Biotechnol 2008; 77(5): 965-73.
[http://dx.doi.org/10.1007/s00253-007-1256-6] [PMID: 18038131]

[97] Frisvad JC, Yilmaz N, Thrane U, Rasmussen KB, Houbraken J, Samson RA. Talaromyces atroroseus, a new species efficiently producing industrially relevant red pigments. PLoS One 2013; 8(12): e84102.
[http://dx.doi.org/10.1371/journal.pone.0084102] [PMID: 24367630]

[98] Singh D, Kaur G. Swainsonine, a novel fungal metabolite: optimization of fermentative production and bioreactor operations using evolutionary programming. Bioprocess Biosyst Eng 2014; 37(8): 1599-607.
[http://dx.doi.org/10.1007/s00449-014-1132-6] [PMID: 24500619]

[99] Mapari SAS, Hansen ME, Meyer AS, Thrane U. Computerized screening for novel producers of Monascus-like food pigments in *Penicillium* species. J Agric Food Chem 2008; 56(21): 9981-9.
[http://dx.doi.org/10.1021/jf801817q] [PMID: 18841978]

[100] Mapari SA, Meyer AS, Thrane U, Frisvad JC. Identification of potentially safe promising fungal cell factories for the production of polyketide natural food colorants using chemotaxonomic rationale. Microb Cell Fact 2009; 8(24): 24.
[http://dx.doi.org/10.1186/1475-2859-8-24] [PMID: 19397825]

[101] Venil CK, Zakaria ZA, Ahmad WA. Bacterial pigments and their applications. Process Biochem 2013; 48(7): 1065-79.
[http://dx.doi.org/10.1016/j.procbio.2013.06.006]

[102] Rodrigues DB, Menezes CR, Mercadante AZ, Jacob-Lopes E, Zepka LQ. Bioactive pigments from microalgae *Phormidium autumnale*. Food Res Int 2015; 77: 273-9.
[http://dx.doi.org/10.1016/j.foodres.2015.04.027]

[103] Becker J, Lange A, Fabarius J, Wittmann C. Top value platform chemicals: bio-based production of organic acids. Curr Opin Biotechnol 2015; 36: 168-75.
[http://dx.doi.org/10.1016/j.copbio.2015.08.022] [PMID: 26360870]

[104] Theron MM, Lues JFR, Eds. Organic acids and food preservation. CRC Press 2010.
[http://dx.doi.org/10.1201/9781420078435]

[105] Sauer M, Porro D, Mattanovich D, Branduardi P. Microbial production of organic acids: expanding the markets. Trends Biotechnol 2008; 26(2): 100-8.
[http://dx.doi.org/10.1016/j.tibtech.2007.11.006] [PMID: 18191255]

[106] Alonso S, Rendueles M, Díaz M. Microbial production of specialty organic acids from renewable and waste materials. Crit Rev Biotechnol 2015; 35(4): 497-513.
[http://dx.doi.org/10.3109/07388551.2014.904269] [PMID: 24754448]

[107] Ricke SC. Perspectives on the use of organic acids and short chain fatty acids as antimicrobials. Poult Sci 2003; 82(4): 632-9.
[http://dx.doi.org/10.1093/ps/82.4.632] [PMID: 12710485]

[108] Zorn H, Czermak P. Biotechnology of food and feed additives. Adv Biochem Eng Biotechnol 2014; 143.

[109] Riemensperger A. 2016. Organic acid-based products – market evaluation and technical comment.

[110] Plumridge A, Hesse SJA, Watson AJ, Lowe KC, Stratford M, Archer DB. The weak acid preservative sorbic acid inhibits conidial germination and mycelial growth of *Aspergillus niger* through intracellular acidification. Appl Environ Microbiol 2004; 70(6): 3506-11.
[http://dx.doi.org/10.1128/AEM.70.6.3506-3511.2004] [PMID: 15184150]

[111] Berovic M, Legisa M. Citric acid production. Biotechnol Annu Rev 2007; 13: 303-43.
[http://dx.doi.org/10.1016/S1387-2656(07)13011-8] [PMID: 17875481]

[112] Dhillon GS, Brar SK, Verma M, Tyagi RD. Recent advances in citric acid bio-production and recovery. Food Bioprocess Technol 2011; 4(4): 505-29.
[http://dx.doi.org/10.1007/s11947-010-0399-0]

[113] Couto SR, Sanroman MA. Application of solid-state fermentation to food industry - A review. J Food Eng 2006; 76(3): 291-302.
[http://dx.doi.org/10.1016/j.jfoodeng.2005.05.022]

[114] Nikbakht R, Sadrzadeh M, Mohammadi T. Effect of operating parameters on concentration of citric acid using electrodialysis. J Food Eng 2007; 83(4): 596-604.
[http://dx.doi.org/10.1016/j.jfoodeng.2007.04.010]

[115] Soccol CR, Vandenberghe LPS, Rodrigues C, Pandey A. New perspectives for citric acid production and application. Food Technol Biotechnol 2006; 44(2): 141-9.

[116] Abdel-Rahman MA, Tashiro Y, Sonomoto K. Recent advances in lactic acid production by microbial fermentation processes. Biotechnol Adv 2013; 31(6): 877-902.
[http://dx.doi.org/10.1016/j.biotechadv.2013.04.002] [PMID: 23624242]

[117] Yadav AK, Chaudhari AB, Kothari RM. Bioconversion of renewable resources into lactic acid: an industrial view. Crit Rev Biotechnol 2011; 31(1): 1-19.
[http://dx.doi.org/10.3109/07388550903420970] [PMID: 20476870]

[118] Zhang ZY, Jin B, Kelly JM. Production of lactic acid from renewable materials by *Rhizopus* fungi. Biochem Eng J 2007; 35(3): 251-63.
[http://dx.doi.org/10.1016/j.bej.2007.01.028]

[119] de la Rosa P, Jordano R, Medina LM. Influence of low concentrations of an acid preservative on sponge cakes under different storage conditions. J Food Sci 2009; 74(2): M80-2.
[http://dx.doi.org/10.1111/j.1750-3841.2009.01059.x] [PMID: 19323762]

[120] Tormo M, Izco JM. Alternative reversed-phase high-performance liquid chromatography method to analyse organic acids in dairy products. J Chromatogr A 2004; 1033(2): 305-10.
[http://dx.doi.org/10.1016/j.chroma.2004.01.043] [PMID: 15088752]

[121] Jin B, Huang LP, Lant P. *Rhizopus arrhizus*-a producer for simultaneous saccharification and fermentation of starch waste materials to L(+)-lactic acid. Biotechnol Lett 2003; 25(23): 1983-7.
[http://dx.doi.org/10.1023/B:BILE.0000004389.53388.d0] [PMID: 14719810]

[122] Bulut S, Elibol M, Ozer D. Effect of different carbon sources on L(+)-lactic acid production by *Rhizopus oryzae*. Biochem Eng J 2004; 21(1): 33-7.
[http://dx.doi.org/10.1016/j.bej.2004.04.006]

[123] Choi SH, Chin KB. Evaluation of sodium lactate as a replacement for conventional chemical preservatives in comminuted sausages inoculated with *Listeria monocytogenes*. Meat Sci 2003; 65(1): 531-7.
[http://dx.doi.org/10.1016/S0309-1740(02)00245-0] [PMID: 22063246]

[124] Crowley S, Mahony J, van Sinderen D. Current perspectives on antifungal lactic acid bacteria as natural bio-preservatives. Trends Food Sci Technol 2013; 33(2): 93-109.
[http://dx.doi.org/10.1016/j.tifs.2013.07.004]

[125] Kumar N, Pruthi V. Potential applications of ferulic acid from natural sources. Biotechnol Rep (Amst) 2014; 4: 86-93.
[http://dx.doi.org/10.1016/j.btre.2014.09.002] [PMID: 28626667]

[126] Overhage J, Steinbüchel A, Priefert H. Biotransformation of eugenol to ferulic acid by a recombinant strain of *Ralstonia eutropha* H16. Appl Environ Microbiol 2002; 68(9): 4315-21.
[http://dx.doi.org/10.1128/AEM.68.9.4315-4321.2002] [PMID: 12200281]

[127] Lambert F, Zucca J, Ness F, Aigle M. Production of ferulic acid and coniferyl alcohol by conversion of eugenol using a recombinant strain of *Saccharomyces cerevisiae*. Flavour Fragrance J 2014; 29(1): 14-21.
[http://dx.doi.org/10.1002/ffj.3173]

[128] Birtić S, Dussort P, Pierre F-X, Bily AC, Roller M. Carnosic acid. Phytochemistry 2015; 115: 9-19.
[http://dx.doi.org/10.1016/j.phytochem.2014.12.026] [PMID: 25639596]

[129] Gutierrez L-F, Hamoudi S, Belkacemi K. Lactobionic acid: A high value-added lactose derivative for food and pharmaceutical applications. Int Dairy J 2012; 26(2): 103-11.
[http://dx.doi.org/10.1016/j.idairyj.2012.05.003]

[130] Steiner P, Sauer U. Overexpression of the ATP-dependent helicase RecG improves resistance to weak organic acids in Escherichia coli. Appl Microbiol Biotechnol 2003; 63(3): 293-9.
[http://dx.doi.org/10.1007/s00253-003-1405-5] [PMID: 12898065]

[131] EUFIC. 2015. Available from: http://www.efsa.europa.eu/en/topics/topic/sweeteners

[132] EFSA. Sweeteners. European Food Safety Authority 2016. [February]

[133] Carocho M, Morales P, Ferreira ICFR. Sweeteners as food additives in the XXI century: A review of what is known, and what is to come. Food Chem Toxicol 2017; 107(Pt A): 302-17.
[http://dx.doi.org/10.1016/j.fct.2017.06.046] [PMID: 28689062]

[134] Philippe RN, De Mey M, Anderson J, Ajikumar PK. Biotechnological production of natural zero-calorie sweeteners. Curr Opin Biotechnol 2014; 26: 155-61.
[http://dx.doi.org/10.1016/j.copbio.2014.01.004] [PMID: 24503452]

[135] Mordor Intelligence. Available from: https://www.mordorintelligence.com/industry-reports/globa--food-sweetener-market-industry

[136] Sylvetsky AC, Rother KI. Trends in the consumption of low-calorie sweeteners. Physiol Behav 2016; 164(Pt B): 446-50.
[http://dx.doi.org/10.1016/j.physbeh.2016.03.030] [PMID: 27039282]

[137] Chattopadhyay S, Raychaudhuri U, Chakraborty R. Artificial sweeteners - a review. J Food Sci Technol 2014; 51(4): 611-21.
[http://dx.doi.org/10.1007/s13197-011-0571-1] [PMID: 24741154]

[138] von Poser Toigo E, Huffell AP, Mota CS, Bertolini D, Pettenuzzo LF, Dalmaz C. Toigo EvP. Metabolic and feeding behavior alterations provoked by prenatal exposure to aspartame. Appetite 2015; 87: 168-74.
[http://dx.doi.org/10.1016/j.appet.2014.12.213] [PMID: 25543075]

[139] Chatsudthipong V, Muanprasat C. Stevioside and related compounds: therapeutic benefits beyond sweetness. Pharmacol Ther 2009; 121(1): 41-54.
[http://dx.doi.org/10.1016/j.pharmthera.2008.09.007] [PMID: 19000919]

[140] Sang Z, Jiang Y, Tsoi Y-K, Leung KS-Y. Evaluating the environmental impact of artificial sweeteners: a study of their distributions, photodegradation and toxicities. Water Res 2014; 52: 260-74.
[http://dx.doi.org/10.1016/j.watres.2013.11.002] [PMID: 24289948]

[141] Hao Y, Wang F, Huanga W, Tanga X, Zoub Q, Lic Z, *et al.* Sucrose substitution by polyols in sponge cake and their effects on the foaming and thermal properties of egg protein. Food Hydrocoll 2016; 57: 153-9.
[http://dx.doi.org/10.1016/j.foodhyd.2016.01.006]

[142] Rywińska A, Marcinkiewicz M, Cibis E, Rymowicz W. Optimization of medium composition for erythritol production from glycerol by *Yarrowia lipolytica* using response surface methodology. Prep Biochem Biotechnol 2015; 45(6): 515-29.
[http://dx.doi.org/10.1080/10826068.2014.940966] [PMID: 25387364]

[143] Sawada K, Taki A, Yamakawa T, Seki M. Key role for transketolase activity in erythritol production by *Trichosporonoides megachiliensis* SN-G42. J Biosci Bioeng 2009; 108(5): 385-90.
[http://dx.doi.org/10.1016/j.jbiosc.2009.05.008] [PMID: 19804861]

[144] Saran S, Mukherjee S, Dalal J, Saxena RK. High production of erythritol from *Candida sorbosivorans* SSE-24 and its inhibitory effect on biofilm formation of *Streptococcus* mutans. Bioresour Technol 2015; 198: 31-8.
[http://dx.doi.org/10.1016/j.biortech.2015.08.146] [PMID: 26363499]

[145] Mirończuk AM, Rakicka M, Biegalska A, Rymowicz W, Dobrowolski A. A two-stage fermentation process of erythritol production by yeast *Y. lipolytica* from molasses and glycerol. Bioresour Technol 2015; 198: 445-55.
[http://dx.doi.org/10.1016/j.biortech.2015.09.008] [PMID: 26409857]

[146] Moon HJ, Jeya M, Kim IW, Lee JK. Biotechnological production of erythritol and its applications. Appl Microbiol Biotechnol 2010; 86(4): 1017-25.
[http://dx.doi.org/10.1007/s00253-010-2496-4] [PMID: 20186409]

[147] Kohl ES, Leet TH, Lee DY, Kim HJ, Ryu YW, Seo JH. Scale-up of erythritol production by an osmophilic mutant of *Candida magnoliae*. Biotechnol Lett 2003; 25(24): 2103-5.
[http://dx.doi.org/10.1023/B:BILE.0000007076.64338.ce] [PMID: 14969417]

[148] Savergave LS, Gadre RV, Vaidya BK, Narayanan K. Strain improvement and statistical media optimization for enhanced erythritol production with minimal by-products from *Candida magnoliae* mutant R23. Biochem Eng J 2011; 55(2): 92-100.
[http://dx.doi.org/10.1016/j.bej.2011.03.009]

[149] Ishizuka H, Wako K, Kasumi T, Sasaki T. Breeding of a mutant of *aureobasidium* sp with high erythritol production. J Ferment Bioeng 1989; 68(5): 310-4.
[http://dx.doi.org/10.1016/0922-338X(89)90003-2]

[150] Ravella SR, Gallagher J, Fish S, Prakasham RS. Overview on Commercial Production of Xylitol, Economic Analysis and Market Trends.D-Xylitol: Fermentative Production, Application and

Commercialization. Berlin, Heidelberg: Springer Berlin Heidelberg 2012; pp. 291-306.
[http://dx.doi.org/10.1007/978-3-642-31887-0_13]

[151] Ur-Rehman S, Mushtaq Z, Zahoor T, Jamil A, Murtaza MA. Xylitol: a review on bioproduction, application, health benefits, and related safety issues. Crit Rev Food Sci Nutr 2015; 55(11): 1514-28.
[http://dx.doi.org/10.1080/10408398.2012.702288] [PMID: 24915309]

[152] Tamburini E, Costa S, Marchetti MG, Pedrini P. Optimized Production of Xylitol from Xylose Using a Hyper-Acidophilic *Candida tropicalis*. Biomolecules 2015; 5(3): 1979-89.
[http://dx.doi.org/10.3390/biom5031979] [PMID: 26295411]

[153] Breuer U, Harms H. *Debaryomyces hansenii*--an extremophilic yeast with biotechnological potential. Yeast 2006; 23(6): 415-37.
[http://dx.doi.org/10.1002/yea.1374] [PMID: 16652409]

[154] Jo JH, Oh SY, Lee HS, Park YC, Seo JH. Dual utilization of NADPH and NADH cofactors enhances xylitol production in engineered *Saccharomyces cerevisiae*. Biotechnol J 2015; 10(12): 1935-43.
[http://dx.doi.org/10.1002/biot.201500068] [PMID: 26470683]

[155] Wannawilai S, Sirisansaneeyakul S, Chisti Y. Benzoate-induced stress enhances xylitol yield in aerobic fed-batch culture of *Candida mogii* TISTR 5892. J Biotechnol 2015; 194: 58-66.
[http://dx.doi.org/10.1016/j.jbiotec.2014.11.037] [PMID: 25499077]

[156] Rodrigues RCLB, Kenealy WR, Jeffries TW. Xylitol production from DEO hydrolysate of corn stover by *Pichia stipitis* YS-30. J Ind Microbiol Biotechnol 2011; 38(10): 1649-55.
[http://dx.doi.org/10.1007/s10295-011-0953-4] [PMID: 21424687]

[157] Rao RS, Jyothi ChP, Prakasham RS, Sarma PN, Rao LV. Xylitol production from corn fiber and sugarcane bagasse hydrolysates by *Candida tropicalis*. Bioresour Technol 2006; 97(15): 1974-8.
[http://dx.doi.org/10.1016/j.biortech.2005.08.015] [PMID: 16242318]

[158] de Albuquerque TL, da Silva IJ Jr, de Macedo GR, Ponte Rocha MV. Biotechnological production of xylitol from lignocellulosic wastes: A review. Process Biochem 2014; 49(11): 1779-89.
[http://dx.doi.org/10.1016/j.procbio.2014.07.010]

[159] Kim YS, Kim SY, Kim JH, Kim SC. Xylitol production using recombinant *Saccharomyces cerevisiae* containing multiple xylose reductase genes at chromosomal delta-sequences. J Biotechnol 1999; 67(2-3): 159-71.
[http://dx.doi.org/10.1016/S0168-1656(98)00172-2] [PMID: 9990733]

[160] Kong MK, Kang H-J, Kim JH, Oh SH, Lee PC. Metabolic engineering of the *Stevia rebaudiana* ent-kaurene biosynthetic pathway in recombinant *Escherichia coli*. J Biotechnol 2015; 214: 95-102.
[http://dx.doi.org/10.1016/j.jbiotec.2015.09.016] [PMID: 26392384]

[161] Rumelhard M, Hosako H, Eurlings IMJ, *et al.* Safety evaluation of rebaudioside A produced by fermentation. Food Chem Toxicol 2016; 89: 73-84.
[http://dx.doi.org/10.1016/j.fct.2016.01.005] [PMID: 26776281]

[162] Lehmann M, Trueheart J, Zwartjens P, Wu L, Boer V, Sagt CMJ, *et al.* Diterpene production. Google Patents 2013.

[163] Kvasnicka F, Voldrich M, Vyhnálek J. Determination of glycyrrhizin in liqueurs by on-line coupled capillary isotachophoresis with capillary zone electrophoresis. J Chromatogr A 2007; 1169(1-2): 239-42.
[http://dx.doi.org/10.1016/j.chroma.2007.08.074] [PMID: 17875310]

[164] Izumori K, Miyoshi T, Tokuda S, Yamabe K. Production of d-Tagatose from Dulcitol by *Arthrobacter globiformis*. Appl Environ Microbiol 1984; 48(5): 1055-7.
[PMID: 16346663]

[165] Xu Z, Xu Z, Tang B, Li S, Gao J, Chi B, *et al.* Construction and co-expression of polycistronic plasmids encoding thermophilic L-arabinose isomerase and hyperthermophilic β-galactosidase for single-step production of D-tagatose. Biochem Eng J 2016; 109: 28-34.

[http://dx.doi.org/10.1016/j.bej.2015.12.015]

[166] Bali V, Panesar PS, Bera MB, Panesar R. Fructo-oligosaccharides: Production, Purification and Potential Applications. Crit Rev Food Sci Nutr 2015; 55(11): 1475-90.
[http://dx.doi.org/10.1080/10408398.2012.694084] [PMID: 24915337]

[167] Gu W, Xia Q, Yao J, Fu S, Guo J, Hu X. Recombinant expressions of sweet plant protein mabinlin II in *Escherichia coli* and food-grade *Lactococcus lactis*. World J Microbiol Biotechnol 2015; 31(4): 557-67.
[http://dx.doi.org/10.1007/s11274-015-1809-2] [PMID: 25649203]

[168] Faus I. Recent developments in the characterization and biotechnological production of sweet-tasting proteins. Appl Microbiol Biotechnol 2000; 53(2): 145-51.
[http://dx.doi.org/10.1007/s002530050001] [PMID: 10709975]

[169] Edens L, Heslinga L, Klok R, *et al.* Cloning of cDNA encoding the sweet-tasting plant protein thaumatin and its expression in *Escherichia coli*. Gene 1982; 18(1): 1-12.
[http://dx.doi.org/10.1016/0378-1119(82)90050-6] [PMID: 7049841]

[170] Faus I, Patiño C, Río JL, del Moral C, Barroso HS, Rubio V. Expression of a synthetic gene encoding the sweet-tasting protein thaumatin in *Escherichia coli*. Biochem Biophys Res Commun 1996; 229(1): 121-7.
[http://dx.doi.org/10.1006/bbrc.1996.1767] [PMID: 8954093]

[171] Daniell S, Mellits KH, Faus I, Connerton I. Refolding the sweet-tasting protein thaumatin II from insoluble inclusion bodies synthesised in *Escherichia coli*. Food Chem 2000; 71(1): 105-10.
[http://dx.doi.org/10.1016/S0308-8146(00)00151-5]

[172] Masuda T, Tamaki S, Kaneko R, *et al.* Cloning, expression and characterization of recombinant sweet-protein thaumatin II using the methylotrophic yeast *Pichia pastoris*. Biotechnol Bioeng 2004; 85(7): 761-9.
[http://dx.doi.org/10.1002/bit.10786] [PMID: 14991654]

[173] Ide N, Kaneko R, Wada R, *et al.* Cloning of the thaumatin I cDNA and characterization of recombinant thaumatin I secreted by *Pichia pastoris*. Biotechnol Prog 2007; 23(5): 1023-30.
[PMID: 17691810]

[174] Ide N, Masuda T, Kitabatake N. Effects of pre- and pro-sequence of thaumatin on the secretion by *Pichia pastoris*. Biochem Biophys Res Commun 2007; 363(3): 708-14.
[http://dx.doi.org/10.1016/j.bbrc.2007.09.021] [PMID: 17897626]

[175] Masuda T, Ide N, Ohta K, Kitabatake N. High-yield Secretion of the Recombinant Sweet-Tasting Protein Thaumatin I. Food Sci Technol Res 2010; 16(6): 585-92.
[http://dx.doi.org/10.3136/fstr.16.585]

[176] Faus I, del Moral C, Adroer N, *et al.* Secretion of the sweet-tasting protein thaumatin by recombinant strains of *Aspergillus niger* var. awamori. Appl Microbiol Biotechnol 1998; 49(4): 393-8.
[http://dx.doi.org/10.1007/s002530051188] [PMID: 9615480]

[177] Moralejo FJ, Cardoza RE, Gutierrez S, Martin JF. Thaumatin production in *Aspergillus awamori* by use of expression cassettes with strong fungal promoters and high gene dosage. Appl Environ Microbiol 1999; 65(3): 1168-74.
[PMID: 10049878]

[178] Luchakivska Y, Komarnitskii I, Kuchuk M. Obtaining of Transgenic Carrot (*Daucus carota* L.) and Celery (*Apium graveolens* L.) Plants Expressing the Recombinant Gene of Thaumatin II Protein. In Vitro Cell Dev Biol Anim 2012; 48: 80.

[179] Pham NB, Schäfer H, Wink M. Production and secretion of recombinant thaumatin in tobacco hairy root cultures. Biotechnol J 2012; 7(4): 537-45.
[http://dx.doi.org/10.1002/biot.201100430] [PMID: 22125283]

[180] Sun H-J, Kataoka H, Yano M, Ezura H. Genetically stable expression of functional miraculin, a new

type of alternative sweetener, in transgenic tomato plants. Plant Biotechnol J 2007; 5(6): 768-77.
[http://dx.doi.org/10.1111/j.1467-7652.2007.00283.x] [PMID: 17692073]

[181] Hirai T, Fukukawa G, Kakuta H, Fukuda N, Ezura H. Production of recombinant miraculin using transgenic tomatoes in a closed cultivation system. J Agric Food Chem 2010; 58(10): 6096-101.
[http://dx.doi.org/10.1021/jf100414v] [PMID: 20426470]

[182] Yano M, Hirai T, Kato K, Hiwasa-Tanase K, Fukuda N, Ezura H. Tomato is a suitable material for producing recombinant miraculin genetically stable manner. Plant Sci 2010; 178(5): 469-73.
[http://dx.doi.org/10.1016/j.plantsci.2010.02.016]

[183] Al Bachchu MA, Jin S-B, Park J-W, Boo K-H, Sun H-J, Kim Y-W, *et al.* Functional Expression of Miraculin, a Taste-modifying Protein, in Transgenic Miyagawa Wase Satsuma Mandarin (*Citrus unshiu* Marc.). J Korean Soc Appl Biol Chem 2011; 54(1): 24-9.
[http://dx.doi.org/10.3839/jksabc.2011.003]

[184] Hirai T, Duhita N, Hiwasa-Tanase K, Ezura H. Cultivation under salt stress increases the concentration of recombinant miraculin in transgenic tomato fruit, resulting in an increase in purification efficiency. Plant Biotechnol 2011; 28(4): 387-92.
[http://dx.doi.org/10.5511/plantbiotechnology.11.0726a]

[185] Hirai T, Kim Y-W, Kato K, Hiwasa-Tanase K, Ezura H. Uniform accumulation of recombinant miraculin protein in transgenic tomato fruit using a fruit-ripening-specific E8 promoter. Transgenic Res 2011; 20(6): 1285-92.
[http://dx.doi.org/10.1007/s11248-011-9495-9] [PMID: 21359850]

[186] Hirai T, Kurokawa N, Duhita N, *et al.* The HSP terminator of *Arabidopsis thaliana* induces a high level of miraculin accumulation in transgenic tomatoes. J Agric Food Chem 2011; 59(18): 9942-9.
[http://dx.doi.org/10.1021/jf202501e] [PMID: 21861502]

[187] Hirai T, Shohael AM, Kim Y-W, Yano M, Ezura H. Ubiquitin promoter-terminator cassette promotes genetically stable expression of the taste-modifying protein miraculin in transgenic lettuce. Plant Cell Rep 2011; 30(12): 2255-65.
[http://dx.doi.org/10.1007/s00299-011-1131-x] [PMID: 21830129]

[188] Hiwasa-Tanase K, Nyarubona M, Hirai T, Kato K, Ichikawa T, Ezura H. High-level accumulation of recombinant miraculin protein in transgenic tomatoes expressing a synthetic miraculin gene with optimized codon usage terminated by the native miraculin terminator. Plant Cell Rep 2011; 30(1): 113-24.
[http://dx.doi.org/10.1007/s00299-010-0949-y] [PMID: 21076835]

[189] Kurokawa N, Hirai T, Takayama M, Hiwasa-Tanase K, Ezura H. An E8 promoter-HSP terminator cassette promotes the high-level accumulation of recombinant protein predominantly in transgenic tomato fruits: a case study of miraculin. Plant Cell Rep 2013; 32(4): 529-36.
[http://dx.doi.org/10.1007/s00299-013-1384-7] [PMID: 23306632]

[190] Takai A, Satoh M, Matsuyama T, *et al.* Secretion of miraculin through the function of a signal peptide conserved in the Kunitz-type soybean trypsin inhibitor family. FEBS Lett 2013; 587(12): 1767-72.
[http://dx.doi.org/10.1016/j.febslet.2013.04.026] [PMID: 23660404]

[191] Jin SB, Sun HJ, Al Bachchu MA, Chung SJ, Lee J, Han S-I, *et al.* Production of Recombinant Miraculin Protein Using Transgenic Citrus Cell Suspension Culture System. J Korean Soc Appl Biol Chem 2013; 56(3): 271-4.
[http://dx.doi.org/10.1007/s13765-013-3074-0]

[192] Matsuyama T, Satoh M, Nakata R, Aoyama T, Inoue H. Functional expression of miraculin, a taste-modifying protein in *Escherichia coli*. J Biochem 2009; 145(4): 445-50.
[http://dx.doi.org/10.1093/jb/mvn184] [PMID: 19122203]

[193] Ito K, Asakura T, Morita Y, *et al.* Microbial production of sensory-active miraculin. Biochem Biophys Res Commun 2007; 360(2): 407-11.
[http://dx.doi.org/10.1016/j.bbrc.2007.06.064] [PMID: 17592723]

[194]   Chen Z, Cai H, Lu F, Du L. High-level expression of a synthetic gene encoding a sweet protein, monellin, in *Escherichia coli*. Biotechnol Lett 2005; 27(22): 1745-9.
[http://dx.doi.org/10.1007/s10529-005-3544-5] [PMID: 16314964]

[195]   Chen Z, Heng C, Li Z, Liang X, Xinchen S. Expression and secretion of a single-chain sweet protein monellin in *Bacillus subtilis* by sacB promoter and signal peptide. Appl Microbiol Biotechnol 2007; 73(6): 1377-81.
[http://dx.doi.org/10.1007/s00253-006-0609-x] [PMID: 17028871]

[196]   Zhang XL, Ito T, Kondo K, Kobayashi T, Honda H. Production of single chain recombinant monellin by high cell density culture of genetically engineered *Candida utilis* using limited feeding of sodium ions. J Chem Eng of Jpn 2002; 35(7): 654-9.
[http://dx.doi.org/10.1252/jcej.35.654]

[197]   Chen Z, Li Z, Yu N, Yan L. Expression and secretion of a single-chain sweet protein, monellin, in *Saccharomyces cerevisiae* by an α-factor signal peptide. Biotechnol Lett 2011; 33(4): 721-5.
[http://dx.doi.org/10.1007/s10529-010-0479-2] [PMID: 21107648]

[198]   Reddy CS, Vijayalakshmi M, Kaul T, Islam T, Reddy MK. Improving flavour and quality of tomatoes by expression of synthetic gene encoding sweet protein monellin. Mol Biotechnol 2015; 57(5): 448-53.
[http://dx.doi.org/10.1007/s12033-015-9838-5] [PMID: 25645814]

[199]   Kim IH, Lim KJ. Large-scale purification of recombinant monellin from yeast. J Ferment Bioeng 1996; 82(2): 180-2.
[http://dx.doi.org/10.1016/0922-338X(96)85046-X]

[200]   Assadi-Porter FM, Aceti DJ, Cheng H, Markley JL. Efficient production of recombinant brazzein, a small, heat-stable, sweet-tasting protein of plant origin. Arch Biochem Biophys 2000; 376(2): 252-8.
[http://dx.doi.org/10.1006/abbi.2000.1725] [PMID: 10775410]

[201]   Berlec A, Tompa G, Slapar N, Fonović UP, Rogelj I, Strukelj B. Optimization of fermentation conditions for the expression of sweet-tasting protein brazzein in *Lactococcus lactis*. Lett Appl Microbiol 2008; 46(2): 227-31.
[http://dx.doi.org/10.1111/j.1472-765X.2007.02297.x] [PMID: 18215220]

[202]   Lee Y-W, Kim K-Y, Han S-H, Kang C-H, So J-S. Expression of the sweet-tasting protein brazzein in *Lactobacillus* spp. Food Sci Biotechnol 2012; 21(3): 895-8.
[http://dx.doi.org/10.1007/s10068-012-0116-z]

[203]   Rachid A, Belloir C, Chevalier J, Desmetz C, Miller M-L, Poirier N, *et al.* Optimization of the Production of Recombinant Brazzein Secreted by the Yeast *Pichia pastoris*. Chemical Senses 2009; 34(7): A80-A.

[204]   Rachid A, Desmetz C, Chevalier J, Briand L. Purification and Characterization of Recombinant Brazzein Secreted by the Yeast *Pichia pastoris*. Chem Senses 2009; 34(3): E66-7.

[205]   Poirier N, Roudnitzky N, Brockhoff A, *et al.* Efficient production and characterization of the sweet-tasting brazzein secreted by the yeast *Pichia pastoris*. J Agric Food Chem 2012; 60(39): 9807-14.
[http://dx.doi.org/10.1021/jf301600m] [PMID: 22958103]

[206]   Jo H-J, Noh J-S, Kong K-H. Efficient secretory expression of the sweet-tasting protein brazzein in the yeast *Kluyveromyces lactis*. Protein Expr Purif 2013; 90(2): 84-9.
[http://dx.doi.org/10.1016/j.pep.2013.05.001] [PMID: 23684772]

[207]   Yan S, Song H, Pang D, *et al.* Expression of plant sweet protein brazzein in the milk of transgenic mice. PLoS One 2013; 8(10): e76769.
[http://dx.doi.org/10.1371/journal.pone.0076769] [PMID: 24155905]

[208]   Suzuki M, Kurimoto E, Nirasawa S, *et al.* Recombinant curculin heterodimer exhibits taste-modifying and sweet-tasting activities. FEBS Lett 2004; 573(1-3): 135-8.
[http://dx.doi.org/10.1016/j.febslet.2004.07.073] [PMID: 15327988]

[209] Kurimoto E, Suzuki M, Amemiya E, *et al.* Curculin exhibits sweet-tasting and taste-modifying activities through its distinct molecular surfaces. J Biol Chem 2007; 282(46): 33252-6.
[http://dx.doi.org/10.1074/jbc.C700174200] [PMID: 17895249]

[210] Nakajima K, Asakura T, Maruyama J, Morita Y, Oike H, Misaka T, *et al.* Recombinant neoculin produced by *Aspergillus oryzae* has the native taste-modifying activity, recognizable by human sweet taste receptor. Chemical Senses 2006; 31(5): A21-A.

[211] Nakajima K, Asakura T, Maruyama J, *et al.* Extracellular production of neoculin, a sweet-tasting heterodimeric protein with taste-modifying activity, by *Aspergillus oryzae.* Appl Environ Microbiol 2006; 72(5): 3716-23.
[http://dx.doi.org/10.1128/AEM.72.5.3716-3723.2006] [PMID: 16672522]

[212] Vonder Haar C, Peterson TC, Martens KM, Hoane MR. Vitamins and nutrients as primary treatments in experimental brain injury: Clinical implications for nutraceutical therapies. Brain Res 2015.
[PMID: 26723564]

[213] Carocho M, Ferreira ICFR. A review on antioxidants, prooxidants and related controversy: natural and synthetic compounds, screening and analysis methodologies and future perspectives. Food Chem Toxicol 2013; 51: 15-25.
[http://dx.doi.org/10.1016/j.fct.2012.09.021] [PMID: 23017782]

[214] Hussain N, Irshad F, Jabeen Z, Shamsi IH, Li Z, Jiang L. Biosynthesis, structural, and functional attributes of tocopherols in planta; past, present, and future perspectives. J Agric Food Chem 2013; 61(26): 6137-49.
[http://dx.doi.org/10.1021/jf4010302] [PMID: 23713813]

[215] Davey MW, Van Montagu M, Inze D, Sanmartin M, Kanellis A, Smirnoff N, *et al.* Plant L-ascorbic acid: chemistry, function, metabolism, bioavailability and effects of processing. J Sci Food Agric 2000; 80(7): 825-60.
[http://dx.doi.org/10.1002/(SICI)1097-0010(20000515)80:7<825::AID-JSFA598>3.0.CO;2-6]

[216] Survase SA, Bajaj IB, Singhal RS. Biotechnological production of vitamins. Food Technol Biotechnol 2006; 44(3): 381-96.

[217] Dong L-F, Neuzil J. Vitamin E Analogues as Prototypic Mitochondria-Targeting Anti-cancer Agents.Mitochondria: The Anti- cancer Target for the Third Millennium. Dordrecht: Springer Netherlands 2014; pp. 151-81.
[http://dx.doi.org/10.1007/978-94-017-8984-4_7]

[218] Palozza P, Krinsky NI. beta-Carotene and alpha-tocopherol are synergistic antioxidants. Arch Biochem Biophys 1992; 297(1): 184-7.
[http://dx.doi.org/10.1016/0003-9861(92)90658-J] [PMID: 1637180]

[219] Takeyama H, Kanamaru A, Yoshino Y, Kakuta H, Kawamura Y, Matsunaga T. Production of antioxidant vitamins, beta-carotene, vitamin C, and vitamin E, by two-step culture of *Euglena gracilis* Z. Biotechnol Bioeng 1997; 53(2): 185-90.
[http://dx.doi.org/10.1002/(SICI)1097-0290(19970120)53:2<185::AID-BIT8>3.0.CO;2-K] [PMID: 18633963]

[220] Abalde J, Fabregas J. B-carotene, vitamin C and vitamin E content of the marine microalgae *Dunaliella tertiolecta* cultured with different nitrogen sources. Bioresour Technol 1991; 38: 121-5.
[http://dx.doi.org/10.1016/0960-8524(91)90142-7]

[221] Carballo-Cárdenas EC, Tuan PM, Janssen M, Wijffels RH. Vitamin E (alpha-tocopherol) production by the marine microalgae *Dunaliella tertiolecta* and *Tetraselmis suecica* in batch cultivation. Biomol Eng 2003; 20(4-6): 139-47.
[http://dx.doi.org/10.1016/S1389-0344(03)00040-6] [PMID: 12919791]

[222] Hultquist DE, Xu F, Quandt KS, *et al.* Evidence that NADPH-dependent methemoglobin reductase and administered riboflavin protect tissues from oxidative injury. Am J Hematol 1993; 42(1): 13-8.

[http://dx.doi.org/10.1002/ajh.2830420105] [PMID: 8416288]

[223] Halliwell B, Gutteridge JMC. Free radicals, lipid peroxidation, and cell damage. Lancet 1984; 2(8411): 1095.
[http://dx.doi.org/10.1016/S0140-6736(84)91530-7] [PMID: 6150163]

[224] Inci S, Ozcan OE, Kilinç K. Time-level relationship for lipid peroxidation and the protective effect of alpha-tocopherol in experimental mild and severe brain injury. Neurosurgery 1998; 43(2): 330-5.
[http://dx.doi.org/10.1097/00006123-199808000-00095] [PMID: 9696087]

[225] Powers HJ. Riboflavin (vitamin B-2) and health. Am J Clin Nutr 2003; 77(6): 1352-60.
[http://dx.doi.org/10.1093/ajcn/77.6.1352] [PMID: 12791609]

[226] Betz AL, Ren XD, Ennis SR, Hultquist DE. Riboflavin reduces edema in focal cerebral ischemia. Acta Neurochir Suppl (Wien) 1994; 60: 314-7.
[PMID: 7976577]

[227] Horiuchi J, Hiraga K. Industrial application of fuzzy control to large-scale recombinant vitamin B2 production. J Biosci Bioeng 1999; 87(3): 365-71.
[http://dx.doi.org/10.1016/S1389-1723(99)80047-4] [PMID: 16232483]

CHAPTER 7

Designing Bioactive Nanoparticles: The Era of Nutraceuticals

Clara Grosso, Patrícia Valentão and **Paula B. Andrade***

REQUIMTE/LAQV, Laboratório de Farmacognosia, Departamento de Química, Faculdade de Farmácia, Universidade do Porto, Rua de Jorge Viterbo Ferreira, no. 228, 4050-313 Porto, Portugal

Abstract: Two thousand and five hundred years ago, Hippocrates stated 'let food be the medicine and medicine be the food', but only in 1989, Stephen DeFelice coined the term nutraceutical to describe the hybrid between 'nutrition' and 'pharmaceutical'. A 'nutraceutical' is defined as a supplement to the diet that is composed of bioactive compounds found in foods and botanicals, vitamins, and minerals. It is formulated and taken under the form of capsules, tablets, etc., resulting in beneficial health impacts. Owing to their biological properties, phenolic compounds are considered nutraceuticals with great potential; however, their effects are limited due to their low bioavailability. The rationale for developing an efficient drug/nutraceutical delivery system is to increase the bioavailability and half-time of the drug in the vicinity of the target cells, reducing at the same time, its exposure to non-target cells. Nutraceutical delivery to the brain is a major challenge imposed by physical barriers, such as the blood-brain barrier (BBB). In the last decade, nanotechnology has become a powerful strategy to enable nutraceuticals' target-delivery to tissues and organs, including the brain. This chapter will present the latest results obtained with phenolic-based nanoparticles, showing the failures, achievements, and most promising routes for future works.

Keywords: Antimicrobial, Antioxidant, Antiproliferative, Anticancer, Catechins, Curcumin, Dendrimers, Flavonoids, Liposomes, Medicinal plants, Metal Nanoparticles, Neuroprotective, Nutraceutical, Polymeric Nanoparticles, Quercetin, Resveratrol, Solid-lipid Nanoparticles, Stilbenes.

INTRODUCTION

The term "nutraceutical" was coined from "nutrition" and "pharmaceutical" in 1989 by Stephen DeFelice, founder and chairman of the Foundation for Innovation in Medicine, Cranford, NJ. Nutraceutical was defined as "a food (or

* **Corresponding author Paula B. Andrade:** REQUIMTE/LAQV, Laboratório de Farmacognosia, Departamento de Química, Faculdade de Farmácia, Universidade do Porto, Rua de Jorge Viterbo Ferreira, no. 228, 4050-313 Porto, Portugal; Tel: +351-220428654; Fax: +351-226093390; Email: pandrade@ff.up.pt

Ali Osman (Ed.)

part of a food) that offers medical or health benefits, including the prevention and/or treatment of a disorder" [1]. Nutraceuticals englobe bioactive compounds that occur in food or are produced *de novo* in human metabolism, botanicals or their constituents, vitamins, and minerals. They are delivered in forms that differ from conventional foods or beverages [2]. Among them, polyphenols standout as multipotent therapeutic agents since they display a plethora of *in vitro* and *in vivo* bioactivities: antioxidant [3 - 9], antimicrobial [10 - 18], neuroprotective [19 - 25], anti-inflammatory [26 - 34], immunomodulatory [35 - 39], anticancer [40 - 48] and cardiovascular protection [31, 49 - 56].

Despite the promising effectiveness of polyphenols in the prevention of several diseases, their low bioavailability is an obstacle. This parameter varies according to their dimensions, level of polymerization, degree and type of glycosylation and hydrophobicity [57]. Many studies have reported the kinetics and extent of polyphenol absorption. The results from 97 intervention studies clearly showed a great variability; isoflavones and gallic acid being well absorbed in humans, followed by catechins, flavanones, and quercetin glucosides, although with different kinetics. The compounds exhibiting a more conditioned absorption pattern were found to be proanthocyanidins, the galloylated tea catechins, and the anthocyanins [57].

While anthocyanins and some aglycones of flavonols (quercetin) and isoflavones (daidzin and genistein) are soon absorbed in the stomach [57 - 60], other polyphenols resist acid hydrolysis and reach the intestine. The aglycones and some glycosides are broken down by the action of phloridzin hydrolase and cytosolic β-glucosidase in the small intestine, facilitating their absorption [61]. The most complex polyphenols arrive to the colon and are hydrolysed by the colonic microflora into aglycones, then into various aromatic acids (*e.g.* hydroxyphenylacetic acids, hydroxyphenylpropionic acids and phenylvale-rolactones). The same happens for the non-esterified and esterified hydroxy-cinnamic acids, with absorption of non-esterified structures being faster than the esterified ones [62, 63]. Right after the absorption step, and before entering the blood stream, aglycones undergo phase II metabolism (in small intestine and liver) forming methylated, sulphated, and glucuronide metabolites by the action of catechol-*O*-methyltransferases, sulfotransferases and uridine-5'-diphosphate glucuronosyltransferases, respectively [61, 63]. Polyphenol metabol- ites do not freely circulate in the blood, but bound to plasma proteins, such as albumin. Those that are not up taken by tissues are eliminated *via* biliary and urinary routes [61, 63]. Overall, the dietary polyphenols undergo several structural changes, and the observed bioactive effect is not due to the parent compound but due to their metabolites [62, 63]. This means that, on the one side studies in the last decades have shown that the bioactive compounds are not the parent compounds acquired from the diet but, instead, the metabolites resulting from their metabolism along the gastrointestinal tract [61]. On the other side, there is a huge amount of information regarding the *in vitro* potentiality of the parent compounds. The only way to analyse the activity of these parent compounds in animal tests is by encapsulating them. *In vivo* parallel experiments analysing the extract of medicinal and edible plants and fruits in both forms (encapsulated and non-encapsulated) will allow comparing their effects and concluding whether it is important or not to increase the concentration of the parent compounds *in vivo*. If the non-encapsulated form proves its stronger effectiveness, it will also be important to assess if the effect observed can be potentiated by the encapsulation of the metabolites resulting from gastrointestinal metabolism.

Taking into account the above-mentioned points, it is obviously very difficult for polyphenols to reach the brain in their primitive form, and if the presence of the blood-brain barrier (BBB) is added to this, then it is worthy to say that the brain-target delivery of polyphenols is one of the most challenging tasks in the modern science. This barrier impairs the accessibility of potential drugs to the central nervous system (CNS) and, therefore, conditioning new pharmacological approaches, once effective concentrations are not reached [64]. As explained above, although the effect observed *in vivo* should mainly be due to the resulting metabolites, we cannot neglect the *in vitro* effect of parent compounds observed using cell-based assays [65].

According to the National Institute of Health (NIH) guidelines, a nanoparticle (NP) is any material used in the formulation of a drug, resulting in a final product smaller than 1 µm in size [66]. NP encapsulation is an effective method that provides not only protection of the desired molecule from environmental degradation agents, but also increases drug stability, controls its rate of release, and increases its bioavailability [67 - 69]. Therefore, NP-based systems have gained popularity due to their ability to overcome biological barriers, allowing a successful delivery of hydrophobic drugs to specific target sites [66, 70]. Unfortunately, the body recognizes hydrophobic particles as foreigners, and therefore, they are quickly taken up by the mononuclear phagocytic system (MPS). This system eliminates NPs from the blood stream unless the particles are designed to escape from its recognition. In general, small particles (<100 nm) with a hydrophilic surface have the greatest ability to avoid the MPS. To overcome opsonisation, *i.e.* removal of nanoparticle drug carriers, several methods have been developed to camouflage NPs from the MPS. For instance, the addition of hydrophilic poly-ethylene glycol (PEG), poly-vinyl pyrrolidone (PVP), poloxamers, polysorbate 80, D-α-tocopheryl polyethylene glycol 1000 succinate (TPGS), polysorbate 20 or polysaccharides, like dextran, to the surface of NPs results in increased half-life blood circulation periods [71 - 73]. Moreover, functionalization of nanoparticles with PEG, amino acids, peptides, antibodies, fluorescent antibodies, etc., for bioimaging and delivering drugs more efficiently to their targets has also been used [74 - 77].

This chapter will review the recent research on the antioxidant, antimicrobial, anticancer and neuroprotective activities of different polyphenolic-based NPs, highlighting the nutraceutical nature of polyphenols.

TYPES OF NANOPARTICLES

Polymeric Nanoparticles

In polymeric NPs, the bioactive molecule can be encapsulated within the polymeric matrix or adsorbed onto the surface. Therapeutic polymeric NPs are composed of biodegradable or biocompatible materials that have a limited period of time *in vivo*, and then gradually degrade into molecules that can become dissolved or metabolized and eliminated from the body. Among the most used ones are poly (lactic-co-glycolic acid) (PLGA), poly (lactic acid) (PLA), poly (ε-caprolactone) (PCL), alginic acid, gelatin and chitosan [76]. The synthetic PLGA and PLA undergo hydrolysis to the biodegradable metabolite lactic acid and glycolic acid or just lactic acid, respectively. PCL, also a synthetic polymer, is hydrolysed at its ester linkages, in physiological conditions. Chitosan, a polymer prepared by *N*-deacetylation of the crustacean derived natural biopolymer chitin, is enzymatically hydrolysed at glucosamine–glucosamine, glucosamine–*N*-acety--glucosamine and *N*-acetyl-glucosamine–*N*-acetyl-glucosamine linkages [71, 78]. Alginate is an anionic polymer produced by brown algae and bacteria and consists of α-$_L$-guluronic acid and β-$_D$-mannuronic acid residues linearly linked by 1,4-glycosidic linkages. The production of alginate nanoparticles is less common than the other polymer-based NPs [79].

Polymeric NPs present some advantages: they are structurally stable and can be synthesized with a sharper size distribution. Their particle properties, such as mean size, zeta potentials and drug release profiles, can be precisely adjusted by selecting different polymer lengths, surfactants, and organic solvents during the synthesis [76]. However, the toxicity of polymeric NPs is not well documented (although they are generally regarded as safe) [80] and the lack of information concerning the scale-up production is limited [81]. Nonetheless, efforts to scale-up the production of polymeric NPs have been made [82, 83].

Dendrimers are branched synthetic polymers. Their structure begins with a central atom or group of atoms called core. From this central structure, the branches of other atoms called 'dendrons', grow through a number of chemical reactions. The end-groups of the dendrons (*i.e.*, the groups reaching the outer periphery), can be functionalized to modify the physicochemical or biological properties of the dendrimers. Drugs can either be attached to dendrimers' end groups or encapsulated in the core [66, 84].

Dendrimers have several biomedical applications due to their chemical homogeneity, the possibility of enhancing their size by repetitive addition of chemical moieties, and finally the high density of surface groups that are suitable for ligand-binding. Compared to traditional linear polymers, dendrimers have also better physical and chemical properties [85]. Again, one major drawback is the difficulty of scaling-up the multi-step synthesis of dendrimers [86, 87]. In recent years, efforts have been made to scale-up the process [88].

Inorganic Nanoparticles

Magnetic NPs are those NPs that show a response to an applied magnetic field. There are five classes of magnetic NPs: ferromagnetic (iron, nickel, and cobalt), paramagnetic (gadolinium, magnesium, lithium, and tantalum), diamagnetic (cooper, silver, gold, and most of the known elements), antiferromagnetic (MnO, CoO, NiO, and $CuCl_2$), and ferrimagnetic (magnetite - Fe_3O_4 and maghemite - γ-Fe_2O_3) [89]. Even though metallic nanoparticles are biocompatible, a significant fraction of metal particles can be retained and accumulated in the body after drug administration, probably causing toxicity, a fact that constitutes a concern [76].

The localized surface plasmon resonance (LSPR) characteristic of noble metal nanoparticles (*e.g.* silver and gold) is responsible for the wide applications in biomedical sensing and imaging [90, 91]. The SPR band intensity and wavelength depend on the metal type, particle size, shape, structure, composition, and the dielectric constant of the surrounding medium [92]. Moreover, the antimicrobial properties of silver and gold NPs find applications in health industry, food storage, textile coatings, and environment. Indeed, they are well-known antimicrobial agents against a wide range of over 650 microorganisms, including Gram-negative and Gram-positive bacteria, fungi, and viruses [93 - 95]. The toxic mechanisms underlying the antimicrobial effect of metals are not yet fully understood. Hypotheses include: (a) the metal reduction potential, which can lead directly or indirectly to the formation of reactive oxygen species; (b) the metal donor atom selectivity and/or speciation, which promote the binding of metal ions to some atoms of donor ligands (O, N and S) leading, for instance, to the destruction of Fe–S clusters from bacterial dehydratases; (c) bacterial membrane disruption promoted by the adsorption of positive metal ions to highly electronegative membrane macromolecules; (d) some metals, mainly Ag, disrupt the activity of the bacterial electron transport chain, causing proton leakage across the membrane [96, 97].

Graphene Nanoparticles

Graphene is a monolayer of sp2-bonded carbon atoms that are closely packed into a 2-dimensional honeycomb crystal lattice. Its exceptional electronic, optical, mechanical, and thermal properties have shown a wide range of applications, including drug-delivery [98, 99]. Materials of the graphene family include few-layer graphene, reduced graphene oxide (rGO), graphene nanosheets, and graphene oxide (GO) [100]. Cell and animal studies have shown that the hydrophobic forms (GO and graphene) are more toxic than rGO because they interact with cell membrane lipids, causing toxicity. Since graphene is not a degradable material, its solubility and biocompatibility can be improved by chemically coating it with hydrophilic polymers, such as PEG. Indeed, PEGylation has been shown to reduce GO cytotoxicity [100].

Graphene-nanoparticle hybrids are becoming very popular. These hybrid structures can be synthesized by combining graphene or its derivatives - GO and reduced graphene oxide rGO - with several types of NPs, such as metal (*e.g.*, noble metal and magnetic), metal oxide, or silica NPs (SiNPs). There are two types of hybrids: graphene–nanoparticle composites, where the NP grows on graphene sheets, and graphene-encapsulated NPs, where NP surface is coated with graphene [98].

Liposomes

Liposomes are bilayered phospholipid vesicles. Their composition includes natural and/or synthetic phospholipids, such as phosphatidylethanolamine, phosphatidylglycerol, phosphatidylcholine (=lecithin), phosphatidylserine, and phosphatidylinositol. Liposome bilayers may also contain cholesterol, hydrophilic polymer conjugated lipids, and water. The addition of cholesterol to the lipid bilayer of liposomes reduces their permeability and increases their stability [101]. By optimizing the lipid composition, liposome size, membrane fluidity, surface charge, and steric stabilization, it is possible to enhance the therapeutic index of liposomes, over those observed for the conventional formulations [77].

Liposomes are used to deliver either hydrophobic or hydrophilic drugs, through incorporation in the lipid bilayer itself or encapsulation in the inner aqueous core, respectively [66].

Many techniques have been reported for liposome preparation, but the major challenge in liposome production continues to be the scale-up. Most of the described preparation techniques are not suitable for scaling up due to their complexity, low reproducibility, and predictability of the preparations [102, 103]. Nevertheless, focus has been given to the scale-up of liposomes production with some successful results [103, 104].

Solid Lipid Nanoparticles and Nanostructured Lipid Carriers

Solid-lipid NPs are made from room temperature-solid lipids (triglycerides, partial glycerides, fatty acids, steroids, and waxes), surfactant(s) (soybean lecithin, egg lecithin, poloxamer 188, and sodium cholate) and water. Thus, they have a hydrophilic surface with a hydrophobic core, allowing the easy loading of hydrophobic drugs into it [105].

The production of solid-lipid NPs is easy to scale-up [106, 107], which is a great advantage. But certain drawbacks have been associated to this type of NPs: (a) they contain high amount of water (70–95%), (b) their drug-loading capacity is limited due to the crystalline structure of solid-lipids, (c) expulsion of encapsulated drugs can occur during storage due to the formation of the perfect crystalline lattice, a thermodynamically more stable form, (d) drug release profile may change with storage time, (e) polymorphic transitions are possible, (f) particle growth is possible during storage, and (g) gelation of the dispersion may take place during storage [108]. Many of these limitations can be overcome by the new generation of nanostructured lipid carriers (NLC). In contrast to solid-lipid NPs, the NLCs are produced by controlled mixing of solid lipids with spatially incompatible liquid lipids, which is particularly useful as they provide structural defects where drugs can be incorporated [109]. Indeed, the majority of drugs have higher solubility in a liquid lipid rather than in a solid lipid [110]. Therefore, NCLs have better long-term drug stability and lower drug expulsion rate during storage than solid lipid NPs [111].

SYNTHESIS OF NANOPARTICLES

Several methods have been used to synthesize NPs. Two types of approaches can be followed, namely bottom-up and top-down. In the first strategy, a structure is built from small units, while in the second a larger unit is reduced in size [112].

Among the methods available to synthesize polymeric NPs, such as PLGA, PLA, and PCL based-NPs, some can be highlighted, such as single and double emulsion-solvent evaporation methods, and nanoprecipitation method, also called solvent displacement method. In oil-water (O/W, single) or in water-oil-water (W/O/W, double) emulsion-solvent evaporation methods, hydrophobic drugs are added directly to the oil phase (O/W), while hydrophilic drugs may be first emulsified with the polymer solution prior to formation of particles (W/O/W). After this step, high intensity sonication facilitates the formation of small polymer droplets. The resulting emulsion is added to a larger aqueous phase and stirred for several hours. The organic solvent is then evaporated under reduced pressure.

Hardened nanoparticles are collected and washed by centrifugation [113, 114] (Fig. **7.1**).

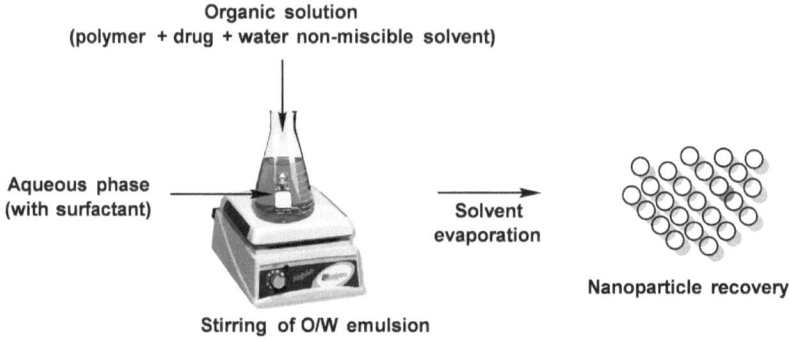

Fig. (7.1). Synthesis of nanoparticles by the single emulsion-solvent evaporation method (adapted from [114, 115]).

In nanoprecipitation method, the preformed polymer and drug are dissolved in a water miscible organic solvent (*e.g.* acetone or methanol). The solution is then injected into an aqueous solution, which may or may not contain a surfactant. Through fast solvent diffusion, the NPs are formed immediately. After that, the solvents are removed under reduced pressure. Surfactants or stabilizers can be included in the process to modify the size and the surface properties, or to ensure the stability of the nanoparticle dispersion [114, 115] (Fig. **7.2**).

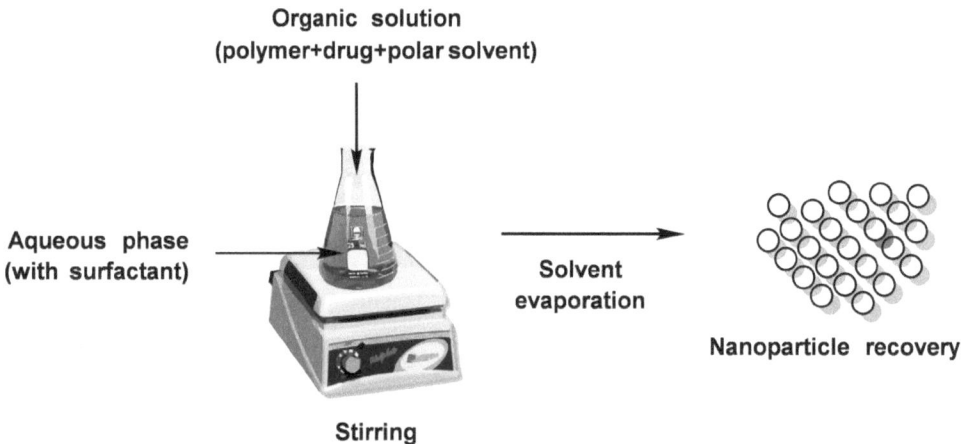

Fig. (7.2). Synthesis of nanoparticles by the nanoprecipitation method (adapted from [114, 115]).

Chitosan polymeric NPs are mainly prepared according to the ionic gelation method. In acidic aqueous solution (*e.g.* acetic acid, pH 3.5), chitosan oligomers can self-assemble into NPs through ionic cross-linking with multivalent anions. The polyanion most commonly used for the ionic crosslinking with the cationic chitosan is tripolyphosphate (TPP), which is nontoxic. Due to the complexation between oppositely charged species, chitosan undergoes ionic gelation and precipitates [116, 117].

Conventionally, silver and gold NPs are synthesized by chemical reduction method involving hazard chemicals as reducing agents. In this bottom-up approach, sodium borohydride, sodium citrate, ascorbate, and others are used to reduce Ag^+ ($AgNO_3$) or Au^+ ($HAuCl_4 \cdot 3H_2O$) or Pd^{2+} ($PdCl_2$) in aqueous and non-aqueous solutions [93]. Conversely, green synthesis of metal NPs has been adopted, since it is environmentally friendly, cost effective, and easy to scale-up. In this type of synthesis, hazard chemicals are substituted by plant, bacteria, or fungal compounds or extracts with reducing capacity [118] (Fig. **7.3**). A green combustion can also be adopted to produce plant-based ZnO NPs at 400 °C, in which plant extracts also work as reducing agents.

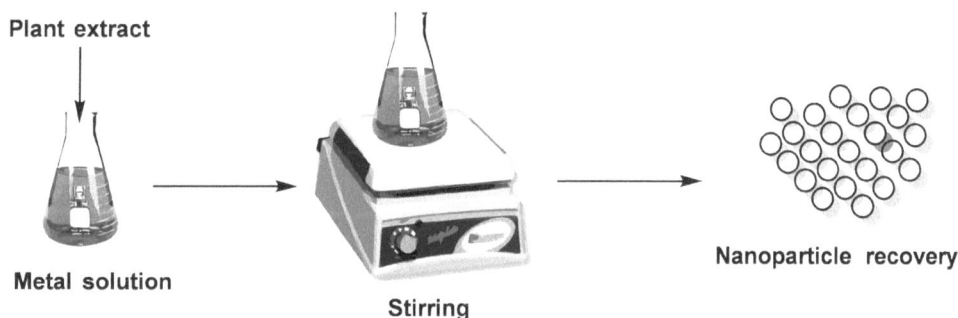

Fig. (7.3). Synthesis of metallic nanoparticles from plant extracts (adapted from [118]).

There are four classical methods to prepare liposomes, namely hydration of a thin lipid film, reverse-phase evaporation technique, solvent injection technique, and detergent dialysis. The difference between them is the way by which the lipids are drying down from organic solvents and then re-dispersed in aqueous media [119]. In hydration of a thin lipid film, a mixture of phospholipid and cholesterol are dispersed in organic solvent. Then, the organic solvent is evaporated at reduced pressure and the dry lipidic film, deposited on the flask wall, is hydrated by adding an aqueous buffer solution under agitation, at temperatures above the lipid transition temperature (Fig. **7.4**) [120].

In the reverse-phase evaporation technique, a lipidic film is prepared by evaporating organic solvent under reduced pressure. The system is purged with nitrogen and the lipids are re-dissolved in a second organic phase. Large vesicles are formed when an aqueous buffer is introduced into this mixture. The organic solvent is then removed and the system is maintained under continuous nitrogen [118]. The solvent injection method involves the dissolution of the lipid into an organic phase (ethanol or ether), followed by the injection of the lipid solution into aqueous media, forming liposomes [118]. Liposomes can also be formed by detergent dialysis. In this method, lipids are dissolved with detergent, yielding defined mixed micelles. Later, the detergent is removed by controlled dialysis. Other less employed techniques for the production of liposomes are nanoprecipitation and emulsion techniques because they require large amounts of organic solvent [118].

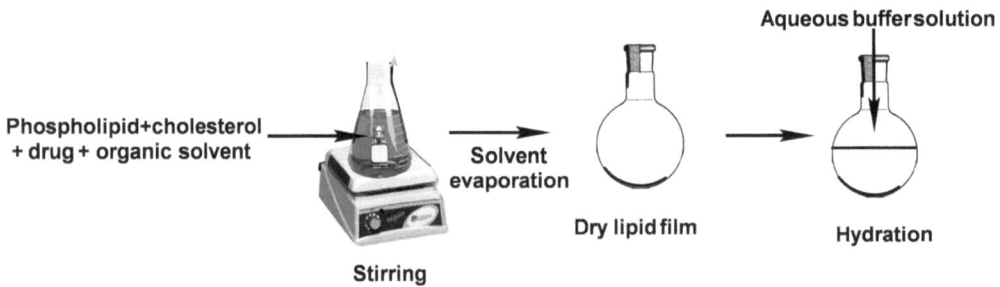

Fig. (7.4). Liposome synthesis by hydration of a thin lipid film.

Concerning other classes of nanoparticles, SLN can be prepared by microemulsion dilution method and rGO NPs by modified Hummers method. This method consists of generating GO through the addition of potassium permanganate to a solution of graphite, sodium nitrate and sulfuric acid [119].

BIOACTIVE POLYPHENOLS

The most studied polyphenolic-based NPs, both concerning processes of synthesis and biological properties, are those containing resveratrol [66, 121 - 170], tea polyphenols [96, 171 - 182], curcumin [180 - 213], and quercetin [214 - 261]. The intensive research on the development of nanostructures bearing these compounds is justified by their biological potential but low bioavailability.

Resveratrol, or 3,4',5-trihydroxystilbene (Fig. **7.5**), is a stilbene present in many natural foods, such as grapes (*Vitis vinifera* L.), purple grape juice and red wine, peanuts (*Arachis hypogaea* L.), blueberries (*Vaccinium* spp.), and blackberries (*Morus* spp.). Numerous studies have reported that *trans*-resveratrol, the most stable isoform, is a preventive agent of several pathologies, namely, neurodegenerative processes, viral and bacterial infections, cardiovascular diseases, and cancer [262 - 264]. The poor bioavailability of resveratrol in humans has been a major concern. Pharmacokinetic studies in humans have shown absorption and a rapid metabolism of resveratrol, but relatively low excretion in urine and faeces. However, maximal plasma concentration (*Cmax*) of free resveratrol is extremely low (<37 nM) after oral ingestion at physiological doses. Resveratrol-*O*-glucuronides and sulphates are the major plasma and urine metabolites, the sulphated forms being predominant [61, 265].

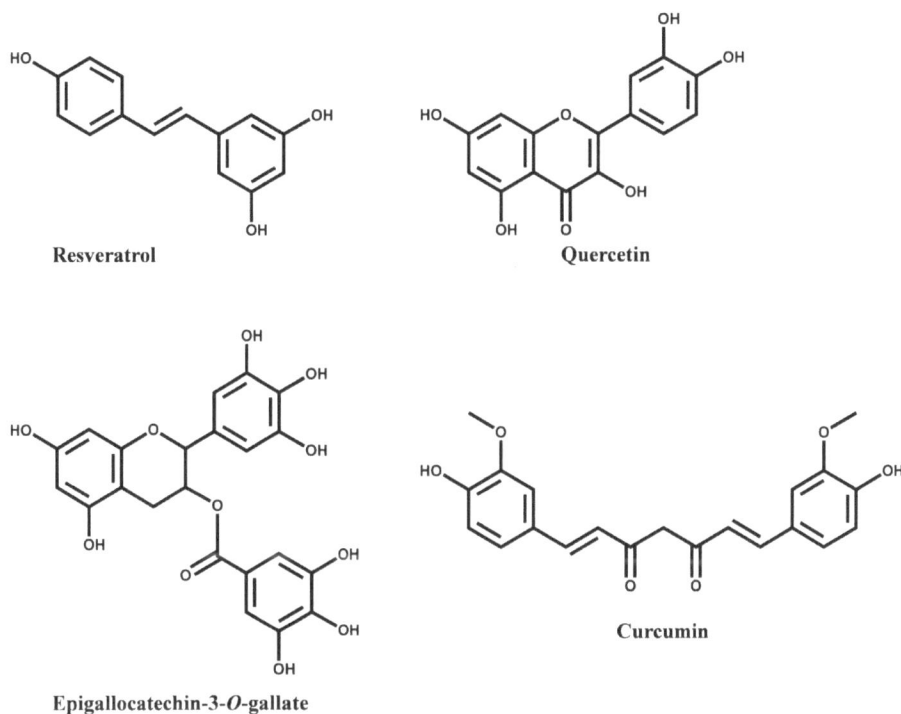

Resveratrol   **Quercetin**

Epigallocatechin-3-*O*-gallate   **Curcumin**

Fig. (7.5). Chemical structure of the most cited encapsulated polyphenols.

Camellia sinensis L. green leaves contain high levels of flavan-3-ol, predominantly (−)-epigallocatechin, (−)-epigallocatechin-3-*O*-gallate (Fig. **7.5**), and (−)-epicatechin-3-*O*-gallate). The levels of these compounds decay during fermentation of the green leaves to produce black tea, by the action of polyphenol oxidase. At the same time, accumulation of theaflavins and thearubigins occurs [61]. After an acute green tea feeding study, the plasma contained a total of 12 metabolites in the form of *O*-methylated, sulphated, and glucuronide conjugates of epicatechin and epigallocatechin along with the native green tea flavan-3-ols. The *Cmax* of epigallocatechin-3-*O*-gallate is 55 nM. The presence of unmetabolized form of epigallocatechin-3-*O*-gallate suggests that the galloyl moiety can be responsible for its escaping from phase II metabolism [61].

Moreover, studies indicate that epigallocatechin-3-*O*-gallate is mainly excreted through the bile, while epigallocatechin and epicatechin are excreted through urine and bile [266].

The link between tea consumption and reduced risk of cardiovascular diseases, prevention of certain types of cancer, dementia, osteoporosis, dental caries, kidney stones, among others, has been extensively explored [266].

Quercetin (Fig. **7.5**) is a flavonol ubiquitously found in vegetables and fruits, including apples (*Malus domestica* Borkh), onions (*Allium cepa* L.), black currant (*Ribes nigrum* L.) and black tea (*Cammelia sinensis* L.) [267]; but it mainly occurs in glycosidic forms. Quercetin possesses several biological effects including antioxidant, anti-inflammatory, anticancer, and antidiabetic activities [268]. After a cleavage of the sugar moieties, quercetin aglycone is then conjugated into its glucuronide, sulphated, and methyl metabolites, in order to facilitate their elimination [61].

Curcumin (Fig. **7.5**) is a yellow pigment present in the spice turmeric (*Curcuma longa* L.). Extensive research has shown its antioxidant, anti-inflammatory, anticancer, antiviral, and antibacterial activities. Despite this broad spectrum of activities, its low bioavailability limits its clinical application [269]. In a pharmacodynamic and pharmacokinetic study with cancer patients, curcumin was detected in the faeces, but not in plasma, blood cells, or urine. Curcumin can undergo metabolic *O*-conjugation, as its sulphate conjugate was detected in the faeces of one patient [270].

BIOACTIVE PHENOLIC-RICH MEDICINAL PLANTS

Several medicinal plants and fruits consumed all over the world are a rich source of polyphenols, and to translate this potential to *in vivo* results, several strategies are being employed. Most of the following examples described below did not report the phenolic composition of the encapsulated form or of the raw material; thus, the identification of the putative compounds, present in the natural matrices described, was obtained from other studies.

Antioxidant Activity

The antioxidant activity of NPs based on several medicinal plant species and fruits has been extensively studied (Table **7.1**).

To understand why the reported species display such high antioxidant activities, we first need to address their chemical profile. An extract prepared from the rhizome of *Acorus calamus* L., containing 25 mg gallic acid equivalent (GAE)/g and 9 mg rutin equivalent (RE)/g, was incorporated in silver NPs and its scavenging activity and its ferric reducing power were evaluated [271].

Table 7.1. Antioxidant activity of polyphenolic-based NPs.

Species	Type of Nanoparticle	Size (nm)	Assay
Acorus calamus L [271]	Silver NP	31.83	Scavenging activity (DPPH, $O_2^{\bullet-}$ and H_2O_2) and ferric reducing power
Brassica oleraceae L. var. *capitata* f. *rubra* DC [272]	SLN	178–437	Scavenging activity (DPPH)
Cammelia sinensis L [171]	PCL[a] and alginate	380	Scavenging activity (DPPH)
Cammelia sinensis L [273]	Chitosan	282–301	Scavenging activity (DPPH) and fish oxidation (peroxide value)
Cammelia sinensis L [274]	Gold NP		Scavenging activity (DPPH, $O_2^{\bullet-}$, $^{\bullet}NO$)
Cassia fistula L. [275, 276]	Zinc oxide NP	5–15	Scavenging activity (DPPH)
Garcinia xanthochymus Roxb [277]	Zinc oxide NP	20–30	Scavenging activity (DPPH)
Mangifera indica L [278]	SAE[b]	<400	Scavenging activity (DPPH)
Mimusops elengi L [279]	Silver NP	12.8–30.48	Scavenging activity (DPPH)
Pterocarpus marsupium Roxb [280]	Gold NP	72–85	Total antioxidant capacity (phosphomolybdate method)
Saururus chinensis (Lour.) Baill [281]	Palladium NP	4	Scavenging activity (DPPH)
Terminalia bellirica (Gaertn.) Roxb [282]	Silver NP	20	Scavenging activity (DPPH)
Compounds			
2,5-Dihydroxybenzoic acid [283]	Low and high molecular weights chitosan NP	300–600	ORAC assay
4-Hydroxybenzaldehyde [284]	Dendrimer with polyether core	–	Scavenging activity (DPPH and $^{\bullet}OH$)
Catechin [285]	PLGA	410.59	Scavenging activity ($O_2^{\bullet-}$), lipid peroxidation assay and metal chelating activity assay
Chlorogenic acid [286]	Chitosan	146–255	Scavenging activity (ABTS$^{\bullet+}$)
Curcumin [143]	PCL lipid-core nanocapsules	190–210	Scavenging activity ($^{\bullet}OH$)
Curcumin [287]	PEG-iron oxide NP	80–90	Scavenging activity (DPPH)
Epigallocatechin-3-*O*-gallate [150]	Gold NP	25.55	Scavenging activity (DPPH and ABTS$^{\bullet+}$)

(Table 7.1) cont.....

Compounds	Type of Nanoparticle	Size (nm)	Assay
Fisetin [150]	Gold NP	9.76	Scavenging activity (DPPH and ABTS$^{\bullet+}$)
Gallic acid [288]	PLGA NP with or without PS80^c	223–228	Scavenging activity (ABTS$^{\bullet+}$)
Gallic acid [233]	Silver- selenium NP	30–35	Scavenging activity (DPPH and ABTS$^{\bullet+}$)
Protocatechuic acid [283]	Low and high molecular weights chitosan NP	300–600	ORAC assay
Quercetin [233]	Silver- selenium NP	30–35	Scavenging activity (DPPH and ABTS$^{\bullet+}$)
Quercetin [236]	Germanium NP	15–45	Scavenging activity ($^{\bullet}$OH)
Quercetin [285]	PLGA	399.67	Scavenging activity ($O_2^{\bullet-}$), lipid peroxidation assay and metal chelating activity assay
Resveratrol [143]	PCL lipid-core nanocapsules	190–210	Scavenging activity ($^{\bullet}$OH)
Resveratrol [164]	SLN	293–586	Anti-lipoperoxidative activity
Resveratrol [150]	Gold NP	14.55	Scavenging activity (DPPH and ABTS$^{\bullet+}$)
Resveratrol + curcumin [143]	PCL lipid-core nanocapsules	190–210	Scavenging activity ($^{\bullet}$OH)
Syringaldehyde [284]	Dendrimer with polyether core	–	Scavenging activity (DPPH and $^{\bullet}$OH)
Syringaldehyde [289]	Dendrimer with tris(2-aminoethyl)amine core	–	Scavenging activity (DPPH)
Syringaldehyde [289]	Dendrimer with aminomethylbenzylamine core	–	Scavenging activity (DPPH)
Vanillin [284]	Dendrimer with polyether core	–	Scavenging activity (DPPH and $^{\bullet}$OH)
Vanillin [289]	Dendrimer with tris(2-aminoethyl)amine core	–	Scavenging activity (DPPH)

^{a} PCL – polycaprolactone;
^{b} SAE – supercritical antisolvent extraction;
^{c} PS80 – polysorbate 80;

The DPPH scavenging activity of SLN, made with the hydroethanol extract from the leaves of *Brassica oleracea* L. var. *capitata* f. *rubra* DC, was assessed [272]. This species is well known for its high content of anthocyanins, mainly acylated anthocyanins (cyanidin-3-O-(sinapoyl)diglucoside-5-O-glucoside, cyanidin-3-O-(glycopyranosyl-sinapoyl)diglucoside-5-O-glucoside, cyanidin-3-O-(glycopyranosyl-p-coumaroyl)diglucoside-5-O-glucoside, cyanidin-3-O-(glycopyranosyl- feruloyl)diglucoside-5-O-glucoside, cyanidin-3-O-(p-coumaroyl)(sinapoyl)triglucoside-5-*O*-glucoside, cyanidin-3-*O*-(feruloyl)(sinapoyl)triglucoside-5-*O*-glucoside, cyanidin-3-*O*-(*p*-coumaroyl)diglucoside-5-*O*-glucoside, cyanidin-3-*O*-(feruloyl) diglucoside-5-*O*-glucoside, cyanidin-3-*O*-(sinapoyl)diglucoside-5-*O*-glucoside, cyanidin-3-*O*-(sinapoyl)(*p*-coumaroyl)diglucoside-5-*O*-glucoside, cyanidin-3-*O*-(sinapoyl)(feruloyl)diglucoside-5-*O*-glucoside, cyanidin-3-*O*-(sinapoyl)(sinapoyl) diglucoside-5-*O*-glucoside, cyanidin-3-*O*-diglucoside-5-*O*-glucoside) [290]. The DPPH scavenging activity was also evaluated for the zinc oxide NPs prepared with the aqueous extracts from the leaves of *Cassia fistula* L [275, 276]. and from the fruits of *Garcinia xanthochymus* Roxb [277], for the powder obtained from the supercritical antisolvent extraction of by-products of canned fruits and juice of *Mangifera indica* L [278], for the silver NPs englobing the aqueous extracts from the fruits of *Mimusops elengi* L [279], from the fruits of *Terminalia bellirica* (Gaertn.) Roxb [282], and for the palladium NPs containing the aqueous extract from the leaves of *Saururus chinensis* (Lour.) Baill [281]. The antioxidant activity observed can be attributed to the phenolic content of these extracts. *C. fistula* possess hydroxyanthraquinones (rhein, emodin, physcion, chrysophanol), phenolic acids, tannins (proanthocyanidins), flavonoids (flavan-3-ols, flavone glycosides) besides triterpene derivatives [291]. *G. xanthochymus* is rich in xanthones, namely 1,3,5,6-tetrahydroxy-4,7,8-(3-methyl-2 butenyl) xanthone, garciniaxanthone, 1,4,5,6-tetrahydroxy-7,8,-di (3-methylbut-2-enyl) xanthone and 1,2,6-trihydroxy-5-methoxy-7-(3-methylbut-2-enyl) xanthone, xanthochymol, isoxanthochymol, 1,5-dihydroxyanthone and 1,7-dihydroxyxanthone and the biflavonoid volkensi flavone [277]. Several phenolics were identified in *M. indica*, namely mangiferin, isomangiferin, quercetin-3-*O*-galactoside, quercetin-3-*O*-glucoside, quercetin-3-*O*-xyloside, quercetin-3-*O*-arabinoside, quercetin and kaempferol [278]. *M. elengi* extract encapsulated in silver NPs was composed of ascorbic acid, gallic acid, pyrogallol and resorcinol [279]. The fruits of *T. bellirica* contain ascorbic acid, hydrolyzable tannins, gallic acid, pyrogallol, methyl gallate, chebulagic acid, ellagic acid, and resorcinol [282, 291]. Total phenolic acid content in *S. chinensis* leaf extract was found to be 1.8 ± 0.08 g of GAE/100 g dried weight of leaf [282].

C. sinensis L. leaves aqueous extract was already encapsulated in chitosan NP [273], polycaprolactone combined with chitosan NPs [171] and gold NPs [274], and their scavenging activity tested against several oxygen and nitrogen radical species (Table **7.1**). As reported above, this species is rich in flavan-3-ols: (+)-epigallocatechin-3-*O*-gallate, (+)-epigallocatechin, (+)-epicatechin-3-*O*-gallate, and (+)-epicatechin [171].

The assessment of the total antioxidant activity of the aqueous extract from the wood of *Pterocarpus marsupium* Roxb. encapsulated in gold NPs was carried out by the phosphomolybdate method [281]. Phenolics isolated from the wood of this species include marsupsin, pterosupin, pterostilbene, and (-)-epicatechin [292].

Antimicrobial Activity

In order to maximize the antibacterial and antifungal activities, polyphenol rich extracts from medicinal plants and fruits have been mainly incorporated in silver NPs. Table **7.2** lists some recent examples of the combination between silver/gold and polyphenols against different strains of Gram-positive and Gram-negative bacteria and fungi.

Table 7.2. Antimicrobial activity of polyphenolic-based NPs.

Species	Type of Nanoparticle	Size (nm)	Assay	Microorganisms
Acalypha indica L [*293*].	Silver NP	20–30	MIC, membrane permeability test and respiration activity	**Bacteria:** *Escherichia coli* (MTCC443) and *Vibrio cholerae* (MTCC3904)
Acorus calamus L [271]	Silver NP	31.83	Disc diffusion method and growth kinetic studies	**Bacteria:** *Bacillus subtilis, Bacillus cereus* and **Staphylococcus***aureus*
Boerhaavia diffusa L [294]	Silver NP	25	Disc diffusion method and MIC	**Bacteria:** *Aeromonas hydrophila* (MTCC646), *Flavobacterium branchiophilum* (ATCC35036) and *Pseudomonas fluorescence* (MTCC671)
Brassica rapa L [295]	Silver NP	5.7–24.4	**Disc diffusion assay**	**Fungi:** *Gloeophyllum abietinum* (KACC51949), *Gloeophyllum trabeum* (KACC43361), *Chaetomium globosum* (KACC42262) and *Phanerochaete sordida* (KACC43367)

(*Table 7.2*) *cont.....*

Species	Type of Nanoparticle	Size (nm)	Assay	Microorganisms
Carica papaya L [296]	Silver NP	50–200	Disc diffusion method and MIC	**Bacteria:** *Micrococcus luteus* (MCC2155), **Staphylococcus***aureus* (MCC2408), *Bacillus subtilis* (MCC2511), **Escherichia***coli* (MCC2412), *Pseudomonas putida* and *Klebsiella pneumoniae* (MCC2451)
Cassia fistula L [275, 276]	Zinc oxide NP	5–15	Well diffusion method	**Bacteria:** *Klebsiella Aerogenes* (NCIM2098), *Escherichia coli* (NCIM5051), *Staphyloccus aureus* (NCIM5022) and *Pseudomonas desmolyticum* (NCIM2028)
Citrus sinensis L [297]	Silver NP	10–35	Well diffusion method	**Bacteria:** *Escherichia coli, Pseudomonas aeruginosa* and *Staphylococcus aureus*
Cymbopogan citratus (DC.) Stapf [298]	Silver NP	32	Well diffusion method	**Bacteria:** *Staphylococcus aureus, Escherichia coli, Proteus mirabilis, Klebsiella pneumoniae* and *Salmonella typhi* **Fungi:***Candida albicans* and *Aspergillus niger* (NCIM616)
Cynodon dactylon (L.) Pers [299]	Silver NP	8–10	Well diffusion method	**Bacteria:** *Escherichia coli* (MTCC4604), *Pseudomonas aeruginosa* (MTCC 4676), *Staphylococcus aureus* (NCIM 2127) and *Salmonella typhimurium* (MTCC 733)
Emblica officinalis Gaertn [300]	Silver NP	10–70	Disc diffusion method	**Bacteria:** *Staphylococcus aureus, Bacillus subtilis, Escherichia coli, Klebsiella pneumoniae*

(Table 7.2) cont.....

Species	Type of Nanoparticle	Size (nm)	Assay	Microorganisms
Garcinia mangostana L [301]	Silver NP	6–57	**Disc diffusion method**	**Bacteria:** *Escherichia coli* and *Staphylococcus aureus*
Malus domestica L [302]	Silver NP	10–40	MIC[a] and well diffusion method	**Bacteria:** *Bacillus cereus* (ATCC11778), *Staphylococcus aureus* (ATCC25175), *Citrobacter koseri* (ATCCBAA-895) and *Pseudomonas aeruginosa* (ATCC10145) **Fungus:***Candida albicans* (ATCC90028)
Mimusops elengi L [279]	Silver NP	12.8–30.48	MIC and well diffusion method	**Bacteria:** *Escherichia coli* (ATCC 25922) and *Staphylococcus aureus* (ATCC 25923)
Moringa oleifera L [303]	Silver NP	57	Well diffusion method	**Bacteria:** *Staphylococcus aureus, Escherichia coli* and *Bacillus cereus* **Fungi:***Candida albicans, Candida tropicalis* and *Candida krusei*
Musa paradisiaca L [304]	Silver NP	–	Well diffusion method	**Bacteria:** *Enterobacter aerogenes* (MTCC 111), *Escherichia coli* (MTCC 728) and *Klebsiella* sp. **Fungi:***Candida albicans* (BH and BX) and *Candida lipolytica* (NCIM 3589)
Psoralea corylifolia L [305]	Silver NP	50–200	**Disc diffusion method**	**Bacteria:** *Klebsiella pneumoniae, Escherichia coli, Bacillus subtilis, Salmonella paratyphi* B, *Pseudomonas aeruginosa* and *Staphylococcus aureus*

(Table 7.2) cont.....

Species	Type of Nanoparticle	Size (nm)	Assay	Microorganisms
Rosmarinus officinalis L [306]	Silver NP	29	MIC and well diffusion method	**Bacteria:** *Staphylococcus aureus* (ATCC25293), *Bacillus subtilis* (ATCC 6633), *Escherichia coli* (ATCC33218) and *Pseudomonas aeruginosa* (ATCC 9027)
Sesuvium portulacastrum L [307]	Silver NP stabilized with PVA[b]	5–20	Disc diffusion method	**Bacteria:** *Pseudomanas aeruginosa, Klebsiella pneumoniae, Staphylococcus aureus, Listeria monocytogenes* and *Micrococcus luteu* **Fungi:***Alternaria alternata, Candida albicans, Penicillium* sp. and *Fusarium* sp.
Terminalia bellirica (Gaertn.) Roxb [282]	Silver NP	20	**Well diffusion method and MIC**	**Bacteria:** *Escherichia coli* (ATCC25922) and *Staphylococcus aureus* (ATCC25923)
Terminalia chebula Retz [308]	Silver NP	<100	**MIC, MBC[a] and well diffusion method**	**Bacteria:** *Escherichia coli* (ATCC25922) and *Staphylococcus aureus* (ATCC 25923)
Vitex negundo L [309]	Silver NP	10–30	**Disc diffusion assay**	**Bacteria:** *Escherichia coli* (ATCC25922) and *Staphylococcus aureus* (ATCC 25923)
Compounds				
2,5-Dihydroxybenzoic acid [310]	Low and high molecular weight chitosan	300–600	MIC and MBC	**Bacteria:** *Bacillus cereus* (ATCC 11778), *Escherichia coli* O157:H7, *Listeria innocua* (NCTC 11288), *Staphylococcus aureus* (ATCC 6538), *Salmonella typhimurium* (ATCC 14028) and *Yersinia enterocolitica* (NCTC 10406)
Epigallocatechin-3-O-gallate [96]	Silver NPs on collagen fiber	22.3	MIC	**Bacteria:** *Escherichia coli*

(Table 7.2) cont.....

Compounds	Type of Nanoparticle	Size (nm)	Assay	Microorganisms
Gallic acid [233]	Silver–Selenium NP	30–35	Well diffusion method	**Bacteria:** *Escherichia coli* (MTCC 433) and *Bacillus subtilis* (MTCC 441)
Protocatechuic acid [310]	Low and high molecular weight chitosan	300–600	MIC and MBC	**Bacteria:** *Bacillus cereus* (ATCC 11778), *Escherichia coli* O157:H7, *Listeria innocua* (NCTC 11288), *Staphylococcus aureus* (ATCC 6538), *Salmonella typhimurium* (ATCC 14028) and *Yersinia enterocolitica* (NCTC 10406)
Quercetin [214]	PCL	251.25–267.76	Well diffusion method, MIC, bacterial growth kinetics	**Bacteria:** *Bacillus subtilis* (MTCC 1790), *Staphylococcus aureus* (MTCC 3160), *Salmonella typhimurium* (MTCC 3224) and *Escherichia coli* (MTCC 1678)
Quercetin [216]	PLGA	100–150	MIC and membrane permeability test and mice models	**Bacteria:** *Escherichia coli* and *Micrococcus tetragenus*
Quercetin [233]	Silver–Selenium NP	30–35	Well diffusion method	**Bacteria:** *Escherichia coli* (MTCC 433) and *Bacillus subtilis* (MTCC 441)
Resveratrol [130]	Silver NP and Gold NP	8.32–21.84	MIC	**Bacteria:** *Staphylococcus aureus* (SG511, 285 and 503), *Streptococcus pyogenes* (308A, T 12 A and 77 A), *Pseudomonas aeruginosa* (9027, 1592E, 1771 and 1771 M), *Escherichia coli* (078, TEM, 1502, DC 0 and DC 2),

(Table 7.2) cont.....

Compounds	Type of Nanoparticle	Size (nm)	Assay	Microorganisms
Resveratrol [130]	Silver NP and Gold NP	8.32–21.84	MIC	*Salmonella typhimurium, Klebsiella oxytoca* 1082E, *Klebsiella aerogenes* 1522E, *Enterobacter cloacae* (P99 and 1321E) and *Streptococcus pneumoniae*
Rosmarinic acid [310]	Low and high molecular weight chitosan	300–600	MIC and MBC	**Bacteria:** *Bacillus cereus* (ATCC 11778), *Escherichia coli* O157:H7, *Listeria innocua* (NCTC 11288), *Staphylococcus aureus* (ATCC 6538), *Salmonella typhimurium* (ATCC 14028) and *Yersinia enterocolitica* (NCTC 10406)

[a] MIC – minimum inhibitory concentration; MBC - minimum bactericidal concentration
[b] PVA - polyvinyl alcohol

Underpinning the antimicrobial activity of plant-based NPs is the chemical profile of the encapsulated material. Silver NPs have been prepared with the aqueous extracts from different parts of plants: the leaves of *Acalypha indica* L [293], *Brassica rapa* L [295], *Carica papaya* L [296], *Cynodon dactylon* (L.) Pers [299], *Cymbopogan citratus* (DC.) Stapf [298], *Garcinia mangostana* L. *[301], Moringa oleifera* L [303], *Rosmarinus officinalis* L. [306], and *Vitex negundo* L [309]; the *callus* and leaves of *Sesuvium portulacastrum* (L.) L [307]; the rhizomes of *A. calamus* [271]; the whole plant of *Boerhaavia diffusa* L [294];{Nakkala, 2014 #169} the peels of *Citrus sinensis* L [297]. and of *Musa paradisiaca* L *[304];* the fruits of *Emblica officinalis* Gaertn [300], *Malus domestica* L [302], *M. elengi* L [279], *T. bellirica* (Gaertn.) Roxb [282]. and *Terminalia chebula* Retz [308]; and the seeds of *Psoralea corylifolia* L [305].

Since these authors have incorporated aqueous extracts, the contribution of the more polar compounds (such as phenolic compounds) to the antimicrobial activity is more significant than that of the more apolar compounds (such as volatile oils), which are also known for their strong effect. *A. indica* L. is a species containing different classes of polyphenols and other compounds, such as tannins (potassium brevifolincarboxylate, acaindinin, 1-*O*-galloyl-glucose, 1,2,3,6-tetra-galloyl-glucose, corilagin, geraniin, acetonylgeraniin A, euphormisin M2, repandusinic acid

A and chebulagic acid), flavonoids (quercetin 3-*O*-glucoside and quercetin-3-*O*-rutinoside, biorobin, nicotiflorin, clitorin, mauritianin, chrysin and galangin), the cyanogenic glucoside acalyphin, amides (acalyphamidc, aurantiamide, succinimide), and the pyranoquinolinone alkaloid flindersin [311, 312]. *Brassica rapa* L. is rich in several flavonoids and hydroxycinnamic acid derivatives (kaempferol-3-*O*-sophoroside-7-*O*-glucoside, kaempferol-3-*O*-sophoroside-7-*O*-sophoroside, kaempferol-3-*O*-(feruloyl/caffeoyl)-sophoroside-7-*O*-glucoside, kaempferol-3,7-*O*-diglucoside, kaempferol-3-*O*-sophoroside, kaempferol-3-*O*-glucoside, isorhamnetin-3,7-*O*-diglucoside, ferulic acid, sinapic acid, caffeic acid, 1,2-disinapoylgentiobiose, 1,2'-disinapoyl-2-feruloylgentiobiose and 3-*p*-coumaroylquinic acid) [313] and glucosinolates [314]. Although *C. papaya* L. is known by its content in alkaloids (carpaine, pseudocarpaine, dehydrocarpaine I and II) and cyanogenic compounds, it also contains phenolic compounds (protocatechuic acid, *p*-coumaric acid, caffeic acid, chlorogenic acid, 5,7-dimethoxycoumarin, kaempferol, quercetin) [315]. The leaves of *C. dactylon* (L.) Pers. contain flavonoids (apigenin, luteolin, orientin and vitexin), other phenolics, carotenoids (β-carotene, neoxanthin, violaxanthin), phytosterols, glycosides, saponins and volatile oils [316]. The chemical profile of the leaves of *C. citratus* (DC.) Stapf comprises luteolin and its 6-C and 7-*O*-glycosides, isoorientin-2'-*O*-rhamnoside, quercetin, kaempferol and apigenin, elimicin, catechol, chlorogenic acid, caffeic acid, and hydroquinone. The essential oil mainly consists of citral [317]. *G. mangostana* L. leaves are known for their xanthone content (1,6-dihydroxy-3-methoxy-2-isoprenyl-xanthone, 1-hydroxy-6-acetoxy-3-methoxy-2-isoprenyl-xanthone and gartanin) [318]. *Moringa oleifera* is characterized by a volatile oil and several phenolic compounds (gallic acid, ellagic acid, chlorogenic acid, ferulic acid, luteolin, quercetin-3-*O*-rutinoside, quercetin, kaempferol, apigenin) [319]. *R. officinalis* L. is a well-studied source of phenolic compounds, including carnosic acid, carnosol, rosmanol, its isomers epiisorosmanol and epirosmanol, methylcarnosate, epirosmanolmethylether, and 5,6,7,10-tetrahydro-7-hydroxyrosmaquinone, rosmarinic acid, rosmarinic acid-3-*O*-glucoside, rosmadial, rosmaridiphenol, syringic acid, and the flavonoids homoplantaginin, cirsimaritin, genkwanin, gallocatechin, nepetrin, hesperidin, 6-hydroxyluteolin-7-*O*-glucoside, luteolin-3'-*O*-glucuronide, and two isomers of luteolin-3'-*O*-(*O*-acetyl)-glucuronide, triterpenes and organic acids (quinic acid) [320]. The phenolic profile of *V. negundo* L. includes the presence of aucubin, agnuside, protocatechuic acid, protocatechualdehyde, 4-hydroxybenzoic acid, caffeic acid, neochlorogenic acid, chlorogenic acid, cryptochlorogenin acid, isochlorogenic acids A, B and C, schaftoside, isoschaftoside, flavosativaside, vitexin, vitexin-2-*O*-rhamnoside, isovitexin, kaaempferol-3-*O*-(6''-malonylglucoside), quercetin-3-*O*-galactoside, luteoloside, kaempferol-3-*O*-rutinoside, apigenin-7-*O*-glucoside, luteolin, quercetin, apigenin and casticin [321]. Several flavonoids were isolated from *S.*

portulacastrum L.: 3,5,4'-trihydroxy-6,7-dimethoxyflavone-3-*O*-glucoside, eupalitin and sesuviosides C, D, E and F [322, 323]. Several flavonoids and cinnamic acids have been identified in *B. diffusa* L., such as flavonols (quercetin, kaempferol), flavonoid glycosides (quercetin-3-*O*-(2″-rhamnosyl)-robinobioside, kaempferol-3-*O*-(2″-rhamnosyl)-robinobioside, quercetin-3-*O*-robinobioside, eupalitin-3-*O*-galactosyl(1→2)-glucoside, kaempferol-3-*O*-robinobioside, eupalitin-3-*O*-galactoside) and cinnamic acids (3,4-dihydroxy-5-methoxycinnamoyl-rhamnoside) [324, 325]. The peels of *Citrus sinensis* L. are characterized by the presence of flavones (nobiletin, tangeretin, 3,5,6,7,8,3',4'-heptamethoxyflavone and 5,6,7,4'-tetramethoxyflavone) and essential oil [326]. The peels of *M. paradisiaca* L. are a rich source of kaempferol-3-*O*-glucoside, quercetin-3-*O*-glucoside, quercetin, quercetin-3-*O*-rutinoside, myricetin, catechin, chlorogenic acid, carvacrol, caffeic acid, eugenol, vanillic acid, protocatechuic acid, *p*-coumaric acid and ferulic acid [327]. *E. officinalis* Gaertn. contains L-ascorbic acid, gallic acid, corillagin (gallic acid derivative), ellagic acid and phyllanthin (lignan) [328]. The phenolic compounds present in *M. domestica* L. are quercetin, kaempferol, myricetin, epicatechin, procyanidin B2, and chorogenic acid [302]. *T. chebula* Retz. contains hydrolyzable tannins (chebulanin, punicalagin, terchebin, chebulinic acid) and tannic, ellagic and gallic acids [291]. One meroterpene (bakuchiol) and four flavonoids (bavachinin, bavachin, isobavachin and isobavachalcone) were isolated from the seeds of *P. corylifolia* L [329]. Moreover, with the purpose of improving the antimicrobial effect, zinc oxide NPs containing the aqueous extract of *C. fistula* L. were developed [275, 276].

Antiproliferative Activity

Different types of NPs have been used to incorporate aqueous or hydromethanol extracts from medicinal plants and fruits, as follows: silver NPs (rhizome of *A. calamus* L [271], leaves of *Clerodendron serratum* (L.) Moon [330]. *and Melia dubia* Cav [331], fruits of *M. domestica* Borkh [302] and *Psidium guajava* L [332].); gold NPs (leaves of *C. sinensis* L [274], the pods of *Illicium verum* Hook.f [333]. and fruits of *V. vinifera* L [334].); chitosan NPs [335] and the reduced graphene oxide NP [124] of the leaves of *C. sinensis* and the PLGA-PEG NPs from the fruits of *Punica granatum* L [336].

The antiproliferative potential of these NPs has been tested in diverse human cancer cell lines, namely HeLa (epithelial cells from cervix), A549 (epithelial cells from lung), MCF-7 (epithelial cells from mammary gland/breast), HepG2 (epithelial cells from liver), HT29 (epithelial cells from colon), SW48 (epithelial cells from colon), Ehrlich ascites carcinoma, HBL-100 (epithelial cells from mammary gland/breast), Hs578T (epithelial cells from mammary gland/breast), HPAF-II (epithelial cells from pancreas), HCT116 (epithelial cells from colon),

MIA Paca-2 (epithelial cells from pancreas), Su86.86 (epithelial cells from pancreas), BxPC3 (epithelial cells from pancreas), Capan 1 (epithelial cells from pancreas), Panc-1 (epithelial cells from pancreas/duct), E3LZ10.7 (epithelial cells from pancreas), PL5 (cells from pancreas), PL8 (cells from pancreas), Dalton lymphoma cells, Eca9706 (epithelial cells from oesophagus), 4T1 (epithelial cells from mammary gland), MDA-MB-231 (epithelial cells from mammary gland/breast), A375 (epithelial cells from skin), WRL-68 (epithelial cells from cervix), C6 glioma cells, SMMC-7721 (cells from liver), SH-SY5Y (epithelial cells from bone marrow), SiHa (epithelial cells from cervix), and THP-1 (monocytes from peripheral blood) (Table **7.3**).

Table 7.3. Anti-proliferative activity of polyphenolic-based NPs.

Species	Type of Nanoparticle	Size (nm)	Assays	Cell Lines/Animal Models
Acorus calamus L [271]	Silver NP	31.83	Cell viability (MTT), estimation of apoptotic effect (propidium iodide (PI) stain assay, DAPI, TUNEL assay)	HeLA and A549 cells
Cammelia sinensis L [274]	Gold NP		Cell viability (MTT, LDH leakage[a]), estimation of lipid peroxidation, of reduced glutathione, of catalase, of superoxide dismutase, of heme oxygenase 1, of NAD(P)H quinone oxidoreductase-1 and estimation of apoptotic effect (Bax[b], Bcl2[b], pAKT[b], caspase 3 and pIκB[b] expression)	MCF-7 cells, mice bearing Ehrlich ascites carcinoma and normal hepatocytes
Cammelia sinensis L [335]	Chitosan	400	Cell viability (MTT), estimation of apoptotic effect (annexin V/PI double stain assay)	HepG2 cells
Cammelia sinensis L [124]	Reduced graphene oxide	–	Cell viability (MTT), photothermal treatment (NIR diode laser)	HT29 and SW48 cells

(Table 7.3) cont.....

Species	Type of Nanoparticle	Size (nm)	Assays	Cell Lines/Animal Models
Clerodendron serratum (L.) Moon [330]	Silver NP		Cell viability (MTT)	Ehrlich ascites carcinoma
Illicium verum Hook.f [333]	Gold NP	20–150	Cell viability, cytotoxicity and caspase 3/7 activation (Apotox-Glo triplex assay kit), oxidative stress (GSH[c]-Glo, GSH assay kit)	A549 cells
Malus domestica L [302]	Silver NP	10–40	Cell viability (MTT)	MCF-7 cells
Melia dubia Cav [331]	Silver NP	7.3	Cell viability (MTT)	MCF-7 cells
Psidium guajava L [332]	Silver NP	2–10	Cell viability (MTT)	HeLa cells
Vitis vinifera L [334]	Gold NP GSH[c]- Gold NP LA[c]- Gold NP	20–45 44–50 60–80	Cell viability (MTT, LDH leakage), apoptotic effect (DAPI staining), DNA fragmentation analysis	HBL-100 cells
Punica granatum L [337]	PLGA-PEG	160	Cell growth (acid phosphatase assay)	MCF-7 and Hs578T cells
Compounds				
Caffeic acid [338]	Silver NP	3–10	Cell viability (MTT) and apoptotic effect (annexin V-FLUOS/ PI assay)	HepG2 cells
Curcumin [213]	Liposome	110–135		MCF-7 cells
Curcumin [338]	β-Cyclodextrin-pluronic modified iron oxide NP	10.5	Anti-proliferation (hematocymeter), estimation of Bcl-xL[b], Mcl-1[b], PCNA[b], β-catenin and MUC1[b] and *in vivo* studies	HPAF-II and Panc-1 cells and athymic nude (nu/nu) mice
Curcumin [339]	Human serum albumin NP	130–150	Cell viability (MTT) and *in vivo* assays	HCT116 and MiaPaCa2 and Balb/c nu/nu male mice

(Table 7.3) cont.....

Compounds	Type of Nanoparticle	Size (nm)	Assays	Cell Lines/Animal Models
Curcumin [180]	NIPAAM+ VP+PEG-A[d]	50	Cell viability (MTT)	MiaPaca2, Su86.86, BxPC3, Capan1, Panc1, E3LZ10.7, PL5 and PL8 cells and athymic (nude) mice
Ellagic acid [337]	PLGA-PEG NP	175	Cell growth (acid phosphatase assay)	MCF-7 and Hs578T cells
Epigallocatechin-3-O-gallate [341]	PLGA NP	239	Cell viability (MTT), inhibition of NF-κB[b] activation, suppression of the expression of cyclin D1, matrix metalloproteinase-9, and vascular endothelial growth factor, enhancement of the apoptotic effect of cisplatin in cell lines and *in vivo* studies	HeLa, THP-1, and A549 cells and mice bearing Ehrlich ascites carcinoma
Epigallocatechin-3-O-gallate [342]	Genipin-cross-linked caseinophosphopeptide–chitosan NP	245.3	Trypan blue dye exclusion assay	HepG2 and BGC823 cells
Epigallocatechin-3-O-gallate [150]	Gold NP	25.55	Cell viability (MTS[a]), apoptotic effect (staurosporine treatment)	SH-SY5Y cells expressing CFP-DEVD-YFP reporter
Epigallocatechin-3-O-gallate [173]	Casein micelles	–	Trypan blue dye exclusion assay, sulforhodamine B cytotoxicity assay	HT-29 cells
Epigallocatechin-3-O-gallate [343]	Chitosan	–	Cell viability (MTT), cell cycle analysis (APO-direct kit), apoptotic effect (expression of Bax[b], Bcl-2[b], caspase 3 and 9) and *in vivo* studies	Mel 928 cells and Athymic (nu/nu) male nude mice

(Table 7.3) cont.....

Compounds	Type of Nanoparticle	Size (nm)	Assays	Cell Lines/Animal Models
Fisetin [150]	Gold NP	9.76	Cell viability (MTS), apoptotic effect (staurosporine treatment)	SH-SY5Y cells expressing CFP-DEVD-YFP reporter
Gallic acid [233]	Silver-selenium NP	30–35	Cell viability (MTT)	Dalton lymphoma cells
Quercetin [344]	Lipossome	–	Cell viability (MTT) and estimation of expressions of DNMT1[b], NF-κBp65[b], HDAC1[b], Cyclin D1, caspase-3 and p16INK4α[b]	Eca9706 cells
Quercetin [224]	MPEG[c]-PLA	100–200	Cell viability (MTT, LDH leakage), apoptotic effect (Annexin V-FITC/PI double staining) and *in vivo* assays	4 T1 and MDA-MB-231 cells and female BABL/c mice
Quercetin [225]	PLGA	113.7	Cell viability (MTT), interaction with nuclear DNA, proliferation assay (haemocytometer), cell cycle analysis, mitochondrial membrane potential, intracellular ROS generation, apoptotic effect, estimation of cytochrome c, rac-1, caspase 3 and histone de-acetylase activities, Western blot analysis (anti-cytochrome c, anti-Bax[b], anti-p53[b], anti-Apaf1[b], anti-PARP[b], anti-HDAC1[b],	HepG2, HeLa, A375, and WRL-68 cells
Quercetin [233]	Silver-selenium NP	30–35	Cell viability (MTT)	Dalton lymphoma cells
Quercetin [236]	Germanium NP	32–33	Cell viability (MTT)	MCF-7 cells

(Table 7.3) cont.....

Compounds	Type of Nanoparticle	Size (nm)	Assays	Cell Lines/Animal Models
Quercetin [241]	PEG2000-lipossomes	50–100	Cytotoxicity (LDH leakage assay), apoptotic/necrotic effect (Annexin V-FITC/PI double staining), measurement of mitochondrial membrane potential and ROS generation, Western blot assay (antibodies to p53[b], p-p53[b], cytochrome c, caspase-3 and -9, and β-actin)	C6 glioma cells
Punicalagin [337]	PLGA-PEG	216	Cell growth (acid phosphatase assay)	MCF-7 and Hs578T cells
Resveratrol [121]	PEG	58	Cell viability (MTT)	HeLa and MCF-7 cells
Resveratrol [150]	Gold NP	14.55	Cell viability (MTS), apoptotic effect (staurosporine treatment)	SH-SY5Y cells expressing CFP-DEVD-YFP reporter
Tea polyphenols [345]	Platinum NP	30–60	Cell viability (MTT), apoptotic effect (propidium iodide staining)	SiHa cells
Theaflavin [341]	PLGA	215	Cell viability (MTT), inhibition of NF-κB[b] activation, suppression of the expression of cyclin D1, matrix metalloproteinase-9, and vascular endothelial growth factor, enhancement of the apoptotic effect of cisplatin in cell lines and *in vivo* studies	HeLa, THP-1 and A549 cells and mice bearing Ehrlich ascites carcinoma

[a] MTT - 3-(4,5-Dimethyl-2-thiazolyl)-2,5-diphenyl-2H-tetrazolium bromide; MTS - 5-(3-carboxymetho-xyphenyl)-2-(4,5-dimethylthiazolyl)-3-(4-sulfophenyl) tetrazolium, inner salt; LDH - lactate dehydrogenase.
[b] Bax - BCL2-associated X protein; Bcl2 - B-cell CLL/lymphoma 2; AKT - Protein kinase B; pAKT - phospho-Akt; pIκB – phospho-IκB; GSH-glutathione; Bcl-xL - BCL2 like ; Mcl-1 - myeloid cell leukemia 1;

PCNA - proliferating cell nuclear antigen; MUC1 - Mucin 1, cell surface associated; NF-κB - nuclear factor-kappa B; DNMT1 - DNA (cytosine-5-)-methyltransferase 1; HDAC1 - histone deacetylase 1; HDAC2 - histone deacetylase 2; p16^{INK4a} - cyclin-dependent kinase inhibitor 2A; p53 - Tumor protein p53; p-p53 - Phospho-p53; Apaf1 - apoptotic peptidase activating factor 1; PARP - Poly ADP ribose polymerase; cdk1 - cyclin-dependent kinase 1; p21 – protein p21.

^{c} GSH – glutathione; LA – lipoic acid.

^{d} NIPAAM - N-isopropylacrylamide; VP - N-vinyl-2-pyrrolidone; PEG-A -poly(ethyleneglycol)monoacrylate.

^{e} Methoxy poly (ethylene glycol)-poly(lactide).

Leaves of *C. serratum* (L.) Moon. contain α-spinasterol, (+) – catechin, luteolin and luteolin-7-*O*-glucuronide, apigenin, baicalein, scutellarein, 6-hydroxyluteolin, a glucoside of 6-hydroxyluteolin, caffeic and ferulic acids and a mixture of glucose, arabinose, and glucuronic acid [346]. The total phenolic and flavonoid contents of *M. dubia* Cav. aqueous extract were found to be 77.3 and 36 mg/g of GAE, respectively [347]. The fruits of *P. guajava* L. are characterized by the presence of anthocyanins (delphinidin-3-*O*-glucoside, cyanidin-3-*O*-glucoside), flavonols (myricetin-3-*O*-glucoside, myricetin-3-*O*-arabinoside, myricetin-3-*O*-xyloside, quercetin-3-*O*-galactoside, quercetin-3-*O*-glucoside quercetin-3-*O*-arabinopyranoside (guaijaverin), quercetin-3-*O*-arabinoside (avicularin), isorhamnetin-3-*O*-glucoside, isorhamnetin-3-*O*-galactoside and quercetin), gallocatechin-(4α-8)-gallocatechol, gallocatechin-(4α-8)-catechin, turpinionosides A and triterpenes (pedunculoside, guavenoic acid, madecassic acid, asiatic acid, abscisic acid and pinfaensin) [348]. A more varied chemical composition was found for *I. verum*, comprising essential oils, prenylated C_6–C_3 compounds, lignans, sesquiterpenes and flavonoids (kaempferol and its glycosides, and quercetin and its glycosides) [349]. *V. vinifera* L. is a rich source of flavan-3-ols (catechin, epicatechin, epicatechin-3-*O*-gallate, epigallocatechin, epigallocatechin-3-*O*-gallate, *etc.*), and anthocyanins (*e.g.* cyanidin-3-*O*-glucoside, malvidin-3-*O*-glucoside, delphinidin-3-*O*-glucoside, and petunidin-3-*O*-glucoside) among other phenolic compounds [334, 350]. *P. granatum* L. contain many polyphenolic compounds, including flavonoids, condensed tannins, and hydrolyzable tannins (ellagitannins and gallotannins). Ellagitannins are considered the most promising bioactive polyphenols of pomegranates, punicalagin being the most abundant one [337].

Neuroprotective Activity

Brain is the ultimate target for drug delivery. Polyphenolic compounds may influence brain function outside the CNS (*e.g.* by improving cerebral blood flow or by modulating signalling pathways from peripheral organs to the brain), at the BBB (*e.g.* by altering multi-drug-resistant protein-dependent influx and efflux mechanisms of various biomolecules), and inside the CNS (*e.g.* by directly modifying the activity of neurons and glial cells). However, reliable data on polyphenolic compounds uptake into the brain of animals is limited [351].

The most challenging task is to deliver bioactive compounds to the brain due to the protection of BBB with the narrow diameter of approximately less than 20 nm. Nanotechnology research plays a pivotal role in improving the permeability, solubility, and stability of bioactive compounds towards a better delivery to specific targets, including the brain [352]. One strategy commonly applied to over cross the BBB is by functionalizing them in order to be recognized by the membrane receptors, such as transferrin receptor, insulin receptor, and low-density lipoprotein receptors [352].

A very limited number of examples is available from literature concerning the encapsulation of neuroprotective medicinal plant extracts (Table **7.4**). The aqueous extract from the leaves of *Bacopa monnieri* (L.) Wettst [353]. and the methanol extract from the leaves of *Hypericum perforatum* L [354]. have been encapsulated in platinum and gold NPs, respectively.

The methanol extract of *B. monnieri* was previously investigated and seventeen compounds were isolated, including phenolic compounds (*p*-hydroxybenzoic acid, 3,4-dimethoxycinnamic acid, quercetin, apigenin, luteolin), triterpenes, and sterols [355]. *H. perforatum* extracts contain several classes of plant phenolics, including phloroglucinols (hyperforin and adhyperforin), naphthodianthrones (hypericin, pseudohypericin, protohypericin and protopseudohypericin), flavonoids (mostly quercetin and kaempferol glycosides and aglycones, and biflavonoids), and phenolic acids (mainly isomeric caffeoylquinic acids) [356].

Table 7.4. Neuroprotective effect of polyphenolic-based NPs.

Species	Type of Nanoparticle	Size (nm)	Assays	Cell Lines/ Animal Models
Bacopa monnieri (L.) Wettst [353]	Platinum NP	5–20	Content/activity of MDA[a], GSH[a], SOD[a], GPx[a], and CAT[a] and contents of dopamine, DOPAC[a], and HVA[a] in the MPTP[a] zebrafish brain, locomotion activity	Zebrafish
Hypericum perforatum L [354]	Gold NP	–	T-maze test, mirror chamber test, estimation of lipid peroxidation, GSH level and SOD and CAT activities	Male Swiss albino mice
Compounds				
Breviscapine [357]	PEG-SLN	21.6-29.9	Assessment of P-glycoprotein ATPase activity, *in vivo* assays	Adult male Sprague Dawley rats

(Table 7.4) cont.....

Compounds	Type of Nanoparticle	Size (nm)	Assays	Cell Lines/ Animal Models
Curcumin [358]	NIPAAM+VP+AA[b]	–	Cell Titer Glo® (CTG) assay, LDH[c] leakage, protection against oxidative stress, ROS content, free and GSSG[a] content, caspase activity and *in vivo* assays	Neuroblastoma cell line (SK--SH cells) and athymic mice
Curcumin [359]	ApoE-Poly(butyl) Cyanoacrylate	197	Cell viability (MTT[c]), estimation of ROS content (H2DCF-DA[c]), antiapoptotic and antinecrotic activity (AnnexinV-FITC and P staining), cell cycle analysis, protection against Aβ, staining with Hoechst 33342, detection of caspase-3	SH-SY5Y
Curcumin [197]	Tet 1 peptide-PLGA	150–200	Cell viability (MTT), anti-amyloid activity	LAG cell line and GI-1 glioma cells
Curcumin [202]	Liposomes	57.7–97.6	Aggregation of the amyloid-β--42 (thioflavin T fluorescence based assay and sandwich immunoassay)	*In vitro*
Curcumin [205]	Liposomes	207.2	Cell viability (MTT), Aβ secretion in cell culture, *post-mortem* and *in vivo* assays	HEK cells, Neuroblastoma cell line (SH-SY5Y cells) and hAPPsw SH-SY5Y cells, post-mortem brain tissue of AD patients and APPxPS1 mice
Curcumin [207]	PEG2000-liposomes	116.1–159.3	Staining of AD tissues, Aβ aggregation assays (thioflavin T assay), cell uptake studies	*In vitro*, hCMEC/D3 cells and *post-mortem* AD brains
Gallic acid [360]	Chitosan	104–166	Despair swim test, tail suspension test, locomotor activity, estimation of MAO-A activity, estimation of lipid peroxidation, GSH level and CAT activity	Swiss male albino mice

(Table 7.4) cont.....

Compounds	Type of Nanoparticle	Size (nm)	Assays	Cell Lines/ Animal Models
α-Mangostin [361]	PEG–PLA	50–300	Brain clearance of ^{125}I-Aβ_{1-42}, activation of microglia and astrocytes (CD45 and GFAP markers), neuronal loss (Nissl staining), spatial learning and memory (Morris water maze test)	Chang liver cells, glial cells (BV-2 cells), brain capillary endothelial cells and SAMP8 and APP/PS1 transgenic mice
Quercetin [362]	PLGA	270	Protection against arsenic, lipid peroxidation assay, reduced glutathione assay, estimation of the activity of CAT, GPx, reduced GSH, GST[a] and glucose-6-phosphate dehydrogenase	Adult female Swiss albino rats
Quercetin [363]	PLGA	10–50	Protection against cerebral ischemia-reperfusion, measurement of ROS level, lipid peroxidation assay, GSH assay, estimation of SOD, iNOS[a] and caspase 3 activities, quantitation of cerebral edema, measurement of the fluidity of mitochondrial membrane	Male Sprague Dawley rats
Quercetin [234]	β-CD-dodecylcarbonate nanoparticles	214.8	MTT, protection against oxysterols, estimation of CD36 and β1-integrin, IL-8[a], MCP-1[a], MMP-9[a] and TLR-4[a] expression	SH-SY5Y cells
Resveratrol [161]	PVP-b-PCL	89.3	MTT, LDH leakage, protection against H_2O_2, measurement of intracellular ROS, MDA level, antiapoptotic effect (Bcl-2/Bax[a], caspase-3	Rat cortical cell culture
Resveratrol [156]	Transferrin-PEG-PLA	153.3	MTT, caspase-3 activity, cellular uptake, *in vivo* assays	Rat C6 glioma cells, rat astrocytes and human U87 glioma cells

[a] MDA – malondialdehyde; GSH – reduced glutathione; GSSG – oxidized glutathione; GPx – glutathione peroxidase; GST - glutathione-S-transferase; SOD – superoxide dismutase; CAT – catalase; DOPAC - 3,4-dihydroxyphenylacetic acid; HVA - homovanillic acid; MPTP - (1-methyl-4-phenyl-1,2-3,6-tetrahydropyridine); iNOS – inducible nitric oxide synthase; IL-8 – interleukin 8; MCP-1 -monocyte Chemoattractant Protein-1; MMP-9 - matrix metallopeptidase 9; TLR-4 - Toll-like receptor 4; Bcl2 - B-cell CLL/lymphoma 2; Bax - BCL2-associated X protein.

[b] N-isopropylacrylamide (NIPAAM), vinylpyrrolidone (VP), and acrylic acid (AA).

[c] LDH – lactate dehydrogenase; MTT - 3-(4,5-dimethyl-2-thiazolyl)-2,5-diphenyl-2H-tetrazolium bromide; H2DCF-DA - 2',7'-dichlorodihydrofluorescein diacetate.

COMPARISON OF THE ACTIVITY OF ENCAPSULATED AND NON-ENCAPSULATED EXTRACTS

The comparison between the effect of pure extracts and encapsulated formulations containing the same extracts allows discussing the effectiveness of nanoformulations. Regarding the antioxidant activity, metallic NPs produced by green synthesis display strong antioxidant activity, although sometimes the activity observed for the non-encapsulated form is stronger. For instance, silver NPs containing *A. calamus* extract showed 10% higher activity than the extract alone against DPPH, while the extract was better scavenger of superoxide anion radical. Both forms exhibited similar reducing power [271]. Nallamuthu *et al.* [286] observed similar IC_{50} values for the encapsulated chlorogenic acid (92 ± 5 µg/ml) and for the pure compound (89 ± 3 µg/ml) in the *in vitro* ABTS assay. Another important factor was that kinetic studies revealed the burst release of 69% chlorogenic acid from nanoparticles only at the end of 100th hour, showing their stability. Pool *et al.* [285] showed that PLGA NPs increased the antiradical and chelating properties of quercetin and catechin. On the other hand, zinc oxide NPs + *Cassia fistula* extract were weaker DPPH scavengers (IC_{50} = 2853 µg/mL) than the extract alone (IC_{50} = 54 µg/mL) [275]. Similarly, chitosan NPs loaded with 2,5-dihydroxybenzoic and protocatechuic acids displayed lower antioxidant activity (ORAC method) when compared to free phenolic compounds [282].

As explained before, silver NPs display strong antibacterial activity since they interact with bacterial membrane and DNA. Banala *et al.* [296] compared the effect of *C. papaya* extract with silver NPs containing the same extract. They observed larger zones of inhibition with the silver NPs against both Gram-positive and Gram-negative bacteria. Sahu *et al.* [299] even noticed the absence of activity for the extract from *Cynodon dactylon*, while $AgNO_3$ and, to a less extent silver NPs, were active against *P. aeruginosa*, *S. aureus*, *E. coli* and *S. typhymurium*.

Nanotechnology can also provide an improved antiproliferative activity of a determined drug. Singh *et al.* [340] reported that theoflavin/EGCG encapsulated PLGA NPs induced an increase up to ~7-fold, when compared with pure theoflavin/EGCG in terms of exerting antiproliferative effects and also in enhancing the anticancer potential of cisplatin in lung carcinoma, cervical carcinoma, and acute monocytic leukemia cells. Theoflavin/EGCG-NPs were more efficient in sensitizing A549 cells to cisplatin-induced apoptosis, and alone or in combination with cisplatin inhibited NF-κB activation more efficiently and supressed the expression of cyclin D1, matrix metalloproteinase-9 and vascular endothelial growth factor. Theoflavin-NPs were also found to be more effective than pure theoflavin/EGCG in inducing the cleavage of caspase-3 and caspase-9 and Bax/Bcl2 ratio in favor of apoptosis.

Apart from testing NPs in cell assays, it is important to assess their activity in more complex systems, such as animal models. Indeed, Prakash *et al.* [354] tested the effect of nanohypericum (*H. perforatum* gold NPs) for the treatment of restraint stress-induced behavioral and biochemical alteration in male albino mice. Using T-maze and Mirror chamber tests they verified that pure extract and NPs (200 and 20 mg/kg) treatment significantly improved spatial memory and anti-anxiety-like behavior as compared to control animals. The NPs-treated group significantly improved the activities as compared to the group treated with pure non-encapsulated extract. In another study, chitosan NPs with gallic acid displayed a more pronounced effect in decreasing the immobility period of mice in forced-swimming test (FST) and tail suspension test (TST) than gallic acid alone. These NPs (10 mg/kg, i.p.) significantly decreased monoamine oxidase-A (MAO-A) activity, malondialdehyde levels, and catalase activity in mice, showing their antidepressant effect [360].

Despite the activities depicted in Tables **7.1-7.4**, it is important to evaluate their toxicity to prevent harmful effects when ingested by humans.

TOXICITY OF NANOPARTICLES

Nanoparticle toxicology, or nanotoxicity, refers to the study of the potential toxic effects of NPs on biological and ecological systems. Despite the thousands of papers dealing with the synthesis and biological potential of NPs, the interaction mechanisms underlying the NPs-living systems are not fully understood. One reason is the discrepancy between the amount of information gathered from *in vitro* studies and the one collected form *in vivo* studies [364]. Moreover, several studies have been performed without a complete characterization and description of the assayed NPs and the solutions used under the experimental conditions. The extensive parametric characterization of the NPs in terms of mean size and size distribution, shape, agglomeration or aggregation state, crystal structure, surface chemistry/charge/area, stability over time, dosing metric, chemical composition and uptake should be done [365]. This lack of information is evidenced in the studies compiled in this chapter, since only few of them reported the chemical composition of the raw material and/or of the end-product (NP).

Several characteristics contribute to the putative toxic effect of NPs [366]:

• The size - the smaller is the NP, the higher is its surface-to-volume ratio, which makes it more reactive and toxic. Secondly, along with size decrease, an increased penetration into animal tissues is observed;

- The shape - spherical NPs are more toxic than rod-shaped NPs, because they can easily cross cell membranes;
- The nature - biocompatible and biodegradable polymeric NPs are supposed to be less toxic than silver NPs;
- The reactivity - the main mechanism of particle-induced damage is the production of reactive oxygen species;
- The mobility - diffusion of NPs increases as their size decreases;
- The stability - at the pH of biological systems, most of the inorganic and organic NPs become soluble. In the case of metals and metal oxide NPs, concentration of metal ions will increase within the cells, which leads to high cell stress;
- The surface chemistry and charge - different functionalities create different charges on NPs surface, which will not only change the charge, but also their action in biological systems;
- Agglomeration/aggregation – agglomerated NPs behave in a different way than the individual dispersed particles mainly because of changes in surface properties;
- Medium and storage time – the medium in which NPs are synthesized or stored in also plays a role in determining their toxic effects. Changes in ionic strength may alter the size and, subsequently, the toxicity of NPs.

According to the Scientific Committee on Emerging and Newly Identified Health Risks (SCENIHR), NPs might have different toxicological properties from the bulk substance, but their risks should be assessed on a case-by-case basis [367]. This risk assessment is very important since consumers are exposed to products based on NPs, either by applying antimicrobial NPs-based cosmetics in the skin or by ingesting NPs in food products and drugs, which directly enter into gastrointestinal tract and interact with lymphatic cell tissues. Thus, it is necessary to find an agreement among researchers and industry in order to develop and standardize *in vivo* cytotoxic assays [367].

In our opinion, the scientific community should reach a consensus decision about the need to guarantee that the encapsulated formulation contains all the compounds present in the native extract and establish that the physical and chemical parameters require mandatory characterization. Only in this way it is possible to compare results among laboratories and correctly assess NPs toxicity.

CONCLUDING REMARKS

This chapter summarizes the advances in drug delivery of bioactive phenolic compounds using nanotechnology, by gathering and interpreting the available data. Four main biological activities characteristics from phenolic compounds were explored, namely antioxidant, antimicrobial, antiproliferative, and neuroprotective.

Despite the hundreds of papers published about the encapsulation of medicinal and edible plants and fruits, very few report the chemical composition of the raw material, as well as the NP composition. This lack of information does not allow assessing the efficiency of entrapment in NP of all the compounds present in the starting material. Moreover, the question still to be answered is if it makes sense to encapsulate the parent phenolic compounds or those resulting from the metabolism along the gastrointestinal tract, since the effects verified in animal studies with pure extracts are due to the metabolites and not to the parent compounds. More *in vivo* parallel experiments analysing extract in both forms (encapsulated and non-encapsulated) will allow to compare their effects.

From the different classes of NPs reviewed, silver NPs are, by far, the most studied one, concerning both their application and toxicity. However, nanotoxicity assessment is still difficult, due to the lack of harmonization among research community/industry about which physical/morphological, chemical, and biological parameters should be fully characterized when formulating a new NP.

CONSENT FOR PUBLICATION

Not applicable.

CONFLICT OF INTEREST

The authors declare no conflict of interest, financial or otherwise.

ACKNOWLEDGEMENT

This work was financed through project UID/QUI/50006/2013, receiving financial support from FCT/MEC through national funds, and co-financed by FEDER, under the Partnership Agreement PT2020. C. Grosso thanks FCT for the FCT Investigator (IF/01332/2014)

ABBREVIATIONS

$^{\bullet}$NO	Nitric oxide
$^{\bullet}$OH	Hydroxyl radical
AA	Acrylic acid
ABTS$^{\bullet+}$	2,2′-Azinobis (3-ethylbenzothiazoline-6-sulfonic acid) cation radical
AKT	Protein kinase B
Apaf1	Apoptotic peptidase activating factor 1

Bax	BCL2-associated X protein
BBB	Blood-brain barrier
Bcl2	B-cell CLL/lymphoma 2
Bcl-xL	BCL2 like
CAT	Catalase
cdk1	Cyclin-dependent kinase 1
CNS	Central nervous system
DNMT1	DNA (cytosine-5)-methyltransferase 1
DOPAC	3,4-Dihydroxyphenylacetic acid
DPPH	2,2-Diphenyl-1-picrylhydrazyl
FST	Forced swimming test
GO	Graphene oxide
GPx	Glutathione peroxidase
GSH	Reduced glutathione
GSSG	Oxidized glutathione
GST	Glutathione-S-transferase
H2DCF-DA	2',7'-Dichlorodihydrofluorescein diacetate
H_2O_2	Hydrogen peroxide
HDAC1	Histone deacetylase 1
HDAC2	Histone deacetylase 2
HVA	Homovanillic acid
IL-8	Interleukin 8
iNOS	Inducible nitric oxide synthase
LA	Lipoic acid
LDH	Lactate dehydrogenase
LSPR	Localized surface plasmon resonance
MBC	Minimum bactericidal concentration
Mcl-1	Myeloid cell leukemia 1
MCP-1	Monocyte chemoattractant protein-1
MDA	Malondialdehyde
MIC	Minimum inhibitory concentration
MMP-9	Matrix metallopeptidase 9
MPEG	Methoxy poly (ethylene glycol)-poly (lactide)
MPS	Mononuclear phagocytic system
MPTP	(1-Methyl-4-phenyl-1, 2, 3, 6-tetrahydropyridine)

MTS	5-(3-Carboxymethoxyphenyl)-2-(4, 5-dimethylthiazoly)-3-(4-sulfophenyl) tetrazolium, inner salt
MTT	3-(4, 5-Dimethyl-2-thiazolyl)-2, 5-diphenyl-2H-tetrazolium bromide
MUC1	Mucin 1, cell surface associated
NF-κB	Nuclear factor-kappa B
NIH	National Institute of Health
NIPAAM	N-Isopropylacrylamide
NLC	Nanostructured lipid carriers
NP	Nanoparticle
$O_2^{\bullet-}$	Superoxide anion radical
ORAC	Oxygen radical absorbance capacity
p16INK4α	Cyclin-dependent kinase inhibitor 2A
p21	Protein p21
p53	Tumor protein p53
pAKT	Phospho-Akt
PARP	Poly ADP ribose polymerase
PCL	Poly (ε-caprolactone)
PCNA	Proliferating cell nuclear antigen
PEG	Poly-ethylene glycol
PEG-A	Poly (ethyleneglycol) monoacrylate
pIκB	Phospho-IκB
PLA	Poly (lactic acid)
PLGA	Poly (lactic-co-glycolic acid)
p-p53	Phospho-p53
PS80	Polysorbate 80
PVA	Polyvinyl alcohol
PVP	Poly-vinyl pyrrolidone
rGO	Reduced graphene oxide
SAE	Supercritical antisolvent extraction
SCENIHR	Scientific Committee on Emerging and Newly Identified Health Risks
SOD	Superoxide dismutase
TLR-4	Toll-like receptor 4
TPGS	Polyethylene glycol 1000 succinate
TST	Tail suspension test
VP	Vinylpyrrolidone

REFERENCES

[1]   Cencic A, Chingwaru W. The role of functional foods, nutraceuticals, and food supplements in intestinal health. Nutrients 2010; 2(6): 611-25.
[http://dx.doi.org/10.3390/nu2060611] [PMID: 22254045]

[2]   Almada AL. Nutraceuticals and functional foods: Aligning with the norm or pioneering through a storm.Nutraceuticals and Functional Foods Regulations in the United States and Around the World. Amsterdam: Elsevier 2014; pp. 3-11.
[http://dx.doi.org/10.1016/B978-0-12-405870-5.00001-3]

[3]   Pandey KB, Rizvi SI. Plant polyphenols as dietary antioxidants in human health and disease. Oxid Med Cell Longev 2009; 2(5): 270-8.
[http://dx.doi.org/10.4161/oxim.2.5.9498] [PMID: 20716914]

[4]   Scalbert A, Johnson IT, Saltmarsh M. Polyphenols: antioxidants and beyond. Am J Clin Nutr 2005; 81(1) (Suppl.): 215S-7S.
[http://dx.doi.org/10.1093/ajcn/81.1.215S] [PMID: 15640483]

[5]   Choi D-Y, Lee Y-J, Hong JT, Lee H-J. Antioxidant properties of natural polyphenols and their therapeutic potentials for Alzheimer's disease. Brain Res Bull 2012; 87(2-3): 144-53.
[http://dx.doi.org/10.1016/j.brainresbull.2011.11.014] [PMID: 22155297]

[6]   Frei B, Higdon JV. Antioxidant activity of tea polyphenols in vivo: evidence from animal studies. J Nutr 2003; 133(10): 3275S-84S.
[http://dx.doi.org/10.1093/jn/133.10.3275S] [PMID: 14519826]

[7]   Lu Y, Yeap Foo L. Antioxidant activities of polyphenols from sage (*Salvia officinalis*). Food Chem 2001; 75: 197-202.
[http://dx.doi.org/10.1016/S0308-8146(01)00198-4]

[8]   Pérez-Jiménez J, Saura-Calixto F. Macromolecular antioxidants or non-extractable polyphenols in fruit and vegetables: Intake in four European countries. Food Res Int 2015; 74: 315-23.
[http://dx.doi.org/10.1016/j.foodres.2015.05.007] [PMID: 28411997]

[9]   Halliwell B. Are polyphenols antioxidants or pro-oxidants? What do we learn from cell culture and *in vivo* studies? Arch Biochem Biophys 2008; 476(2): 107-12.
[http://dx.doi.org/10.1016/j.abb.2008.01.028] [PMID: 18284912]

[10]   Daglia M. Polyphenols as antimicrobial agents. Curr Opin Biotechnol 2012; 23(2): 174-81.
[http://dx.doi.org/10.1016/j.copbio.2011.08.007] [PMID: 21925860]

[11]   Taguri T, Tanaka T, Kouno I. Antimicrobial activity of 10 different plant polyphenols against bacteria causing food-borne disease. Biol Pharm Bull 2004; 27(12): 1965-9.
[http://dx.doi.org/10.1248/bpb.27.1965] [PMID: 15577214]

[12]   Coppo E, Marchese A. Antibacterial activity of polyphenols. Curr Pharm Biotechnol 2014; 15(4): 380-90.
[http://dx.doi.org/10.2174/138920101504140825121142] [PMID: 25312620]

[13]   Moreno S, Scheyer T, Romano CS, Vojnov AA. Antioxidant and antimicrobial activities of rosemary extracts linked to their polyphenol composition. Free Radic Res 2006; 40(2): 223-31.
[http://dx.doi.org/10.1080/10715760500473834] [PMID: 16390832]

[14]   Katalinić V, Možina SS, Skroza D, *et al.* Polyphenolic profile, antioxidant properties and antimicrobial activity of grape skin extracts of 14 *Vitis vinifera* varieties grown in Dalmatia (Croatia). Food Chem 2010; 119: 715-23.
[http://dx.doi.org/10.1016/j.foodchem.2009.07.019]

[15]   Sakanaka S, Juneja LR, Taniguchi M. Antimicrobial effects of green tea polyphenols on thermophilic spore-forming bacteria. J Biosci Bioeng 2000; 90(1): 81-5.
[http://dx.doi.org/10.1016/S1389-1723(00)80038-9] [PMID: 16232822]

[16]   Hamilton-Miller JM. Antimicrobial properties of tea (*Camellia sinensis* L.). Antimicrob Agents Chemother 1995; 39(11): 2375-7.
[http://dx.doi.org/10.1128/AAC.39.11.2375] [PMID: 8585711]

[17]   Proestos C, Chorianopoulos N, Nychas GJE, Komaitis M. RP-HPLC analysis of the phenolic compounds of plant extracts. investigation of their antioxidant capacity and antimicrobial activity. J Agric Food Chem 2005; 53(4): 1190-5.
[http://dx.doi.org/10.1021/jf040083t] [PMID: 15713039]

[18]   Yamamoto H, Ogawa T. Antimicrobial activity of perilla seed polyphenols against oral pathogenic bacteria. Biosci Biotechnol Biochem 2002; 66(4): 921-4.
[http://dx.doi.org/10.1271/bbb.66.921] [PMID: 12036078]

[19]   Menard C, Bastianetto S, Quirion R. Neuroprotective effects of resveratrol and epigallocatechin gallate polyphenols are mediated by the activation of protein kinase C gamma. Front Cell Neurosci 2013; 7: 281.
[http://dx.doi.org/10.3389/fncel.2013.00281] [PMID: 24421757]

[20]   Daglia M, Di Lorenzo A, Nabavi SF, Talas ZS, Nabavi SM. Polyphenols: well beyond the antioxidant capacity: gallic acid and related compounds as neuroprotective agents: you are what you eat! Curr Pharm Biotechnol 2014; 15(4): 362-72.
[http://dx.doi.org/10.2174/1389201015041408252120737] [PMID: 24938889]

[21]   Tavares L, Figueira I, McDougall GJ, *et al.* Neuroprotective effects of digested polyphenols from wild blackberry species. Eur J Nutr 2013; 52(1): 225-36.
[http://dx.doi.org/10.1007/s00394-012-0307-7] [PMID: 22314351]

[22]   Reglodi D, Renaud J, Tamas A, *et al.* Novel tactics for neuroprotection in Parkinson's disease: Role of antibiotics, polyphenols and neuropeptides. Prog Neurobiol 2017; 155: 120-48.
[http://dx.doi.org/10.1016/j.pneurobio.2015.10.004] [PMID: 26542398]

[23]   Bhullar KS, Rupasinghe HPV. Polyphenols: multipotent therapeutic agents in neurodegenerative diseases. Oxid Med Cell Longev 2013; 2013: 891748.
[http://dx.doi.org/10.1155/2013/891748] [PMID: 23840922]

[24]   Mandel S, Youdim MBH. Catechin polyphenols: neurodegeneration and neuroprotection in neurodegenerative diseases. Free Radic Biol Med 2004; 37(3): 304-17.
[http://dx.doi.org/10.1016/j.freeradbiomed.2004.04.012] [PMID: 15223064]

[25]   Campos-Esparza MR, Sánchez-Gómez MV, Matute C. Molecular mechanisms of neuroprotection by two natural antioxidant polyphenols. Cell Calcium 2009; 45(4): 358-68.
[http://dx.doi.org/10.1016/j.ceca.2008.12.007] [PMID: 19201465]

[26]   González R, Ballester I, López-Posadas R, *et al.* Effects of flavonoids and other polyphenols on inflammation. Crit Rev Food Sci Nutr 2011; 51(4): 331-62.
[http://dx.doi.org/10.1080/10408390903584094] [PMID: 21432698]

[27]   Chen B-T, Li W-X, He R-R, *et al.* Anti-inflammatory effects of a polyphenols-rich extract from tea (*Camellia sinensis*) flowers in acute and chronic mice models. Oxid Med Cell Longev 2012; 2012: 537923.
[http://dx.doi.org/10.1155/2012/537923] [PMID: 22900128]

[28]   Joseph SV, Edirisinghe I, Burton-Freeman BM. Fruit polyphenols: A review of anti-inflammatory effects in humans. Crit Rev Food Sci Nutr 2016; 56(3): 419-44.
[http://dx.doi.org/10.1080/10408398.2013.767221] [PMID: 25616409]

[29]   Jean-Gilles D, Li L, Ma H, Yuan T, Chichester CO III, Seeram NP. Anti-inflammatory effects of polyphenolic-enriched red raspberry extract in an antigen-induced arthritis rat model. J Agric Food Chem 2012; 60(23): 5755-62.
[http://dx.doi.org/10.1021/jf203456w] [PMID: 22111586]

[30]   Michel P, Dobrowolska A, Kicel A, *et al.* Polyphenolic profile, antioxidant and anti-inflammatory

activity of eastern teaberry (*Gaultheria procumbens* L.) leaf extracts. Molecules 2014; 19(12): 20498-520.
[http://dx.doi.org/10.3390/molecules191220498] [PMID: 25493634]

[31]   Tipoe GL, Leung T-M, Hung M-W, Fung M-L. Green tea polyphenols as an anti-oxidant and anti-inflammatory agent for cardiovascular protection. Cardiovasc Hematol Disord Drug Targets 2007; 7(2): 135-44.
[http://dx.doi.org/10.2174/187152907780830905] [PMID: 17584048]

[32]   Aquilano K, Baldelli S, Rotilio G, Ciriolo MR. Role of nitric oxide synthases in Parkinson's disease: a review on the antioxidant and anti-inflammatory activity of polyphenols. Neurochem Res 2008; 33(12): 2416-26.
[http://dx.doi.org/10.1007/s11064-008-9697-6] [PMID: 18415676]

[33]   de la Lastra CA, Villegas I. Resveratrol as an anti-inflammatory and anti-aging agent: mechanisms and clinical implications. Mol Nutr Food Res 2005; 49(5): 405-30.
[http://dx.doi.org/10.1002/mnfr.200500022] [PMID: 15832402]

[34]   Lee M, Kim S, Kwon O-K, Oh S-R, Lee H-K, Ahn K. Anti-inflammatory and anti-asthmatic effects of resveratrol, a polyphenolic stilbene, in a mouse model of allergic asthma. Int Immunopharmacol 2009; 9(4): 418-24.
[http://dx.doi.org/10.1016/j.intimp.2009.01.005] [PMID: 19185061]

[35]   John CM, Sandrasaigaran P, Tong CK, Adam A, Ramasamy R. Immunomodulatory activity of polyphenols derived from *Cassia auriculata* flowers in aged rats. Cell Immunol 2011; 271(2): 474-9.
[http://dx.doi.org/10.1016/j.cellimm.2011.08.017] [PMID: 21924708]

[36]   Makare N, Bodhankar S, Rangari V. Immunomodulatory activity of alcoholic extract of *Mangifera indica* L. in mice. J Ethnopharmacol 2001; 78(2-3): 133-7.
[http://dx.doi.org/10.1016/S0378-8741(01)00326-9] [PMID: 11694357]

[37]   Mitjans M, del Campo J, Abajo C, *et al.* Immunomodulatory activity of a new family of antioxidants obtained from grape polyphenols. J Agric Food Chem 2004; 52(24): 7297-9.
[http://dx.doi.org/10.1021/jf049403z] [PMID: 15563210]

[38]   Abajo C, Boffill MA, del Campo J, *et al. In vitro* study of the antioxidant and immunomodulatory activity of aqueous infusion of *Bidens pilosa.* J Ethnopharmacol 2004; 93(2-3): 319-23.
[http://dx.doi.org/10.1016/j.jep.2004.03.050] [PMID: 15234771]

[39]   Gao X, Deeb D, Media J, *et al.* Immunomodulatory activity of resveratrol: discrepant *in vitro* and *in vivo* immunological effects. Biochem Pharmacol 2003; 66(12): 2427-35.
[http://dx.doi.org/10.1016/j.bcp.2003.08.008] [PMID: 14637200]

[40]   Fresco P, Borges F, Diniz C, Marques MPM. New insights on the anticancer properties of dietary polyphenols. Med Res Rev 2006; 26(6): 747-66.
[http://dx.doi.org/10.1002/med.20060] [PMID: 16710860]

[41]   Yang GY, Liao J, Kim K, Yurkow EJ, Yang CS. Inhibition of growth and induction of apoptosis in human cancer cell lines by tea polyphenols. Carcinogenesis 1998; 19(4): 611-6.
[http://dx.doi.org/10.1093/carcin/19.4.611] [PMID: 9600345]

[42]   Lamoral-Theys D, Pottier L, Dufrasne F, *et al.* Natural polyphenols that display anticancer properties through inhibition of kinase activity. Curr Med Chem 2010; 17(9): 812-25.
[http://dx.doi.org/10.2174/092986710790712183] [PMID: 20156174]

[43]   Landis-Piwowar KR, Huo C, Chen D, *et al.* A novel prodrug of the green tea polyphenol (---epigallocatechin-3-gallate as a potential anticancer agent. Cancer Res 2007; 67(9): 4303-10.
[http://dx.doi.org/10.1158/0008-5472.CAN-06-4699] [PMID: 17483343]

[44]   Hadi SM, Bhat SH, Azmi AS, Hanif S, Shamim U, Ullah MF. Oxidative breakage of cellular DNA by plant polyphenols: a putative mechanism for anticancer properties. Semin Cancer Biol 2007; 17(5): 370-6.

[http://dx.doi.org/10.1016/j.semcancer.2007.04.002] [PMID: 17572102]

[45]   Gomes CA, da Cruz TG, Andrade JL, Milhazes N, Borges F, Marques MPM. Anticancer activity of phenolic acids of natural or synthetic origin: a structure-activity study. J Med Chem 2003; 46(25): 5395-401.
[http://dx.doi.org/10.1021/jm030956v] [PMID: 14640548]

[46]   Colomer R, Sarrats A, Lupu R, Puig T. Natural polyphenols and their synthetic analogs as emerging anticancer agents. Curr Drug Targets 2017; 18(2): 147-59.
[http://dx.doi.org/10.2174/1389450117666160112113930] [PMID: 26758667]

[47]   Fantini M, Benvenuto M, Masuelli L, *et al. In vitro* and *in vivo* antitumoral effects of combinations of polyphenols, or polyphenols and anticancer drugs: perspectives on cancer treatment. Int J Mol Sci 2015; 16(5): 9236-82.
[http://dx.doi.org/10.3390/ijms16059236] [PMID: 25918934]

[48]   Koňariková K, Ježovičová M, Keresteš J, Gbelcová H, Ďuračková Z, Žitňanová I. Anticancer effect of black tea extract in human cancer cell lines. Springerplus 2015; 4: 127.
[http://dx.doi.org/10.1186/s40064-015-0871-4] [PMID: 25825685]

[49]   Grassi D, Desideri G, Croce G, Tiberti S, Aggio A, Ferri C. Flavonoids, vascular function and cardiovascular protection. Curr Pharm Des 2009; 15(10): 1072-84.
[http://dx.doi.org/10.2174/138161209787846982] [PMID: 19355949]

[50]   Parks DA, Booyse FM. Cardiovascular protection by alcohol and polyphenols: role of nitric oxide. Ann N Y Acad Sci 2002; 957: 115-21.
[http://dx.doi.org/10.1111/j.1749-6632.2002.tb02910.x] [PMID: 12074966]

[51]   Pasten C, Olave NC, Zhou L, Tabengwa EM, Wolkowicz PE, Grenett HE. Polyphenols downregulate PAI-1 gene expression in cultured human coronary artery endothelial cells: molecular contributor to cardiovascular protection. Thromb Res 2007; 121(1): 59-65.
[http://dx.doi.org/10.1016/j.thromres.2007.02.001] [PMID: 17379280]

[52]   Bradamante S, Barenghi L, Villa A. Cardiovascular protective effects of resveratrol. Cardiovasc Drug Rev 2004; 22(3): 169-88.
[http://dx.doi.org/10.1111/j.1527-3466.2004.tb00139.x] [PMID: 15492766]

[53]   Chiva-Blanch G, Arranz S, Lamuela-Raventos RM, Estruch R. Effects of wine, alcohol and polyphenols on cardiovascular disease risk factors: evidences from human studies. Alcohol Alcohol 2013; 48(3): 270-7.
[http://dx.doi.org/10.1093/alcalc/agt007] [PMID: 23408240]

[54]   Khurana S, Venkataraman K, Hollingsworth A, Piche M, Tai TC. Polyphenols: benefits to the cardiovascular system in health and in aging. Nutrients 2013; 5(10): 3779-827.
[http://dx.doi.org/10.3390/nu5103779] [PMID: 24077237]

[55]   Guo X, Tresserra-Rimbau A, Estruch R, *et al.* Effects of polyphenol, measured by a biomarker of total polyphenols in urine, on cardiovascular risk factors after a long-term follow-up in the PREDIMED study. Oxid Med Cell Longev 2016; 2016: 2572606.
[http://dx.doi.org/10.1155/2016/2572606] [PMID: 26881019]

[56]   Covas M-I, de la Torre R, Fitó M. Virgin olive oil: a key food for cardiovascular risk protection. Br J Nutr 2015; 113 (Suppl. 2): S19-28.
[http://dx.doi.org/10.1017/S0007114515000136] [PMID: 26148918]

[57]   Manach C, Williamson G, Morand C, Scalbert A, Rémésy C. Bioavailability and bioefficacy of polyphenols in humans. I. Review of 97 bioavailability studies. Am J Clin Nutr 2005; 81(1) (Suppl.): 230S-42S.
[http://dx.doi.org/10.1093/ajcn/81.1.230S] [PMID: 15640486]

[58]   Talavéra S, Felgines C, Texier O, Besson C, Lamaison J-L, Rémésy C. Anthocyanins are efficiently absorbed from the stomach in anesthetized rats. J Nutr 2003; 133(12): 4178-82.

[http://dx.doi.org/10.1093/jn/133.12.4178] [PMID: 14652368]

[59]   Crespy V, Morand C, Besson C, Manach C, Demigne C, Remesy C. Quercetin, but not its glycosides, is absorbed from the rat stomach. J Agric Food Chem 2002; 50(3): 618-21.
[http://dx.doi.org/10.1021/jf010919h] [PMID: 11804539]

[60]   Piskula MK, Yamakoshi J, Iwai Y. Daidzein and genistein but not their glucosides are absorbed from the rat stomach. FEBS Lett 1999; 447(2-3): 287-91.
[http://dx.doi.org/10.1016/S0014-5793(99)00307-5] [PMID: 10214963]

[61]   Del Rio D, Rodriguez-Mateos A, Spencer JPE, Tognolini M, Borges G, Crozier A. Dietary (poly)phenolics in human health: structures, bioavailability, and evidence of protective effects against chronic diseases. Antioxid Redox Signal 2013; 18(14): 1818-92.
[http://dx.doi.org/10.1089/ars.2012.4581] [PMID: 22794138]

[62]   Duda-Chodak A. The inhibitory effect of polyphenols on human gut microbiota. J Physiol Pharmacol 2012; 63(5): 497-503.
[PMID: 23211303]

[63]   Manach C, Scalbert A, Morand C, Rémésy C, Jiménez L. Polyphenols: food sources and bioavailability. Am J Clin Nutr 2004; 79(5): 727-47.
[http://dx.doi.org/10.1093/ajcn/79.5.727] [PMID: 15113710]

[64]   Wohlfart S, Gelperina S, Kreuter J. Transport of drugs across the blood-brain barrier by nanoparticles. J Control Release 2012; 161(2): 264-73.
[http://dx.doi.org/10.1016/j.jconrel.2011.08.017] [PMID: 21872624]

[65]   Kelsey NA, Wilkins HM, Linseman DA. Nutraceutical antioxidants as novel neuroprotective agents. Molecules 2010; 15(11): 7792-814.
[http://dx.doi.org/10.3390/molecules15117792] [PMID: 21060289]

[66]   Babu A, Templeton AK, Munshi A, Ramesh R. Nanoparticle-based drug delivery for therapy of lung cancer: progress and challenges. J Nanomater 2013; 863951.

[67]   Joye IJ, Davidov-Pardo G, McClements DJ. Encapsulation of resveratrol in biopolymer particles produced using liquid antisolvent precipitation. Part 2: Stability and functionality. Food Hydrocoll 2015; 49: 127-34.
[http://dx.doi.org/10.1016/j.foodhyd.2015.02.038]

[68]   Joye IJ, Davidov-Pardo G, McClements DJ. Nanotechnology for increased micronutrient bioavailability. Trends Food Sci Technol 2014; 40: 168-82.
[http://dx.doi.org/10.1016/j.tifs.2014.08.006]

[69]   Malam Y, Loizidou M, Seifalian AM. Liposomes and nanoparticles: nanosized vehicles for drug delivery in cancer. Trends Pharmacol Sci 2009; 30(11): 592-9.
[http://dx.doi.org/10.1016/j.tips.2009.08.004] [PMID: 19837467]

[70]   Dan N. Transport and release in nano-carriers for food applications. J Food Eng 2016; 175: 136-44.
[http://dx.doi.org/10.1016/j.jfoodeng.2015.12.017]

[71]   Kumari A, Yadav SK, Yadav SC. Biodegradable polymeric nanoparticles based drug delivery systems. Colloids Surf B Biointerfaces 2010; 75(1): 1-18.
[http://dx.doi.org/10.1016/j.colsurfb.2009.09.001] [PMID: 19782542]

[72]   Immordino ML, Dosio F, Cattel L. Stealth liposomes: review of the basic science, rationale, and clinical applications, existing and potential. Int J Nanomedicine 2006; 1(3): 297-315.
[PMID: 17717971]

[73]   Karakoti AS, Das S, Thevuthasan S, Seal S. PEGylated inorganic nanoparticles. Angew Chem Int Ed Engl 2011; 50(9): 1980-94.
[http://dx.doi.org/10.1002/anie.201002969] [PMID: 21275011]

[74]   Tiwari PM, Vig K, Dennis VA, Singh SR. Functionalized gold nanoparticles and their biomedical

applications. Nanomaterials (Basel) 2011; 1(1): 31-63.
[http://dx.doi.org/10.3390/nano1010031] [PMID: 28348279]

[75]   Allen TM, Cullis PR. Liposomal drug delivery systems: from concept to clinical applications. Adv Drug Deliv Rev 2013; 65(1): 36-48.
[http://dx.doi.org/10.1016/j.addr.2012.09.037] [PMID: 23036225]

[76]   Salouti M, Ahangari A. Nanoparticle based Drug Delivery Systems for Treatment of Infectious Diseases. Application of Nanotechnology in Drug Delivery. InTech 2014; pp. 155-92.
[http://dx.doi.org/10.5772/58423]

[77]   Ulrich AS. Biophysical aspects of using liposomes as delivery vehicles. Biosci Rep 2002; 22(2): 129-50.
[http://dx.doi.org/10.1023/A:1020178304031] [PMID: 12428898]

[78]   Kean T, Thanou M. Biodegradation, biodistribution and toxicity of chitosan. Adv Drug Deliv Rev 2010; 62(1): 3-11.
[http://dx.doi.org/10.1016/j.addr.2009.09.004] [PMID: 19800377]

[79]   Paques JP, van der Linden E, van Rijn CJM, Sagis LMC. Preparation methods of alginate nanoparticles. Adv Colloid Interface Sci 2014; 209: 163-71.
[http://dx.doi.org/10.1016/j.cis.2014.03.009] [PMID: 24745976]

[80]   Singh RP, Ramarao P. Accumulated polymer degradation products as effector molecules in cytotoxicity of polymeric nanoparticles. Toxicol Sci 2013; 136(1): 131-43.
[http://dx.doi.org/10.1093/toxsci/kft179] [PMID: 23976781]

[81]   Wissing SA, Kayser O, Müller RH. Solid lipid nanoparticles for parenteral drug delivery. Adv Drug Deliv Rev 2004; 56(9): 1257-72.
[http://dx.doi.org/10.1016/j.addr.2003.12.002] [PMID: 15109768]

[82]   Galindo-Rodríguez SA, Puel F, Briançon S, Allémann E, Doelker E, Fessi H. Comparative scale-up of three methods for producing ibuprofen-loaded nanoparticles. Eur J Pharm Sci 2005; 25(4-5): 357-67.
[http://dx.doi.org/10.1016/j.ejps.2005.03.013] [PMID: 15916889]

[83]   Ranjan AP, Mukerjee A, Helson L, Vishwanatha JK. Scale up, optimization and stability analysis of Curcumin C3 complex-loaded nanoparticles for cancer therapy. J Nanobiotechnology 2012; 10: 38.
[http://dx.doi.org/10.1186/1477-3155-10-38] [PMID: 22937885]

[84]   Abbasi E, Aval SF, Akbarzadeh A, *et al.* Dendrimers: synthesis, applications, and properties. Nanoscale Res Lett 2014; 9(1): 247.
[http://dx.doi.org/10.1186/1556-276X-9-247] [PMID: 24994950]

[85]   Kesharwani P, Banerjee S, Gupta U, *et al.* PAMAM dendrimers as promising nanocarriers for RNAi therapeutics. Mater Today 2015; 18: 565-72.
[http://dx.doi.org/10.1016/j.mattod.2015.06.003]

[86]   Majoros IJ, Williams CR, Tomalia DA, Baker JR Jr. New Dendrimers: Synthesis and characterization of POPAM−PAMAM hybrid dendrimers. Macromolecules 2008; 41(22): 8372-9.
[http://dx.doi.org/10.1021/ma801843a] [PMID: 21258604]

[87]   Pearson RM, Bae JW, Hong S. Multifunctional Dendritic Nanocarriers: The Architecture and Applications in Targeted Drug Delivery. Yeo Y, E Nanoparticulate Drug Delivery Systems: Strategies, Technologies, and Applications. New Jersey: John Wiley & Sons, Inc. 2013.
[http://dx.doi.org/10.1002/9780470571224.pse492]

[88]   Murota M, Sato S, Tsubokawa N. Scale-up synthesis of hyperbranched poly(amidoamine)-grafted ultrafine silica using dendrimer synthesis methodology in solvent-free dry-system. Polym Adv Technol 2002; 13: 144-50.
[http://dx.doi.org/10.1002/pat.194]

[89]   Issa B, Obaidat IM, Albiss BA, Haik Y. Magnetic nanoparticles: surface effects and properties related to biomedicine applications. Int J Mol Sci 2013; 14(11): 21266-305.

[http://dx.doi.org/10.3390/ijms141121266] [PMID: 24232575]

[90]   Fong KE, Yung L-YL. Localized surface plasmon resonance: a unique property of plasmonic nanoparticles for nucleic acid detection. Nanoscale 2013; 5(24): 12043-71.
[http://dx.doi.org/10.1039/c3nr02257a] [PMID: 24166199]

[91]   Khlebtsov NG, Dykman LA. Optical properties and biomedical applications of plasmonic nanoparticles. J Quant Spectrosc Radiat Transf 2010; 111: 1-35.
[http://dx.doi.org/10.1016/j.jqsrt.2009.07.012]

[92]   Huang X, El-Sayed MA. Gold nanoparticles: Optical properties and implementations in cancer diagnosis and photothermal therapy. J Adv Res 2010; 1: 13-28.
[http://dx.doi.org/10.1016/j.jare.2010.02.002]

[93]   Ahmed S, Ahmad M, Swami BL, Ikram S. A review on plants extract mediated synthesis of silver nanoparticles for antimicrobial applications: A green expertise. J Adv Res 2016; 7(1): 17-28.
[http://dx.doi.org/10.1016/j.jare.2015.02.007] [PMID: 26843966]

[94]   Bindhu MR, Umadevi M. Antibacterial activities of green synthesized gold nanoparticles. Mater Lett 2014; 120: 122-5.
[http://dx.doi.org/10.1016/j.matlet.2014.01.108]

[95]   Zhang Y, Shareena Dasari TP, Deng H, Yu H. Antimicrobial activity of gold nanoparticles and ionic gold. J Environ Sci Health C Environ Carcinog Ecotoxicol Rev 2015; 33(3): 286-327.
[http://dx.doi.org/10.1080/10590501.2015.1055161] [PMID: 26072980]

[96]   Palza H. Antimicrobial polymers with metal nanoparticles. Int J Mol Sci 2015; 16(1): 2099-116.
[http://dx.doi.org/10.3390/ijms16012099] [PMID: 25607734]

[97]   Wu H, He L, Gao M, Gao S, Liao X, Shi B. One-step *in situ* assembly of size-controlled silver nanoparticles on polyphenol-grafted collagen fiber with enhanced antibacterial properties. New J Chem 2011; 35: 2902-9.
[http://dx.doi.org/10.1039/c1nj20674e]

[98]   Yin PT, Shah S, Chhowalla M, Lee K-B. Design, synthesis, and characterization of graphene-nanoparticle hybrid materials for bioapplications. Chem Rev 2015; 115(7): 2483-531.
[http://dx.doi.org/10.1021/cr500537t] [PMID: 25692385]

[99]   Wu Q-H. Synthesis of graphene nanoparticles and their application in surface Raman enhancement. Spectrosc Lett 2014; 47: 704-9.
[http://dx.doi.org/10.1080/00387010.2013.834936]

[100] Seabra AB, Paula AJ, de Lima R, Alves OL, Durán N. Nanotoxicity of graphene and graphene oxide. Chem Res Toxicol 2014; 27(2): 159-68.
[http://dx.doi.org/10.1021/tx400385x] [PMID: 24422439]

[101] Laouini A, Jaafar-Maalej C, Limayem-Blouza I, Sfar S, Charcosset C, Fessi H. Preparation, characterization and applications of liposomes: state of the art. J Colloid Sci Biotechnol 2012; 1: 147-68.
[http://dx.doi.org/10.1166/jcsb.2012.1020]

[102] Charcosset C, Juban A, Valour J-P, Urbaniak S, Fessi H. Preparation of liposomes at large scale using the ethanol injection method: Effect of scale-up and injection devices. Chem Eng Res Des 2015; 94: 508-15.
[http://dx.doi.org/10.1016/j.cherd.2014.09.008]

[103] Laouini A, Charcosset C, Fessi H, Holdich RG, Vladisavljević GT. Preparation of liposomes: a novel application of microengineered membranes--from laboratory scale to large scale. Colloids Surf B Biointerfaces 2013; 112: 272-8.
[http://dx.doi.org/10.1016/j.colsurfb.2013.07.066] [PMID: 23999143]

[104] Pham TT, Jaafar-Maalej C, Charcosset C, Fessi H. Liposome and niosome preparation using a membrane contactor for scale-up. Colloids Surf B Biointerfaces 2012; 94: 15-21.

[http://dx.doi.org/10.1016/j.colsurfb.2011.12.036] [PMID: 22326648]

[105] Saupe A, Rades T. Solid lipid nanoparticles.Mozafari MR, E Nanocarrier technologies: Frontiers of Nanotherapy. The Netherland: Springer 2006; pp. 41-50.
[http://dx.doi.org/10.1007/978-1-4020-5041-1_3]

[106] Dingler A, Gohla S. Production of solid lipid nanoparticles (SLN): scaling up feasibilities. J Microencapsul 2002; 19(1): 11-6.
[http://dx.doi.org/10.1080/02652040010018056] [PMID: 11811752]

[107] Kakkar V, Kaur IP. Preparation, characterization and scale-up of sesamol loaded solid lipid nanoparticles. Nanotechnology Development 2012; 2: 40-5.
[http://dx.doi.org/10.4081/nd.2012.e8]

[108] Das S, Chaudhury A. Recent advances in lipid nanoparticle formulations with solid matrix for oral drug delivery. AAPS PharmSciTech 2011; 12(1): 62-76.
[http://dx.doi.org/10.1208/s12249-010-9563-0] [PMID: 21174180]

[109] Müller RH, Radtke M, Wissing SA. Nanostructured lipid matrices for improved microencapsulation of drugs. Int J Pharm 2002; 242(1-2): 121-8.
[http://dx.doi.org/10.1016/S0378-5173(02)00180-1] [PMID: 12176234]

[110] Tamjidi F, Shahedi M, Varshosaz J, Nasirpour A. Nanostructured lipid carriers (NLC): A potential delivery system for bioactive food molecules. Innov Food Sci Emerg Technol 2013; 19: 29-43.
[http://dx.doi.org/10.1016/j.ifset.2013.03.002]

[111] Beloqui A, Solinís MÁ, Rodríguez-Gascón A, Almeida AJ, Préat V. Nanostructured lipid carriers: Promising drug delivery systems for future clinics. Nanomedicine (Lond) 2016; 12(1): 143-61.
[http://dx.doi.org/10.1016/j.nano.2015.09.004] [PMID: 26410277]

[112] William RS, Medeiros-Ribeiro G. Chapter 2. Size and shape of epitaxial nanostructures. In: Tsakalakos T, Ovid'ko IA, Vasudevan AK, Eds. Nanostructures: Synthesis, Functional Properties and Application. Springer Science & Business Media 2012; 128.

[113] McCall RL, Sirianni RW. PLGA nanoparticles formed by single- or double-emulsion with vitamin E-TPGS. J Vis Exp 2013; 82(82): 51015.
[PMID: 24429733]

[114] Wang Y, Li P, Truong-Dinh Tran T, Zhang J, Kong L. Manufacturing techniques and surface engineering of polymer based nanoparticles for targeted drug delivery to cancer. Nanomaterials (Basel) 2016; 6(2): 26.
[http://dx.doi.org/10.3390/nano6020026] [PMID: 28344283]

[115] Reis CP, Neufeld RJ, Ribeiro AJ, Veiga F, Nanoencapsulation I. Nanoencapsulation I. Methods for preparation of drug-loaded polymeric nanoparticles. Nanomedicine (Lond) 2006; 2(1): 8-21.
[http://dx.doi.org/10.1016/j.nano.2005.12.003] [PMID: 17292111]

[116] Koukaras EN, Papadimitriou SA, Bikiaris DN, Froudakis GE. Insight on the formation of chitosan nanoparticles through ionotropic gelation with tripolyphosphate. Mol Pharm 2012; 9(10): 2856-62.
[http://dx.doi.org/10.1021/mp300162j] [PMID: 22845012]

[117] Mitra A, Dey B. Chitosan microspheres in novel drug delivery systems. Indian J Pharm Sci 2011; 73(4): 355-66.
[PMID: 22707817]

[118] Ingale AG, Chaudhari AN. Biogenic synthesis of nanoparticles and potential applications: An eco-friendly approach. J Nanomed Nanotechnol 2013; 4: 165.
[http://dx.doi.org/10.4172/2157-7439.1000165]

[119] Laouini A, Jaafar-Maalej C, Limayem-Blouza I, Sfar S, Charcosset C, Fessi H. Preparation, characterization and applications of liposomes: State of the art. Journal of Colloid Science and Biotechnology 2012; 1: 147-68.
[http://dx.doi.org/10.1166/jcsb.2012.1020]

[120] Abdolahad M, Janmaleki M, Mohajerzadeh S, Akhavan O, Abbasi S. Polyphenols attached graphene nanosheets for high efficiency NIR mediated photodestruction of cancer cells. Mater Sci Eng C 2013; 33(3): 1498-505.
[http://dx.doi.org/10.1016/j.msec.2012.12.052] [PMID: 23827601]

[121] Wang W, Zhang L, Le Y, Chen J-F, Wang J, Yun J. Synergistic effect of PEGylated resveratrol on delivery of anticancer drugs. Int J Pharm 2016; 498(1-2): 134-41.
[http://dx.doi.org/10.1016/j.ijpharm.2015.12.016] [PMID: 26685725]

[122] Penalva R, Esparza I, Larraneta E, González-Navarro CJ, Gamazo C, Irache JM. Gonz⊐lez-Navarro CJ, Gamazo C, Irache JM. Zein-based nanoparticles improve the oral bioavailability of resveratrol and its anti-inflammatory effects in a mouse model of endotoxic shock. J Agric Food Chem 2015; 63(23): 5603-11.
[http://dx.doi.org/10.1021/jf505694e] [PMID: 26027429]

[123] Davidov-Pardo G, Joye IJ, McClements DJ. Encapsulation of resveratrol in biopolymer particles produced using liquid antisolvent precipitation. Part 1: Preparation and characterization. Food Hydrocoll 2015; 45: 309-16.
[http://dx.doi.org/10.1016/j.foodhyd.2014.11.023]

[124] Pandita D, Kumar S, Poonia N, Lather V. Solid lipid nanoparticles enhance oral bioavailability of resveratrol, a natural polyphenol. Food Res Int 2014; 62: 1165-74.
[http://dx.doi.org/10.1016/j.foodres.2014.05.059]

[125] Neves AR, Lúcio M, Martins S, Lima JL, Reis S. Novel resveratrol nanodelivery systems based on lipid nanoparticles to enhance its oral bioavailability. Int J Nanomedicine 2013; 8: 177-87.
[PMID: 23326193]

[126] Sanna V, Roggio AM, Siliani S, *et al.* Development of novel cationic chitosan-and anionic alginate-coated poly(D,L-lactide-co-glycolide) nanoparticles for controlled release and light protection of resveratrol. Int J Nanomedicine 2012; 7: 5501-16.
[PMID: 23093904]

[127] Ramalingam P, Ko YT. Improved oral delivery of resveratrol from N-trimethyl chitosan-g-palmitic acid surface-modified solid lipid nanoparticles. Colloids Surf B Biointerfaces 2016; 139: 52-61.
[http://dx.doi.org/10.1016/j.colsurfb.2015.11.050] [PMID: 26700233]

[128] Kumar S, Lather V, Pandita D. A facile green approach to prepare core-shell hybrid PLGA nanoparticles for resveratrol delivery. Int J Biol Macromol 2016; 84: 380-4.
[http://dx.doi.org/10.1016/j.ijbiomac.2015.12.036] [PMID: 26708438]

[129] Soo E, Thakur S, Qu Z, Jambhrunkar S, Parekh HS, Popat A. Enhancing delivery and cytotoxicity of resveratrol through a dual nanoencapsulation approach. J Colloid Interface Sci 2016; 462: 368-74.
[http://dx.doi.org/10.1016/j.jcis.2015.10.022] [PMID: 26479200]

[130] Park S, Cha S-H, Cho I, *et al.* Antibacterial nanocarriers of resveratrol with gold and silver nanoparticles. Mater Sci Eng C 2016; 58: 1160-9.
[http://dx.doi.org/10.1016/j.msec.2015.09.068] [PMID: 26478416]

[131] Zhang J, Chen Y, Li D, Cao Y, Wang Z, Li G. Colorimetric determination of islet amyloid polypeptide fibrils and their inhibitors using resveratrol functionalized gold nanoparticles. Mikrochim Acta 2016; 183: 659-65.
[http://dx.doi.org/10.1007/s00604-015-1687-1]

[132] Lindner GdaR, Dalmolin LF, Khalil NM, Mainardes RM. Influence of the formulation parameters on the particle size and encapsulation efficiency of resveratrol in PLA and PLA-PEG blend nanoparticles: a factorial design. J Nanosci Nanotechnol 2015; 15(12): 10173-82.
[http://dx.doi.org/10.1166/jnn.2015.11687] [PMID: 26682464]

[133] Kumar S, Sangwan P, Lather V, Pandita D. Biocompatible PLGA-oil hybrid nanoparticles for high loading and controlled delivery of resveratrol. J Drug Deliv Sci Technol 2015; 30: 54-62.

[http://dx.doi.org/10.1016/j.jddst.2015.09.016]

[134] Tomoaia G, Horovitz O, Mocanu A, *et al.* Effects of doxorubicin mediated by gold nanoparticles and resveratrol in two human cervical tumor cell lines. Colloids Surf B Biointerfaces 2015; 135: 726-34.
[http://dx.doi.org/10.1016/j.colsurfb.2015.08.036] [PMID: 26340362]

[135] Neves AR, Reis S, Segundo MA. Development and validation of a HPLC method using a monolithic column for quantification of *trans*-resveratrol in lipid nanoparticles for intestinal permeability studies. J Agric Food Chem 2015; 63(12): 3114-20.
[http://dx.doi.org/10.1021/acs.jafc.5b00390] [PMID: 25764378]

[136] Jung K-H, Lee JH, Park JW, *et al.* Resveratrol-loaded polymeric nanoparticles suppress glucose metabolism and tumor growth *in vitro* and *in vivo.* Int J Pharm 2015; 478(1): 251-7.
[http://dx.doi.org/10.1016/j.ijpharm.2014.11.049] [PMID: 25445992]

[137] Karthikeyan S, Hoti SL, Prasad NR. Resveratrol loaded gelatin nanoparticles synergistically inhibits cell cycle progression and constitutive NF-kappaB activation, and induces apoptosis in non-small cell lung cancer cells. Biomed Pharmacother 2015; 70: 274-82.
[http://dx.doi.org/10.1016/j.biopha.2015.02.006] [PMID: 25776512]

[138] Hailun L, Yu H, Yong X, *et al.* The protective effects and mechanism of resveratrol-loaded nanoparticles on HK-2 cells suffering from hypoxia-reoxygenayion. Curr Signal Transduct Ther 2015; 10: 104-11.
[http://dx.doi.org/10.2174/1574362410666150625190132]

[139] Guo L, Peng Y, Li Y, *et al.* Cell death pathway induced by resveratrol-bovine serum albumin nanoparticles in a human ovarian cell line. Oncol Lett 2015; 9(3): 1359-63.
[http://dx.doi.org/10.3892/ol.2015.2851] [PMID: 25663913]

[140] Hardy N, Viola HM, Johnstone VPA, *et al.* Nanoparticle-mediated dual delivery of an antioxidant and a peptide against the L-Type Ca^{2+} channel enables simultaneous reduction of cardiac ischemia-reperfusion injury. ACS Nano 2015; 9(1): 279-89.
[http://dx.doi.org/10.1021/nn5061404] [PMID: 25493575]

[141] Pangeni R, Sharma S, Mustafa G, Ali J, Baboota S. Vitamin E loaded resveratrol nanoemulsion for brain targeting for the treatment of Parkinson's disease by reducing oxidative stress. Nanotechnology 2014; 25(48): 485102.
[http://dx.doi.org/10.1088/0957-4484/25/48/485102] [PMID: 25392203]

[142] Jose S, Anju SS, Cinu TA, Aleykutty NA, Thomas S, Souto EB. *In vivo* pharmacokinetics and biodistribution of resveratrol-loaded solid lipid nanoparticles for brain delivery. Int J Pharm 2014; 474(1-2): 6-13.
[http://dx.doi.org/10.1016/j.ijpharm.2014.08.003] [PMID: 25102112]

[143] Coradini K, Lima FO, Oliveira CM, *et al.* Co-encapsulation of resveratrol and curcumin in lipid-core nanocapsules improves their *in vitro* antioxidant effects. Eur J Pharm Biopharm 2014; 88(1): 178-85.
[http://dx.doi.org/10.1016/j.ejpb.2014.04.009] [PMID: 24780440]

[144] Mohanty RK, Thennarasu S, Mandal AB. Resveratrol stabilized gold nanoparticles enable surface loading of doxorubicin and anticancer activity. Colloids Surf B Biointerfaces 2014; 114: 138-43.
[http://dx.doi.org/10.1016/j.colsurfb.2013.09.057] [PMID: 24176891]

[145] Yin H, Si J, Xu H, *et al.* Resveratrol-loaded nanoparticles reduce oxidative stress induced by radiation or amyloid-beta in transgenic Caenorhabditis elegans. J Biomed Nanotechnol 2014; 10(8): 1536-44.
[http://dx.doi.org/10.1166/jbn.2014.1897] [PMID: 25016653]

[146] Lozano-Pérez AA, Rodriguez-Nogales A, Ortiz-Cullera V, *et al.* Silk fibroin nanoparticles constitute a vector for controlled release of resveratrol in an experimental model of inflammatory bowel disease in rats. Int J Nanomedicine 2014; 9: 4507-20.
[PMID: 25285004]

[147] Sun C, Qian Q, Yu H, Chen L, Yang X, Zhou M, *et al.* Incorporation of water-insoluble resveratrol

into nanoscale poly(lactic-co-glycolic acid) matrix and their bio-properties. Sci Adv Mater 2014; 6: 1900-6.
[http://dx.doi.org/10.1166/sam.2014.1917]

[148]   Singh G, Pai RS. *In-vitro/in-vivo* characterization of *trans*-resveratrol-loaded nanoparticulate drug delivery system for oral administration. J Pharm Pharmacol 2014; 66(8): 1062-76.
[PMID: 24779896]

[149]   Singh G, Pai RS. Optimized PLGA nanoparticle platform for orally dosed *trans*-resveratrol with enhanced bioavailability potential. Expert Opin Drug Deliv 2014; 11(5): 647-59.
[http://dx.doi.org/10.1517/17425247.2014.890588] [PMID: 24661109]

[150]   Sanna V, Pala N, Dessì G, *et al.* Single-step green synthesis and characterization of gold-conjugated polyphenol nanoparticles with antioxidant and biological activities. Int J Nanomedicine 2014; 9: 4935-51.
[PMID: 25364251]

[151]   Hao J, Gao Y, Zhao J, *et al.* Preparation and optimization of resveratrol nanosuspensions by antisolvent precipitation using Box-Behnken design. AAPS PharmSciTech 2015; 16(1): 118-28.
[http://dx.doi.org/10.1208/s12249-014-0211-y] [PMID: 25209687]

[152]   Mohan A, Narayanan S, Sethuraman S, Krishnan UM. Novel resveratrol and 5-fluorouracil coencapsulated in PEGylated nanoliposomes improve chemotherapeutic efficacy of combination against head and neck squamous cell carcinoma. BioMed Res Int 2014; 2014: 424239.
[http://dx.doi.org/10.1155/2014/424239] [PMID: 25114900]

[153]   Carlson LJ, Cote B, Alani AWG, Rao DA. Polymeric micellar co-delivery of resveratrol and curcumin to mitigate *in vitro* doxorubicin-induced cardiotoxicity. J Pharm Sci 2014; 103(8): 2315-22.
[http://dx.doi.org/10.1002/jps.24042] [PMID: 24914015]

[154]   Sun R, Zhao G, Ni S, Xia Q. Lipid based nanocarriers with different lipid compositions for topical delivery of resveratrol: comparative analysis of characteristics and performance. J Drug Deliv Sci Technol 2014; 24: 591-600.
[http://dx.doi.org/10.1016/S1773-2247(14)50124-4]

[155]   Kamath MS, Ahmed SS, Dhanasekaran M, Santosh SW. Polycaprolactone scaffold engineered for sustained release of resveratrol: therapeutic enhancement in bone tissue engineering. Int J Nanomedicine 2014; 9: 183-95.
[PMID: 24399875]

[156]   Guo W, Li A, Jia Z, Yuan Y, Dai H, Li H. Transferrin modified PEG-PLA-resveratrol conjugates: *in vitro* and *in vivo* studies for glioma. Eur J Pharmacol 2013; 718(1-3): 41-7.
[http://dx.doi.org/10.1016/j.ejphar.2013.09.034] [PMID: 24070814]

[157]   Sanna V, Siddiqui IA, Sechi M, Mukhtar H. Resveratrol-loaded nanoparticles based on poly(epsilon-caprolactone) and poly(D,L-lactic-co-glycolic acid)-poly(ethylene glycol) blend for prostate cancer treatment. Mol Pharm 2013; 10(10): 3871-81.
[http://dx.doi.org/10.1021/mp400342f] [PMID: 23968375]

[158]   Pando D, Caddeo C, Manconi M, Fadda AM, Pazos C. Nanodesign of olein vesicles for the topical delivery of the antioxidant resveratrol. J Pharm Pharmacol 2013; 65(8): 1158-67.
[http://dx.doi.org/10.1111/jphp.12093] [PMID: 23837583]

[159]   Bu L, Gan L-C, Guo X-Q, *et al. Tran*s-resveratrol loaded chitosan nanoparticles modified with biotin and avidin to target hepatic carcinoma. Int J Pharm 2013; 452(1-2): 355-62.
[http://dx.doi.org/10.1016/j.ijpharm.2013.05.007] [PMID: 23685116]

[160]   Figueiró F, Bernardi A, Frozza RL, *et al.* Resveratrol-loaded lipid-core nanocapsules treatment reduces *in vitro* and *in vivo* glioma growth. J Biomed Nanotechnol 2013; 9(3): 516-26.
[http://dx.doi.org/10.1166/jbn.2013.1547] [PMID: 23621009]

[161]   Lu X, Xu H, Sun B, Zhu Z, Zheng D, Li X. Enhanced neuroprotective effects of resveratrol delivered

by nanoparticles on hydrogen peroxide-induced oxidative stress in rat cortical cell culture. Mol Pharm 2013; 10(5): 2045-53.
[http://dx.doi.org/10.1021/mp400056c] [PMID: 23534345]

[162]   Karthikeyan S, Rajendra Prasad N, Ganamani A, Balamurugan E. Anticancer activity of resveratrol-loaded gelatin nanoparticles on NCI-H460 non-small cell lung cancer cells. Biomed Prev Nutr 2013; 3: 64-73.
[http://dx.doi.org/10.1016/j.bionut.2012.10.009]

[163]   Gokce EH, Korkmaz E, Dellera E, Sandri G, Bonferoni MC, Ozer O. Resveratrol-loaded solid lipid nanoparticles *versus* nanostructured lipid carriers: evaluation of antioxidant potential for dermal applications. Int J Nanomedicine 2012; 7: 1841-50.
[http://dx.doi.org/10.2147/IJN.S29710] [PMID: 22605933]

[164]   Carlotti ME, Sapino S, Ugazio E, Gallarate M, Morel S. Resveratrol in solid lipid nanoparticles. J Dispers Sci Technol 2012; 33: 465-71.
[http://dx.doi.org/10.1080/01932691.2010.548274]

[165]   Kim S, Ng WK, Dong Y, Das S, Tan RBH. Preparation and physicochemical characterization of *trans*-resveratrol nanoparticles by temperature-controlled antisolvent precipitation. J Food Eng 2012; 108: 37-42.
[http://dx.doi.org/10.1016/j.jfoodeng.2011.07.034]

[166]   Zhang QH, Xiong QP, Shi YY, Zhang DY. Study on preparation and characterization of resveratrol solid lipid nanoparticles and its anticancer effects *in vitro*. Zhong Yao Cai 2010; 33(12): 1929-32.
[PMID: 21548373]

[167]   Shao J, Li X, Lu X, *et al.* Enhanced growth inhibition effect of resveratrol incorporated into biodegradable nanoparticles against glioma cells is mediated by the induction of intracellular reactive oxygen species levels. Colloids Surf B Biointerfaces 2009; 72(1): 40-7.
[http://dx.doi.org/10.1016/j.colsurfb.2009.03.010] [PMID: 19395246]

[168]   Lu X, Ji C, Xu H, *et al.* Resveratrol-loaded polymeric micelles protect cells from Abeta-induced oxidative stress. Int J Pharm 2009; 375(1-2): 89-96.
[http://dx.doi.org/10.1016/j.ijpharm.2009.03.021] [PMID: 19481694]

[169]   Guo L-Y, Yao J-P, Sui L-H. Preparation and effects of resveratrol bovine serum albumin nanoparticles on proliferation of human ovarian carcinoma cell SKOV3. Chem J Chin Univ 2009; 30: 474-7.

[170]   Kim B-K, Lee J-S, Oh J-K, Park D-J. Preparation of resveratrol-loaded poly(ε-caprolactone) nanoparticles by oil-in-water emulsion solvent evaporation method. Food Sci Biotechnol 2009; 18: 157-61.

[171]   Sanna V, Lubinu G, Madau P, *et al.* Polymeric nanoparticles encapsulating white tea extract for nutraceutical application. J Agric Food Chem 2015; 63(7): 2026-32.
[http://dx.doi.org/10.1021/jf505850q] [PMID: 25599125]

[172]   Lu Y-C, Luo P-C, Huang C-W, *et al.* Augmented cellular uptake of nanoparticles using tea catechins: effect of surface modification on nanoparticle-cell interaction. Nanoscale 2014; 6(17): 10297-306.
[http://dx.doi.org/10.1039/C4NR00617H] [PMID: 25069428]

[173]   Haratifar S, Meckling KA, Corredig M. Antiproliferative activity of tea catechins associated with casein micelles, using HT29 colon cancer cells. J Dairy Sci 2014; 97(2): 672-8.
[http://dx.doi.org/10.3168/jds.2013-7263] [PMID: 24359816]

[174]   Tang D-W, Yu S-H, Ho Y-C, Huang B-Q, Tsai G-J, Hsieh H-Y, *et al.* Characterization of tea catechins-loaded nanoparticles prepared from chitosan and an edible polypeptide. Food Hydrocoll 2013; 30: 33-41.
[http://dx.doi.org/10.1016/j.foodhyd.2012.04.014]

[175]   Wisuitiprot W, Somsiri A, Ingkaninan K, Waranuch N. A novel technique for chitosan microparticle preparation using a water/silicone emulsion: green tea model. Int J Cosmet Sci 2011; 33(4): 351-8.

[http://dx.doi.org/10.1111/j.1468-2494.2010.00635.x] [PMID: 21323933]

[176] Rocha S, Generalov R, Pereira MdoC, Peres I, Juzenas P, Coelho MA. Epigallocatechin gallate-loaded polysaccharide nanoparticles for prostate cancer chemoprevention. Nanomedicine (Lond) 2011; 6(1): 79-87.
[http://dx.doi.org/10.2217/nnm.10.101] [PMID: 21182420]

[177] Dube A, Nicolazzo JA, Larson I. Chitosan nanoparticles enhance the intestinal absorption of the green tea catechins (+)-catechin and (-)-epigallocatechin gallate. Eur J Pharm Sci 2010; 41(2): 219-25.
[http://dx.doi.org/10.1016/j.ejps.2010.06.010] [PMID: 20600878]

[178] Chen Y-C, Yu S-H, Tsai G-J, Tang D-W, Mi F-L, Peng Y-P. Novel technology for the preparation of self-assembled catechin/gelatin nanoparticles and their characterization. J Agric Food Chem 2010; 58(11): 6728-34.
[http://dx.doi.org/10.1021/jf1005116] [PMID: 20476739]

[179] Hu B, Pan C, Sun Y, Hou Z, Ye H, Zeng X. Optimization of fabrication parameters to produce chitosan-tripolyphosphate nanoparticles for delivery of tea catechins. J Agric Food Chem 2008; 56(16): 7451-8.
[http://dx.doi.org/10.1021/jf801111c] [PMID: 18627163]

[180] Bisht S, Feldmann G, Soni S, *et al.* Polymeric nanoparticle-encapsulated curcumin ("nanocurcumin"): a novel strategy for human cancer therapy. J Nanobiotechnology 2007; 5: 3.
[http://dx.doi.org/10.1186/1477-3155-5-3] [PMID: 17439648]

[181] Tiyaboonchai W, Tungpradit W, Plianbangchang P. Formulation and characterization of curcuminoids loaded solid lipid nanoparticles. Int J Pharm 2007; 337(1-2): 299-306.
[http://dx.doi.org/10.1016/j.ijpharm.2006.12.043] [PMID: 17287099]

[182] Das RK, Kasoju N, Bora U. Encapsulation of curcumin in alginate-chitosan-pluronic composite nanoparticles for delivery to cancer cells. Nanomedicine (Lond) 2010; 6(1): 153-60.
[http://dx.doi.org/10.1016/j.nano.2009.05.009] [PMID: 19616123]

[183] Sun D, Zhuang X, Xiang X, *et al.* A novel nanoparticle drug delivery system: the anti-inflammatory activity of curcumin is enhanced when encapsulated in exosomes. Mol Ther 2010; 18(9): 1606-14.
[http://dx.doi.org/10.1038/mt.2010.105] [PMID: 20571541]

[184] Sahu A, Kasoju N, Bora U. Fluorescence study of the curcumin-casein micelle complexation and its application as a drug nanocarrier to cancer cells. Biomacromolecules 2008; 9(10): 2905-12.
[http://dx.doi.org/10.1021/bm800683f] [PMID: 18785706]

[185] Yallapu MM, Gupta BK, Jaggi M, Chauhan SC. Fabrication of curcumin encapsulated PLGA nanoparticles for improved therapeutic effects in metastatic cancer cells. J Colloid Interface Sci 2010; 351(1): 19-29.
[http://dx.doi.org/10.1016/j.jcis.2010.05.022] [PMID: 20627257]

[186] Bhawana , Basniwal RK, Buttar HS, Jain VK, Jain N. Curcumin nanoparticles: preparation, characterization, and antimicrobial study. J Agric Food Chem 2011; 59(5): 2056-61.
[http://dx.doi.org/10.1021/jf104402t] [PMID: 21322563]

[187] Tsai Y-M, Chien C-F, Lin L-C, Tsai T-H. Curcumin and its nano-formulation: the kinetics of tissue distribution and blood-brain barrier penetration. Int J Pharm 2011; 416(1): 331-8.
[http://dx.doi.org/10.1016/j.ijpharm.2011.06.030] [PMID: 21729743]

[188] Anitha A, Deepagan VG, Divya Rani VV, Menon D, Nair SV, Jayakumar R. Preparation, characterization, *in vitro* drug release and biological studies of curcumin loaded dextran sulphate–chitosan nanoparticles. Carbohydr Polym 2011; 84: 1158-64.
[http://dx.doi.org/10.1016/j.carbpol.2011.01.005]

[189] Sou K, Inenaga S, Takeoka S, Tsuchida E. Loading of curcumin into macrophages using lipid-based nanoparticles. Int J Pharm 2008; 352(1-2): 287-93.
[http://dx.doi.org/10.1016/j.ijpharm.2007.10.033] [PMID: 18063327]

[190] Lim KJ, Bisht S, Bar EE, Maitra A, Eberhart CG. A polymeric nanoparticle formulation of curcumin inhibits growth, clonogenicity and stem-like fraction in malignant brain tumors. Cancer Biol Ther 2011; 11(5): 464-73.
[http://dx.doi.org/10.4161/cbt.11.5.14410] [PMID: 21193839]

[191] Kakkar V, Singh S, Singla D, Kaur IP. Exploring solid lipid nanoparticles to enhance the oral bioavailability of curcumin. Mol Nutr Food Res 2011; 55(3): 495-503.
[http://dx.doi.org/10.1002/mnfr.201000310] [PMID: 20938993]

[192] Mourtas S, Canovi M, Zona C, *et al.* Curcumin-decorated nanoliposomes with very high affinity for amyloid-β1-42 peptide. Biomaterials 2011; 32(6): 1635-45.
[http://dx.doi.org/10.1016/j.biomaterials.2010.10.027] [PMID: 21131044]

[193] Misra R, Sahoo SK. Coformulation of doxorubicin and curcumin in poly(D,L-lactide-co-glycolide) nanoparticles suppresses the development of multidrug resistance in K562 cells. Mol Pharm 2011; 8(3): 852-66.
[http://dx.doi.org/10.1021/mp100455h] [PMID: 21480667]

[194] Mulik RS, Mönkkönen J, Juvonen RO, Mahadik KR, Paradkar AR. Transferrin mediated solid lipid nanoparticles containing curcumin: enhanced *in vitro* anticancer activity by induction of apoptosis. Int J Pharm 2010; 398(1-2): 190-203.
[http://dx.doi.org/10.1016/j.ijpharm.2010.07.021] [PMID: 20655375]

[195] Dhule SS, Penfornis P, Frazier T, *et al.* Curcumin-loaded γ-cyclodextrin liposomal nanoparticles as delivery vehicles for osteosarcoma. Nanomedicine (Lond) 2012; 8(4): 440-51.
[http://dx.doi.org/10.1016/j.nano.2011.07.011] [PMID: 21839055]

[196] Nayak AP, Tiyaboonchai W, Patankar S, Madhusudhan B, Souto EB. Curcuminoids-loaded lipid nanoparticles: novel approach towards malaria treatment. Colloids Surf B Biointerfaces 2010; 81(1): 263-73.
[http://dx.doi.org/10.1016/j.colsurfb.2010.07.020] [PMID: 20688493]

[197] Mathew A, Fukuda T, Nagaoka Y, *et al.* Curcumin loaded-PLGA nanoparticles conjugated with Tet-1 peptide for potential use in Alzheimer's disease. PLoS One 2012; 7(3): e32616.
[http://dx.doi.org/10.1371/journal.pone.0032616] [PMID: 22403681]

[198] Yen F-L, Wu T-H, Tzeng C-W, Lin L-T, Lin C-C. Curcumin nanoparticles improve the physicochemical properties of curcumin and effectively enhance its antioxidant and antihepatoma activities. J Agric Food Chem 2010; 58(12): 7376-82.
[http://dx.doi.org/10.1021/jf100135h] [PMID: 20486686]

[199] Akhtar F, Rizvi MMA, Kar SK. Oral delivery of curcumin bound to chitosan nanoparticles cured *Plasmodium yoelii* infected mice. Biotechnol Adv 2012; 30(1): 310-20.
[http://dx.doi.org/10.1016/j.biotechadv.2011.05.009] [PMID: 21619927]

[200] Manju S, Sreenivasan K. Gold nanoparticles generated and stabilized by water soluble curcumin-polymer conjugate: blood compatibility evaluation and targeted drug delivery onto cancer cells. J Colloid Interface Sci 2012; 368(1): 144-51.
[http://dx.doi.org/10.1016/j.jcis.2011.11.024] [PMID: 22200330]

[201] Re F, Cambianica I, Zona C, *et al.* Functionalization of liposomes with ApoE-derived peptides at different density affects cellular uptake and drug transport across a blood-brain barrier model. Nanomedicine (Lond) 2011; 7(5): 551-9.
[http://dx.doi.org/10.1016/j.nano.2011.05.004] [PMID: 21658472]

[202] Taylor M, Moore S, Mourtas S, *et al.* Effect of curcumin-associated and lipid ligand-functionalized nanoliposomes on aggregation of the Alzheimer's Aβ peptide. Nanomedicine (Lond) 2011; 7(5): 541-50.
[http://dx.doi.org/10.1016/j.nano.2011.06.015] [PMID: 21722618]

[203] Sun J, Bi C, Chan HM, Sun S, Zhang Q, Zheng Y. Curcumin-loaded solid lipid nanoparticles have

prolonged *in vitro* antitumour activity, cellular uptake and improved *in vivo* bioavailability. Colloids Surf B Biointerfaces 2013; 111: 367-75.
[http://dx.doi.org/10.1016/j.colsurfb.2013.06.032] [PMID: 23856543]

[204] Rejinold NS, Sreerekha PR, Chennazhi KP, Nair SV, Jayakumar R. Biocompatible, biodegradable and thermo-sensitive chitosan-g-poly (N-isopropylacrylamide) nanocarrier for curcumin drug delivery. Int J Biol Macromol 2011; 49(2): 161-72.
[http://dx.doi.org/10.1016/j.ijbiomac.2011.04.008] [PMID: 21536066]

[205] Basnet P, Hussain H, Tho I, Skalko-Basnet N. Liposomal delivery system enhances anti-inflammatory properties of curcumin. J Pharm Sci 2012; 101(2): 598-609.
[http://dx.doi.org/10.1002/jps.22785] [PMID: 21989712]

[206] Lazar AN, Mourtas S, Youssef I, *et al.* Curcumin-conjugated nanoliposomes with high affinity for Aβ deposits: possible applications to Alzheimer disease. Nanomedicine (Lond) 2013; 9(5): 712-21.
[http://dx.doi.org/10.1016/j.nano.2012.11.004] [PMID: 23220328]

[207] Mourtas S, Lazar AN, Markoutsa E, Duyckaerts C, Antimisiaris SG. Multifunctional nanoliposomes with curcumin-lipid derivative and brain targeting functionality with potential applications for Alzheimer disease. Eur J Med Chem 2014; 80: 175-83.
[http://dx.doi.org/10.1016/j.ejmech.2014.04.050] [PMID: 24780594]

[208] Shao J, Zheng D, Jiang Z, *et al.* Curcumin delivery by methoxy polyethylene glycol-poly(caprolactone) nanoparticles inhibits the growth of C6 glioma cells. Acta Biochim Biophys Sin (Shanghai) 2011; 43(4): 267-74.
[http://dx.doi.org/10.1093/abbs/gmr011] [PMID: 21349881]

[209] Gangwar RK, Dhumale VA, Kumari D, Nakate UT, Gosavi SW, Sharma RB, *et al.* Conjugation of curcumin with PVP capped gold nanoparticles for improving bioavailability. Mater Sci Eng C 2012; 32: 2659-63.
[http://dx.doi.org/10.1016/j.msec.2012.07.022]

[210] Li X, Nan K, Li L, Zhang Z, Chen H. *In vivo* evaluation of curcumin nanoformulation loaded methoxy poly(ethylene glycol)-graft-chitosan composite film for wound healing application. Carbohydr Polym 2012; 88: 84-90.
[http://dx.doi.org/10.1016/j.carbpol.2011.11.068]

[211] Hasan M, Belhaj N, Benachour H, *et al.* Liposome encapsulation of curcumin: physico-chemical characterizations and effects on MCF7 cancer cell proliferation. Int J Pharm 2014; 461(1-2): 519-28.
[http://dx.doi.org/10.1016/j.ijpharm.2013.12.007] [PMID: 24355620]

[212] Mulik RS, Mönkkönen J, Juvonen RO, Mahadik KR, Paradkar AR. ApoE3 mediated polymeric nanoparticles containing curcumin: apoptosis induced *in vitro* anticancer activity against neuroblastoma cells. Int J Pharm 2012; 437(1-2): 29-41.
[http://dx.doi.org/10.1016/j.ijpharm.2012.07.062] [PMID: 22890189]

[213] Ma'mani L, Nikzad S, Kheiri-Manjili H, *et al.* Curcumin-loaded guanidine functionalized PEGylated I3ad mesoporous silica nanoparticles KIT-6: practical strategy for the breast cancer therapy. Eur J Med Chem 2014; 83: 646-54.
[http://dx.doi.org/10.1016/j.ejmech.2014.06.069] [PMID: 25014638]

[214] Dinesh Kumar V, Verma PRP, Singh SK. Morphological and *in vitro* antibacterial efficacy of quercetin loaded nanoparticles against food-borne microorganisms. LWT 2016; 66: 638-50.
[http://dx.doi.org/10.1016/j.lwt.2015.11.004]

[215] Minaei A, Sabzichi M, Ramezani F, Hamishehkar H, Samadi N. Co-delivery with nano-quercetin enhances doxorubicin-mediated cytotoxicity against MCF-7 cells. Mol Biol Rep 2016; 43(2): 99-105.
[http://dx.doi.org/10.1007/s11033-016-3942-x] [PMID: 26748999]

[216] Sun D, Li N, Zhang W, Yang E, Mou Z, Zhao Z, *et al.* Quercetin-loaded PLGA nanoparticles: a highly effective antibacterial agent *in vitro* and anti-infection application *in vivo*. J Nanopart Res 2016; 18: 3.
[http://dx.doi.org/10.1007/s11051-015-3310-0]

[217]   Qi Y, Jiang M, Cui Y-L, Zhao L, Zhou X. Synthesis of quercetin loaded nanoparticles based on alginate for Pb(II) adsorption in aqueous solution. Nanoscale Res Lett 2015; 10(1): 408.
[http://dx.doi.org/10.1186/s11671-015-1117-7] [PMID: 26474889]

[218]   Bagad M, Khan ZA. Poly(n-butylcyanoacrylate) nanoparticles for oral delivery of quercetin: preparation, characterization, and pharmacokinetics and biodistribution studies in Wistar rats. Int J Nanomedicine 2015; 10: 3921-35.
[PMID: 26089668]

[219]   Lai F, Franceschini I, Corrias F, *et al.* Maltodextrin fast dissolving films for quercetin nanocrystal delivery. A feasibility study. Carbohydr Polym 2015; 121: 217-23.
[http://dx.doi.org/10.1016/j.carbpol.2014.11.070] [PMID: 25659692]

[220]   Pandey SK, Patel DK, Thakur R, Mishra DP, Maiti P, Haldar C. Anti-cancer evaluation of quercetin embedded PLA nanoparticles synthesized by emulsified nanoprecipitation. Int J Biol Macromol 2015; 75: 521-9.
[http://dx.doi.org/10.1016/j.ijbiomac.2015.02.011] [PMID: 25701491]

[221]   Linkevičiūtė A, Misiūnas A, Naujalis E, Barauskas J. Preparation and characterization of quercetin-loaded lipid liquid crystalline systems. Colloids Surf B Biointerfaces 2015; 128: 296-303.
[http://dx.doi.org/10.1016/j.colsurfb.2015.02.001] [PMID: 25701115]

[222]   Sambandam B, Kumar SS, Ayyaswamy A, Yadav Bv N, Thiyagarajan D. Synthesis and characterization of poly D-L lactide (PLA) nanoparticles for the delivery of quercetin. Int J Pharm Pharm Sci 2015; 7: 42-9.

[223]   Ding H, Zhu L, Wei X-K, Shen Q. Solid lipid nanoparticles of quercetin (a flavonoid) in recovery of motor function after spinal injuries. J Biomater Tissue Eng 2015; 5: 509-13.
[http://dx.doi.org/10.1166/jbt.2015.1337]

[224]   Sharma G, Park J, Sharma AR, *et al.* Methoxy poly(ethylene glycol)-poly(lactide) nanoparticles encapsulating quercetin act as an effective anticancer agent by inducing apoptosis in breast cancer. Pharm Res 2015; 32(2): 723-35.
[http://dx.doi.org/10.1007/s11095-014-1504-2] [PMID: 25186442]

[225]   Bishayee K, Khuda-Bukhsh AR, Huh SO. PLGA-loaded gold-nanoparticles precipitated with quercetin downregulate HDAC-Akt activities controlling proliferation and activate p53-ROS crosstalk to induce apoptosis in hepatocarcinoma cells. Mol Cells 2015; 38(6): 518-27.
[http://dx.doi.org/10.14348/molcells.2015.2339] [PMID: 25947292]

[226]   Abdel-Wahhab MA, Aljawish A, El-Nekeety AA, *et al.* Chitosan nanoparticles and quercetin modulate gene expression and prevent the genotoxicity of aflatoxin B_1 in rat liver. Toxicol Rep 2015; 2: 737-47.
[http://dx.doi.org/10.1016/j.toxrep.2015.05.007] [PMID: 28962409]

[227]   González-Esquivel AE, Charles-Niño CL, Pacheco-Moisés FP, Ortiz GG, Jaramillo-Juárez F, Rincón-Sánchez AR. Beneficial effects of quercetin on oxidative stress in liver and kidney induced by titanium dioxide (TiO2) nanoparticles in rats. Toxicol Mech Methods 2015; 25(3): 166-75.
[http://dx.doi.org/10.3109/15376516.2015.1006491] [PMID: 25578686]

[228]   Nday CM, Halevas E, Jackson GE, Salifoglou A. Quercetin encapsulation in modified silica nanoparticles: potential use against Cu(II)-induced oxidative stress in neurodegeneration. J Inorg Biochem 2015; 145: 51-64.
[http://dx.doi.org/10.1016/j.jinorgbio.2015.01.001] [PMID: 25634813]

[229]   Sapino S, Ugazio E, Gastaldi L, *et al.* Mesoporous silica as topical nanocarriers for quercetin: characterization and *in vitro* studies. Eur J Pharm Biopharm 2015; 89: 116-25.
[http://dx.doi.org/10.1016/j.ejpb.2014.11.022] [PMID: 25478737]

[230]   Liu C-H, Huang Y-C, Jhang J-W, Liu Y-H, Wu W-C. Quercetin delivery to porcine cornea and sclera by solid lipid nanoparticles and nanoemulsion. RSC Advances 2015; 5: 100923-33.

[http://dx.doi.org/10.1039/C5RA17423F]

[231] Hassanzadeh S, Khoee S, Beheshti A, Hakkarainen M. Release of quercetin from micellar nanoparticles with saturated and unsaturated core forming polyesters--a combined computational and experimental study. Mater Sci Eng C 2015; 46: 417-26.
[http://dx.doi.org/10.1016/j.msec.2014.10.059] [PMID: 25492006]

[232] Kumar SR, Priyatharshni S, Babu VN, *et al.* Quercetin conjugated superparamagnetic magnetite nanoparticles for *in-vitro* analysis of breast cancer cell lines for chemotherapy applications. J Colloid Interface Sci 2014; 436: 234-42.
[http://dx.doi.org/10.1016/j.jcis.2014.08.064] [PMID: 25278361]

[233] Mittal AK, Kumar S, Banerjee UC. Quercetin and gallic acid mediated synthesis of bimetallic (silver and selenium) nanoparticles and their antitumor and antimicrobial potential. J Colloid Interface Sci 2014; 431: 194-9.
[http://dx.doi.org/10.1016/j.jcis.2014.06.030] [PMID: 25000181]

[234] Testa G, Gamba P, Badilli U, *et al.* Loading into nanoparticles improves quercetin's efficacy in preventing neuroinflammation induced by oxysterols. PLoS One 2014; 9(5): e96795.
[http://dx.doi.org/10.1371/journal.pone.0096795] [PMID: 24802026]

[235] Castangia I, Nácher A, Caddeo C, *et al.* Fabrication of quercetin and curcumin bionanovesicles for the prevention and rapid regeneration of full-thickness skin defects on mice. Acta Biomater 2014; 10(3): 1292-300.
[http://dx.doi.org/10.1016/j.actbio.2013.11.005] [PMID: 24239901]

[236] Guo Y-J, Yang F, Zhang L, Pi J, Cai J-Y, Yang P-H. Facile synthesis of multifunctional germanium nanoparticles as a carrier of quercetin to achieve enhanced biological activity. Chem Asian J 2014; 9(8): 2272-80.
[http://dx.doi.org/10.1002/asia.201402227] [PMID: 24958675]

[237] Varshosaz J, Jafarian A, Salehi G, Zolfaghari B. Comparing different sterol containing solid lipid nanoparticles for targeted delivery of quercetin in hepatocellular carcinoma. J Liposome Res 2014; 24(3): 191-203.
[http://dx.doi.org/10.3109/08982104.2013.868476] [PMID: 24354715]

[238] Han SB, Kwon SS, Jeong YM, Yu ER, Park SN. Physical characterization and *in vitro* skin permeation of solid lipid nanoparticles for transdermal delivery of quercetin. Int J Cosmet Sci 2014; 36(6): 588-97.
[http://dx.doi.org/10.1111/ics.12160] [PMID: 25220288]

[239] Caddeo C, Díez-Sales O, Pons R, Fernàndez-Busquets X, Fadda AM, Manconi M. Topical anti-inflammatory potential of quercetin in lipid-based nanosystems: *in vivo* and *in vitro* evaluation. Pharm Res 2014; 31(4): 959-68.
[http://dx.doi.org/10.1007/s11095-013-1215-0] [PMID: 24297068]

[240] Aditya NP, Macedo AS, Doktorovova S, Souto EB, Kim S, Chang P-S, *et al.* Development and evaluation of lipid nanocarriers for quercetin delivery: A comparative study of solid lipid nanoparticles (SLN), nanostructured lipid carriers (NLC), and lipid nanoemulsions (LNE). LWT 2014; 59: 115-21.
[http://dx.doi.org/10.1016/j.lwt.2014.04.058]

[241] Wang G, Wang J, Luo J, *et al.* PEG2000-DPSE-coated quercetin nanoparticles remarkably enhanced anticancer effects through induced programed cell death on C6 glioma cells. J Biomed Mater Res A 2013; 101(11): 3076-85.
[PMID: 23529952]

[242] Jain AS, Shah SM, Nagarsenker MS, *et al.* Lipid colloidal carriers for improvement of anticancer activity of orally delivered quercetin: formulation, characterization and establishing *in vitro-in vivo* advantage. J Biomed Nanotechnol 2013; 9(7): 1230-40.
[http://dx.doi.org/10.1166/jbn.2013.1636] [PMID: 23909137]

[243] Ha H-K, Kim JW, Lee M-R, Lee W-J. Formation and characterization of quercetin-loaded chitosan

oligosaccharide/β-lactoglobulin nanoparticle. Food Res Int 2013; 52: 82-90.
[http://dx.doi.org/10.1016/j.foodres.2013.02.021]

[244]   Das DK, Chakraborty A, Bhattacharjee S, Dey S. Biosynthesis of stabilised gold nanoparticle using an aglycone flavonoid, quercetin. J Exp Nanosci 2013; 8: 649-55.
[http://dx.doi.org/10.1080/17458080.2011.591001]

[245]   Scalia S, Franceschinis E, Bertelli D, Iannuccelli V. Comparative evaluation of the effect of permeation enhancers, lipid nanoparticles and colloidal silica on *in vivo* human skin penetration of quercetin. Skin Pharmacol Physiol 2013; 26(2): 57-67.
[http://dx.doi.org/10.1159/000345210] [PMID: 23207877]

[246]   Faddah LM, Baky NAA, Mohamed AM, Al-Rasheed NM, Al-Rasheed NM. Protective effect of quercetin and/or l-arginine against nano-zinc oxide-induced cardiotoxicity in rats. J Nanopart Res 2013; 15: 1520-33.
[http://dx.doi.org/10.1007/s11051-013-1520-x]

[247]   Bose S, Du Y, Takhistov P, Michniak-Kohn B. Formulation optimization and topical delivery of quercetin from solid lipid based nanosystems. Int J Pharm 2013; 441(1-2): 56-66.
[http://dx.doi.org/10.1016/j.ijpharm.2012.12.013] [PMID: 23262430]

[248]   Pimple S, Manjappa AS, Ukawala M, Murthy RSR. PLGA nanoparticles loaded with etoposide and quercetin dihydrate individually: *in vitro* cell line study to ensure advantage of combination therapy. Cancer Nanotechnol 2012; 3(1-6): 25-36.
[http://dx.doi.org/10.1007/s12645-012-0027-y] [PMID: 26069494]

[249]   Gao X, Wang B, Wei X, *et al.* Anticancer effect and mechanism of polymer micelle-encapsulated quercetin on ovarian cancer. Nanoscale 2012; 4(22): 7021-30.
[http://dx.doi.org/10.1039/c2nr32181e] [PMID: 23044718]

[250]   Khoei S, Azarian M, Khoee S. Effect of hyperthermia and triblock copolymeric nanoparticles as quercetin carrier on DU145 prostate cancer cells. Curr Nanosci 2012; 8: 690-6.
[http://dx.doi.org/10.2174/157341312802884355]

[251]   Kumari A, Kumar V, Yadav SK. Plant extract synthesized PLA nanoparticles for controlled and sustained release of quercetin: a green approach. PLoS One 2012; 7(7): e41230.
[http://dx.doi.org/10.1371/journal.pone.0041230] [PMID: 22844443]

[252]   Pal R, Roy M, Chakraborti AS. Preparation and characterization of quercetin-loaded poly(lactide-co-glycolide) nanoparticles. Adv Sci Lett 2012; 10: 27-32.
[http://dx.doi.org/10.1166/asl.2012.2149]

[253]   Chen-yu G, Chun-fen Y, Qi-lu L, *et al.* Development of a quercetin-loaded nanostructured lipid carrier formulation for topical delivery. Int J Pharm 2012; 430(1-2): 292-8.
[http://dx.doi.org/10.1016/j.ijpharm.2012.03.042] [PMID: 22486962]

[254]   Chakraborty S, Stalin S, Das N, Choudhury ST, Ghosh S, Swarnakar S. The use of nano-quercetin to arrest mitochondrial damage and MMP-9 upregulation during prevention of gastric inflammation induced by ethanol in rat. Biomaterials 2012; 33(10): 2991-3001.
[http://dx.doi.org/10.1016/j.biomaterials.2011.12.037] [PMID: 22257724]

[255]   Wang G, Wang JJ, Yang GY, *et al.* Effects of quercetin nanoliposomes on C6 glioma cells through induction of type III programmed cell death. Int J Nanomedicine 2012; 7: 271-80.
[PMID: 22275840]

[256]   Fang R, Hao R, Wu X, Li Q, Leng X, Jing H. Bovine serum albumin nanoparticle promotes the stability of quercetin in simulated intestinal fluid. J Agric Food Chem 2011; 59(11): 6292-8.
[http://dx.doi.org/10.1021/jf200718j] [PMID: 21542648]

[257]   Bernardy N, Romio AP, Barcelos EI, *et al.* Nanoencapsulation of quercetin via miniemulsion polymerization. J Biomed Nanotechnol 2010; 6(2): 181-6.
[http://dx.doi.org/10.1166/jbn.2010.1107] [PMID: 20738073]

[258] Li H, Zhao X, Ma Y, Zhai G, Li L, Lou H. Enhancement of gastrointestinal absorption of quercetin by solid lipid nanoparticles. J Control Release 2009; 133(3): 238-44.
[http://dx.doi.org/10.1016/j.jconrel.2008.10.002] [PMID: 18951932]

[259] Zhang Y, Yang Y, Tang K, Hu X, Zou G. Physicochemical characterization and antioxidant activity of quercetin-loaded chitosan nanoparticles. J Appl Polym Sci 2008; 107: 891-7.
[http://dx.doi.org/10.1002/app.26402]

[260] Wu T-H, Yen F-L, Lin L-T, Tsai T-R, Lin C-C, Cham T-M. Preparation, physicochemical characterization, and antioxidant effects of quercetin nanoparticles. Int J Pharm 2008; 346(1-2): 160-8.
[http://dx.doi.org/10.1016/j.ijpharm.2007.06.036] [PMID: 17689897]

[261] Egorova EM, Revina AA. Synthesis of metallic nanoparticles in reverse micelles in the presence of quercetin. Colloids Surf A Physicochem Eng Asp 2000; 168: 87-96.
[http://dx.doi.org/10.1016/S0927-7757(99)00513-0]

[262] Delmas D, Aires V, Limagne E, *et al.* Transport, stability, and biological activity of resveratrol. Ann N Y Acad Sci 2011; 1215: 48-59.
[http://dx.doi.org/10.1111/j.1749-6632.2010.05871.x] [PMID: 21261641]

[263] Hsieh T-C, Wu JM. Resveratrol: Biological and pharmaceutical properties as anticancer molecule. Biofactors 2010; 36(5): 360-9.
[http://dx.doi.org/10.1002/biof.105] [PMID: 20623546]

[264] Gambini J, Inglés M, Olaso G, *et al.* Properties of resveratrol: *in vitro* and *in vivo* studies about metabolism, bioavailability, and biological effects in animal models and humans. Oxid Med Cell Longev 2015; 2015: 837042.
[http://dx.doi.org/10.1155/2015/837042] [PMID: 26221416]

[265] Smoliga JM, Blanchard O. Enhancing the delivery of resveratrol in humans: if low bioavailability is the problem, what is the solution? Molecules 2014; 19(11): 17154-72.
[http://dx.doi.org/10.3390/molecules191117154] [PMID: 25347459]

[266] Higdon JV, Frei B. Tea catechins and polyphenols: health effects, metabolism, and antioxidant functions. Crit Rev Food Sci Nutr 2003; 43(1): 89-143.
[http://dx.doi.org/10.1080/10408690390826464] [PMID: 12587987]

[267] Erlund I. Review of the flavonoids quercetin, hesperetin, and naringenin. Dietary sources, bioactivities, bioavailability, and epidemiology. Nutr Res 2004; 24: 851-74.
[http://dx.doi.org/10.1016/j.nutres.2004.07.005]

[268] Kawabata K, Mukai R, Ishisaka A. Quercetin and related polyphenols: new insights and implications for their bioactivity and bioavailability. Food Funct 2015; 6(5): 1399-417.
[http://dx.doi.org/10.1039/C4FO01178C] [PMID: 25761771]

[269] Prasad S, Tyagi AK, Aggarwal BB. Recent developments in delivery, bioavailability, absorption and metabolism of curcumin: the golden pigment from golden spice. Cancer Res Treat 2014; 46(1): 2-18.
[http://dx.doi.org/10.4143/crt.2014.46.1.2] [PMID: 24520218]

[270] Sharma RA, McLelland HR, Hill KA, *et al.* Pharmacodynamic and pharmacokinetic study of oral Curcuma extract in patients with colorectal cancer. Clin Cancer Res 2001; 7(7): 1894-900.
[PMID: 11448902]

[271] Nakkala JR, Mata R, Gupta AK, Sadras SR. Biological activities of green silver nanoparticles synthesized with *Acorous calamus* rhizome extract. Eur J Med Chem 2014; 85: 784-94.
[http://dx.doi.org/10.1016/j.ejmech.2014.08.024] [PMID: 25147142]

[272] Ravanfar R, Tamaddon AM, Niakousari M, Moein MR. Preservation of anthocyanins in solid lipid nanoparticles: Optimization of a microemulsion dilution method using the Placket-Burman and Box-Behnken designs. Food Chem 2016; 199: 573-80.
[http://dx.doi.org/10.1016/j.foodchem.2015.12.061] [PMID: 26776010]

[273]   Bao S, Xu S, Wang Z. Antioxidant activity and properties of gelatin films incorporated with tea polyphenol-loaded chitosan nanoparticles. J Sci Food Agric 2009; 89: 2692-700.
[http://dx.doi.org/10.1002/jsfa.3775]

[274]   Mukherjee S, Ghosh S, Das DK, *et al.* Gold-conjugated green tea nanoparticles for enhanced anti-tumor activities and hepatoprotection-synthesis, characterization and *in vitro* evaluation. J Nutr Biochem 2015; 26(11): 1283-97.
[http://dx.doi.org/10.1016/j.jnutbio.2015.06.003] [PMID: 26310506]

[275]   Suresh D, Nethravathi PC. Udayabhanu, Rajanaika H, Nagabhushana H, Sharma SC. Green synthesis of multifunctional zinc oxide (ZnO) nanoparticles using *Cassia fistula* plant extract and their photodegradative, antioxidant and antibacterial activities. Mater Sci Semicond Process 2015; 31: 446-54.
[http://dx.doi.org/10.1016/j.mssp.2014.12.023]

[276]   Suresh D, Nethravathi PC. Chironji mediated facile green synthesis of ZnO nanoparticles and their photoluminescence, photodegradative, antimicrobial and antioxidant activities. Mater Sci Semicond Process 2015; 40: 759-65.
[http://dx.doi.org/10.1016/j.mssp.2015.06.088]

[277]   Nethravathi PC, Shruthi GS, Suresh D. Udayabhanu, Nagabhushana H, Sharma SC. *Garcinia xanthochymus* mediated green synthesis of ZnO nanoparticles: Photoluminescence, photocatalytic and antioxidant activity studies. Ceram Int 2015; 41: 8680-7.
[http://dx.doi.org/10.1016/j.ceramint.2015.03.084]

[278]   Meneses MA, Caputo G, Scognamiglio M, Reverchon E, Adami R. Antioxidant phenolic compounds recovery from *Mangifera indica* L. by-products by supercritical antisolvent extraction. J Food Eng 2015; 163: 45-53.
[http://dx.doi.org/10.1016/j.jfoodeng.2015.04.025]

[279]   Kiran Kumar HA, Mandal BK, Mohan Kumar K, *et al.* Antimicrobial and antioxidant activities of *Mimusops elengi* seed extract mediated isotropic silver nanoparticles. Spectrochim Acta A Mol Biomol Spectrosc 2014; 130: 13-8.
[http://dx.doi.org/10.1016/j.saa.2014.03.024] [PMID: 24759779]

[280]   Dhamecha D, Jalalpure S, Jadhav K, Sajjan D. Green synthesis of gold nanoparticles using *Pterocarpus marsupium*: characterization and biocompatibility studies. Particul Sci Technol 2015; 2015: 1-9.

[281]   Basavegowda N, Mishra K, Lee YR, Kim SH. Antioxidant and anti-tyrosinase activities of palladium nanoparticles synthesized using *Saururus chinensis*. J Cluster Sci 2016; 27: 733-44.
[http://dx.doi.org/10.1007/s10876-016-0984-0]

[282]   Hoskote Anand KK, Mandal BK. Activity study of biogenic spherical silver nanoparticles towards microbes and oxidants. Spectrochim Acta A Mol Biomol Spectrosc 2015; 135: 639-45.
[http://dx.doi.org/10.1016/j.saa.2014.07.013] [PMID: 25128676]

[283]   Madureira AR, Pereira A, Pintado M. Chitosan nanoparticles loaded with 2,5-dihydroxybenzoic acid and protocatechuic acid: Properties and digestion. J Food Eng 2016; 174: 8-14.
[http://dx.doi.org/10.1016/j.jfoodeng.2015.11.007]

[284]   Lee CY, Nanah CN, Held RA, *et al.* Effect of electron donating groups on polyphenol-based antioxidant dendrimers. Biochimie 2015; 111: 125-34.
[http://dx.doi.org/10.1016/j.biochi.2015.02.001] [PMID: 25668210]

[285]   Pool H, Quintanar D, Figueroa JD, Marinho Mano C, Bechara JEH. Antioxidant effects of quercetin and catechin encapsulated into PLGA nanoparticles. J Nanomater 2012; 145380.

[286]   Nallamuthu I, Devi A, Khanum F. Chlorogenic acid loaded chitosan nanoparticles with sustained release property, retained antioxidant activity and enhanced bioavailability. Asian Journal of Pharmaceutical Sciences 2015; 10: 203-11.

[http://dx.doi.org/10.1016/j.ajps.2014.09.005]

[287] Konwarh R, Saikia JP, Karak N, Konwar BK. 'Poly(ethylene glycol)-magnetic nanoparticles-curcumin' trio: directed morphogenesis and synergistic free-radical scavenging. Colloids Surf B Biointerfaces 2010; 81(2): 578-86.
       [http://dx.doi.org/10.1016/j.colsurfb.2010.07.062] [PMID: 20729041]

[288] de Cristo Soares Alves A, Mainardes RM, Khalil NM. Nanoencapsulation of gallic acid and evaluation of its cytotoxicity and antioxidant activity. Mater Sci Eng C 2016; 60: 126-34.
       [http://dx.doi.org/10.1016/j.msec.2015.11.014] [PMID: 26706515]

[289] Lee CY, Sharma A, Uzarski RL, *et al.* Potent antioxidant dendrimers lacking pro-oxidant activity. Free Radic Biol Med 2011; 50(8): 918-25.
       [http://dx.doi.org/10.1016/j.freeradbiomed.2010.10.699] [PMID: 20977937]

[290] Podsędek A, Redzynia M, Klewicka E, Koziołkiewicz M. Matrix effects on the stability and antioxidant activity of red cabbage anthocyanins under simulated gastrointestinal digestion. BioMed Res Int 2014; 2014: 365738.
       [http://dx.doi.org/10.1155/2014/365738] [PMID: 24575407]

[291] Surveswaran S, Cai Y-Z, Corke H, Sun M. Systematic evaluation of natural phenolic antioxidants from 133 Indian medicinal plants. Food Chem 2007; 102: 938-53.
       [http://dx.doi.org/10.1016/j.foodchem.2006.06.033]

[292] Manickam M, Ramanathan M, Jahromi MA, Chansouria JPN, Ray AB. Antihyperglycemic activity of phenolics from *Pterocarpus marsupium.* J Nat Prod 1997; 60(6): 609-10.
       [http://dx.doi.org/10.1021/np9607013] [PMID: 9214733]

[293] Krishnaraj C, Jagan EG, Rajasekar S, Selvakumar P, Kalaichelvan PT, Mohan N. Synthesis of silver nanoparticles using *Acalypha indica* leaf extracts and its antibacterial activity against water borne pathogens. Colloids Surf B Biointerfaces 2010; 76(1): 50-6.
       [http://dx.doi.org/10.1016/j.colsurfb.2009.10.008] [PMID: 19896347]

[294] Vijay Kumar PPN, Pammi SVN, Kollu P, Satyanarayana KVV, Shameem U. Green synthesis and characterization of silver nanoparticles using *Boerhaavia diffusa* plant extract and their anti bacterial activity. Ind Crops Prod 2014; 52: 562-6.
       [http://dx.doi.org/10.1016/j.indcrop.2013.10.050]

[295] Narayanan KB, Park HH. Antifungal activity of silver nanoparticles synthesized using turnip leaf extract (*Brassica rapa* L.) against wood rotting pathogens. Eur J Plant Pathol 2014; 140: 185-92.
       [http://dx.doi.org/10.1007/s10658-014-0399-4]

[296] Banala RR, Nagati VB, Karnati PR. Green synthesis and characterization of *Carica papaya* leaf extract coated silver nanoparticles through X-ray diffraction, electron microscopy and evaluation of bactericidal properties. Saudi J Biol Sci 2015; 22(5): 637-44.
       [http://dx.doi.org/10.1016/j.sjbs.2015.01.007] [PMID: 26288570]

[297] Kaviya S, Santhanalakshmi J, Viswanathan B, Muthumary J, Srinivasan K. Biosynthesis of silver nanoparticles using *citrus sinensis* peel extract and its antibacterial activity. Spectrochim Acta A Mol Biomol Spectrosc 2011; 79(3): 594-8.
       [http://dx.doi.org/10.1016/j.saa.2011.03.040] [PMID: 21536485]

[298] Shalaka AM, Pratik RC, Vrishali BS, Suresh PK. Rapid biosynthesis of silver nanoparticles using *Cymbopogan Citratus* (Lemongrass) and its antimicrobial activity. Micro & Nano Lett 2011; 3: 189-94.
       [http://dx.doi.org/10.1007/BF03353671]

[299] Sahu N, Soni D, Chandrashekhar B, Sarangi BK, Satpute D, Pandey RA. Synthesis and characterization of silver nanoparticles using *Cynodon dactylon* leaves and assessment of their antibacterial activity. Bioprocess Biosyst Eng 2013; 36(7): 999-1004.
       [http://dx.doi.org/10.1007/s00449-012-0841-y] [PMID: 23111848]

[300]   Ramesh PS, Kokila T, Geetha D. Plant mediated green synthesis and antibacterial activity of silver nanoparticles using *Emblica officinalis* fruit extract. Spectrochim Acta A Mol Biomol Spectrosc 2015; 142: 339-43.
[http://dx.doi.org/10.1016/j.saa.2015.01.062] [PMID: 25710891]

[301]   Veerasamy R, Xin TZ, Gunasagaran S, Xiang TFW, Yang EFC, Jeyakumar N, *et al.* Biosynthesis of silver nanoparticles using mangosteen leaf extract and evaluation of their antimicrobial activities. J Saudi Chem Soc 2011; 15: 113-20.
[http://dx.doi.org/10.1016/j.jscs.2010.06.004]

[302]   Lokina S, Stephen A, Kaviyarasan V, Arulvasu C, Narayanan V. Cytotoxicity and antimicrobial activities of green synthesized silver nanoparticles. Eur J Med Chem 2014; 76: 256-63.
[http://dx.doi.org/10.1016/j.ejmech.2014.02.010] [PMID: 24583606]

[303]   Prasad TN, Elumalai EK. Biofabrication of Ag nanoparticles using *Moringa oleifera* leaf extract and their antimicrobial activity. Asian Pac J Trop Biomed 2011; 1(6): 439-42.
[http://dx.doi.org/10.1016/S2221-1691(11)60096-8] [PMID: 23569809]

[304]   Bankar A, Joshi B, Kumar AR, Zinjarde S. Banana peel extract mediated novel route for the synthesis of silver nanoparticles. Colloids Surf A Physicochem Eng Asp 2010; 368: 58-63.
[http://dx.doi.org/10.1016/j.colsurfa.2010.07.024]

[305]   Danai-Tambhale SD, Adhyapak PV. A facile green synthesis of silver nanoparticles using *Psoralea corylifolia* L. seed extract and their *in vitro* antimicrobial activities. Int J Pharma Bio Sci 2014; 5: 457-67.

[306]   Ghaedi M, Yousefinejad M, Safarpoor M, Khafri HZ, Purkait MK. *Rosmarinus officinalis* leaf extract mediated green synthesis of silver nanoparticles and investigation of its antimicrobial properties. J Ind Eng Chem 2015; 31: 167-72.
[http://dx.doi.org/10.1016/j.jiec.2015.06.020]

[307]   Nabikhan A, Kandasamy K, Raj A, Alikunhi NM. Synthesis of antimicrobial silver nanoparticles by callus and leaf extracts from saltmarsh plant, *Sesuvium portulacastrum* L. Colloids Surf B Biointerfaces 2010; 79(2): 488-93.
[http://dx.doi.org/10.1016/j.colsurfb.2010.05.018] [PMID: 20627485]

[308]   Mohan Kumar K, Sinha M, Mandal BK, Ghosh AR, Siva Kumar K, Sreedhara Reddy P. Green synthesis of silver nanoparticles using *Terminalia chebula* extract at room temperature and their antimicrobial studies. Spectrochim Acta A Mol Biomol Spectrosc 2012; 91: 228-33.
[http://dx.doi.org/10.1016/j.saa.2012.02.001] [PMID: 22381795]

[309]   Zargar M, Hamid AA, Bakar FA, *et al.* Green synthesis and antibacterial effect of silver nanoparticles using *Vitex negundo* L. Molecules 2011; 16(8): 6667-76.
[http://dx.doi.org/10.3390/molecules16086667] [PMID: 25134770]

[310]   Madureira AR, Pereira A, Castro PM, Pintado M. Production of antimicrobial chitosan nanoparticles against food pathogens. J Food Eng 2015; 167: 210-6.
[http://dx.doi.org/10.1016/j.jfoodeng.2015.06.010]

[311]   Ma Y-T, Chuang J-I, Lin J-H, Hsu F-L. Phenolics from *Acalypha indica.* J Chin Chem Soc (Taipei) 1997; 44: 499-502.
[http://dx.doi.org/10.1002/jccs.199700075]

[312]   Nahrstedt A, Hungeling M, Petereit F. Flavonoids from *Acalypha indica.* Fitoterapia 2006; 77(6): 484-6.
[http://dx.doi.org/10.1016/j.fitote.2006.04.007] [PMID: 16828241]

[313]   Fernandes F, Valentão P, Sousa C, Pereira JA, Seabra RM, Andrade PB. Chemical and antioxidative assessment of dietary turnip (*Brassica rapa* var. *rapa* L.). Food Chem 2007; 105: 1003-10.
[http://dx.doi.org/10.1016/j.foodchem.2007.04.063]

[314]   Cartea ME, de Haro A, Obregón S, Soengas P, Velasco P. Glucosinolate variation in leaves of

Brassica rapa crops. Plant Foods Hum Nutr 2012; 67(3): 283-8.
[http://dx.doi.org/10.1007/s11130-012-0300-6] [PMID: 23001436]

[315]   Canini A, Alesiani D, D'Arcangelo G, Tagliatesta P. Gas chromatography–mass spectrometry analysis of phenolic compounds from *Carica papaya* L. leaf. J Food Compos Anal 2007; 20: 584-90.
[http://dx.doi.org/10.1016/j.jfca.2007.03.009]

[316]   Ashokkumar K, Selvaraj K, Muthukrishnan SD. *Cynodon dactylon* (L.) Pers.: An updated review of its phytochemistry and pharmacology. J Med Plants Res 2013; 7: 3477-83.

[317]   Shah G, Shri R, Panchal V, Sharma N, Singh B, Mann AS. Scientific basis for the therapeutic use of *Cymbopogon citratus*, stapf (Lemon grass). J Adv Pharm Technol Res 2011; 2(1): 3-8.
[http://dx.doi.org/10.4103/2231-4040.79796] [PMID: 22171285]

[318]   Pedraza-Chaverri J, Cárdenas-Rodríguez N, Orozco-Ibarra M, Pérez-Rojas JM. Medicinal properties of mangosteen (*Garcinia mangostana*). Food Chem Toxicol 2008; 46(10): 3227-39.
[http://dx.doi.org/10.1016/j.fct.2008.07.024] [PMID: 18725264]

[319]   Valdez-Solana MA, Mejía-García VY, Téllez-Valencia A, García-Arenas G, Salas-Pacheco J, Alba-Romero JJ, *et al*. Nutritional content and elemental and phytochemical analyses of *Moringa oleifera* grown in Mexico. J Chem 2015; 860381.

[320]   Borrás-Linares I, Stojanović Z, Quirantes-Piné R, *et al. Rosmarinus officinalis* leaves as a natural source of bioactive compounds. Int J Mol Sci 2014; 15(11): 20585-606.
[http://dx.doi.org/10.3390/ijms151120585] [PMID: 25391044]

[321]   Huang M, Zhang Y, Xu S, *et al.* Identification and quantification of phenolic compounds in *Vitex negundo* L. var. *cannabifolia* (Siebold et Zucc.) Hand.-Mazz. using liquid chromatography combined with quadrupole time-of-flight and triple quadrupole mass spectrometers. J Pharm Biomed Anal 2015; 108: 11-20.
[http://dx.doi.org/10.1016/j.jpba.2015.01.049] [PMID: 25703235]

[322]   Khajuria RK, Suri KA, Suri OP, Atal CK. 3,5,4′-Trihydroxy-6,7-dimethoxyflavone 3-glucoside from *Sesuvium portulacastrum.* Phytochemistry 1982; 21: 1179-80.
[http://dx.doi.org/10.1016/S0031-9422(00)82450-4]

[323]   Buckingham J, Munasinghe VRN. Dictionary of flavonoids. Boca Raton: CRC Press, Taylor & Francis Group 2015; pp. H-125.
[http://dx.doi.org/10.1201/b18170]

[324]   Mishra S, Aeri V, Gaur PK, Jachak SM. Phytochemical, therapeutic, and ethnopharmacological overview for a traditionally important herb: *Boerhavia diffusa* Linn. BioMed Res Int 2014; 2014: 808302.
[http://dx.doi.org/10.1155/2014/808302] [PMID: 24949473]

[325]   Ferreres F, Sousa C, Justin M, *et al.* Characterisation of the phenolic profile of *Boerhaavia diffusa* L. by HPLC-PAD-MS/MS as a tool for quality control. Phytochem Anal 2005; 16(6): 451-8.
[http://dx.doi.org/10.1002/pca.869] [PMID: 16315490]

[326]   Li S, Lambros T, Wang Z, Goodnow R, Ho C-T. Efficient and scalable method in isolation of polymethoxyflavones from orange peel extract by supercritical fluid chromatography. J Chromatogr B Analyt Technol Biomed Life Sci 2007; 846(1-2): 291-7.
[http://dx.doi.org/10.1016/j.jchromb.2006.09.010] [PMID: 17035106]

[327]   Oboh G, Akinsanmi OA, Adefegha SA, Akinyemi AJ. Interaction of plantain (*Musa paradisiaca*) peel extracts (unripe, ripe and over ripe) with key enzymes linked to hypertension (angiotensin-I converting enzyme) and their antioxidant activities (*in vitro*): A nutraceutical approach. Adv Food Sci 2014; 37: 50-7.

[328]   Poltanov EA, Shikov AN, Dorman HJD, *et al.* Chemical and antioxidant evaluation of Indian gooseberry (*Emblica officinalis* Gaertn., syn. *Phyllanthus emblica* L.) supplements. Phytother Res 2009; 23(9): 1309-15.

[http://dx.doi.org/10.1002/ptr.2775] [PMID: 19172666]

[329] Haraguchi H, Inoue J, Tamura Y, Mizutani K. Antioxidative components of *Psoralea corylifolia* (Leguminosae). Phytother Res 2002; 16(6): 539-44.
[http://dx.doi.org/10.1002/ptr.972] [PMID: 12237811]

[330] Priyadharshini Raman R, Parthiban S, Srinithya B, Vinod Kumar V, Philip Anthony S, Sivasubramanian A, *et al.* Biogenic silver nanoparticles synthesis using the extract of the medicinal plant *Clerodendron serratum* and its *in vitro* antiproliferative activity. Mater Lett 2015; 160: 400-3.
[http://dx.doi.org/10.1016/j.matlet.2015.08.009]

[331] Kathiravan V, Ravi S, Ashokkumar S. Synthesis of silver nanoparticles from *Melia dubia* leaf extract and their *in vitro* anticancer activity. Spectrochim Acta A Mol Biomol Spectrosc 2014; 130: 116-21.
[http://dx.doi.org/10.1016/j.saa.2014.03.107] [PMID: 24769382]

[332] Lokina S, Stephen A, Kaviyarasan V, Arulvasu C, Narayanan V. Cytotoxicity and antimicrobial studies of silver nanoparticles synthesized using *Psidium guajava* L. extract. Synth React Inorg Met-Org Nano-Met Chem 2015; 45: 426-32.
[http://dx.doi.org/10.1080/15533174.2013.831881]

[333] Sathishkumar M, Pavagadhi S, Mahadevan A, Balasubramanian R. Biosynthesis of gold nanoparticles and related cytotoxicity evaluation using A549 cells. Ecotoxicol Environ Saf 2015; 114: 232-40.
[http://dx.doi.org/10.1016/j.ecoenv.2014.03.020] [PMID: 24835429]

[334] Amarnath K, Mathew NL, Nellore J, Siddarth CRV, Kumar J. Facile synthesis of biocompatible gold nanoparticles from *Vites vinefera* and its cellular internalization against HBL-100 cells. Cancer Nanotechnol 2011; 2(1-6): 121-32.
[http://dx.doi.org/10.1007/s12645-011-0022-8] [PMID: 26316896]

[335] Liang J, Li F, Fang Y, *et al.* Cytotoxicity and apoptotic effects of tea polyphenol-loaded chitosan nanoparticles on human hepatoma HepG2 cells. Mater Sci Eng C 2014; 36: 7-13.
[http://dx.doi.org/10.1016/j.msec.2013.11.039] [PMID: 24433880]

[336] Shirode AB, Bharali DJ, Nallanthighal S, Coon JK, Mousa SA, Reliene R. Nanoencapsulation of pomegranate bioactive compounds for breast cancer chemoprevention. Int J Nanomedicine 2015; 10: 475-84.
[PMID: 25624761]

[337] Guo D, Dou D, Ge L, Huang Z, Wang L, Gu N. A caffeic acid mediated facile synthesis of silver nanoparticles with powerful anti-cancer activity. Colloids Surf B Biointerfaces 2015; 134: 229-34.
[http://dx.doi.org/10.1016/j.colsurfb.2015.06.070] [PMID: 26208293]

[338] Yallapu MM, Ebeling MC, Khan S, *et al.* Novel curcumin-loaded magnetic nanoparticles for pancreatic cancer treatment. Mol Cancer Ther 2013; 12(8): 1471-80.
[http://dx.doi.org/10.1158/1535-7163.MCT-12-1227] [PMID: 23704793]

[339] Kim TH, Jiang HH, Youn YS, *et al.* Preparation and characterization of water-soluble albumin-bound curcumin nanoparticles with improved antitumor activity. Int J Pharm 2011; 403(1-2): 285-91.
[http://dx.doi.org/10.1016/j.ijpharm.2010.10.041] [PMID: 21035530]

[340] Singh M, Bhatnagar P, Mishra S, Kumar P, Shukla Y, Gupta KC. PLGA-encapsulated tea polyphenols enhance the chemotherapeutic efficacy of cisplatin against human cancer cells and mice bearing Ehrlich ascites carcinoma. Int J Nanomedicine 2015; 10: 6789-809.
[http://dx.doi.org/10.2147/IJN.S79489] [PMID: 26586942]

[341] Hu B, Xie M, Zhang C, Zeng X. Genipin-structured peptide-polysaccharide nanoparticles with significantly improved resistance to harsh gastrointestinal environments and their potential for oral delivery of polyphenols. J Agric Food Chem 2014; 62(51): 12443-52.
[http://dx.doi.org/10.1021/jf5046766] [PMID: 25479066]

[342] Siddiqui IA, Bharali DJ, Nihal M, *et al.* Excellent anti-proliferative and pro-apoptotic effects of (---epigallocatechin-3-gallate encapsulated in chitosan nanoparticles on human melanoma cell growth

both *in vitro* and *in vivo.* Nanomedicine (Lond) 2014; 10(8): 1619-26.
[http://dx.doi.org/10.1016/j.nano.2014.05.007] [PMID: 24965756]

[343] Zheng N-G, Wang J-L, Yang S-L, Wu J-L. Aberrant epigenetic alteration in Eca9706 cells modulated by nanoliposomal quercetin combined with butyrate mediated via epigenetic-NF-κB signaling. Asian Pac J Cancer Prev 2014; 15(11): 4539-43.
[http://dx.doi.org/10.7314/APJCP.2014.15.11.4539] [PMID: 24969881]

[344] Alshatwi AA, Athinarayanan J, Vaiyapuri Subbarayan P. Green synthesis of platinum nanoparticles that induce cell death and G2/M-phase cell cycle arrest in human cervical cancer cells. J Mater Sci Mater Med 2015; 26(1): 5330.
[http://dx.doi.org/10.1007/s10856-014-5330-1] [PMID: 25577212]

[345] Poornima BS, Prakash LH. Pradeep, Harini A. Pharmacological review on *Clerodendrum serratum* Linn. Moon. J Pharmacogn Phytochem 2015; 3: 126-30.

[346] Valentina P, Kaliappan I, Kiruthiga B, Parimala MJ. Preliminary phytochemical analysis and biological screening of extracts of leaves of *Melia dubia* Cav. Int J Res Ayurveda Pharm 2013; 4: 417-9.
[http://dx.doi.org/10.7897/2277-4343.04322]

[347] Flores G, Wu S-B, Negrin A, Kennelly EJ. Chemical composition and antioxidant activity of seven cultivars of guava (*Psidium guajava*) fruits. Food Chem 2015; 170: 327-35.
[http://dx.doi.org/10.1016/j.foodchem.2014.08.076] [PMID: 25306353]

[348] Wang G-W, Hu W-T, Huang B-K, Qin L-P. *Illicium verum*: a review on its botany, traditional use, chemistry and pharmacology. J Ethnopharmacol 2011; 136(1): 10-20.
[http://dx.doi.org/10.1016/j.jep.2011.04.051] [PMID: 21549817]

[349] Mateus N, Machado JM, de Freitas V. Development changes of anthocyanins in *Vitis vinifera* grapes grown in the Douro Valley and concentration in respective wines. J Sci Food Agric 2002; 82: 1689-95.
[http://dx.doi.org/10.1002/jsfa.1237]

[350] Schaffer S, Halliwell B. Do polyphenols enter the brain and does it matter? Some theoretical and practical considerations. Genes Nutr 2012; 7(2): 99-109.
[http://dx.doi.org/10.1007/s12263-011-0255-5] [PMID: 22012276]

[351] Ganesan P, Ko H-M, Kim I-S, Choi D-K. Recent trends in the development of nanophytobioactive compounds and delivery systems for their possible role in reducing oxidative stress in Parkinson's disease models. Int J Nanomedicine 2015; 10: 6757-72.
[http://dx.doi.org/10.2147/IJN.S93918] [PMID: 26604750]

[352] Lajoie JM, Shusta EV. Targeting receptor-mediated transport for delivery of biologics across the blood-brain barrier. Annu Rev Pharmacol Toxicol 2015; 55: 613-31.
[http://dx.doi.org/10.1146/annurev-pharmtox-010814-124852] [PMID: 25340933]

[353] Nellore J, Pauline C, Amarnath K. *Bacopa monnieri* phytochemicals mediated synthesis of platinum nanoparticles and its neurorescue effect on 1-methyl 4-phenyl 1,2,3,6 tetrahydropyridine-induced experimental parkinsonism in zebrafish. J Neurodegener Dis 2013; 2013: 972391.
[http://dx.doi.org/10.1155/2013/972391] [PMID: 26317003]

[354] Prakash DJ, Arulkumar S, Sabesan M. Effect of nanohypericum (*Hypericum perforatum* gold nanoparticles) treatment on restraint stressinduced behavioral and biochemical alteration in male albino mice. Pharmacognosy Res 2010; 2(6): 330-4.
[http://dx.doi.org/10.4103/0974-8490.75450] [PMID: 21713134]

[355] Zhou Y, Shen Y-H, Zhang C, Zhang W-D. Chemical constituents of *Bacopa monnieri.* Chem Nat Compd 2007; 43: 355-7.
[http://dx.doi.org/10.1007/s10600-007-0133-y]

[356] Orčić DZ, Mimica-Dukić NM, Francišković MM, Petrović SS, Jovin EĐ. Antioxidant activity relationship of phenolic compounds in *Hypericum perforatum* L. Chem Cent J 2011; 5: 34.

[http://dx.doi.org/10.1186/1752-153X-5-34] [PMID: 21702979]

[357]   Liu Z, Okeke CI, Zhang L, *et al.* Mixed polyethylene glycol-modified breviscapine-loaded solid lipid nanoparticles for improved brain bioavailability: preparation, characterization, and *in vivo* cerebral microdialysis evaluation in adult Sprague Dawley rats. AAPS PharmSciTech 2014; 15(2): 483-96.
[http://dx.doi.org/10.1208/s12249-014-0080-4] [PMID: 24482026]

[358]   Ray B, Bisht S, Maitra A, Maitra A, Lahiri DK. Neuroprotective and neurorescue effects of a novel polymeric nanoparticle formulation of curcumin (NanoCurc™) in the neuronal cell culture and animal model: implications for Alzheimer's disease. J Alzheimers Dis 2011; 23(1): 61-77.
[http://dx.doi.org/10.3233/JAD-2010-101374] [PMID: 20930270]

[359]   Mulik RS, Mönkkönen J, Juvonen RO, Mahadik KR, Paradkar AR. ApoE3 mediated poly(butyl) cyanoacrylate nanoparticles containing curcumin: study of enhanced activity of curcumin against beta amyloid induced cytotoxicity using *in vitro* cell culture model. Mol Pharm 2010; 7(3): 815-25.
[http://dx.doi.org/10.1021/mp900306x] [PMID: 20230014]

[360]   Nagpal K, Singh SK, Mishra DN. Nanoparticle mediated brain targeted delivery of gallic acid: *in vivo* behavioral and biochemical studies for improved antioxidant and antidepressant-like activity. Drug Deliv 2012; 19(8): 378-91.
[http://dx.doi.org/10.3109/10717544.2012.738437] [PMID: 23173579]

[361]   Yao L, Gu X, Song Q, *et al.* Nanoformulated alpha-mangostin ameliorates Alzheimer's disease neuropathology by elevating LDLR expression and accelerating amyloid-beta clearance. J Control Release 2016; 226: 1-14.
[http://dx.doi.org/10.1016/j.jconrel.2016.01.055] [PMID: 26836197]

[362]   Ghosh A, Mandal AK, Sarkar S, Panda S, Das N. Nanoencapsulation of quercetin enhances its dietary efficacy in combating arsenic-induced oxidative damage in liver and brain of rats. Life Sci 2009; 84(3-4): 75-80.
[http://dx.doi.org/10.1016/j.lfs.2008.11.001] [PMID: 19036345]

[363]   Ghosh A, Sarkar S, Mandal AK, Das N. Neuroprotective role of nanoencapsulated quercetin in combating ischemia-reperfusion induced neuronal damage in young and aged rats. PLoS One 2013; 8(4): e57735.
[http://dx.doi.org/10.1371/journal.pone.0057735] [PMID: 23620721]

[364]   Elsaesser A, Howard CV. Toxicology of nanoparticles. Adv Drug Deliv Rev 2012; 64(2): 129-37.
[http://dx.doi.org/10.1016/j.addr.2011.09.001] [PMID: 21925220]

[365]   Love SA, Maurer-Jones MA, Thompson JW, Lin Y-S, Haynes CL. Assessing nanoparticle toxicity. Annu Rev Anal Chem (Palo Alto, Calif) 2012; 5: 181-205.
[http://dx.doi.org/10.1146/annurev-anchem-062011-143134] [PMID: 22524221]

[366]   Sajid M, Ilyas M, Basheer C, *et al.* Impact of nanoparticles on human and environment: review of toxicity factors, exposures, control strategies, and future prospects. Environ Sci Pollut Res Int 2015; 22(6): 4122-43.
[http://dx.doi.org/10.1007/s11356-014-3994-1] [PMID: 25548015]

[367]   Beer C, Foldbjerg R, Hayashi Y, Sutherland DS, Autrup H. Toxicity of silver nanoparticles - nanoparticle or silver ion? Toxicol Lett 2012; 208(3): 286-92.
[http://dx.doi.org/10.1016/j.toxlet.2011.11.002] [PMID: 22101214]

Recent Advances in Biotechnology, 2018, Vol. 4, 341-382　　　　　341

CHAPTER 8

Dairy Enzyme Discovery and Technology

Jan Kjølhede Vester[1], Jeppe Wegener Tams[1] and Ali Osman[2,*]

[1] *Novozymes A/S, Krogshøjvej 36, 2880 Bagsværd, Denmark*

[2] *DSM Food Specialties, Alexander Felminglaan 1, 2613 Delft, The Netherlands*

Abstract: Enzymes have a vital role in adding value to milk, and their function varies widely from coagulants used to make cheese, to bioprotective enzymes used to enhance the shelf-life of dairy products, proteases used for acceleration of cheese ripening and modification of functional properties of milk proteins, lipases used to develop lipolytic flavors in cheese ripening, and lactases used to hydrolyze lactose to alleviate lactose intolerance and produce galactooligosaccharides as dietary fibers. This chapter (i) presents the recent advances in enzyme discovery approaches used to search for novel enzymes with interesting features, supported by few examples of relevance to the dairy industry, and (ii) discusses the up-to-date developments in industrial dairy enzyme applications, with particular focus on lactose bioconversion by lactose hydrolysis and transgalactosylation.

Keywords: Cultivation Techniques, Extremophiles, Galactooligosaccharides, Lactose Hydrolysis, Lactose Oxidation, Metagenomics, Protein Engineering.

INTRODUCTION

Enzymes are proteins that act as catalysts. When one substance needs to be transformed into another, nature uses enzymes to speed up the process. Enzymes are stable and biodegradable. They also work at low temperatures and moderate pH levels. This makes them the most environmentally-friendly catalysts. In the dairy industry, the application of enzymes is well established. Rennets (rennin, a mixture of chymosin and pepsin obtained mainly from animal and microbial sources) are used for coagulation of milk in the first stage of cheese production. Proteases of various kinds are used for acceleration of cheese ripening, modification of protein functional properties, and reducing the allergenic properties of cow milk proteins for infants. Lipases are used mainly in cheese ripening for the development of lipolytic flavors. Lactases (β- galactosidases) are used to hydrolyze lactose to glucose and galactose as a digestive aid and to

* **Corresponding author Ali Osman**: DSM Food Specialties, Alexander Felminglaan 1, 2613 Delft, The Netherlands; Tel: +4523202272; Email: ali.osman@dsm.com

improve the solubility and sweetness in various dairy products. Lysozyme and lactoperoxidase are used to control microbial contamination and improve the shelf-life of dairy products.

The above examples of applications have reached their current industrial state, partially due to the tremendous work done by enzyme manufacturers to discover new enzymes or tailor-make the features of the existing enzymes for specific applications. The wonderful diversity of organisms in nature together with their enzymatic arsenal have often been the starting point for the discovery and identification of novel enzymes. Natural diversity approaches and optimization strategies are complementary routes and both are equally important in developing a high-quality diversity of enzymes. Today, the discovery of enzymes for the food industry is not only a multidisciplinary effort that involves a wide array of different screening technologies but is also based on close collaboration between food scientists, who understand food chemistry and the intended industrial application, and biotechnologies who can deliver the right enzymes for these specific applications.

This chapter briefly reviews (i) enzyme discovery approaches, with a focus on improved cultivation techniques, metagenomic methods, extremophiles, and screening for enzymatic activities, (ii) novel systems in enzyme expression and production, and (iii) the recent advances in applying enzymes in the dairy industry with a focus on lactose-free products, and synthesis of galactooligosaccharides.

DISCOVERY OF NEW ENZYMES

When searching for new enzymes, it is important to keep the final application requirements in mind. It can be possible to engineer the desired characteristics into an enzyme backbone (*e.g.* increase stability, modify optimum temperature), but there is no guarantee that this will work and it can be tedious and require advanced setups to generate the optimal variants. Therefore, it is highly recommended to take these requirements into consideration already when initiating enzyme discovery efforts, and activity assays for screening should focus on this. A straight forward approach is to look for enzymes from microorganisms from natural environments with characteristics similar to the application, if such environments can be found. Alternatively, these characteristics can be mimicked in the laboratory *e.g.* by establishing appropriate enrichment cultures. In this section, various approaches for enzyme discovery will be discussed with focus on methods to access the vast majority of microorganisms that cannot be cultured in the laboratory by standard techniques as well as systems for enzyme discovery based on extremophiles.

Approaches for Enzyme Discovery

Most of the enzymes being used commercially today originate from microorganisms that have been cultured. Having a natural enzyme producer in culture is desirable, since it allows for easy initial characterization of enzyme activity as well as evaluation of potential risks of the host organism. However, it is generally accepted that less than 1% of the total cell counts of an environmental sample can be easily cultivated in the laboratory [1], leaving the vast majority of microbial biodiversity unreachable by traditional cultivation based methods. This uncultured and poorly explored microbial diversity harbors an enormous potential for the discovery of new enzymes with industrial potential, including enzymes for the dairy industry. Methods to explore this enzymatic diversity exist and can be divided into two groups: *improved cultivation techniques* and *metagenomics methods* (Fig. **8.1**), which will be discussed in the following paragraphs [2].

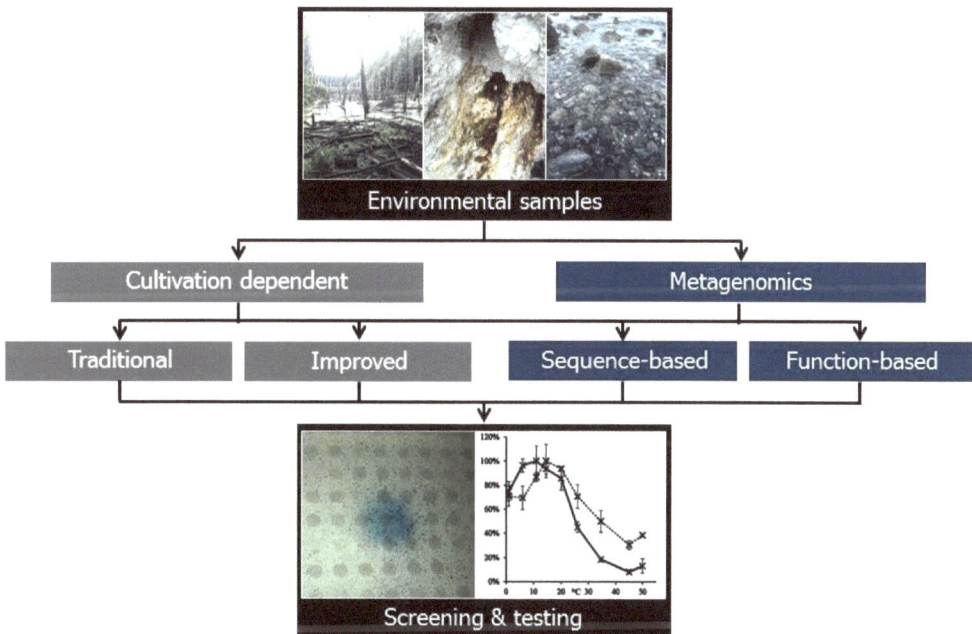

Fig. (8.1). General overview of approaches for enzyme discovery from environmental samples. Cultivation dependent methods (gray boxes) consist of traditional as well as improved cultivation methods, and metagenomics methods (based on DNA, blue boxes) are sequence- or function-based. See text for details on the different methods. Modified from Vester *et al.* [2].

Improved Cultivation Techniques

There are many reasons to why most bacteria evade cultivation efforts, including missing growth factors, lack of specific nutrients, overgrowth by others, toxic nutrient levels, special requirements regarding pH, temperature, osmolality, pressure, and many more [3]. Several techniques with the aim of culturing previously uncultured microorganisms have been developed. A general strategy is to mimic the natural environment or even to bring this environment into the laboratory. One of the first methods was the *diffusion chamber* [4], which was later modified into a high-throughput version, the *isolation chip* (iChip) [5]. The principle of these techniques is that cells are separated from the environment by permeable membranes, which allow for diffusion of chemical components including growth factors and nutrients. By placing them in natural or engineered environments, the growth of cells can be established, and once growth is established, pure isolates can often be obtained. These can then be screened for enzymatic activities of interest. The *hollow fiber membrane chambers* [6] is also build on this principle, and variations like encapsulation of microorganisms in *gel-microdroplets* in a continuous fed-batch system [7], as well as *in-situ* incubation of *encapsulated microorganisms* [8] have successfully been used to establish cultures of novel microorganisms.

Less advanced techniques have also been applied with success to increase the cultivability of microorganisms. These include the use of *gellan gum* instead of agar in growth media [9], significantly increasing the *incubation time* [10, 11], and addition of *resuscitation-promoting factors* [12], or *siderophores* [13] to promote growth. *Dilution to extinction* is a prominent technique [14], which was used to successfully culture members of the ubiquitous marine SAR11 clade for the first time [15]. The principle of this is to dilute samples down to single cells before cultivation in isolation. Recently, large-scale dilution to extinction was used to isolate numerous previously uncultured strains from sea-water [16], and Sakai and Kurosawa [17] coupled enrichment cultures with 16S diversity analysis to *dilution to extinction* to isolate novel thermophilic archaea.

Metagenomics Methods

Despite the efforts in improved cultivation techniques, isolates of most microorganisms cannot be obtained and thereby their enzymes remain unexplored by cultivation dependent efforts. However, these can be accessed by metagenomics methods, which rely on DNA instead of viable cells. DNA extraction is inheritably biased and careful considerations must be made on which method(s) to apply and whether to use *direct* (*in situ* lysis of cells) or *indirect* (*ex situ* lysis of cells) DNA extraction. The choice of which methods to apply depends on the sample matrix, biomass level, presence of unwanted DNA (*e.g.* from biofilm, plant or algae), and no general rule can be applied. A low amount of DNA is also a common problem, not least for extreme environments. This can be solved by *multiple displacement amplification* (MDA), where the φ29 DNA polymerase is used to amplify any DNA present [18]. However, MDA is heavily biased, especially for complex samples, and one should realize that the resulting DNA will not represent the natural situation [19, 20]. An alternative to MDA is *enrichment cultures*, which can be used to establish cultures with increased biomass allowing for the extraction of higher amount of DNA. Carefully designed enrichment cultures can bias the microbial community in a desired direction, ideally enriching for enzymatic activities of interest increasing the hit-rate for these.

When DNA has been obtained, there are two options: *sequence-* or *function-based* metagenomics (Fig. **8.1**). In the *sequence-based* approach, metagenomics sequence is established and annotated using bioinformatics tools. The identified genes of interest then must be cloned and expressed in a suitable host to verify and characterize activities. This approach has a very high-throughput and generates a lot of potential genes of interest that can be analyzed further. However, the functional annotation using bioinformatics is generally based on sequence homology [21], meaning that truly novel genes will not be annotated due to low homology to previously characterized genes. Sequence-based metagenomics has been used in numerous studies; one example with relevance for the dairy industry is a comparison of the microbiomes of adult and a breast-fed baby elephant by Ilmberger *et al.* [22], which showed that the majority of glycosyl hydrolase (GH) families in the baby elephant were β-galactosidases, illustrating this environment as a natural enrichment for lactose degradation. Searching for enzymes with specific characteristics (*e.g.* cold-adaption) through sequence-based metagenomics is not trivial, but machine- learning based on general traits can be applied. For example, psychrophilic enzymes are often found to have character-istic features including decreased core hydrophobicity, lower arginine/ lysine ratio, more and longer loops, decreased number of proline residues in loops, as well as fewer disulfide bridges and hydrogen bonds [23]. These characteristics could be implemented in bioinformatics tools for gene annotation to increase the chance of identifying cold-adapted enzymes, and this idea could be applied for thermostable, halophilic, *etc.* enzymes as well.

Function-based metagenomics offer the chance of finding completely novel enzymes, since it does not rely on previously characterized genes. The principle of this method is to randomly fragment DNA and insert it into a vector for functional expression in a suitable host. Although the potential for discovery of truly novel enzymes is good, the associated hit-rate is very low and many difficult decisions and compromises regarding choice of vector, expression host, screening conditions *etc.* have to be made [24, 25]. Some of the non-standard systems available for enzyme expression are discussed later in this chapter, and these could be adopted for function-based metagenomics to increase the hit-rate for specific enzymes (*e.g.* archaeal).

Cultivation and metagenomics based methods for enzyme discovery were compared in a study by Vester *et al.* [20]. The microbial community of a cold and alkaline environment in Greenland was analyzed for cold-adapted enzymes by a combination of cultivation dependent (classical strain collection) and independent (function- and sequence-based metagenomics) approaches. The results demonstrated the benefits and drawbacks of the different approaches. The cultivation based approach led to a high number of enzyme producing strains; however, there was a high level of redundancy and low novelty, suggesting that traditional cultivation based methods are best suited for previously uncharacterized environments. The functional expression had a low hit-rate, but the discovered enzymes were novel and easily produced in the chosen host, *Escherichia coli*. Sequencing of the metagenome revealed, as expected, that most of the potential enzymes were not discovered by functional expression. More details on cultivation, sequence- and function-based metagenomics can be found in a review by Vester *et al.* [2].

Extremophiles

When searching for new enzymes of industrial relevance, extremophilic microorganisms are of special interest. These organisms, including their enzymes, have evolved to function at extreme conditions, of which many resemble the conditions applied for various industrial processes. Cultivation of extremophilic organisms can be very challenging. Growth conditions can be difficult to mimic in the laboratory, and *in situ* incubation can be impossible due to several factors including hostile conditions, inaccessible locations, and prolonged incubation time [2]. Many extreme environments are, therefore, fairly under-explored, increasing the chance of finding truly novel enzymes. Extremophiles are generally defined as microorganisms that can live and reproduce at extremes of temperature ($<15°C$, $>45°C$), pH (<5, >8.5), salinity ($>6\%$ NaCl (w/v)), pressure (>50 MPa), as well as in the presence of high levels of heavy metals and radiation, and most of the identified extremophiles are Archaea [26]. The prokaryotic limits of life, as determined by pure isolates, are currently -12 to $122°C$, pH 0 to 14, 0 to 33% NaCl (w/v) and 0 to 120 MPa [27] (Table **8.1**). However, these limits are most likely underestimates, since they are based on growth of isolates rather than *in situ* growth, where the combination of multiple extremes can push the limits further as exemplified by the discovery of active bacterial cells in Arctic sea ice brines at -20°C [28]. These limits of life also define the limits where one can expect to find functional enzymes. The applications of some of these "extremozymes" are discussed below.

Table 8.1. Prokaryotic limits of life with examples of strains. These limits are determined by pure prokaryotic isolates and are likely underestimates of true limits (see text for details). Modified from Harrison *et al.* [27].

	Limits	Strain	Domain	Source
Temperature (°C)	-12	*Psychromonas ingrahamii*	Bacteria	Sea ice
	122	*Methanopyrus kandleri*	Archaea	Hydrothermal sediment
Salinity (% NaCl (w/v))	0	*Caldivirga maquilingensis*	Archaea	Acidic hot spring
		Telmatobacter bradus	Bacteria	*Sphagnum* peat
	33	*Bacillus aidingensis*	Bacteria	Salt-lake sediment
pH	0	*Acidiplasma aeolicum*	Archaea	Hydrothermal pool
		Picrophilus oshimae	Archaea	Hydrothermal systems in solfataric fields
	14	*Nocardiopsis valliformis*	Bacteria	Alkali lake
Pressure (MPa)	0	*Desulfovibrio piezophilus*	Bacteria	Wood falls
	120	*Pyrococcus yayanosii*	Archaea	Deep-sea hydrothermal vent

Psychrophilic enzymes are optimized to function at low temperature, which is partly obtained by a highly flexible structure. They are appealing to the dairy industry, since they allow for processes running at low temperatures, which can reduce the risk of spoilage, preserve flavor, and prevent change of nutritional value. In addition, psychrophilic enzymes have high specific activity reducing the amount of enzyme required and they are easily inactivated. Cold-adapted microorganisms are found in many places including high mountains, polar regions, glaciers, deep-oceans, caves, the upper atmosphere, as well as refrigerated appliances and on the surface of plants and animals living in cold environments. In the dairy industry, cold-active proteases can be used to increase cheese maturation. Microorganisms producing cold-adapted proteases have been found in multiple places including soils, glaciers, sea ice, deep-sea, fish, and fish microbiome [29]. Also, cold-adapted β-galactosidases that hydrolyze lactose at low temperature have been found. Examples are β-galactosidase from the Antarctic psychrophile *Pseudoalteromonas haloplanktis* [30], the Arctic bacterium *Alkalilactibacillus ikkense* [31], and a β-galactosidase from a cold and alkaline Arctic metagenome [20]. Psychrophilic enzymes could also be used for cleaning applications in the dairy industry, where they could reduce the need to cycle between cool operating and hot cleaning temperatures. In addition, the heat-lability often found for the psychrophilic enzymes can be an advantage to ease the inactivation by applying moderate heat after cleaning [23]. It has also been shown that cold-adaption can be engineered into existing enzymes, as exemplified by site-directed modifications on a subtilisin protease performed by Tindbaek *et al.* [32], and increased heat-lability can be introduced by removing glycosylations

[33]. The heat-lability might not always be beneficial, and cold-adapted enzymes from deep-sea environments have been found to have increased thermal stability [34], making deep-sea environments a good place to search for stable cold-active enzymes.

Like psychrophilic enzymes, the use of *thermophilic* enzymes for high-temperature bioprocessing has many advantages including reduced risk of contamination, continues recovery of volatile products, increased solubility of substrates, decreased viscosity, and reduced cooling costs [35]. Thermophilic enzymes are not only providing improved performance at high temperature, but are often also associated with solvent tolerance and stability [36]. The low activity at low temperatures of thermostable enzymes can be exploited to inactivate the enzymes by cooling [37]. Ultra high temperature (UHT) treatment in which temperature is raised to 135-150°C for a few seconds, is a preservation technique often used for drinks, such as milk [38]. Thermophilic enzymes could potentially be active after UHT treatment, which could be used to generate lactose free milk. Another example of using thermophilic enzymes in the dairy industry is a β-galactosidase from the archaea *Solfolobus solfataricus* that has been used to produce galactooligosaccharides (GOS), and there was an increase in GOS yield when temperature was increased from 60 to 75°C. This was due to increase in the lactose concentration, which is required to prevent water from being the accepter in the reaction, which results in hydrolysis of lactose instead of transgalactosylation [39].

Enzymes that tolerate high pressure, *piezophilic* enzymes, have several possible applications in the food industry, especially within high-pressure food pasteurization and processing [40], but the use of these have been limited. This is mainly due to the difficulties in cultivating piezophiles and therefore leaves an open opportunity to explore *e.g.* by metagenomics approaches [26].

Finally, it is worth mentioning that a general increased knowledge on extremozymes can be used as inspiration for engineering optimized versions of existing enzymes that perform well but lack a certain characteristic.

Screening for Enzymatic Activities

It is critical to have a robust and reliable screening system for the enzymatic activities of interest, and preferably it should have high-throughput capacity, especially if a low hit-rate method like function-based metagenomics is used. Screening systems are constantly being developed and can be very specialized. It is beyond the scope of this chapter to provide a full overview, and instead a few examples and general comments are presented.

Plate based screenings using *e.g.* chromogenic substrates can often be applied directly at the desired conditions of interest (*e.g.* high/low temperature and/or pH), and multiple substrates can be combined to reduce the workload and allow for more high-throughput screening [41, 42]. If the growth requirements of a host organism are different from those desired for the enzyme activity, conditions can be switched from growth to screening after the biomass and enzyme production has been established [23]. This approach was used to be able to screen for proteolytic activities in *E. coli* at pH10 by using a high pH top-agar with substrate [43], and to screen for cold-adapted β-galactosidase, α-amylase, protease, and phosphatase activities at low temperature in *E. coli* by transferring expression libraries to a lower temperature after growth at 37°C [20].

Ingham *et al.* [44] have developed the 'micro-Petri dish, a million-well growth chip' which allows high-throughput cultivation and screening of microorganisms. With this system, it is possible to screen both engineered libraries as well as environmental samples directly for cells with desired enzymatic activities, requiring that a suitable assay is available for those activities. Two interesting approaches for high-throughput screening of function-based metagenomics libraries are *substrate-induced gene expression* (SIGEX) and *product-induced gene expression* (PIGEX). In SIGEX, DNA of interest is cloned upstream of a promoter-less fluorescent reporter, and cells with DNA harboring promoters induced under the conditions applied, can be selectively sorted using fluorescence-activated cell sorting (FACS) [45]. In PIGEX, a library is co-cultured with a reporter strain with a product sensitive promoter, allowing for detection of cells producing the desired product [46]. Finally, an extremely high-throughput method for screening for microbial enzymes using gel microdroplets (GMD) and microfluidics has been developed by Hosokawa *et al.* [47] for function-based metagenomics, and by Nakamura *et al.* [48] for environmental samples directly.

Novel Systems in Enzyme Expression and Production

The most common expression and production strains are based on standard and well characterized hosts with a substantial genetic toolbox available (*e.g. E. coli*, *Bacillus subtilis*, *Saccharomyces cerevisiae*, and *Pichia pastoris*). However, as described previously, novel enzymes with industrial potential are likely to be found in non-standard environments, and it is therefore likely that they can be problematic to express using standard hosts. Several systems for enzyme expression and production have been developed using non-standard hosts (*e.g.* extremophiles) or no host at all (*in vitro* expression). These will be discussed below.

Expression Systems Based on Extremophiles

Low-temperature expression systems offer the possibility to maintain stability of heat-labile enzymes, increase the chance of producing soluble proteins instead of inclusion bodies, and be able to thermally suppress enzyme activity, *e.g.* by producing a thermophilic protease with toxic effects at low temperature [23]. Several systems for *low-temperature* expression have been developed. For gram-negative bacteria, these include expression in *Pseudoalteromonas haloplanktis* [49], *Shewanella livingstonensis* [50], and an *E. coli* strain optimized for low temperature expression. The latter was generated by including two chaperones from the psychrophilic Antarctic bacterium *Oleispira antarctica* [51], a system commercially available as ArcticExpress (Agilent Technologies, Santa Clara, CA). For psychrophilic gram-positive bacteria, a shuttle vector for expression in *Arthrobacter*, *Microbacterium*, *Curtobacterium*, and *Rhodoglobus* has been developed [52].

As discussed previously, deep-sea organisms could be a source of stable cold-adapted enzymes, and vectors for cloning and expression in the psychro-*piezophilic* bacteria *Photobacterium profundum* SS9 and *Shewanella piezoto-lerans* WP3 are available. These bacteria originate from samples taken at 1,914 m and 2,551 m for *P. profundum* SS9 and *S. piezotolerans* WP3, respectively [40]. Recently, the low-temperature-inducible gene expression system developed for *S. piezotolerans*, was shown also to be working in other *Shewanella* species, including the fresh water lake bacterium *S. oneidensis* and psychrophilic *S. psychrophila* isolated at 1,910 m depth [53].

For *high-temperature* expression, multiple organisms are available including both bacteria and archaea. *Thermus thermophiles* is a bacterium with optimum growth at 70-80°C, that has been shown to be able to express active tagged versions of homologous proteins [35]. *Thermotoga* are bacteria with a growth optimum around 80°C of which *Thermotoga* sp. strain RQ2 has been engineered to overexpress heterologous active cellulases [54]. *Thermococcus kodakarensis* is an archaeon with optimal growth at 85°C [35], which has been developed to enable expression of heterologous genes with strong constitutive promoters as well as a signal peptide for secretion of enzymes [55]. *Pyrococcus furiosus* is the most developed hyperthermophile host for protein expression, including both constitutive and inducible promoters. It has an optimum growth temperature at 100°C. The cold-inducible promoter available for *P. furiosus* allows for temperature shift production, where growth- and production-phase can be separated by lowering the temperature from 80°C to 72°C [35]. For the hyperthermophilic archaeon *S. solfataricus*, which shows optimal growth at 80°C and pH 3, a virus based shuttle vector has been developed for production of recombinant and tagged proteins by genome integration [56]. Both homologous and heterologous proteins have been expressed with success in *S. solfataricus* and the use of both His and Strep tags have been demonstrated. Small non-integrative multicopy shuttle vectors are also available for *Sulfolobus* [57], and shuttle vectors for *Sulfolobus acidocaldarius* have recently been optimized [58]. Genetic systems for expression in the hyperthermo-*piezo*-philic archaea *Pyrococcus abyssi* GE5 and *Thermococcus barophilus* MP are also available. These strains were originally sampled at 2,200 m and 3,550 m for *P. abyssi* GE5 and *T. barophilus* MP, respectively [39].

Halophilic expression system is available for the halophilic archaea *Halobacterium salinarum* by a shuttle-vector with inducible K_+ dependent promoter [59]. Proteins from *H. salinarum* are generally acidic and negative in charge, which allows them to remain in solution at high salt concentrations [60]. In addition, a series of tryptophan inducible expression plasmids has been developed for expression of proteins in the archaea *Halobacterium volcanii*, and a strain has been optimized for His tag purification and transformation of methylated DNA [61].

In vitro Expression Systems

The basic principle of *in vitro* expression (also known as cell-free protein synthesis) systems is to generate a crude extract from cells, remove endogenous DNA and mRNA, add free amino acids and energy components, and initiate protein synthesis by adding template DNA or mRNA. The use of DNA as template is referred to as *coupled* reactions since transcription and translation are coupled, whereas the use of mRNA as template is called *linked* reactions. The main benefits of *in vitro* expression systems are 1) high-throughput potential, 2) no need for time-consuming cloning, and 3) the fact that they are open systems allowing for easy manipulation of reaction conditions [62]. *In vitro* expression systems are available from a range of organisms including bacteria, archaea, fungi, plants, insects, and mammals; each has its benefits and limitations. Some of these are discussed briefly in the following.

In vitro expression based on *E. coli* has been studied and applied intensively since it was first reported more than 50 years ago [63]. It benefits from the extensive toolbox available from this model organism and can produce high protein yields in a simple, fast, and cost-efficient manner. However, being a prokaryote, it suffers from limited capability of post-translational modifications required for many eukaryotic proteins as well as lack of endogenous membrane structures for synthesis of membrane proteins. A series of commercial *in vitro* expression kits based on *E. coli* are available, including EasyXpress and RTS (biotechrabbit

GmbH, Hennigsdorf, Germany), PURExpress® (New England Biolabs, Ipswich, MA), and Expressway™ (Thermo Fischer Scientific, Waltman, MA). *E. coli* based systems are optimized for expression at 37°C, which can be a limitation for both psychro- and thermophilic proteins as discussed previously. The optimal temperature for an *E. coli* based system was lowered to 25°C by using a 5'UTR from a cold-shock gene [64], and the ArcticExpress strain containing chaperones from a psychrophilic bacterium could potentially be used for low-temperature *in vitro* expression based on *E. coli*.

Several systems are available for thermophilic proteins. These proteins might require high temperature for correct folding, and the increased temperature could also be beneficial in decreasing the formation of secondary mRNA structures that inhibit translation [62]. A high temperature *in vitro* expression system has been established based on the thermophilic bacteria *Thermus thermophilus* supplemented with recombinant expressed *E. coli* proteins [65]. In this system, mRNA was successfully generated for five of six heterologous genes, and translation worked from 37°C to 65°C. The lower growth limit of *T. thermophilus* is 47°C, illustrating the wider operation range of cell free expression systems. However, 65°C is well below the upper growth limit of *T. thermophilus* (~80°C), which is most likely due to missing co-factors (*e.g.* chaperones) or lower heat-stability by recombinant expressed proteins. The most thermophilic organisms belong to the archaea, and *in vitro* expression systems have been developed using archaea. The thermophilic archaea *Sulfolobus* sp. B12 has been used to establish an *in vitro* transcription system with an optimal temperature at 75°C [66], and an *in vitro* expression system with a temperature range of 40-80°C has been established and optimized for the thermophilic archaea *Thermococcus kodakaraensis* [67].

A highly developed *in vitro* expression system is based on the protozoan *Leishmania tarentolea*. The benefits of *L. tarentolea* compared to other eukaryotic cell-free expression systems are that it is amendable to large scale fermentation and can easily be genetically modified. At the same time, it does not suffer the problems with batch-to-batch variations experienced in organisms with long lifecycles due to influence of environmental factors [68]. *L. tarentolea* has been shown to be superior to *E. coli* based cell-free expression systems in expressing mammalian and protozoan GTPases functionally [69]. Interestingly, species-independent translational leader sequences (SITS) have been developed and shown to facilitate expression on both prokaryotic (*E. coli*) and eukaryotic (rabbit reticulocyte lysate, insect cell extract, wheat germ extract, *S. cerevisiae*, and *L. tarentolea*) cell-free expression systems. This circumvents the requirement of *in vitro* capping or addition of UTRs to mRNA before expression in eukaryotic systems like *L. tarentolea* [70]. The *L. tarentolea* expression system is commercially available as LEXSY (Jena Bioscience GmbH, Jena, Germany), which works best at 20-27°C. This includes an antisense oligonucleotide to specifically block expression of endogenous *L. tarentolea* mRNAs.

Several other well-established systems for *in vitro* expression based on plants (wheat germ extract, tobacco BY-2 extract), fungi (yeast extract), insects and mammals (rabbit reticulocyte extract, CHO cell extract, HeLa cell lysates) are commercially available. More details can be found in a recent thorough evaluation of the potentials and limitations of currently available *in vitro* expression systems by Zemella *et al.* [62].

Improving Yield and Functionality of Newly Discovered Enzymes

Besides having the required activity, enzymes must fulfill certain criteria to make it into industry. Industrial production is a scientific field on its own, so only very general considerations are presented here. Production should be economically feasible, products must be sufficiently stable and easy to purify, and production system and strains must be robust. To keep production costs low, the media used for production should be cheap, safe, consistent and easily accessible. This means that chromosomal integration is generally preferred for production, since the use of plasmids adds an additional expense in terms of components that needs to be added for selection pressure. Downstream processing is also very important in terms of keeping production costs low. Often it is highly preferable to have the enzymes secreted into the culture broth which makes downstream purification less challenging and thereby less expensive. This can often be achieved by introducing a strong signal peptide from the host in front of the gene encoding the enzyme of interest. In the same way, solubility of enzymes can be increased by fusion to a small soluble protein. When choosing host for production, it can be relevant to go for a host which has 'generally regarded as safe' (GRAS) status to allow for easier approval of products. Also, strain engineering can be applied to significantly increase the yield. This could be by expressing co-factors like chaperones to assist in correct folding of the enzymes or to remove endogenous proteases with activity on the protein produced.

FORMULATION OF ENZYMES FOR FINAL USE

Formulation of enzymes is a requirement of all enzyme products to ensure that the enzyme is active in use after storage at the enzyme- and food production-site. Solid and liquid formulations are the two major formulation categories; which formulation to use depends on enzyme stability and the application.

Solid Formulation

Solid formulation is an obvious choice if dry blending of enzyme and substrate is preferred before application. However, for most dairy applications the enzyme is added to liquid milk and therefore a proper mixing system should be present in applications of solid enzyme products. The most common reason for selecting solid formulation is its superior stability on the shelf at ambient temperature for years. The most challenging aspect for the enzyme producer is to (i) find the right condition to dry the enzyme so it can be re-solubilized in a fully active form and (ii) avoid dust formation in handling. Beside stability reasons, solid state formulation can be selected if a high enzyme concentrate product is needed or if no additional additives are wanted in the final application (additives is always included in liquid formulation).

Liquid Formulation

Liquid formulation has the advantage that it is easily applicable to milk using tubes/pump and also offers a more efficient mixing than solid formulation. The main problem for a liquid formulated enzyme product is stability during storage due to loss of activity *e.g.* proteolysis, denaturation, and precipitation. Furthermore, microbial stability should also be controlled which is a noticeable difference between solid and liquid formulation, as microbial growth is efficiently diminished in solid formulation.

Typically, only a few formulation additives are allowed for a relevant food segment and the enzyme producer should be aware if any carry-over issue exists *e.g.* (i) from ingredients used for enzyme fermentation, (ii) if the additives used for enzyme formulation exceed the maximum levels in the final food, for example if the addition of NaCl in the enzyme formulation exceeds the maximum sodium level in the final food, as there are other sources of sodium in the food ingredients list. Common liquid formulations of enzymes are glycerol, sorbitol, sucrose, NaCl, KCl, and if preservatives are needed for microbial stability then Na-benzoate and K-sorbate are also options.

ENZYMATIC PROCESSES IN THE DAIRY INDUSTRY

As milk consists almost exclusively of proteins, lipids, lactose, and water, the potential enzymes used commercially in the dairy industry are limited and at present the most relevant enzymes are lactases, proteases, peptidases, and lipases. As lactose is only found in milk, the application of lactases is unique for the dairy industry.

Lactose Hydrolysis

Lactose constitutes the largest part of the dry matter in milk, and lactose hydrolysis is one of the most advanced enzymatic applications in the dairy industry, if not in the whole food industry. The benefits of lactose hydrolysis in milk and other dairy streams, for example in whey and whey permeate, are numerous, such as (1) the production of sweetening syrups for use in ice cream and other dairy and non-dairy foods, (2) enabling fermentation by lactose negative microorganisms which has benefits in producing certain types of biomass, such as *saccharomyces cerevisiae* and other lactose negative yeasts, (3) the avoidance of lactose crystallization in products, such as ice cream and dulce de leche, (4) increasing the availability of dairy nutrients worldwide by alleviating lactose intolerance, and (5) reducing added sugars in dairy products by naturally unlocking the sweetness *via* lactose hydrolysis into glucose and galactose.

Lactose hydrolysis is carried out commercially by β-galactosidases (E.C. 3.2.1.23), usually known as lactases, which are widely ubiquitous enzymes in nature and are probably one of the most studied enzymes in the literature. Gist Brocades nv, a Dutch company (acquired by DSM in the late 1990s), launched the first commercially available yeast lactase derived from *Kluyveromyces lactis* in the late 1960s. However, the move towards industrial applications of the lactose hydrolysis process gained momentum in mid-1980s, when more attention was given to lactose intolerance in the dairy industry. Today, lactose free products are not only seen as products suitable for lactose intolerant consumers but are also positioned as healthy products in the market, together with other healthy market trends and concepts.

Potential Modes of Lactose Hydrolysis

The various potential modes by which lactose hydrolysis can be achieved in the dairy industry are:

Membrane Reactor Processes

This can be basically applied in protein-free streams, such as whey permeate. In this case, the enzyme is recovered from the reaction mixture with a second UF (ultra-filtration) equipment and the permeate, which contains the hydrolysed lactose and does not contain the enzyme, is remixed with the whey retentate, which is produced by a first UF process. However, the economic feasibility and complexity of this process is not convincing on the industrial scale, due to reasons related to membrane fouling, membrane selectivity, effects of some solutes on flux, the need for a high membrane area to overcome the low mass transfer rate, gradual loss of enzyme activity, potential of microbial growth during continuous operation at ambient or higher temperatures, and the fact that it is very challenging to apply this technology in protein rich media, such as milk [71 - 73]. In addition, the reduced price of commercial soluble lactases decreased the interest in the potential application of membrane reactors for lactose hydrolysis, on the commercial scale. The use of soluble enzymes has been more economically feasible than investments in complex reactor processes.

Immobilized Lactase Systems

This appears to have a great potential for large scale application for lactose hydrolysis in milk, permeate, or whey, for reasons related to cost-saving, process continuity without pre-incubation steps, and not mixing the lactase with the final dairy product. The useful lifetime of an immobilized system can be usually between hundred and several thousand hours, which significantly reduces the cost compared with the use of soluble enzymes. However, in case of milk, the microbial stability of the reactor is very difficult to control since the whole process of lactose hydrolysis needs to be carried out at neutral pH values. Probably the only exception is operation at very low or very high temperatures to avoid microbial issues in milk. However, prolonged incubation of milk at high temperature has adverse effects on the milk nutritional and organoleptic quality. Another major disadvantage of immobilized lactase systems is the fact that milk proteins tend to adsorb onto the surface of the immobilized enzyme surface, *i.e.* the inert carrier of the immobilization system, and foul the reactor. Additionally, neutral β-galactosidases, which are mostly used for lactose hydrolysis in milk, are not very stable when immobilized with classical techniques, such as adsorption or covalent linkage, unlike acid lactases which show more stability upon immobilization [71 - 73]. To our knowledge, the only known commercial application of an immobilized lactase for lactose hydrolysis in milk was based on immobilizing a neutral-pH enzyme of *Kluveromyces lactis* by entrapping it in porous cellulose acetate fibres. Immobilization gave the enzyme a shelf life of about 100 days. Low-lactose milk has been produced using this technology by SNAM Progetti at the Centrale di Latte in Milan.

Mechanically Disrupted, and Immobilized Non-Viable Microbial Cells Containing High Lactase Activity

This allows lactose diffusion into the cells and thus enhances the efficiency of the lactose hydrolysis process. This approach might sound economically and technically feasible, as there is no need to purify and formulate the lactase enzyme and since non-viable cells will not affect the milk quality by their metabolic activity [71 - 73]. However, there are many disadvantages related to the fact that other enzyme activities present in these non-viable cells might cause off-flavour in the product. Lactose hydrolysis nowadays covers a wide range of dairy products of different compositions and shelf-lives, and purity of even commercial lactase preparations (*e.g.* free from side activities, such as arylsulfatase, amylase, protease, and invertase) is a vital requirement in the dairy industry to ensure the production of high quality lactose free dairy products. Furthermore, the reproducibility of the hydrolysis process, which involves the use of permeabilized, mechanically disrupted, and immobilized non-viable microbial cells containing high lactase activity, is very challenging as the stability of lactase activity contained in these cells is not guaranteed over a long time, *e.g.* hundreds of hours, to ensure a continuous hydrolysis process. Last but not least, the complexity of this mode of lactose hydrolysis, compared to adding soluble commercial lactase into milk, makes it not attractive on the large industrial scale.

Use of Free Soluble Lactases

Adding soluble lactases into milk and other dairy streams to produce lactose-free products is the major commercial mode of lactose hydrolysis. The simplicity of this process, the continuous advances in the purity of commercial lactase preparations, and the significant decrease in the price of commercial lactase will keep this mode of lactose hydrolysis as the preferred mode by industry for many coming years. In this mode, lactase is mixed with *e.g.* milk either before the process or at the end of the process. So far, the majority of applications include adding lactase into milk and incubating the milk at cold temperatures (mostly < 10 °C) for enough time to reduce the lactose content to < 0.1% or < 0.01% of the milk, in most countries, followed by processing, such as pasteurization and UHT treatment. In this way, the added lactase is denatured and considered only as a processing aid. The drawbacks related to this type of application include the use of considerable amount of the enzyme, and the fact that upon severe heat treatment (*e.g.* UHT) Maillard reaction takes place at a fast rate due to (i) the presence of highly reactive reducing monosaccharides in the milk, and (ii) the fact that reducing sugars are present at a double molar concentration compared to normal untreated milk. Recently, process engineering technologies, such as Aldose and Flexdose systems developed by Tetra-pak and Varidose system developed by GEA has made it possible to dose very small quantities of sterile lactases with high precision into the product stream after heat treatment and just before the time of aseptic packaging. This is very beneficial in the case of UHT and ESL (extended shelf-life) milk drinks, and lactose hydrolysis in this case takes place during the first few days of storage. In addition to significantly saving on the cost of enzyme, the major benefits of using Aldose, Flexdose, and Varidose systems over the traditional batch incubation of the enzyme with milk are (i) decreasing the extent and level of Maillard reactions, browning, and Advanced Glycation End products (AGEs) in the final products, due to the fact that the reducing monosaccharides are not present during the heat treatment, (ii) saving a complete process step, *i.e.* incubating the enzyme with milk; this improves the production capacity at dairies and positively contributes to sustainability. On the other hand, when commercial lactases are added at the end of heat treatment just before aseptic filling, they are considered as ingredients. Besides, they need to be sterile in the case of Flexdose and Varidose systems from a microbial perspective, as well as very pure from all deleterious side activities, as the lactase is in continuous contact with the milk during the entire shelf life of the product. In terms of lactase side activities, Nielsen *et al.* [74] studied the storage-induced changes in lactose free UHT milk produced using 5 commercial lactase preparations (termed A1, A2, B1, B2, and B3), compared with conventional UHT milk. The highest level of free amino acids during storage was found when using A1 and B1, while it was significantly lower when using A2, B2, and B3. Compared to conventional UHT milk, lactose free milk prepared using all 5 lactase preparations showed significantly higher levels of free amino acids, which was partly due to release of C-terminal amino acid residues from intact β- and α_{s1}-casein, indicating carboxypeptidase activity. Moreover, using A1, A2, and B1 lactase preparations increased the release of peptides derived from the hydrophobic regions of β- and α_{s1}-casein, which could potentially contribute to development of bitterness in milk during storage, while these peptides were not observed when using B2 and B3 lactase preparations. This study clearly showed that the residual proteolytic side-activity differs significantly between commercial lactase preparations, and has a vital role in determining the quality and shelf-life of lactose free UHT milk. In a previous study by the same research group, sensory properties of lactose free UHT milk, particularly bitterness, stale aroma, and yellowness of milk, were correlated with the level of peptides and free amino acids observed as a result of proteolytic side activities in commercial lactase preparations [75]; a fact which suggests that elimination of residual proteolytic activity in lactase preparations may therefore be of interest in relation to increasing the shelf-life of lactose free UHT milk.

Recent Advances in Research- and Commercial-Lactases

Most of the commercially available lactases in the market are neutral lactases, derived mainly from *Kluyveromyces lactis*, *Kluyveromyces fragilis*, and *Bacillus circulans*. There are many formulations of these enzymes marketed by many suppliers, which differ in the enzymatic strength of the lactase and in the several grades of purity of the preparations suited for different dairy applications. These enzymes have found applications in various dairy products, such as liquid milk, fermented products, milk powder, fresh cheese, cream cheese, and many other dairy products. Another class of commercially available lactases is the acid lactase, derived mainly from *Aspergillus oryzae*, which is suitable for fermented dairy products as well as for use in acid whey applications. Recently, new commercial lactases, derived from *Bifidobacterium bifidum,* have been introduced to the market By Novozymes and Chr. Hansen. These lactases have the benefits of (i) retaining most of their activity at a pH range of 5-7, making them suitable for applications in both liquid milk and fermented dairy products, and (ii) reaching very low lactose levels faster than the known acid lactases from *A. oryzae*, which makes them a very cost-effective choice, particularly for fermented dairy products. However, in a country like France, these lactases need to be labeled as ingredients because they are not completely denatured by the final pH value of fermented dairy products. Moreover, DSM has just launched a new lactase derived from *Kluyveromyces lactis*, which is claimed to be the fastest lactase in the market, *i.e.* able to accomplish the lactose hydrolysis process in milk faster than any other commercial lactase. In addition to improving production capacity and sustainability, this will certainly enable the dairy industry to process more lactose free milk without requiring further capital expenditure (CAPEX) in incubation tanks.

The vast expansion of the lactose free market, the need for more streamlined processes in the dairy industry, and the continuous requirements with regards to decreasing the cost-in-use of lactase applications on large industrial scale are all factors that will shape new innovations in commercial enzyme preparations. Of great interest are lactases which retain a substantial part of their activity at temperatures < 10 °C, as the lactose hydrolysis processes are performed at cold temperatures. This can allow a faster hydrolysis process, the use of much lower lactase dosages than what is currently used, reduced energy consumption, and easy inactivation. These lactases can also find applications in processes which involve the addition of lactase before aseptic filling using Aldose, Flexdose, and Varidose systems, because dairy products produced using these systems are usually stored at ambient temperatures (on average 20 °C), which allows dosing less lactase, and thus having less deleterious side-activities, as has been previously described.

Examples of these cold-active lactases are numerous, but no commercial applications of these cold-active lactases yet exist on an industrial scale. For example, a novel β-galactosidase, derived from a novel arctic bacterium, *Alkalilactibacillus ikkense*, was found to be highly active at low temperatures with more than 60% of its maximal activity maintained at 0 °C. The apparent optimal activity was observed at temperatures between 20 °C and 30 °C and at pH 8. Comparison of this β-galactosidase with a commercially available enzyme showed that the conversion rate of the *A. ikkense* enzyme was approximately two-fold higher at temperatures between 0 °C and 20 °C [76]. Furthermore, a new β-galactosidase, derived from the Antarctic bacterium *Arthrobacter* sp. 32c, maintained 60% of its maximum activity at 25 °C and 15% of its maximum activity at 0 °C [77]. Likewise, Mahdian *et al.* [78] found out that the cold active β-galactosidase derived from *Planococcus* sp-L4 maintained ca. 20% of its maximum activity at 5 °C. Moreover, Vester *et al.* [20], identified 2 novel β-galactosidases (BGal $_{117E2}$ and BGal $_{13H5}$) from cold-adapted arctic metagenome, using a functional expression library established in *Escherichia coli*. Both enzymes retained 20-30% activity at 10°C, and displayed typical adaptation characteristics to low temperature (*e.g.* low arginine, low proline, and low ratio of arginine/arginine + lysine).

Before cold-active lactases find commercial applications, various criteria should be met, such as cost-effective production, stability and solubility of the enzymes, robust expression and purification systems, as well as high catalytic efficiency (kcat/Km) of these newly discovered enzymes compared to the currently used commercial lactases.

Synthesis of Galactooligosaccharides (GOS)

In addition to lactose hydrolysis, converting lactose into prebiotic galactooligosaccharides (GOS) has recently attracted very much attention and shown a lot of progress, with focus mainly on using lactose in concentrated solutions to produce high yields of GOS as functional food ingredients. Besides, successful attempts of *in-situ* production of GOS at the natural lactose levels in milk have been recently reported.

Prebiotics are selectively fermented ingredients that allow specific changes in the composition and/or activity of the gastrointestinal microbiota, which confer benefits upon the host health and well-being [79, 80]. Galactooligosaccharides (GOS), oligomers of galactose linked to a terminal end of glucose or galactose [81], are considered as established prebiotics, due to the wealth of scientific studies showing how GOS fulfil the several criteria required for ingredients to be defined as prebiotics [79, 80]. Commercially speaking, GOS production is carried out in aqueous solutions using β-galactosidases, which catalyze the hydrolysis of terminal non-reducing β-D-galactose residues in β-D-galactosides (*e.g.* lactose). During lactose hydrolysis, β-galactosidases cleave the β-(1→4) glycosidic linkage between galactose and glucose, which releases glucose into the medium. Then, the enzyme transfers the galactosyl moiety into acceptor molecules containing hydroxyl groups. Galactose is formed when the acceptor molecule is water (hydrolysis pathway), while a new glycoside or oligosaccharide is formed when other acceptor molecules are present in the reaction medium (transgalactosylation pathway) [82, 83] (Fig. **8.2**). Lactose itself can act as both substrate and acceptor of the galactosyl moiety, and under conditions of high lactose concentration, a substantial mixture of oligosaccharides containing di-, tri-, tetra-, and higher oligosaccharides is produced from lactose.

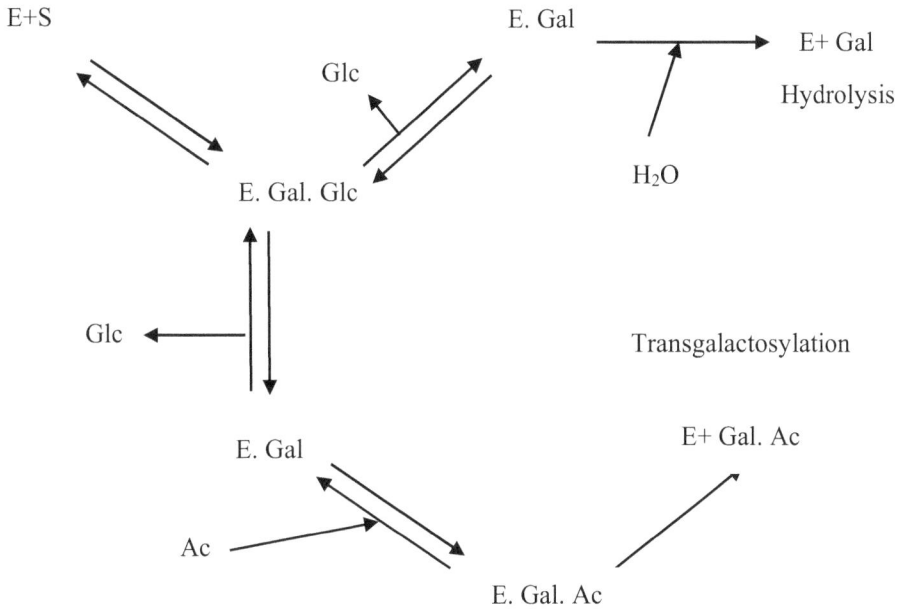

Fig. (8.2). A schematic representation of transgalactosylation during the lactose hydrolysis. E, S, Glc, Gal, and AC stand for β-galactosidase, lactose, glucose, galactose and a galactosyl acceptor respectively.

The type and the relative composition of the formed oligosaccharides depend on the ratio of the transferase activity of the used β-galactosidase to its hydrolytic activity, which is influenced by the source and characteristics of the enzyme, the relative concentration of different galactosyl acceptors in the reaction medium, and the reaction conditions (pH, temperature, time, *etc…*) [82, 84 - 86]. Despite the commercial use of β-galactosidases in GOS synthesis, they have various drawbacks, such as the low yield of GOS and the less regio- and stereo-specific type of catalysis performed by β-galactosidases. The low yield, however, can be enhanced by (1) performing GOS synthesis at a lower water activity of the reaction medium and at high concentrations of donor and acceptor molecules (*e.g.* using high concentrations of lactose), (2) removing the products from the reaction medium, and (3) controlling the synthesis conditions [83, 87].

Factors Influencing GOS Production from Lactose

Initial Lactose Concentration

Increasing the lactose concentration ensures that less water molecules act as acceptors of the galactose moiety, which favors transgalactosylation over hydrolysis, and results in the transfer of galactosyl moieties into lactose, and thus the production of high GOS yields [82, 84, 86, 88, 89]. Moreover, increasing the initial lactose concentration usually leads to obtaining GOS mixtures with higher percentage of DP ≥ 3 (DP = degree of polymerization) compared to transgalactosylated disaccharides, as in the case of β-galactosidase from *Bacillus circulans*, *Aspergillus oryzae* [84] and *Bifidobacterium bifidum* NCIMB 41171 [86]. GOS yields of > 50% (% of total carbohydrates) have been already obtained using highly concentrated lactose solutions [90].

Temperature

Temperature influences (1) lactose solubility, (2) lactose conversion rate, (3) transgalactosylation reaction rate, (4) hydrolysis reaction rate, and (5) β-galactosidase stability. Conducting GOS synthesis at high temperatures allows the use of highly concentrated lactose solutions, which favors transgalactosylation over hydrolysis. The exact effect of temperature on the hydrolysis rate and the transgalactosylation rate individually is not very well examined. Yet, few studies have generally indicated that high GOS yields (a higher transgalactosylation rate compared to hydrolysis rate) were obtained by increasing the reaction temperature at a fixed initial lactose concentration [90, 91]. In addition, Maillard reactions, taking place between the amino side chains of the β-galactosidase and the reducing sugars in the reaction medium, are accelerated at high temperatures, which can significantly deactivate the used β-galactosidase [92, 93]. Furthermore, the inactivation of β-galactosidase at high temperatures negatively affect GOS synthesis. Overcoming this particular issue is usually carried out by using thermophilic β-galactosidases with high transgalactosylation activity or by using immobilization to stabilize the used β-galactosidase. Temperature has also shown an effect on the DP in GOS mixtures. For example, the content of oligosaccharides with DP ≥ 3 increased in the GOS mixture, obtained using β-galactosidases from *Bifidobacterium bifidum* NCIMB 41171, as the reaction temperature increased from 40 to 60 °C [86]. Moreover, the content of DP 3 and DP 4 also increased as a function of temperature (up till 47 °C) using the β-galactosidase from *Aspergillus oryzae*. Besides, a β-galactosidase from *Aspergillus aculeatus* produced high content of DP ≥ 3 and low content of DP 2 by increasing the reaction temperature [94].

Enzyme Sources

The different sources of β-galactosidases also determine the diverse optimum process parameters (temperature, pH, *etc*…) required to conduct GOS synthesis. β-Galactosidases from different sources have different degrees of selectivity in terms of using water or other molecules as galactosyl acceptors, and thus produce different GOS yields with different degrees of polymerization and types of glycosidic linkages [89, 90, 95, 96]. For instance, the GOS produced by a β-galactosidase from *Bacillus circulans* consisted of di- to hexa-saccharides [97 - 99], while a β-galactosidase from *Aspergillus oryzae* produced mainly trisaccharides [100, 101] and a small amount of tetrasaccharides [102]. Also, mainly trisaccharides and, to a lesser extent, tetrasaccharides were obtained using a β-galactosidase from *Bifidobacterium longum* BRCRC 15708 [103]. Moreover, Osman *et al*. [90] reported that β-galactosidases from *Bifidobacterium bifidum* NCIMB 41171 produced mainly transgalactosylated disaccharides ($\sim$ 46 – 55% of the GOS mixture) and trisaccharides ($\sim$ 36 – 42% of the GOS mixture), whereas the content of GOS with DP $\geq$ 4 was low ($\sim$ 9 – 13% of the GOS mixture). In terms of the glycosidic linkages, different β-galactosidases usually produce GOS containing different glycosidic linkages. For instance, the β-galactosidase from *Lactobacillus reuteri* preferred to form β-(1→3) and β-(1→6) linkages [104], while mainly β-(1→4) linkages were formed using the β-galactosidase from *Cryptococcus laurentii* [105] and *Bacillus circulans* [98]. Moreover, *Aspergillus oryzae* β-galactosidase preferred to form β (1→6) linkages mainly, while the GOS produced by *Bifidobacterium bifidum* NCIMB 41171 consisted primarily of β-(1→3) linkages, with less amounts of β-(1→4) and β-(1→6).

GOS Synthesis Time

In the beginning of the GOS synthesis reaction, lactose concentration decreases rapidly with a simultaneous increase in the yield of GOS until a maximum GOS yield is obtained at a certain percentage of lactose conversion, which differs from one β-galactosidase to another. Following this, the reaction enters a stage where a balance between transgalactosylation and hydrolysis is observed [86] leading to the obtainment of a stable GOS yield. What follows this stage includes the hydrolysis of GOS into the more thermodynamically favored products glucose and galactose. Therefore, the GOS synthesis should be stopped at the time point where the maximum GOS yield is obtained [86, 106]. The reaction time also influences the DP in GOS mixtures. Usually, trisaccharides are the first formed GOS, followed by tetrasaccharides and then higher oligosaccharides. This is because trisaccharides need to be present to act as galactosyl acceptors to form tetrasaccharides, which also act as galactosyl acceptors to form pentasaccharides. Therefore, the maximum yields of DP 3, DP 4 and DP > 4 are usually obtained at different time points during the synthesis reaction [106, 107]. The reaction time has also showed an effect on the formed glycosidic bonds in very few studies; for example, the percent of β-D-Galp-(1→4)-β-D-Galp-(1→4)-D-Glc decreased from $\sim$ 95% to $\sim$ 30 – 35% of the total trisaccharides between 1 and 23 h, while the percent of β-D-Galp-(1→4)-β-D-Galp-(1→3)-D-Glc increased to $\sim$ 18% of the total trisaccharides after 23 h of the reaction performed, using the β-galactosidase from *Bacillus circulans* [101].

Other Factors (pH and inhibitors)

Since the kinetics of lactose hydrolysis and GOS synthesis might be affected by pH [108], there is a potential of controlling GOS synthesis by varying the pH of the reaction. Akiyama *et al.* [109] reported that pH 6 was optimum for lactose hydrolysis while pH 7 was optimum for the formation of trisaccharides (DP3), using a β-galactosidase from *Thermus* sp. Z-1. Moreover, a β-galactosidase from *Aspergillus aculeatus* produced higher content of DP 3 at pH 6.5 compared to pH 4.5, 5.5 and 7.5, while the DP 2 and the DP > 3 content did not change with the change in the pH, outlining the potential effects of pH on the DP in GOS mixtures [94]. On the other hand, many studies did not show any effect of pH on the synthesis of GOS. The major effects of pH on the GOS synthesis are so far related to the source of the enzyme (fungal, yeast, bacterial) which determines, the optimum pH at which the GOS synthesis reaction is carried out while maintaining the activity of the enzyme.

In terms of inhibitors, galactose typically has more inhibitory effects than glucose, as galactose can form the galactosyl-enzyme intermediate which can displace lactose from the active site [110, 111]. Neri *et al.* [107] indicated that galactose reduced the rate of lactose conversion by *Aspergillus oryzae* β-galactosidase more than glucose, but both sugars negatively affected the obtained GOS yield. On the other hand, both monosaccharides reduced lactose conversion using *Kluyveromyces lactis* β-galactosidase, but only galactose reduced the maximum GOS yield [112]. In another study, glucose was the main inhibitor for GOS synthesis using a β-galactosidase from *Sterigmatomyces elviae* CBS8119 [113]. The effect of galactose in this case was not examined due to the very low galactose yield during transgalactosylation. Removal of these monosaccharides by nanofiltration or by using specific fermentation cultures, which use glucose and galactose and not GOS are approaches used to decrease the inhibition effects [113, 114].

Improving the GOS Synthesis Process

The recent improvements in the GOS synthesis process covered mainly (1) the increase in the GOS yield, and (2) the production of substantial GOS yields under cold temperatures using the lactose concentrations naturally present in milk, *i.e.in-situ* GOS production. Succeeding in the last point particularly would allow the commercial *in-situ* conversion of lactose in milk into GOS, which will open new opportunities for the dairy industry.

Protein Engineering

Protein engineering aims mainly at increasing the transgalactosylation activity of the used β-galactosidase at the expense of its hydrolytic activity by changing specific amino acids at the active site of the enzyme. For instance, the change of the arginine residue at position 109 to tryptophan in the β-galactosidase (BgaB) from *Geobacillus stearothermophilus* KVE39 increased the GOS yield and productivity by 12 and 17-fold, respectively [115]. Besides, the deletion of approximately 580 amino acid residues from the C-terminal end of the β-galactosidase (BIF3) from *Bifidobacterium bifidum* DSM 20215 increased the transgalactosylation activity of the enzyme. The truncated β-galactosidase could convert ~ 90% of the reacted lactose into GOS, while only 10% of the reacted lactose was hydrolysed [116]. Furthermore, changing the phenylalanine residue at position 426 to tyrosine in the β-glucosidase (CelB) of the hyperthermophilic *Pyrococcus furiosus* increased the GOS yield from 40 to 45% [117]. Also, changing both the phenylalanine residue at position 426 to tyrosine and the methionine residue at position 424 to lysine of the same enzyme showed better transgalactosylation properties at low lactose concentrations compared to the wild-type; the GOS yield was 40% using the engineered β-galactosidase compared to only 18% using the wild type enzyme from 10% initial lactose concentration [117]. This example clearly shows how the authors attempted to make it possible to produce substantial GOS yields at lactose concentrations slightly higher than those concentrations naturally found in milk. Moreover, the β-galactosidase (LacS) from *Sulfolobus solfataricus* P2 was subjected to site-directed mutagenesis and two mutants were obtained. In the first mutant, the phenylalanine residue at position 441 was changed to tyrosine, while in the second mutant the phenylalanine residue at position 359 was changed to glutamine [118]. The GOS yield was 50.9% using the wild type, 61.7% using the first mutant and 58.3% using the second mutant [118]. Protein engineering might also be used to reduce the inhibition effects exerted by galactose and glucose, increase the thermal stability of mesophilic β-galactosidases, and alter the substrate and acceptor specificity to produce novel GOS structures. However, these benefits which may be potentially obtained by protein engineering have not been under investigation like the increase in transgalactosylation activity. Despite the expected benefits using protein engineering, changing, and/or removing many amino acids to obtain the desired improvement in GOS synthesis might negatively change the catalytic efficiency of the engineered enzyme and alter the folding and/or the structure of the used β-galactosidase on the secondary, tertiary, and quaternary levels. These possible changes might affect the eventual performance of the enzyme used in GOS synthesis. Additionally, a debate can always arise with regards to using or banning engineered enzymes, *e.g.* β-galactosidases, in the food industry.

Reaction Medium Engineering

The main aim here is to decrease the availability of water molecules to act as acceptors of the galactosyl moieties, particularly in organic solvent systems, as these systems favor transgalactosylation over hydrolysis compared to aqueous systems, at the same initial lactose concentration [119, 120]. Moreover, the presence of organic solvents limits the potential for microbial contamination in the GOS synthesis process. Wang *et al.* [121] performed GOS synthesis in organic-aqueous bi-phasic media using a novel metagenome derived β-galactosidase BgaP412. They obtained a maximum GOS yield of 46.6% (w/w) at 75.4% lactose conversion in cyclohexane/buffer system (95:5) under the optimum conditions; the GOS yield was higher than that obtained in aqueous medium. Bankova *et al.* [122] also reported improved transgalactosylation in organic-aqueous systems, compared to aqueous systems only, using a β-galactosidase from *Aspergillus oryzae*. Moreover, GOS synthesis using a β-galactosidase from *Aspergillus oryzae* in aqueous-organic co-solvent using enzyme-cyclodextrin co-lyophilizate was 1.8 times more than it was in the aqueous system [123].

Ionic liquids have been also used for the synthesis of GOS and GOS derivatives. Kaftzik *et al.* [124] produced *N*-acetyllactosamine using the transgalactosylation activity of the β-galactosidase from *Bacillus circulans* using 25% v/v of 1,3-d--methyl-imidazolmethyl sulfate as a water-miscible ionic liquid. The use of ionic liquids, in this example, suppressed the hydrolysis of the formed product and doubled the yield to ≈ 60%.

Reverse micelles, which are self-assembling structures of water pools in organic solutions produced by the aid of surfactants, have been also used for GOS synthesis. The catalytic behavior of enzymes entrapped in the water pool of reverse micelles is quite different from that in aqueous media. The water in the pools shows also different properties from the water in bulk aqueous solutions. Another advantage is that the polar core and the surrounding hydrophobic boundary of reverse micelles can include many kinds of substrate molecules, *i.e.* hydrophilic, hydrophobic, or amphiphilic [125]. Reverse micelles can be used without maintaining high lactose concentrations or high temperatures with minimum loss of β-galactosidase activity. For instance, Chen *et al.* [119] stated that GOS synthesis was enhanced in reverse micelles using dioctyl sodium sulfosuccinate/isooctane reverse micelles, *i.e.* 51.2% (w/w) GOS compared to 31% GOS (% w/w) in aqueous systems using a β-galactosidase from *Aspergillus oryzae*. The same effect was found in GOS synthesis by a β-galactosidase from *E. coli* under controllable water concentration in reverse micelles [125].

Nevertheless, there are several disadvantages of using non-aqueous systems for

GOS synthesis, such as (1) some organic solvents have limited use in food applications where concerns can be raised if they are used for GOS production, (2) low quantitative final GOS yields and low volumetric productivities are usually obtained in organic solvent-, ionic liquid-, and reverse micelles systems, as lactose solubility is low in such systems compared to aqueous systems, (3) additional complicated purification steps are required, which adds a lot to the cost of production, (4) non-aqueous systems might not be the choice for complex waste streams, such as whey permeate, (5) the effect of organic systems on improving transgalactosylation at the expense of hydrolysis is not linear, and (6) not all enzymes have tolerance to non-aqueous systems in terms of maintaining their activity and selectivity [120, 124].

In-situ Production of GOS Using Natural Lactose Concentrations in Milk Under Cold Temperatures

The study performed by Jørgensen *et al.* 2001 was probably the first study which reported the synthesis of a substantial GOS yield (ca. 40%) using solutions of low lactose concentrations (10%). Jørgensen *et al.* [116] deleted approximately 580 amino acid residues from the C-terminal end of the β-galactosidase BIF3 from *Bifidobacterium bifidum* DSM20215; this deletion converted the enzyme from a normal hydrolytic β-galactosidase into a highly efficient transgalactosylating enzyme. The truncated β-galactosidase utilized approximately 90% of the reacted lactose to produce galactooligosaccharides, while hydrolysis constituted a 10% side reaction. This high ratio of transgalactosylation to hydrolysis (9:1) was very surprisingly maintained even at lactose concentrations of 10%, implying that the truncated enzyme behaved as a "true" transgalactosylase even at low lactose concentrations.

This inspiring study drew the attention of other researchers to the possibility of producing substantial yields of GOS *in-situ* in milk using the lactose concentrations naturally present in milk. Subsequently, the real initial attempt to produce GOS directly in skimmed milk has been reported by Rodrigues-Colinas *et. al.* [126] using a commercial β-galactosidase from *Bacillus circulans* (Biolactase) at 40 °C; in this study, however, only an acceptable total GOS yield of approximately 0.7% in milk, including transgalactosylated disaccharides and oligosaccharides having DP ≥ 3, was obtained and corresponded to ca. 15% of the total carbohydrates. This yield, however, does not allow rich-in-fiber or source-of-fiber claims in the final dairy products according to FDA and EFSA regulations, which define fibers as those carbohydrate molecules, which are resistant to digestion and absorption, reach the colon intact and have a DP ≥ 3. Rodrigues-Colinas *et al.* [127] described a similar approach using commercial β-galactosidases derived from *Bacillus circulans*, *Kluyveromyces lactis*, and

Aspergillus Oryzae which were added to skimmed milk and incubated at 4 °C. Interestingly, at this low temperature, the obtained GOS yield was 0.81%, 0.48%, and 0.42% in milk, using β-galactosidases from *B. circulans*, *K. lactis*, and *A. oryzae*, respectively. The obtained GOS yield included transgalactosylated disaccharides and oligosaccharides having DP ≥ 3. These yields also do not allow rich-in-fiber or source-of-fiber claims. The advantages of obtaining rich-in-fiber or source-of-fiber claims by the *in-situ* application of β-galactosidases in milk are (1) no fiber ingredients need to be added to dairy products, (2) low lactose levels can be obtained by converting lactose into GOS, glucose and galactose, (3) the calorie content of the final product is considerably lower, (4) the total carbohydrate level in the final products are lower in case lactose is converted to GOS, compared to the case when fiber ingredients are added. To our knowledge, the only study which reported the obtainment of significant GOS yield produced *in-situ* in milk is the study performed by Ray and Osman [128]. In this study, GOS yields (DP ≥ 3) reached up to 1.36% (ca. 27% total carbohydrates), while when transgalactosylated disaccharides (DP 2) were also added to the GOS yield, the final obtained GOS yield was 2.59% (ca. 52% total carbohydrates). This yield was obtained in milk under cold temperatures of < 10 °C and at the natural lactose concentration levels, *i.e.* 4.8%. The results obtained by Ray and Osman [128] allows for rich-in-fiber and/or source-of-fiber claims, particularly in skimmed and semi-skimmed dairy products. The advantages of this inspiring process in the last study, *i.e.in-situ* high yield GOS production in milk under cold temperatures, are that (1) it does not require significant energy input, (2) it is not susceptible to microbial contamination, (3) it allows rich-in-fiber and source-of-fiber claims, and (4) it finds applications at low temperatures to maintain the organoleptic properties of dairy products. It is vital to mention that the enzyme used in this study is already commercially available.

THE USE OF OTHER DAIRY ENZYMES

Peptidases (Proteases)

Peptidases (EC 3.4) can been divided into enzyme classes based on sequence homology (see MEROPS database), and the different enzyme classes have generally distinctive catalytic mechanism/functional properties, such as endo/exo activity, pH optimum, co-factor requirements, and amino acid specificity [129]. The various presented proteases are described here according to their enzyme class.

A1 Peptidase

Aspartic acid peptidases (EC 3.4.23) are endo-peptidases that have two aspartic acids in its catalytic site and have typically a low pH optimum (pH 2-4).

Furthermore, some of these enzymes are expressed as pro-forms and when activated they become unstable at neutral pH [129 - 131].

Chymosin

One of the most well-known industrial dairy enzymes is rennet (enzymes produced in the stomachs of ruminant mammals), where chymosin (3.4.23.4) is the most important enzyme for cheese making. An extended review of structural/function relationship of these cheese-making enzymes have been made previously [130]. Cheese making is the oldest industrial usage of enzymes which dates back to ~5000 BC, and today chymosin is still one of the most dominating industrial enzyme with respect to market size. Traditionally, calf rennet has been used, but today GM-based produced bovine chymosin is the major coagulant enzymes in cheese making, The GM-based produced chymosin is a purer enzyme product which gives a better yield/control of the process without concerns related to diseases of animal origin *e.g.* mad cow disease. Beside bovine chymosin, a GM-based camel chymosin has been introduced in the market that have a higher C/P ratio (clotting unit per protease) *i.e.* hydrolyzing Phe^{105}-Met^{106} bond in kappa casein versus general proteolytic activity. Furthermore, a protein engineered version of bovine chymosin has been introduced which also offers extended textural shelf life in cheese while maintaining low proteolysis.

Pepsin

Bovine pepsin (3.4.23.1) has very low C/P ratio and when present in rennet it gives lower cheese yield and special flavors in the final cheese originating from "unspecific" proteolysis. As pepsin has been present in traditional rennet before the introduction of pure GMM chymosin, flavors originating from general proteolysis (*i.e.* besides hydrolyzing Phe^{105}-Met^{106} bond in kappa casein) are a characteristic of many traditionally produced cheeses. Therefore, peptidases other than GM-based chymosin are typically added, together with GM-based chymosin, to cheese to give unique flavor profiles to various cheese types [129 - 131].

Microbial Rennet

The fungal enzyme from *Rhizomucor miehei* (EC 3.4.23.23) is a non-GM alternative to animal rennet and can be used for halal, kosher, and vegetarian products, but as for pepsin the use of microbial rennet gives a lower cheese yield and generates more flavors.

S1 Peptidases

The peptidases of family S1 contain the catalytic triad His, Asp, and Ser and is a family with a wide range of specificities [129 - 131].

Trypsin

Trypsin (EC 3.4.21.4) is one of the most well characterized peptidases for both analytic and industrial applications. It has a strong preference towards Lys and Arg at the carboxy site of the cleavage site, giving a distinct proteolytic pattern. The industrial production of trypsin is an extraction from porcine pancreas (originally a by-product of insulin production), and it is mainly used to produce low allergenic infant formulae from bovine casein or whey [129 - 131].

Chymotrypsin

Chymotrypsin (EC 3.4.21.1) is extracted together with trypsin from porcine pancreas and its proteolytic preferences are large hydrophobic amino acids at the carboxy site of the cleavage site. Together with trypsin, chymotrypsin gives a protein hydrolysate which mimics peptides generated by mammalian digestion in the intestine [129 - 131].

Microbial Trypsin

Fusarium oxysporum peptidase (EC 3.4.21.4) has also strong preferences for Lys/Arg at the carboxy site at cleave, and can be used as pure enzyme product for kosher/halal/vegetarian products, or if diseases from animal origin should be a concern [129 - 131].

Microbial Chymotrypsin

Nocardiopsis prasina peptidase (EC 3.4.21.1) has a comparable specificity as chymotrypsin and is also an alternative for an animal enzyme [129 - 131].

S8 Peptidases

The serine peptidase family S8 have the catalytic triad in the order Asp, His, and Ser in the sequence, which is a different order to that of families S1 (see MEROPS database) and one of the examples of the rare convergent evolution of a catalytic mechanism [131].

Subtilisin

This endo-peptidase has a broad specificity for peptide bonds, and the commercial products of subtilisin (EC 3.4.21.62) have high catalytic activity up to 70°C. In the dairy industry, it is used to make hydrolysate of *e.g.* whey for sports drinks.

M4 Peptidases

The M4 metallopeptidase family consists of endo-peptidases, which has a HEXXH motif that binds a single zinc ion, and a unique signature identifies a family of zinc-dependent metallopeptidases [129].

Bacillolysin

Bacillus amyloliquefaciens peptidase (EC 3.4.24.28) prefers large hydrophobic amino acids in the carboxy site of cleavage, but can also works at slight acidic pH and is used for cheese ripening and production of casein/whey hydrolysates.

Exo-peptidases

The exo-peptidases present in the market for dairy applications are mixtures of many peptidase activities. For example, exo-peptidases from *Aspergillus oryzae* and *Aspergillus niger* can be used to modify cheese ripening, change flavor profile, and decrease bitterness of casein/whey hydrolysate.

Crosslinking Enzymes

Beside protein hydrolysis, protein crosslinking offers other rheological properties and several enzyme solutions have been mentioned *e.g.* transglutaminase, tyrosinase, laccase, and peroxidase, where the leading industrial crosslinking enzyme is transglutaminase. Transglutaminase from *S. mobareansis* (EC 2.3.2.13) can increase viscosity/thickening and reduce syneresis of yogurt. This is especially relevant for low fat yoghurt to reduce addition of stabilizers (starch, pectin and skim milk powder) which, besides cost reduction, also simplifies ingredient labelling. Furthermore, this enzyme is able to increase yield in cheese production by fixing peptides and proteins in the curd which otherwise would have been in the whey. A more in-depth description of the application of transglutaminase can be found in chapter 1 in this book as well as elsewhere in the literature.

Lipid Active Enzymes

Fusarium venenatum phospholipase (EC 3.1.1.32) can hydrolyze phospholipids to lysophospholipids, and these lysophospholipids increase fat retention during cutting, stirring, and stretching in the cheese making process. For certain cheeses, *e.g.* the pasta filata segment, a 3.5% extra cheese yield is obtained. Furthermore, the release of free fatty acids can contribute to improved flavor profile and faster ripening for cheeses where this is favorable. In addition, triacylglycerol acylhydrolase, such as *Rhizomucor miehei* lipase (EC 3.1.1.3) is also able to generate comparable fatty acids if this is needed for flavor profile and faster ripening. As there are more triglycerides than phospholipids in milk/cheese, the *Rhizomucor miehei* lipase can generate more flavors, but this process should be controlled to avoid excessive amount of free fatty acids. The *Fusarium venenatum* phospholipase gives a more controlled flavor generation due to the limited number of phospholipids. More insights into lipid active enzymes, particularly phospholipases, are discussed in more details in Chapter 3 in this book.

Oxidoreductases

The appearance of a cooked off-flavor in heat-treated milk products is one of the common problems in the dairy industry and is a strong barrier to consumer acceptance of severely heat-treated milk, such as UHT milk. Cooked off-flavour is due to the presence of volatile sulfides and thiols that arise from thermal breakdown of whey proteins, primarily β-lactoglobulin, and the proteinaceous material associated with the fat globule membrane. Of the interesting approaches used to alleviate cooked off-flavour is the work performed by Swaisgood [132] who disclosed a process of removing cooked flavour by treating milk with immobilized sulfhydryl oxidase, which oxidized the sulfhydryl groups in heat treated milk into disulfides. Others, such as Tsuchiya & Petersen [133] used one or more oxidoreductases, in particular laccase, to remove burned flavour from foods, such as milk. Moreover, Sundgren *et al.* [134] managed to reduce the cooked off-flavour obtained in heat-treated milk products by using a carbohydrate oxidase to oxidize at least 0.1% of the lactose into lactobionic acid; it has been found that to maintain the flavour and shelf-life of the milk-related product, lactose oxidizing should be kept low. Although the use of oxidoreductases is very interesting in the dairy industry, no widespread commercial application exists beyond using peroxidases for whey bleaching purposes.

CONCLUDING REMARKS

The area of dairy enzyme science and technology has been rapidly growing in the last few decades, driven by the discovery of many new enzymes with interesting features which add value and positively influence the final characteristics of dairy products and processes. Apart from the advances took place on the research and laboratory scale, little has been applied on the industrial scale with examples limited to new lactases, coagulants, and cheese ripening enzymes. Whether the new interesting applications of enzymes (*e.g. in-situ* GOS synthesis, removing cooked off-flavours of heat treated milk products, protein function modification by cross linking, amidation, de-phosphorylation, controlled partial hydrolysis) will be realized on a large industrial scale or not is linked to (i) the potential commercial value of these new applications, (ii) the required investment, (iii) the availability of other non-enzymatic solutions which provide a similar functionality in the final products, and (iv) how much the dairy industry dares to apply innovative enzymatic solutions beyond the known applications of lactose-free products, cheese making by coagulants, and cheese ripening by peptidases and lipases.

CONSENT FOR PUBLICATION

Not applicable.

CONFLICT OF INTEREST

The authors declare no conflict of interest, financial or otherwise.

ACKNOWLEDGEMENTS

Declared none.

REFERENCES

[1]　Amann RI, Ludwig W, Schleifer KH. Phylogenetic identification and *in situ* detection of individual microbial cells without cultivation. Microbiol Rev 1995; 59(1): 143-69.
[PMID: 7535888]

[2]　Vester JK, Glaring MA, Stougaard P. Improved cultivation and metagenomics as new tools for bioprospecting in cold environments. Extremophiles 2015; 19(1): 17-29.
[http://dx.doi.org/10.1007/s00792-014-0704-3] [PMID: 25399309]

[3]　Vartoukian SR, Palmer RM, Wade WG. Strategies for culture of 'unculturable' bacteria. FEMS Microbiol Lett 2010; 309(1): 1-7.
[PMID: 20487025]

[4]　Kaeberlein T, Lewis K, Epstein SS. Isolating "uncultivable" microorganisms in pure culture in a simulated natural environment. Science 2002; 296(5570): 1127-9.
[http://dx.doi.org/10.1126/science.1070633] [PMID: 12004133]

[5]　Nichols D, Cahoon N, Trakhtenberg EM, *et al.* Use of ichip for high-throughput in situ cultivation of "uncultivable" microbial species. Appl Environ Microbiol 2010; 76(8): 2445-50.
[http://dx.doi.org/10.1128/AEM.01754-09] [PMID: 20173072]

[6]　Aoi Y, Kinoshita T, Hata T, Ohta H, Obokata H, Tsuneda S. Hollow-fiber membrane chamber as a device for *in situ* environmental cultivation. Appl Environ Microbiol 2009; 75(11): 3826-33.
[http://dx.doi.org/10.1128/AEM.02542-08] [PMID: 19329655]

[7]　Zengler K, Toledo G, Rappé M, *et al.* Cultivating the uncultured. Proc Natl Acad Sci USA 2002; 99(24): 15681-6.
[http://dx.doi.org/10.1073/pnas.252630999] [PMID: 12438682]

[8]　Ben-Dov E, Kramarsky-Winter E, Kushmaro A. An *in situ* method for cultivating microorganisms using a double encapsulation technique. FEMS Microbiol Ecol 2009; 68(3): 363-71.
[http://dx.doi.org/10.1111/j.1574-6941.2009.00682.x] [PMID: 19453493]

[9]　Tamaki H, Hanada S, Sekiguchi Y, Tanaka Y, Kamagata Y. Effect of gelling agent on colony formation in solid cultivation of microbial community in lake sediment. Environ Microbiol 2009; 11(7): 1827-34.
[http://dx.doi.org/10.1111/j.1462-2920.2009.01907.x] [PMID: 19302540]

[10]   Vester JK, Glaring MA, Stougaard P. Improving diversity in cultures of bacteria from an extreme environment. Can J Microbiol 2013; 59(8): 581-6.
[http://dx.doi.org/10.1139/cjm-2013-0087] [PMID: 23899002]

[11]   Pulschen AA, Bendia AG, Fricker AD, Pellizari VH, Galante D, Rodrigues F. Isolation of uncultured bacteria from antarctica using long incubation periods and low nutritional media. Front Microbiol 2017; 8: 1346.
[http://dx.doi.org/10.3389/fmicb.2017.01346] [PMID: 28769908]

[12]   Nichols D, Lewis K, Orjala J, *et al.* Short peptide induces an "uncultivable" microorganism to grow *in vitro*. Appl Environ Microbiol 2008; 74(15): 4889-97.
[http://dx.doi.org/10.1128/AEM.00393-08] [PMID: 18515474]

[13]   D'Onofrio A, Crawford JM, Stewart EJ, *et al.* Siderophores from neighboring organisms promote the growth of uncultured bacteria. Chem Biol 2010; 17(3): 254-64.
[http://dx.doi.org/10.1016/j.chembiol.2010.02.010] [PMID: 20338517]

[14]   Connon SA, Giovannoni SJ. High-throughput methods for culturing microorganisms in very-lo--nutrient media yield diverse new marine isolates. Appl Environ Microbiol 2002; 68(8): 3878-85.
[http://dx.doi.org/10.1128/AEM.68.8.3878-3885.2002] [PMID: 12147485]

[15]   Rappé MS, Connon SA, Vergin KL, Giovannoni SJ. Cultivation of the ubiquitous SAR11 marine bacterioplankton clade. Nature 2002; 418(6898): 630-3.
[http://dx.doi.org/10.1038/nature00917] [PMID: 12167859]

[16]   Yang SJ, Kang I, Cho JC. Expansion of Cultured Bacterial Diversity by Large-Scale Dilution-to-Extinction Culturing from a Single Seawater Sample. Microb Ecol 2016; 71(1): 29-43.
[http://dx.doi.org/10.1007/s00248-015-0695-3] [PMID: 26573832]

[17]   Sakai HD, Kurosawa N. Exploration and isolation of novel thermophiles in frozen enrichment cultures derived from a terrestrial acidic hot spring. Extremophiles 2016; 20(2): 207-14.
[http://dx.doi.org/10.1007/s00792-016-0815-0] [PMID: 26860120]

[18]   Spits C, Le Caignec C, De Rycke M, *et al.* Whole-genome multiple displacement amplification from single cells. Nat Protoc 2006; 1(4): 1965-70.
[http://dx.doi.org/10.1038/nprot.2006.326] [PMID: 17487184]

[19]   Yilmaz S, Allgaier M, Hugenholtz P. Multiple displacement amplification compromises quantitative analysis of metagenomes. Nat Methods 2010; 7(12): 943-4.
[http://dx.doi.org/10.1038/nmeth1210-943] [PMID: 21116242]

[20]   Vester JK, Glaring MA, Stougaard P. Discovery of novel enzymes with industrial potential from a cold and alkaline environment by a combination of functional metagenomics and culturing. Microb Cell Fact 2014; 13: 72.
[http://dx.doi.org/10.1186/1475-2859-13-72] [PMID: 24886068]

[21]   Kim M, Lee K-H, Yoon S-W, Kim B-S, Chun J, Yi H. Analytical tools and databases for metagenomics in the next-generation sequencing era. Genomics Inform 2013; 11(3): 102-13.
[http://dx.doi.org/10.5808/GI.2013.11.3.102] [PMID: 24124405]

[22]   Ilmberger N, Güllert S, Dannenberg J, *et al.* A comparative metagenome survey of the fecal microbiota of a breast- and a plant-fed Asian elephant reveals an unexpectedly high diversity of glycoside hydrolase family enzymes. PLoS One 2014; 9(9): e106707.
[http://dx.doi.org/10.1371/journal.pone.0106707] [PMID: 25208077]

[23]   Cavicchioli R, Charlton T, Ertan H, Mohd Omar S, Siddiqui KS, Williams TJ. Biotechnological uses of enzymes from psychrophiles. Microb Biotechnol 2011; 4(4): 449-60.
[http://dx.doi.org/10.1111/j.1751-7915.2011.00258.x] [PMID: 21733127]

[24]   Ekkers DM, Cretoiu MS, Kielak AM, Elsas JD. The great screen anomaly--a new frontier in product discovery through functional metagenomics. Appl Microbiol Biotechnol 2012; 93(3): 1005-20.
[http://dx.doi.org/10.1007/s00253-011-3804-3] [PMID: 22189864]

[25]   Popovic A, Tchigvintsev A, Tran H, *et al.* Metagenomics as a tool for enzyme discovery: Hydrolytic enzymes from marine-related metagenomes. Adv Exp Med Biol 2015; 883: 1-20.
[http://dx.doi.org/10.1007/978-3-319-23603-2_1] [PMID: 26621459]

[26]   Neifar M, Maktouf S, Ghorbel RE, Jaouani A, Cherif A. Extremophiles as source of novel bioactive compounds with industrial potential.Biotechnology of Bioactive Compounds: Sources and Applications. 2015; pp. 245-67.
[http://dx.doi.org/10.1002/9781118733103.ch10]

[27]   Harrison JP, Gheeraert N, Tsigelnitskiy D, Cockell CS. The limits for life under multiple extremes. Trends Microbiol 2013; 21(4): 204-12.
[http://dx.doi.org/10.1016/j.tim.2013.01.006] [PMID: 23453124]

[28]   Junge K, Eicken H, Deming JW. Bacterial Activity at -2 to -20 ° C in Arctic wintertime sea ice. Appl Environ Microbiol 2004; 70(1): 550-7.
[http://dx.doi.org/10.1128/AEM.70.1.550-557.2004] [PMID: 14711687]

[29]   Kuddus M, Ramteke PW. Recent developments in production and biotechnological applications of cold-active microbial proteases. Crit Rev Microbiol 2012; 38(4): 330-8.
[http://dx.doi.org/10.3109/1040841X.2012.678477] [PMID: 22849713]

[30]   Hoyoux A, Jennes I, Dubois P, *et al.* Cold-adapted β-galactosidase from the Antarctic psychrophile Pseudoalteromonas haloplanktis. Appl Environ Microbiol 2001; 67(4): 1529-35.
[http://dx.doi.org/10.1128/AEM.67.4.1529-1535.2001] [PMID: 11282601]

[31]   Schmidt M, Stougaard P. Identification, cloning and expression of a cold-active β-galactosidase from a novel Arctic bacterium, Alkalilactibacillus ikkense. Environ Technol 2010; 31(10): 1107-14.
[http://dx.doi.org/10.1080/09593331003677872] [PMID: 20718293]

[32]   Tindbaek N, Svendsen A, Oestergaard PR, Draborg H. Engineering a substrate-specific cold-adapted subtilisin. Protein Eng Des Sel 2004; 17(2): 149-56.
[http://dx.doi.org/10.1093/protein/gzh019] [PMID: 15047911]

[33]   Feijoo-Siota L, Blasco L, Rodríguez-Rama JL, *et al.* Recent patents on microbial proteases for the dairy industry. Recent Adv DNA Gene Seq 2014; 8(1): 44-55.
[PMID: 25564028]

[34]   Saito R, Nakayama A. Differences in malate dehydrogenases from the obligately piezophilic deep-sea bacterium *Moritella* sp. strain 2D2 and the psychrophilic bacterium *Moritella* sp. strain 5710. FEMS Microbiol Lett 2004; 233(1): 165-72.
[http://dx.doi.org/10.1016/j.femsle.2004.02.004] [PMID: 15043884]

[35]   Zeldes BM, Keller MW, Loder AJ, Straub CT, Adams MWW, Kelly RM. Extremely thermophilic microorganisms as metabolic engineering platforms for production of fuels and industrial chemicals. Front Microbiol 2015; 6: 1209.
[http://dx.doi.org/10.3389/fmicb.2015.01209] [PMID: 26594201]

[36]   Elleuche S, Schäfers C, Blank S, Schröder C, Antranikian G. Exploration of extremophiles for high temperature biotechnological processes. Curr Opin Microbiol 2015; 25: 113-9.
[http://dx.doi.org/10.1016/j.mib.2015.05.011] [PMID: 26066287]

[37]   Synowiecki J, Grzybowska B, Zdziebło A. Sources, properties and suitability of new thermostable enzymes in food processing. Crit Rev Food Sci Nutr 2006; 46(3): 197-205.
[http://dx.doi.org/10.1080/10408690590957296] [PMID: 16527752]

[38]   Conte A, Lacivita V, Esposto D, Saccotelli MA, Del Nobile MA. Patents on the advances in dairy industry. Recent Pat Eng 2014; 8: 41-9.
[http://dx.doi.org/10.2174/1872212108666140212211807]

[39]   Sun X, Duan X, Wu D, Chen J, Wu J. Characterization of Sulfolobus solfataricus β-galactosidase mutant F441Y expressed in Pichia pastoris. J Sci Food Agric 2014; 94(7): 1359-65.
[http://dx.doi.org/10.1002/jsfa.6419] [PMID: 24114556]

[40]   Zhang Y, Li X, Bartlett DH, Xiao X. Current developments in marine microbiology: high-pressure biotechnology and the genetic engineering of piezophiles. Curr Opin Biotechnol 2015; 33: 157-64.
       [http://dx.doi.org/10.1016/j.copbio.2015.02.013] [PMID: 25776196]

[41]   Ten LN, Im WT, Kim MK, Kang MS, Lee ST. Development of a plate technique for screening of polysaccharide-degrading microorganisms by using a mixture of insoluble chromogenic substrates. J Microbiol Methods 2004; 56(3): 375-82.
       [http://dx.doi.org/10.1016/j.mimet.2003.11.008] [PMID: 14967229]

[42]   Ten LN, Im WT, Kim MK, Lee ST. A plate assay for simultaneous screening of polysaccharide- and protein-degrading micro-organisms. Lett Appl Microbiol 2005; 40(2): 92-8.
       [http://dx.doi.org/10.1111/j.1472-765X.2004.01637.x] [PMID: 15644106]

[43]   Lylloff JE, Hansen LBS, Jepsen M, *et al.* Genomic and exoproteomic analyses of cold- and alkaline-adapted bacteria reveal an abundance of secreted subtilisin-like proteases. Microb Biotechnol 2016; 9(2): 245-56.
       [http://dx.doi.org/10.1111/1751-7915.12343] [PMID: 26834075]

[44]   Ingham CJ, Sprenkels A, Bomer J, *et al.* The micro-Petri dish, a million-well growth chip for the culture and high-throughput screening of microorganisms. Proc Natl Acad Sci USA 2007; 104(46): 18217-22.
       [http://dx.doi.org/10.1073/pnas.0701693104] [PMID: 17989237]

[45]   Uchiyama T, Watanabe K. Substrate-induced gene expression (SIGEX) screening of metagenome libraries. Nat Protoc 2008; 3(7): 1202-12.
       [http://dx.doi.org/10.1038/nprot.2008.96] [PMID: 18600226]

[46]   Uchiyama T, Miyazaki K. Product-induced gene expression, a product-responsive reporter assay used to screen metagenomic libraries for enzyme-encoding genes. Appl Environ Microbiol 2010; 76(21): 7029-35.
       [http://dx.doi.org/10.1128/AEM.00464-10] [PMID: 20833789]

[47]   Hosokawa M, Hoshino Y, Nishikawa Y, *et al.* Droplet-based microfluidics for high-throughput screening of a metagenomic library for isolation of microbial enzymes. Biosens Bioelectron 2015; 67: 379-85.
       [http://dx.doi.org/10.1016/j.bios.2014.08.059] [PMID: 25194237]

[48]   Nakamura K, Iizuka R, Nishi S, *et al.* Culture-independent method for identification of microbial enzyme-encoding genes by activity-based single-cell sequencing using a water-in-oil microdroplet platform. Sci Rep-UK 2016; p. 6.

[49]   Tutino ML, Duilio A, Parrilli R, Remaut E, Sannia G, Marino G. A novel replication element from an Antarctic plasmid as a tool for the expression of proteins at low temperature. Extremophiles 2001; 5(4): 257-64.
       [http://dx.doi.org/10.1007/s007920100203] [PMID: 11523895]

[50]   Miyake R, Kawamoto J, Wei YL, *et al.* Construction of a low-temperature protein expression system using a cold-adapted bacterium, Shewanella sp. strain Ac10, as the host. Appl Environ Microbiol 2007; 73(15): 4849-56.
       [http://dx.doi.org/10.1128/AEM.00824-07] [PMID: 17526788]

[51]   Ferrer M, Chernikova TN, Timmis KN, Golyshin PN. Expression of a temperature-sensitive esterase in a novel chaperone-based Escherichia coli strain. Appl Environ Microbiol 2004; 70(8): 4499-504.
       [http://dx.doi.org/10.1128/AEM.70.8.4499-4504.2004] [PMID: 15294778]

[52]   Miteva V, Lantz S, Brenchley J. Characterization of a cryptic plasmid from a Greenland ice core Arthrobacter isolate and construction of a shuttle vector that replicates in psychrophilic high G+C Gram-positive recipients. Extremophiles 2008; 12(3): 441-9.
       [http://dx.doi.org/10.1007/s00792-008-0149-7] [PMID: 18335166]

[53]   Yang XW, Jian HH, Wang FP. pSW2, a novel low-temperature-inducible gene expression vector

based on a filamentous phage of the deep-sea bacterium Shewanella piezotolerans WP3. Appl Environ Microbiol 2015; 81(16): 5519-26.
[http://dx.doi.org/10.1128/AEM.00906-15] [PMID: 26048946]

[54]   Xu H, Han D, Xu Z. Expression of heterologous cellulases in thermotoga sp. Strain RQ2. BioMed Res Int 2015; 2015: 304523.
[http://dx.doi.org/10.1155/2015/304523] [PMID: 26273605]

[55]   Takemasa R, Yokooji Y, Yamatsu A, Atomi H, Imanaka T. Thermococcus kodakarensis as a host for gene expression and protein secretion. Appl Environ Microbiol 2011; 77(7): 2392-8.
[http://dx.doi.org/10.1128/AEM.01005-10] [PMID: 21278271]

[56]   Albers SV, Jonuscheit M, Dinkelaker S, *et al.* Production of recombinant and tagged proteins in the hyperthermophilic archaeon Sulfolobus solfataricus. Appl Environ Microbiol 2006; 72(1): 102-11.
[http://dx.doi.org/10.1128/AEM.72.1.102-111.2006] [PMID: 16391031]

[57]   Berkner S, Grogan D, Albers SV, Lipps G. Small multicopy, non-integrative shuttle vectors based on the plasmid pRN1 for Sulfolobus acidocaldarius and Sulfolobus solfataricus, model organisms of the (cren-)archaea. Nucleic Acids Res 2007; 35(12): e88.
[http://dx.doi.org/10.1093/nar/gkm449] [PMID: 17576673]

[58]   Hwang S, Choi KH, Yoon N, Cha J. Improvement of a Sulfolobus-E. coli shuttle vector for heterologous gene expression in Sulfolobus acidocaldarius. J Microbiol Biotechnol 2015; 25(2): 196-205.
[http://dx.doi.org/10.4014/jmb.1407.07043] [PMID: 25293629]

[59]   Kixmüller D, Greie JC. Construction and characterization of a gradually inducible expression vector for Halobacterium salinarum, based on the kdp promoter. Appl Environ Microbiol 2012; 78(7): 2100-5.
[http://dx.doi.org/10.1128/AEM.07155-11] [PMID: 22287001]

[60]   Mevarech M, Frolow F, Gloss LM. Halophilic enzymes: proteins with a grain of salt. Biophys Chem 2000; 86(2-3): 155-64.
[http://dx.doi.org/10.1016/S0301-4622(00)00126-5] [PMID: 11026680]

[61]   Allers T, Barak S, Liddell S, Wardell K, Mevarech M. Improved strains and plasmid vectors for conditional overexpression of His-tagged proteins in *Haloferax volcanii*. Appl Environ Microbiol 2010; 76(6): 1759-69.
[http://dx.doi.org/10.1128/AEM.02670-09] [PMID: 20097827]

[62]   Zemella A, Thoring L, Hoffmeister C, Kubick S. Cell-free protein synthesis: Pros and cons of prokaryotic and eukaryotic systems. ChemBioChem 2015; 16(17): 2420-31.
[http://dx.doi.org/10.1002/cbic.201500340] [PMID: 26478227]

[63]   Matthaei H, Nirenberg MW. The dependence of cell-free protein synthesis in E. coli upon RNA prepared from ribosomes. Biochem Biophys Res Commun 1961; 4: 404-8.
[http://dx.doi.org/10.1016/0006-291X(61)90298-4] [PMID: 13768264]

[64]   Freischmidt A, Hiltl J, Kalbitzer HR, Horn-Katting G. Enhanced in vitro translation at reduced temperatures using a cold-shock RNA motif. Biotechnol Lett 2013; 35(3): 389-95.
[http://dx.doi.org/10.1007/s10529-012-1091-4] [PMID: 23143178]

[65]   Zhou Y, Asahara H, Gaucher EA, Chong S. Reconstitution of translation from Thermus thermophilus reveals a minimal set of components sufficient for protein synthesis at high temperatures and functional conservation of modern and ancient translation components. Nucleic Acids Res 2012; 40(16): 7932-45.
[http://dx.doi.org/10.1093/nar/gks568] [PMID: 22723376]

[66]   Hüdepohl U, Reiter WD, Zillig W. In vitro transcription of two rRNA genes of the archaebacterium Sulfolobus sp. B12 indicates a factor requirement for specific initiation. Proc Natl Acad Sci USA 1990; 87(15): 5851-5.
[http://dx.doi.org/10.1073/pnas.87.15.5851] [PMID: 2116009]

[67]   Endoh T, Kanai T, Imanaka T. A highly productive system for cell-free protein synthesis using a lysate of the hyperthermophilic archaeon, Thermococcus kodakaraensis. Appl Microbiol Biotechnol 2007; 74(5): 1153-61.
[http://dx.doi.org/10.1007/s00253-006-0753-3] [PMID: 17165083]

[68]   Kovtun O, Mureev S, Jung W, Kubala MH, Johnston W, Alexandrov K. Leishmania cell-free protein expression system. Methods 2011; 55(1): 58-64.
[http://dx.doi.org/10.1016/j.ymeth.2011.06.006] [PMID: 21704167]

[69]   Kovtun O, Mureev S, Johnston W, Alexandrov K. Towards the construction of expressed proteomes using a Leishmania tarentolae based cell-free expression system. PLoS One 2010; 5(12): e14388.
[http://dx.doi.org/10.1371/journal.pone.0014388] [PMID: 21203562]

[70]   Mureev S, Kovtun O, Nguyen UTT, Alexandrov K. Species-independent translational leaders facilitate cell-free expression. Nat Biotechnol 2009; 27(8): 747-52.
[http://dx.doi.org/10.1038/nbt.1556] [PMID: 19648909]

[71]   Ga¨nzle MG, Hasse G, Jelen P. Lactose: Crystallization, hydrolysis and value-added derivatives. Int Dairy J 2008; 18: 685-94.
[http://dx.doi.org/10.1016/j.idairyj.2008.03.003]

[72]   Harju M, Kallioinen H, Tossavainen O. Lactose hydrolysis and other conversions in dairy products: Technological aspects. Int Dairy J 2012; 22: 104-9.
[http://dx.doi.org/10.1016/j.idairyj.2011.09.011]

[73]   Jelen P, Tossavainen O. Low lactose and lactose-free milk and dairy products – prospects, technologies and applications. Aus J Dairy Tec 2003; 58: 161-5.

[74]   Nielsen SD, Zhao D, Le TT, *et al.* Proteolytic side-activity of lactase preparations. Int Dairy J 2018; 78: 159-68.
[http://dx.doi.org/10.1016/j.idairyj.2017.12.001]

[75]   Nielsen SD, Jansson T, Le TT, *et al.* Correlation between sensory properties and peptides derived from hydrolysed-lactose UHT milk during storage. Int Dairy J 2017; 68: 23-31.
[http://dx.doi.org/10.1016/j.idairyj.2016.12.013]

[76]   Schmidt M, Stougaard P. Identification, cloning and expression of a cold-active β-galactosidase from a novel Arctic bacterium, *Alkalilactibacillus ikkense*. Environ Technol 2010; 31(10): 1107-14.
[http://dx.doi.org/10.1080/09593331003677872] [PMID: 20718293]

[77]   Hildebrandt P, Wanarska M, Kur J. A new cold-adapted β-D-galactosidase from the Antarctic *Arthrobacter* sp. 32c - gene cloning, overexpression, purification and properties. BMC Microbiol 2009; 9: 151.
[http://dx.doi.org/10.1186/1471-2180-9-151] [PMID: 19631003]

[78]   Mahdian SMAM, Karimi E, Tanipour MH, *et al.* Expression of a functional cold active β-galactosidase from Planococcus sp-L4 in Pichia pastoris. Protein Expr Purif 2016; 125: 19-25.
[http://dx.doi.org/10.1016/j.pep.2015.09.008] [PMID: 26361980]

[79]   Gibson GR, Probert HM, Loo JV, Rastall RA, Roberfroid MB. Dietary modulation of the human colonic microbiota: updating the concept of prebiotics. Nutr Res Rev 2004; 17(2): 259-75.
[http://dx.doi.org/10.1079/NRR200479] [PMID: 19079930]

[80]   Gibson GR, Scott KP, Rastall RA, *et al.* Dietary prebiotics: current status and new definition. Food Sci Technol Bull: Funct Foods 2010; 7: 1-19.

[81]   Mussatto SI, Mancilha IM. Non-digestible oligosaccharides: A review. Carbohydr Polym 2007; 68: 587-97.
[http://dx.doi.org/10.1016/j.carbpol.2006.12.011]

[82]   Mahoney RR. Galactosyl-oligosaccharide formation during lactose hydrolysis. Food Chem 1998; 63: 147-54.

[http://dx.doi.org/10.1016/S0308-8146(98)00020-X]

[83] Monsan P, Paul F. Enzymatic synthesis of Oligosaccharides. FEMS Microbiol Rev 1995; 16: 187-92.
[http://dx.doi.org/10.1111/j.1574-6976.1995.tb00165.x]

[84] Boon MA, Janssen AEM. Effect of temperature and enzyme origin on the enzymatic synthesis of oligosaccharides. Enzyme Microb Technol 2000; 26(2-4): 271-81.
[http://dx.doi.org/10.1016/S0141-0229(99)00167-2] [PMID: 10689088]

[85] Martinez-Villaluenga C, Cardelle-Cobas A, Corzo N, Olano A, Villamiel M. Optimization of conditions for galactooligosaccharides synthesis during lactose hydrolysis by β-galactosidase from *Kluyveromyces lactis* (Lactozym 3000 L HP G). Food Chem 2008; 107: 258-64.
[http://dx.doi.org/10.1016/j.foodchem.2007.08.011]

[86] Osman A, Tzortzis G, Rastall RA, Charalampopoulos D. A comprehensive investigation of the synthesis of prebiotic galactooligosaccharides by whole cells of *Bifidobacterium bifidum* NCIMB 41171. J Biotechnol 2010; 150(1): 140-8.
[http://dx.doi.org/10.1016/j.jbiotec.2010.08.008] [PMID: 20728480]

[87] Palcic MM. Biocatalytic synthesis of oligosaccharides. Curr Opin Biotechnol 1999; 10(6): 616-24.
[http://dx.doi.org/10.1016/S0958-1669(99)00044-0] [PMID: 10600700]

[88] Gosling A, Stevens GW, Barber AR, Kentish SE, Gras SL. Recent advances refining galactooligosaccharide production from lactose. Food Chem 2010; 121: 307-18.
[http://dx.doi.org/10.1016/j.foodchem.2009.12.063]

[89] Rabiu BA, Jay AJ, Gibson GR, Rastall RA. Synthesis and fermentation properties of novel galacto-oligosaccharides by *β*-galactosidases from *Bifidobacterium* species. Appl Environ Microbiol 2001; 67(6): 2526-30.
[http://dx.doi.org/10.1128/AEM.67.6.2526-2530.2001] [PMID: 11375159]

[90] Osman A, Tzortzis G, Rastall RA, Charalampopoulos D, Bbg IV. BbgIV Is an Important Bifidobacterium β-Galactosidase for the Synthesis of Prebiotic Galactooligosaccharides at High Temperatures. J Agric Food Chem 2012; 60(3): 740-8.
[http://dx.doi.org/10.1021/jf204719w] [PMID: 22148735]

[91] Hung MN, Lee BH. Purification and characterization of a recombinant β-galactosidase with transgalactosylation activity from *Bifidobacterium infantis* HL96. Appl Microbiol Biotechnol 2002; 58(4): 439-45.
[http://dx.doi.org/10.1007/s00253-001-0911-6] [PMID: 11954789]

[92] Bruins ME, Strubel M, van Lieshout JFT, Janssen AEM, Boom RM. Oligosaccharide synthesis by the hyperthermostable β-glucosidase from Pyrococcus furiosus: kinetics and modelling. Enzyme Microb Technol 2003; 33: 3-11.
[http://dx.doi.org/10.1016/S0141-0229(03)00096-6]

[93] Maitin V, Rastall RA. Enzyme glycation influences product yields during oligosaccharides synthesis by reverse hydrolysis. J Mol Catal, B Enzym 2004; 30: 195-202.
[http://dx.doi.org/10.1016/j.molcatb.2004.05.004]

[94] Cardelle-Cobas A, Villamiel M, Olano A, Corzo N. Study of galacto-oligosaccharide formation from lactose using Pectinex Ultra SP-L. J Sci Food Agric 2008; 88: 954-61.
[http://dx.doi.org/10.1002/jsfa.3173]

[95] Prenosil JE, Stuker E, Bourne JR. Formation of oligosaccharides during enzymatic lactose: Part I: State of art. Biotechnol Bioeng 1987; 30(9): 1019-25.
[http://dx.doi.org/10.1002/bit.260300904] [PMID: 18581545]

[96] Zarate S, Lopez-Leiva MH. Oligosaccharide formation during enzymatic lactose hydrolysis: A literature review. J Food Prot 1990; 53: 262-8.
[http://dx.doi.org/10.4315/0362-028X-53.3.262]

[97] Mozaffar Z, Nakanishi K, Matsuno R, Kamikubo T. Purification and properties of β-galactosidases

from *Bacillus circulans.* Agric Biol Chem 1984; 48: 3053-61.

[98]   Mozaffar Z, Nakanishi K, Matsuno R. Continuous production of galactooligosaccharides from lactose using immobilized *β*-galactosidase from *Bacillus circulans.* Appl Microbiol Biotechnol 1986; 25: 224-8.

[99]   Sako T, Matsumoto K, Tanaka R. Recent progress on research and applications of non-digestible galacto-oligosaccharides. Int Dairy J 1999; 9: 69-80.
[http://dx.doi.org/10.1016/S0958-6946(99)00046-1]

[100]   Lopez-Leiva MH, Guzman M. Formation of oligosaccharides during enzymatic hydrolysis of milk whey permeates. Process Biochem 1995; 30: 757-62.
[http://dx.doi.org/10.1016/0032-9592(95)00006-2]

[101]   Yanahira S, Suguri T, Yakabe T, Ikeuchi Y, Hanagata G, Deya E. Formation of oligosaccharides from lactitol by Aspergillus oryzae β-D-galactosidase. Carbohydr Res 1992; 232(1): 151-9.
[http://dx.doi.org/10.1016/S0008-6215(00)91002-4] [PMID: 1423346]

[102]   Iwasaki K, Nakajima M, Nakao S. Galactooligosaccharide production from lactose by an enzymatic batch reaction using β-galactosidase. Process Biochem 1996; 31: 69-76.
[http://dx.doi.org/10.1016/0032-9592(94)00067-0]

[103]   Hsu CA, Lee SL, Chou CC. Enzymatic production of galactooligosaccharides by β-galactosidase from *Bifidobacterium longum* BCRC 15708. J Agric Food Chem 2007; 55(6): 2225-30.
[http://dx.doi.org/10.1021/jf063126+] [PMID: 17316019]

[104]   Maischberger T, Nguyen TH, Sukyai P, *et al.* Production of lactose-free galacto-oligosaccharide mixtures: comparison of two cellobiose dehydrogenases for the selective oxidation of lactose to lactobionic acid. Carbohydr Res 2008; 343(12): 2140-7.
[http://dx.doi.org/10.1016/j.carres.2008.01.040] [PMID: 18353295]

[105]   Ohtsuka K, Tanoh A, Ozawa O, Kanematsu T, Uchida T, Shinke R. Purification and properties of a β-galactosidase with high galactosyl transfer activity from *Cryptococcus laurentii* OKN-4. J Ferment Bioeng 1990; 70: 301-7.
[http://dx.doi.org/10.1016/0922-338X(90)90138-M]

[106]   Albayrak N, Yang ST. Production of galacto-oligosaccharides from lactose by *Aspergillus oryzae* β-galactosidase immobilized on cotton cloth. Biotechnol Bioeng 2002; 77(1): 8-19.
[http://dx.doi.org/10.1002/bit.1195] [PMID: 11745169]

[107]   Neri DFM, Balcão VM, Costa RS, *et al.* Galacto-oligosaccharides production during lactose hydrolysis by free *Aspergillus oryzae* β-galactosidase and immobilized on magnetic polysiloxane-polyvinyl alcohol. Food Chem 2009; 115: 92-9.
[http://dx.doi.org/10.1016/j.foodchem.2008.11.068]

[108]   Huber RE, Kurz G, Wallenfels K. A quantitation of the factors which affect the hydrolase and transgalactosylase activities of β-galactosidase (*E. coli*) on lactose. Biochemistry 1976; 15(9): 1994-2001.
[http://dx.doi.org/10.1021/bi00654a029] [PMID: 5122]

[109]   Akiyama K, Takase M, Horikoshi K, Okonogi S. Production of galactooligosaccharides from lactose using a β-glucosidase from *Thermus* sp. Z-1. Biosci Biotechnol Biochem 2001; 65(2): 438-41.
[http://dx.doi.org/10.1271/bbb.65.438] [PMID: 11302184]

[110]   Hatzinikolaou DG, Katsifas E, Mamma D, Karagouni AD, Christakopoulos P, Kekos D. Modeling of the simultaneous hydrolysis-ultrafiltration of whey permeate by a thermostable beta-galactosidase from *Aspergillus niger.* Biochem Eng J 2005; 24: 161-72.
[http://dx.doi.org/10.1016/j.bej.2005.02.011]

[111]   Jurado E, Camacho F, Luzón G, Vicaria JM. Kinetic models of activity for beta-galactosidases: influence of pH, ionic concentration and temperature. Enzyme Microb Technol 2004; 34: 33-40.
[http://dx.doi.org/10.1016/j.enzmictec.2003.07.004]

[112] Chockchaisawasdee S, Athanasopoulos VI, Niranjan K, Rastall RA. Synthesis of galacto-oligosaccharide from lactose using beta-galactosidase from *Kluyveromyces lactis*: Studies on batch and continuous UF membrane-fitted bioreactors. Biotechnol Bioeng 2005; 89(4): 434-43.
      [http://dx.doi.org/10.1002/bit.20357] [PMID: 15627251]

[113] Onishi N, Tanaka T. Purification and properties of a novel thermostable galacto-oligosaccharid--producing beta-galactosidase from Sterigmatomyces elviae CBS8119. Appl Environ Microbiol 1995; 61(11): 4026-30.
      [PMID: 8526517]

[114] Cheng CC, Yu MC, Cheng TC, Sheu DC, Duan KJ, Tai WL. Production of high-content galacto-oligosaccharide by enzyme catalysis and fermentation with *Kluyveromyces marxianus.* Biotechnol Lett 2006; 28(11): 793-7.
      [http://dx.doi.org/10.1007/s10529-006-9002-1] [PMID: 16786243]

[115] Placier G, Watzlawick H, Rabiller C, Mattes R. Evolved β-galactosidases from *Geobacillus stearothermophilus* with improved transgalactosylation yield for galacto-oligosaccharide production. Appl Environ Microbiol 2009; 75(19): 6312-21.
      [http://dx.doi.org/10.1128/AEM.00714-09] [PMID: 19666723]

[116] Jørgensen F, Hansen OC, Stougaard P. High-efficiency synthesis of oligosaccharides with a truncated β-galactosidase from *Bifidobacterium bifidum.* Appl Microbiol Biotechnol 2001; 57(5-6): 647-52.
      [http://dx.doi.org/10.1007/s00253-001-0845-z] [PMID: 11778873]

[117] Hansson T, Kaper T, van Der Oost J, de Vos WM, Adlercreutz P. Improved oligosaccharide synthesis by protein engineering of beta-glucosidase CelB from hyperthermophilic *Pyrococcus furiosus.* Biotechnol Bioeng 2001; 73(3): 203-10.
      [http://dx.doi.org/10.1002/bit.1052] [PMID: 11257602]

[118] Wu Y, Yuan S, Chen S, Wu D, Chen J, Wu J. Enhancing the production of galacto-oligosaccharides by mutagenesis of *Sulfolobus solfataricus* β-galactosidase. Food Chem 2013; 138(2-3): 1588-95.
      [http://dx.doi.org/10.1016/j.foodchem.2012.11.052] [PMID: 23411285]

[119] Chen SX, Wei DZ, Hu ZH. Synthesis of galacto-oligosaccharides in AOT/isooctane reverse micelles by beta-galactosidase. J Mol Catal, B Enzym 2001; 16: 109-14.
      [http://dx.doi.org/10.1016/S1381-1177(01)00051-0]

[120] Cruz-Guerrero AE, Gómez-Ruiz L, Viniegra-González G, Bárzana E, García-Garibay M. Influence of water activity in the synthesis of galactooligosaccharides produced by a hyperthermophilic beta-glycosidase in an organic medium. Biotechnol Bioeng 2006; 93(6): 1123-9.
      [http://dx.doi.org/10.1002/bit.20824] [PMID: 16470870]

[121] Wang K, Lu Y, Liang WQ, *et al.* Enzymatic synthesis of galacto-oligosaccharides in an organic-aqueous biphasic system by a novel β-galactosidase from a metagenomic library. J Agric Food Chem 2012; 60(15): 3940-6.
      [http://dx.doi.org/10.1021/jf300890d] [PMID: 22443294]

[122] Bankova E, Bakalova N, Petrova S, Kolev D. Enzymatic Synthesis of Oligosaccharides and Alkylglycosides in Water-Organic Media Via Transglycosylation of Lactose. Biotechnol Biotechnol Equip 2006; 20: 114-9.
      [http://dx.doi.org/10.1080/13102818.2006.10817387]

[123] Srisimarat W, Pongsawasdi P. Enhancement of the oligosaccharide synthetic activity of β-galactosidase in organic solvents by cyclodextrin. Enzyme Microb Technol 2008; 43: 436-41.
      [http://dx.doi.org/10.1016/j.enzmictec.2008.06.007]

[124] Kaftzik N, Wasserscheid P, Kragl U. Use of Ionic Liquids to Increase the Yield and Enzyme Stability in the β-Galactosidase Catalysed Synthesis of N-Acetyllactosamine. Org Process Res Dev 2002; 6: 553-7.
      [http://dx.doi.org/10.1021/op0255231]

[125] Chen CW, Ou-Yang C-C, Yeh C-W. Synthesis of galactooligosaccharides and transgalactosylation modeling in reverse micelles. Enzyme Microb Technol 2003; 33: 497-507.
[http://dx.doi.org/10.1016/S0141-0229(03)00155-8]

[126] Rodriguez-Colinas B, Poveda A, Jimenez-Barbero J, Ballesteros AO, Plou FJ. Galacto-oligosaccharide synthesis from lactose solution or skim milk using the β-galactosidase from *Bacillus circulans.* J Agric Food Chem 2012; 60(25): 6391-8.
[http://dx.doi.org/10.1021/jf301156v] [PMID: 22676418]

[127] Rodriguez-Colinas B, Fernandez-Arrojo L, Ballesteros AO, Plou FJ. Galactooligosaccharides formation during enzymatic hydrolysis of lactose: towards a prebiotic-enriched milk. Food Chem 2014; 145: 388-94.
[http://dx.doi.org/10.1016/j.foodchem.2013.08.060] [PMID: 24128493]

[128] Ray CA, Osman A. Lactose-reduced milk products containing galacto-oligosaccharides and monosaccharides and a method of production. US patent 15123560 2015.

[129] Jongeneel CV, Bouvier J, Bairoch A. Discussion letter: A unique signature identifies a family of zinc-dependent metallopeptidases. FEBS Lett 1989; 242211-4.

[130] Yegin S, Dekker P. Progress in the field of aspartic proteinases in cheese manufacturing: structures, functions, catalytic mechanism, inhibition, and engineering. Dairy Sci Technol 2013; 93: 565-94.
[http://dx.doi.org/10.1007/s13594-013-0137-2]

[131] Rawlings ND, Barrett AJ, Finn R. Twenty years of the MEROPS database of proteolytic enzymes, their substrates and inhibitors. Nucleic Acids Res 2016; 44(D1): D343-50.
[http://dx.doi.org/10.1093/nar/gkv1118] [PMID: 26527717]

[132] Swaisgood HE. Process of removing the cooked flavour from milk. US patent 4053644 1977.

[133] Tsuchiya R, Petersen BR. Reduction of malodour. US patent 6080391 1998.

[134] Sundgren A, Ray CA, Nielsen JH. Improved milk and milk related products. WO patent 14785919 2014.

CHAPTER 9

Food Waste Management: The Role of Biotechnology

Dimitris Sarris[1,3,*], Christina N. Economou[2] and Seraphim Papanikolaou[3]

[1] *Department of Food Science and Nutrition, School of the Environment, University of the Aegean, 81400 Myrina-Lemnos, Greece*

[2] *Laboratory of Biochemical Engineering & Environmental Technology (LBEET), Department of Chemical Engineering, University of Patras, 26504 Patras, Greece*

[3] *Laboratory of Food Microbiology and Biotechnology, Department of Food Science and Human Nutrition, Agricultural University of Athens, Athens 11855, Greece*

Abstract: In recent years, the application of biotechnology on residues or by-products of food processing industries has received great interest from researchers, since the production of bulk chemicals and high value-added compounds, such as ethanol, biogas, organic acids, enzymes, mushrooms, *etc.* has been investigated in detail. The utilization of these abundant residues as alternative nutrient sources for microorganisms' growth is also expected to minimize both environmental pollution from their disposal and the final production cost of bio-products. The main objective of the present chapter is to highlight the role of biotechnology on the bioconversion of major industrial and agro-industrial by-products, that have zero acquisition cost, to produce various bio-products.

Keywords: Added-value Products, Bioremediation, Corn Stover, Food Wastes, FOG Wastes, Flour-rich Wastes, Lignocellulosic, Molasses, Mushroom Spent, Olive Mill Wastewaters, Rice Straw, Wheat Straw.

LIGNOCELLULOSIC RESIDUES

In recent years, there has been a great interest in the biotechnological utilization of lignocellulosic residues, such as wheat straw, rice straw/hulls, corn stover, citrus and cotton wastes, spent mushroom substrate, grape pomace *etc.* as cheap substrates for cultivation of various microorganisms with the aim to produce valuable products [1 - 4]. The major components of these residues are cellulose, hemicellulose, and lignin. Lignin is a complex polymer, composed of three aromatic alcohols (p-coumaryl, sinapyl, and coniferyl alcohols). Lignin also

---

* **Corresponding Author Dimitris Sarris:** Department of Food Science and Nutrition, School of the Environment, University of the Aegean, 81400 Myrina-Lemnos, Greece; Tel: +302254083100; Fax: +302254083109; E-mail: dsarris@aegean.gr

Ali Osman (Ed.)

surrounds the cellulose and hemicellulose molecules, so is quite resistant to pre-treatment methods like enzymatic and chemical hydrolysis. Hemicellulose is a heterogeneous polymer constructed of xylose, arabinose, glucose, galactose, mannose and sugar acids, while cellulose is a homopolymer of glucose units linked by a β-1,4-glycosidic bond [5]. It is worth noting that the percentages of these molecules vary between different lignocellulosic materials and depend on weather and growth conditions, age of the plant, and many other factors. Burning is the common approach taken by farmers for removal of lignocellulosic residues from the fields, causing serious environmental problems [6]. The utilization of these abundant residues as an alternative nutrient source for microorganisms' growth is expected to minimize both environmental pollution from their disposal and the final production cost of various bioproducts. Some examples of the biotechnological applications of lignocellulosic by-products are presented below.

FERMENTATION SYSTEMS

Solid-state fermentation (SSF) as well as submerged fermentation (SmF) processes have been used extensively to produce chemicals and high value-added products, *e.g.* ethanol, biodiesel, methane, single cell protein, single cell oil, enzymes, organic acids (lactic and citric acid), mushrooms, *etc.* from various types of substrates [7 - 10].

SSF is defined as any fermentation process where the growth of microorganisms takes place on a solid medium in the absence of free liquid. SSF systems have been applied to utilize low-cost materials, such as lignocellulosic remains, agro-industrial wastes, and by-products of the food industry [8, 11 - 15]. The important advantages of SSF systems are mainly the simple process and the low cost, because the substrates can be utilized without any further pretreatment. In addition, the possibility of substrate contamination is also low, due to the lack of free water in these bioreactor systems [16]. It is well-known from the scientific literature that the yield of enzyme production in SSF is higher than SmF. Nevertheless, controlling the operating parameters, such as pH, temperature, and moisture, still remains a serious problem in these cultivation systems [16].

On the other hand, SmF processes can utilize liquid substrates, such as molasses, agro-industrial wastewaters, fat/oil/grease (FOG) containing wastes, flour-rich waste (FRW) streams *etc.* for microorganisms' cultivation [17 - 22]. Consequently, the mass and heat transfer as well as the kinetic parameters' estimation of microorganisms are more effective in comparison with SSF. Furthermore, controlling the operating conditions in SmF is easier than in SSF systems. However, in most cases, a stage of liquid substrate pre-treatment is required to facilitate the utilization of nutrients by microorganisms. Thus, the best fermentation conditions of microorganisms, concerning the concentration of chemicals and high value-added products, should be estimated in each case to achieve higher productivity per bioreactor volume. The discrimination between SSF and SmF systems is depicted in Table **9.1**.

Table 9.1. Differences between SSF and SmF systems.

	SSF	SmF
Microorganisms used	Yeasts and fungi (and bacteria in some cases)	Yeasts Fungi Bacteria
Products	Flavors Enzymes (i.e. lipase, pectinase etc.) Organic acids (i.e. lactic and citric acid) Xanthan gum	Biomass Lipids Organic acids Other bulk chemicals Ethanol Polyols Polysaccharides etc.
Advantages	Low cost materials (no pre-treatment) Low contamination risk (lack of free water)	Estimation of kinetic & heat & mass transfer parameters Easy control of operating parameters
Disadvantages	Control of operating parameters Difficulties on scale-up	Treatment of raw material (substrate) prior to fermentation Contamination risk

Wheat Straw

Wheat is the most important graminaceous plant cultivated around the world for human nutrition [14]. According to Food and Agriculture Organization (FAO), approximately 77% of global wheat production came mainly from Asian and European countries in 2013/2014 [23]. After harvesting wheat seeds, wheat straw, the remaining part of the plant, constitutes an agricultural by-product [24]. This lignocellulosic material consists of 8-15% lignin, 35-45% cellulose, 20-30% hemicelluloses, 3% proteins, and 10% ash on dry solid basis [9].

Huge amounts of wheat straw are usually burned in the fields increasing air pollution and endangering human health [6]. Wheat straw, like wood, may be used directly through combustion to generate electricity. Sastre *et al.* [25] showed that the combustion of wheat straw produced less emission of greenhouse gases and consumed less fossil fuel compared to electricity from natural gas, while Nguyen *et al.* [26] demonstrated that wheat straw could be an alternative energy source for electricity generation with low emission of nitrogen oxides (NO_x).

Bioethanol production using wheat straw as feedstock has been investigated by many researchers in the scientific literature as an alternative way for renewable energy production [14]. In this case, pretreatment of wheat straw is a necessary stage for liberation of cellulose and hemicelluloses followed by hydrolysis, in order to produce fermentable sugars (*e.g.* glucose, xylose, arabinose, galactose and mannose) for efficient utilization by microorganisms [9]. Among pretreatment methods, chemical techniques are quite simple and fast compared to physical, physicochemical, and biological methods [9, 14]; in this regard, acid hydrolysis is the most common chemical pretreatment method used to obtain high yields of sugars from lignocellulosic materials, like wheat straw [14, 27]. Saccharification can be achieved by chemical or enzymatic hydrolysis. Enzymatic hydrolysis requires less energy and chemicals in comparison with chemical hydrolysis, while there is no formation of inhibitory compounds, like furfural, 5-hydroxymethyl furfural (HMF), and acetic acid, for microorganism's growth [27, 28]. For these reasons, efforts are being made to reduce the cost of enzymes used and improve their performance [9, 29]. Sugar fermentation can be achieved by yeast, bacteria, and fungi strains [14]. Some of the applied microorganisms are *Saccharomyces cerevisiae, Escherichia coli, Zymomonas mobilis, Pachysolen tannophilus, C. shehatae, Pichia stipitis, Candida brassicae, Mucor indicus etc* [9]. Among them, the yeast *S. cerevisiae* is the most widely studied microorganism for ethanol production from wheat straw, achieving high volumetric production yields [30].

Wheat straw, thanks to its zero-acquisition cost and easy availability, has been also used as raw material for lignocellulosic enzymes production (*e.g.* cellulases). The major groups of cellulase enzymes are endo-(1,4)-β-D-glucanase (EC 3.2.1.4), exo-(1,4)-β-D-glucanase (EC 3.2.1.91), and β-glucosidases (EC 3.2.1.21) [31]. These enzymes are synthesized by fungi, bacteria, and actinomycetes strains during their cultivation on lignocellulosic substrates [32], and find applications in various industries, such as agricultural industry, textile industry, pulp and paper industry, wine and brewery industry, animal feed industry, *etc* [31, 32]. Wheat straw has been studied for the production of cellulolytic enzymes by the thermophilic fungus *Thermoascus aurantiacus* using a solid-state bioreactor. Under optimum culture conditions, endoglucanase and β-glucosidase activities obtained were 1709U and 79U per g of dry wheat straw, respectively [33]. In another study, wheat straw was investigated as a raw material for cellulase production by *Aspergillus heteromorphus* in submerged fermentation. The optimum saccharification conditions of temperature and pH were 30 °C and 5, respectively, and the maximum endoglucanase activity (83 IU/ml) was observed after five days of incubation [34]. The fungus *Trichoderma viride* has also been examined for cellulase production in solid state fermentation using wheat straw as substrate [35]. In this case, the maximum cellulase activity of 398 U/mL occurred after acid pretreatment of wheat straw with 2% HCl. Following experiments showed that the purified cellulase had a maximum activity of 148 U/mL. Also, the purified cellulase was compatible with local detergent brands up to twenty days at room temperature, indicating its potential as suitable detergent additive for improved washing performance [36]. Therefore, wheat straw is an attractive feedstock to make these industrial processes economically feasible.

Basidiomycetes, which also show high levels of lignocellulolytic enzyme activities, have the capacity to degrade the lignin and hemicellulose of lignocellulosic materials and produce basidiocarps (mushrooms) [37]. Wheat straws are widely used from mushroom industry as a lignocellulosic material for the cultivation of *Pleurotus* and *Lentinula* strains as single substrate or in combination with other substrates, achieving high biological efficiency (fresh weight of mushroom per dry weight of substrate) and mycelia colonization time as well as high mushroom yield and quality. As examples, Philippoussis *et al.* [37] used wheat straw in combination with wheat bran, soybean flour, and calcium carbonate at 80:12:7:1 ratio, respectively, in order to study the biomass, and the hydrolytic and oxidative enzyme production by *Lentinula edodes*. The results obtained showed that high mycelium growth rates and laccase production can be achieved during cultivation of *L. edodes* on wheat straw. Gaitán-Hernández *et al.* [38] used wheat straw as single substrate to produce Shiitake (*L. edodes*) mushrooms. In this study, the total fresh mushroom production ranged between 370 to 1830 g depending on fungal strain, while the maximum biological efficiency achieved was 78.5%. Koutrotsios *et al.* [39] demonstrated that the biological efficiency of *Pleurotus ostreatus* cultivated on wheat straw without supplements was about 53%. Also, it was observed that wheat straw produces heavier mushrooms in comparison with other substrates, *e.g.* palm tree leaves, pine needles, almond and walnut shells, beech sawdust, corn cobs, and olive mill by-products.

Corn Stover

Corn stover, which includes cobs, leaves, husk and stalk, is an abundant agricultural residue produced in both Europe and America every year [40, 41]. The amount of corn stover remaining in the field has increased significantly during the past years due to increased production of corn crops. It has been estimated that approximately 60–80 million dry tons of corn stover can be available annually for fermentation technology [42]. A part of corn stover is used as supplemental feed for livestocks [43], while burning in the fields is the common approach taken by farmers causing serious environmental problems [44].

On the other hand, various biotechnological methods have been proposed by many researchers in order to minimize the problem of corn wastes disposal. The composition of corn stover depends on climatic parameters, geographic features, phenotype, *etc.* and consists mainly of lignin, cellulose and hemicellulose. The above characteristics make corn stover an attractive source for cellulosic ethanol production [41]. Lignocellulosic constituents are hydrolyzed into simple sugars (hexoses and pentoses) and then fermented into ethanol using microorganisms [45]. Simultaneous saccharification and fermentation, and separate hydrolysis and fermentation have been proposed concerning ethanol production from steam-pretreated corn stover [40, 45]. In this case, separate hydrolysis and fermentation achieved a 13% lower ethanol yield than simultaneous saccharification and fermentation technique, due to the presence of inhibitors, like acetic acid, in the liquid fraction of the pretreated corn stover [40]. In order to obtain higher yields of fermentable sugars with the aim to improve ethanol yields, ammonia fiber explosion pretreatment has been proposed to increase enzymatic digestibility. Ammonia fiber explosion is an alkaline pretreatment process with liquid ammonia that treats the physico-chemical characteristics of lignocellulosic materials under pressure, in order to enhance the conversion of cellulose and hemicellulose to fermentable sugars [41]. Other pretreatment methods of corn stover used to facilitate the enzymatic hydrolysis are ammonia recycled percolation (pretreatment of lignocellulosic substrates with aqueous ammonia using a percolation reactor in recirculation mode), and controlled pH and dilute sulfuric acid pretreatment [46, 47]. The above studies conclusively demonstrated that the utilization of corn stover as alternative feedstock for cellulosic ethanol production could reduce the dependence on corn grain.

Also, corn stover has been targeted as a cheap substrate for biotechnological production of lactic acid which finds applications in the light, food, and healthcare industries [44]. Among many studies conducted for this propose, Xue *et al.* [1] used corn stover hydrolyzate, as a sole carbon source, and a thermophilic strain *Bacillus* sp. XZL4 and achieved the highest L-lactic acid concentration (81 g/L) reported in the literature. Additionally, corn cobs hydrolyzate has been used as raw material for the simultaneous saccharification and lactic acid fermentation by strains of *Lactobacillus delbrukii* [48]. In this study, the final concentration of lactic acid obtained was 17.73 g/L based on the glucose extracted from the saccharified corn cobs. Also, lactic acid production from corncobs has been examined by *Rhizopus oryzae* NRRL-395 achieving a yield of 299.4 g per kg dry corncobs after 48 h fermentation [49]. In order to improve the sugar utilization derived from corn stover for lactic acid production, mixed cultures of *Lactobacillus rhamnosus* and *Lactobacillus brevis* have been studied [50]. These results indicated that mixed cultures of lactic acid microorganisms, that have the ability to increase the substrate conversion efficiency, can produce higher yields of lactic acid compared to single cultures. Therefore, corn by-products could be potential feedstock for efficient production of lactic acid by microbes.

Corn Stover has been also reviewed as an alternative cheap material for citric acid production. Ashour *et al.* [51] have mentioned that *Aspergillus niger* could produce 48.4 g of citric acid per kg dry corn cobs after 8 days of fermentation under the optimized conditions. In another study, *Aspergillus* sp. produced significantly higher amounts of citric acid (243 g per kg dry matter of corncobs) in a shorter time (3 days) after addition of 3% methanol [52]. Note that the yields of citric acid were less than 60 g per kg dry weight of corncobs in the absence of methanol [53]. Pretreatment of corn cobs with dilute sodium hydroxide and Rapidase Pomaliq, a commercial enzyme came from *A. niger* and *T. reesei*, can also improve the yield of citric acid produced (603.5 g per kg dry corn cobs after 3 days of cultivation) [53]. Corn husks could also serve as raw materials for citric acid production by *Aspergillus niger* [53]. Similarly, the pretreatment of corn husks with dilute sodium hydroxide and commercial enzymes, like Rapidase Pomaliq, also significantly increased the yield of citric acid which reached 259 g per kg of dry weight of corn husks.

Corn stover could also be used to obtain other value-added products, such as xylooligosaccharides, which have significant beneficial effects on human health and are usually produced by enzymatic hydrolysis of xylan [54]. Yang *et al.* [54] performed aqueous xylan extraction of corn cobs using diluted sulfuric acid and steaming combined with enzymatic hydrolysis of the extracted xylan aiming for xylooligosaccharides production. The xylooligosaccharides yield after enzymatic hydrolysis was about 67.7% by steaming based on xylan in the raw material. Aachary *et al.* [55] compared mild alkali/acid treatments and pressure cooking of corncob to extract its lignin-saccharide complex in order to enhance enzymatic hydrolysis of xylan to xylooligosaccharides. The results indicated that mild alkali pretreatment efficiently extracted the xylan for commercial enzymes action, resulting in high yield of xylooligosaccharides (81%) in the hydrolyzate.

Rice Straw

Rice straw is an abundant lignocellulosic material produced in large amounts globally, and is discarded as a by-product. It has been estimated that production of 1 kg of grain generates 1 - 1.5 kg of rice straw. Burning of rice straw is the common practice for its removal from the field, while large volume of this is used in animal feed. The lignocellulosic composition of rice straw (cellulose 32–47%, hemicelluloses 19–27% and lignin 5–24%) makes it an excellent material for the production of fermentable sugars, which can find applications as feedstock in several biotechnological fields [11].

In recent years, the production of bioethanol from rice straw either by simultaneous saccharification and fermentation or separate enzymatic hydrolysis and fermentation has been extensively studied [11]. More recently, Sindhu *et al.* [2], utilized the hydrolyzate of rice straw obtained after combined pretreatment and hydrolysis for the production of bioethanol, using the yeast *Saccharomyces cerevisiae,* and the production of poly-3-hydroxybutyrate (biopolymer) using the bacterium *Comamonas* sp.. The maximum ethanol yield was 1.48%, while the maximum biopolymer yield was 35.86% without nutrients addition. Rice straw hydrolyzate has also been studied for the production of ethanol and xylitol by applying pretreatment with moderate aqueous ammonia and sequential fermentation using the yeast *Candida tropicalis* [56]. The obtained results indicated that 220 L of ethanol and 91.5 kg of xylitol can be produced from a ton of rice straw. In another study, the residue of fermented rice straw (glucose and ethanol) for ethanol production was then fed to black soldier fly in order to accumulate lipids for biodiesel production. It was estimated that 4.3 g of biodiesel and 10.9 g of bioethanol could be obtained from the above process per 200 g rice straw [57]. The biotechnological production of biodiesel through fermentation of rice straw hydrolyzate was also studied with the use of the oleaginous yeast *Rhodotorula glutinis* in an airlift bioreactor [58]. In this case, the percentage of lipids accumulated in the biomass was 34%, while the predominant fatty acids were oleic and linoleic.

In addition, rice straw has been used as supplemental material for anaerobic digestion of goat and chicken manure in order to improve biogas production *via* the reduction of the ratio of carbon to nitrogen in the substrates [59, 60]. The obtained results showed that higher biogas and methane yields can be achieved after rice straw addition, while initial carbon to nitrogen ratio between 15 and 25 has been proposed as the optimal ratio for the co-fermentation of goat and chicken manure with rice straw. Also, a low carbon to nitrogen ratio (20) could be beneficial in the composting of swine manure using rice straw as evident by the study of Zhu [61]. Moreover, rice straw has been investigated as single substrate for methane production. Chen *et al.* [62] showed that the method of extrusion pretreatment of rice straw can be effectively applied to maximize the volumetric methane production. It was also observed that extrusion-pretreated rice straw achieved a shorter digestion time and higher degradation yields of cellulose and hemicellulose, thanks to its smaller particle size and larger specific surface area.

Rice straw has been identified as a suitable substrate to produce industrial enzymes, *i.e.* cellulases, xylanases, and laccase. *Niladevi et al.* [63] demonstrated that 17.3 U of laccase can be produced under SSF per g of rice straw, using the actinobacterial strain *Streptomyces psammoticus. Aspergillus niger* was also cultivated on rice straw for the production of cellulases and hemicellulases [64]. The maximum enzyme activity (U per g of dry substrate) obtained were 19, 130,

94, 5070 and 176 for filter paper activity, endoglucanase, β-glucosidase, xylanase and β-xylosidase, respectively. More recently, *Aspergillus niger, Penicillium oxalicum, Colletotrichum gloeosporioides*, and *Pycnoporus sanguineus* were studied to produce xylanase from rice straw [15]. It was observed that *P. oxalicum* showed the highest xylanase activity, about 66 U/mL, compared to other strains.

CITRUS WASTES

Oranges, mandarins, lemons, and grapefruits are the main raw materials for the production of juices in the citrus industry around the world [13]. However, after juice extraction a large volume of solid residues (peels, pulps, and seeds), corresponding to 50% of the initial fruit weight, is generated [65]. The disposal of these wastes on landfill constitutes a serious environmental problem, due to their odor and leachate [3].

The composition of citrus residues allows them to be used in different applications [13]. Conventionally, pectin production for use in food, pharmaceutical, dental, and cosmetic industries is a common practice for utilization of citrus peels [66]. Nevertheless, in order to avoid the high cost of chemical and mechanical methods applied for the above technology, various biotechnological methods aiming at the bioconversion of citrus wastes into high value-added products have been proposed [13].

Among many biotechnological applications of citrus wastes, Torrado *et al.* [3] have studied orange peels as single substrate for citric acid production by using the fungus *Aspergillus niger,* achieving a maximum citric acid yield of 193 g per kg dry weight of orange peel. These values could be significantly improved when combining orange peels and cane molasses (industrial waste) for *Aspergillus niger* cultivation. In this case, the maximum yield of citric acid was 640 g per kg orange peel [67]. Also, citrus peels have been studied for production of various industrially important enzymes. For example, orange and lemon peels have been proposed for polygalacturonase production by *Aspergillus* sp. under submerged and SSF systems achieving remarkable enzyme activities [68, 69]. Irshad *et al.* [70] cultivated the fungus *Trichoderma viridi* under SSF using orange peels as feedstock in order to produce Exo 1, 4-β glucanase. It was observed that the maximum enzyme activity of 412 U/mL was achieved after four days of incubation under optimized fermentation conditions. Mrudula and Anitharaj [71] demonstrated that the fungus *A. niger,* cultivated under SSF using orange peels as substrate, achieved higher yields of pectinase (5283 U per g of dry weight of substrate) in comparison with submerged fermentation. Moreover, Seyis and Aksoz [72] have utilized orange and lemon pulp as well as orange and lemon peels as alternative sources for xylanase production under SSF. In this study, the xylanase activity ranged between 15 to 18 U per mg of protein.

The enrichment of orange peels with γ-linolenic acid by the oleaginous fungus *Cunninghamella echinulata* with the aim of utilization them in animal feed has also been investigated [73]. After ten days of fermentation, the maximum oil content was about 17 mg per g dry weight of orange peels, while the maximum γ-linolenic acid content was about 1.2 mg per g of dry weight of fermented orange peels. Moreover, the fungi *A. niger* and *T. viride* have been studied for production of single cell protein in order to enhance the protein content of lemon pulp in animal feed. The maximum protein content was obtained in *T. viride* cultures (32%) after 25 days of fermentation, while the maximum protein level obtained from *A. niger* was 27% after 15 days [10].

Renewable energy production using environmentally friendly methods is another biotechnological field of citrus waste utilization. Furthermore, the energy production from the above residues can reduce the cost and waste of food industry [74]. Citrus waste has been investigated for the production of both methanol and biogas, thanks to its high carbohydrate content [75]. Citrus peel waste (peel, seeds, and membranes) and mandarin waste (composed of peel and pulp) have been used *via* simultaneous saccharification and fermentation for ethanol production by *Saccharomyces cerevisiae* [75 - 77]. Recently, Wikandari *et al.* [78] examined the leaching technique for orange peel waste pretreatment and demonstrated that citrus waste sludge can be digested to biogas, while the application of membrane bioreactors in the digestion process could improve the theoretical methane yields by more than six times, compared to the conventional digestion system [79]. However, in the above case citrus waste has to be pretreated in order to reduce the concentration of d-limonene below the inhibitory levels for microorganisms' growth prior to the fermentation process [75, 78].

GRAPE POMACE

The wine industry produces millions of kilograms of grape pomace or grape marc, a lignocellulosic residue which mainly consists of seeds and skins. Grape pomace has been utilized as animal feed, although the presence of antinutritional factors (*i.e.* polyphenols) and its low nutritional value reduce digestibility and inhibit the ruminal symbionts. It has also been used as fertilizer, but its phenolic components inhibit the germination of seeds [80].

Few biotechnological methods have been proposed for using grape pomace as a feedstock for the production of valuable bioproducts. Grape pomace has been used as raw material for the production of hydrolytic enzymes, such as xylanase and exo-polygalacturonase by *Aspergillus awamori* in SSF [80]. In this study, the effect of particle size and the initial moisture of grape pomace as well as the extra addition of different carbon sources (apple pectin, glucose, and birchwood xylan) were studied in detail. The findings showed that the particle size did not appear to adversely affect the production of enzymes, in contrast to the initial moisture which can reduce the production of enzymes at very low or high values. In addition, the supplementation of grape pomace with 6% glucose increased both enzymatic activities, while the supplementation with xylan or pectin increased the exo-polygalacturonase production and decreased the production of xylanase. Diaz *et al.* [4] demonstrated that higher values of xylanase, exo-polygalacturonase, and cellulase activities could be achieved by the fungus *A. awamori* in SSF using washed grape pomace supplemented with orange peels as solid substrates. In this case, there has been a four-fold increase in exo-polygalacturonase production and a twenty-fold increase in the production of xylanase and cellulase. On the other hand, the supplementation of grape pomace with orange peels can also reduce the fermentation time and the inhibitory effect of high sugar concentration on the productivity of enzymes.

Moreover, Rodrıguez *et al.* [81] investigated ethanol production by *Saccharomyces cerevisiae* PM-16 in SSF using grape pomace as a source of fermentable sugars. The maximum ethanol content obtained was after 48 h of fermentation, and the ethanol yield on sugars consumed was more than 82%. This study also demonstrated that ethanol production from grape and sugar beet pomace is mostly higher in SSF in comparison to SmF by using sugar beet juice. In a recent study, Corbin *et al.* [7] determined that 31–54% w/w of the dried grape pomace contains 47–80% water-soluble carbohydrates, which can be converted to ethanol attaining a production level of 270 L per ton. Moreover, it was observed that sulfuric acid pre-treatment of grape pomace liberated higher concentration of glucose in comparison with thermal treatments.

The production of citric acid from grape pomace by *Aspergillus niger* in SSF has also been investigated [82]. In this case, the maximum citric acid yield was 60% based on the fermentable carbohydrates consumed. The above value was obtained at a moisture content of 65%, pH 3.8, and fermentation time of four days. These results also showed that the addition of methanol (3%) into grape pomace before fermentation increased the amount of the produced citric acid compared to substrate without the presence of methanol.

In addition, Stredansky and Conti [83] studied the potential of several agro-industrial by-products with high soluble carbohydrate content, including grape pomace, as a source of the exopolysaccharide xanthan, which can find applications as stabilizer in food, pharmaceutical, and petrochemical industries. These SSF experiments were conducted using two strains of the bacterium *Xanthomonas campestris*, able to produce xanthan. The obtained results showed that grape pomace represents a less attractive substrate compared to other cheap materials, like apple pomace and citrus peels, for producing the exopolysaccharide xanthan, due to lower amounts of carbohydrates and absorption efficiency. Nevertheless, the maximum concentration of xanthan achieved using grape pomace was 10 g per liter.

SPENT MUSHROOM SUBSTRATE

Cultivation of edible mushrooms is a process that converts various lignocellulosic materials into a tasty protein-rich food. However, it has been estimated that one kilogram of produced mushroom generates five kilograms of a by-product known as spent mushroom substrate (SMS) [8]. SMS is a kind of lignocellulosic material which contains cellulose, hemicellulose, lignin as well as enzymes, vitamins, polysaccharides, and some trace elements, such as P, K, Ca, Na, Fe, Zn, and Mg [8, 84]. In recent years, significant effort has been placed into valorization of SMS through production of high value-added products in order to consider SMS as a renewable resource of the food industry.

Enzymes, such as laccase (Lac, E.C. 1.10.3.2), manganese peroxidase (MnP, E.C. 1.11.1.13), and lignin peroxidase (LiP, E.C. 1.11.1.14), which are mainly extracted from SMS, can find applications in the biodegradation of toxic pollutants [8]. Laccase obtained from crude extracts of SMS from *Agaricus bisporus* was identified as the main enzyme responsible for oxidation of phenol and polyphenolic compounds [85]. Moreover, immobilization of the crude laccase from the basidiomycete *Pleurotus ostreatus* was used effectively for the continuous elimination of phenolic pollutants, such as 2,6-dimethoxyphenol [86]. Also, SMS from *Pleurotus ostreatus* was able to remove polycyclic aromatic hydrocarbons (PAH) of untreated drill cuttings after 56 days of composting [87]. In addition, the crude enzyme extract containing laccase and MnP from *Pleurotus pulmonarius* SMS was capable of performing a 100% reduction in the initial concentration of chlorothalonil (CTN, an organochlorine fungicide) after 45 min of reaction [88]. In another study, SMS of *Lentinus edodes* was used as a bio-sorbent of heavy metals, such as cadmium, lead, and chromium under batch conditions. It was demonstrated that 25g/L SMS resulted in a removal efficiency of about 80% and 60% for cadmium, lead, and chromium, respectively [89]. Also, SMS of *Pleurotus sajor caju* has been examined for dyes decolourisation [90]. In this study, it was demonstrated that LiP activity, the main enzyme extracted from SMS, was able to decolorize crystal violet, tryphan blue, amido black, congo red, bromophenol blue, methyl green, and remazol brilliant blue R in high percentages (*e.g.* from 40% to 100% after a 24 h incubation period).

Re-utilization of SMS *via* supplementation with other agricultural residues for new cultivation cycles of mushrooms was also studied in the scientific literature. Royse [91] cultivated *P. sajor-caju* on SMS of *L. edodes*. The results showed that the highest biological efficiency of *P. sajor-caju* was obtained by supplementing SMS with 12% soybean and 1% calcium carbonate (CaCO3). Also, SMS from oyster mushroom supplemented with 60% sawdust and 20% wheat bran resulted in the highest biological yield and efficiency for edible mushrooms *P. ostreatus* and *P. florida* compared to other proportions [92]. Additionally, *Pleurotus* SMS in combination with sunflower seed hulls and vermicompost have been used as non-composted material for cultivation of *Agaricus blazei*; it increased the productivity of mushroom cultivation and decreased the contamination level of the substrate [93]. More recently, SMS of *Hypsizigus marmoreus* combined with cottonseed hulls and wheat bran was used for *P. ostreatus* cultivation [94]. The results indicated that substrate supplementation with 12% SMS showed the highest biological efficiency (about 61%) compared with the other supplementation proportions of SMS. Similarly, the best quality traits of mushrooms were obtained by 12% and 25% SMS supplementation.

It has been demonstrated that polysaccharides derived from both the fruiting bodies of higher basidiomycetes and the mycelium biomass appear to have promising antibacterial and antioxidant properties [95]. Based on these findings, SMS has also been investigated as a new source of mushroom polysaccharides. Zhu *et al.* [96] isolated and purified a water-soluble polysaccharide, named PL, from SMS of *L. edodes*. The Results showed that the polysaccharide was mainly composed of mannose, glucose, and rhamnose. The crude polysaccharide also showed significant antibacterial activity against *Escherichia coli*, *Staphylococcus aureus,* and *Sarcina lutea*. Summarizing the above points, the aforementioned studies indicated that novel uses of SMS, and the enzymes and polysaccharides derived from it could provide an economically viable solution for the disposal problem of SMS, decreasing the costs of mushroom industry.

The aforementioned information regarding the valorization of lignocellulosic residues are presented and briefly gathered in Table **9.2**.

OLIVE MILL WASTEWATER (OMW)

Olive mill wastewater (OMW) is the principal waste stream that is derived from the olive fruit processing by mechanical means during olive oil production. In OMW, biological oxygen demand (BOD) and chemical oxygen demand (COD) concentration values can be 200-400 times higher than those in typical municipal sewage [97]. The OMW treatment is considered to be a very difficult task because such effluents are produced in large quantities in a quite short period of time and in scattered small or medium sized facilities [98]. The OMW is generally characterized by intensive violet-dark brown to black color, strong specific olive oil smell, high degree of organic pollution, pH between 3 and 6 (slightly acid), high electrical conductivity, and high content of polyphenols (0.5 – 24.0 g/L) and solid matter [99].

Table 9.2. Raw materials used as substrate, microorganism used for growth, and final product resulting from the valorization of lignocellulosic residues by biotechnological methods.

Raw Material / Substrate	Microorganism	Product	Reference
Wheat straw	*Saccharomyces cerevisiae*	Bioethanol	Jørgensen *et al.* [30]
	Thermoascus aurantiacus	endoglucanase	Kalogeris *et al.* [33]
		β-glucosidase	
	Aspergillus heteromorphus	endoglucanase	Singh *et al.* [34]
	Trichoderma viride	cellulase	Ahmed *et al.* [35]
	Lentinula edodes	mushrooms	Philippoussis *et al.* [37]
			Gaitán-Hernández *et al.* [38]
	Pleurotus ostreatus	mushrooms	Koutrotsios *et al.* [39]
Corn stover	*Bacillus* sp. XZL4	lactic acid	[1]
	Lactobacillus delbrukii		Zulfiqar *et al.* [48]
	Aspergillus niger	citric acid	Ashour [51]
			Hang *et al.* [52]
			Hang *et al.* [53]
Rice straw	*Comamonas* sp.	bioethanol	Sindhu [2]
		poly-3-hydroxybutyrate	
	Candida tropicalis	xylitol	Swain *et al.* [56]
		bioethanol	
	Rhodotorula glutinis	biodiesel	Yen *et al.* [58]
	Streptomyces psammoticus	laccase	Niladevi [63]
	Aspergillus niger	cellulases	Kang *et al.* [64]
		hemicellulases	
Citrus wastes	*Aspergillus niger*	citric acid	
		pectinase	Mrudula *et al.* [71]
	Cunninghamella echinulata	γ-linolenic acid	Gema *et al.* [73]
	Saccharomyces cerevisiae	bioethanol	Wilkins *et al.* [75], Oberoi *et al.* [76], Widmer *et al.* [77]
	Trichoderma viride	single cell protein	De Gregorio *et al.* [10]

(Table 9.2) cont.....

Raw Material / Substrate	Microorganism	Product	Reference
Grape pomace	*Aspergillus awamori*	xylanase	Botella *et al.* [80], Díaz *et al.* [4]
		exo-polygalacturonase	
		cellulase	Díaz *et al.* 2012 [4]
	Saccharomyces cerevisiae	bioethanol	Rodrıguez [81]
	AspergIllus niger	citric acid	Hang *et al.* [82]
	Xanthomonas campestris	xanthan	Stredansky and Conti [83]
SMS	*Pleurotus ostreatus*	laccase	Hublik *et al.* [86], Juárez *et al.* [88]
		manganese peroxidase	Juárez *et al.* [88]
	Pleurotus sajor caju	lignin peroxidase	Singh *et al.* [90]
		mushrooms	Royse *et al.* [91]
	Agaricus blazei	mushrooms	Matute *et al.* [93]
	Lentinoula edodes	polysaccharides	Zhu [96]

The OMW organic fraction is principally composed of sugars, cellulose, pectin, (poly-)phenolic compounds responsible for the dark color, phytotoxic and antimicrobial compounds, simple phenolic acids, phenolic alcohols, polyalcohols, various amino acids, proteins, organic acids, and residual oil [97, 100 - 108]. Regarding the mineral fraction, all minerals are presented in various fresh OMW samples [100, 109 - 113]. In some cases and besides phenolic compounds, OMW derived from press extraction systems also contains reducing carbohydrates (principally glucose) in very high quantities (*i.e.* >70 g/L) [101], which pose significant problems related with their treatment.

Phenolic and fatty compounds, found in OMW, may inhibit the growth of several types of microorganisms and stop the conventional secondary and anaerobic treatments in municipal treatment plants [102]. Such compounds cannot be biodegraded easily [97, 101, 104, 105, 113 - 115].

Treatment of OMWs

Both the variety of OMW components and the tremendous seasonal volume production make their treatment difficult, and therefore their elimination and disposal is one of the main critical environmental problems related to the olive oil industry.

Recent developments indicate that OMW should be valorized as a medium for fermentation purposes rather than discharged as a waste [101]. Therefore, research should focus on both waste bioremediation and production of high-added value products, simultaneously, under cost-effective technologies. The breakdown of phenolic compounds should be considered as the limiting step in OMW treatment by biotechnological processes [103 - 106], as such compounds are not easily biodegradable [97, 101, 105, 113 - 115].

The biotechnological approach requires deep knowledge of the biochemical routes used by microorganisms for the different compounds of OMW in order to select the most appropriate species or "design" new strains that effectively degrade the wide variety of these substances. It should be stressed out that biological processes (especially anaerobic ones) have been found to be more economic and efficient than physical/chemical processes [99].

The structure of the aromatic compounds present in OMW resembles many of the components of lignin. The ability of higher fungi to break down phenolic compounds is based on the secretion of extra-cellular oxidases (ligninolytic enzymes), laccases, lignin peroxidases, and manganese dependent (or independent) peroxidases [101, 105, 116, 117]. The secretion of these enzymes is strain-dependent and is influenced by various culture conditions [101, 104, 105, 118, 119]. The use of filamentous fungi for OMW pretreatment has been shown to reduce toxicity and improve the biodegradability in aerobic degradation. In particular, the pretreatment of OMW with higher fungi, which produce poly-aromatic hydrocarbon-degrading enzymes, has been used to detoxify and decolorize them. However, the application of such processes on large scale compared to bacteria presents limitations due to the difficulty of achieving continuous culture because of the formation of filamentous pellets and mycelia. Such pellets and mycelia might either increase the density of the medium or be destroyed (torn down and therefore terminate fermentation). Another process drawback could be the variation in COD reduction and color removal values obtained after OMW bio-treatment, even when using the same microorganism and operating conditions [99].

Non-genetically modified yeast strains, in general, do not contain in their genetic arsenal the possibility of producing such types of enzymes [118] and thus, the removal of phenolic compounds and the decolorization of OMW by means of fermentation by these yeast species, through the use of the above-mentioned enzymes, should be excluded. On the other hand, Rizzo *et al.* [120] suggested a potential exclusive physical mechanism involving the establishment of weak and reversible interactions, mainly between anthocyanins and yeast walls, by means of adsorption. Moreover, (potentially very weak) assimilation of several phenolic compounds by yeasts should, also, not be excluded [121]. Indeed, there are studies in accordance with the above mentioned studies showing that when non-genetically modified yeast strains were cultivated in OMW-based media, remarkable decolorization and non-negligible removal of phenolic compounds occurred [17, 19, 20].

OMW-based media enriched with other carbon sources have been used for the cultivation of molds, prokaryotic microorganisms, yeasts, and yeast-like species leading not only to the remediation of the waste but also to the production of value added compounds, such as yeast and fungal biomass [17, 19, 20, 115, 122 - 124], exopolysaccharide [101], various enzymes [101, 105, 107, 122, 125], citric acid [20, 115], and bioethanol [17, 19, 119, 126, 127].

Utilization and Applications of OMW

Amongst the different approaches for the utilization and application of OMW (see Fig. **9.1**), these effluents could be also used as substrate for the growth of microorganisms and the production of new (potentially high-added value) products [biosurfactants, biopolymers, activated carbon, enzymes, and production/generation of bioenergy/biofuels (such as alcohols-bioethanol, biohydrogen, biomethane, biodiesel)] [99, 128]. Due to their rich content in sugars and minerals, OMW could be used as an ideal substrate for yeast or other fungi growth, leading to the production of high digestible microbial mass which includes carbohydrates, lipids, minerals, and vitamins. Not only edible (or other) fungi, especially *Pleurotus* or *Lentinula* species, but also *Agaricus bisporus* and *Geotrichum candidum* are able to grow using olive oil by-products (such as blends of OMW with other waste streams, such as cheese whey) [99, 117, 125, 128, 129]. As mentioned above, the content of OMW includes, amongst others, sugars, phenolic compounds, and potentially polyols and lipids. Thus, such effluents could be used as a source of bioethanol (recovered by distillation), biohydrogen, biodiesel precursors (*e.g.* triglycerides), and biomethane production.

Even though there are studies focusing on biomethane production obtained through the anaerobic digestion of OMW substrates, work has also been performed in bioethanol generation [17, 19] and biohydrogen production by photo-fermentative processes and dark fermentation [99, 128]. Biohydrogen can be produced either by direct or indirect biophotolysis, photo-fermentation, and dark fermentation (the latter does not require light energy) [130]. Anaerobic digestion can convert organic substrates to CH_4 and CO_2 (biogas) through the concerted action of a mixture of microbes (consortia). A large number of microbial species with significant taxonomic and physiological differences can produce biohydrogen through single or combined metabolic pathways. Several studies though emphasize that systems producing photo-biological hydrogen using photosynthetic bacteria and OMW will clearly require improvements, due to their dark color and inhibitory effects, either by using high dilution rates or by blending with other waste streams, such as cheese whey. However, the use of high dilution rates requires the addition of nitrogen supplementation and adjustment of OMW's pH; this proved to be impracticable on large scale, as the volume of the effluent is significantly increased [99, 128, 129, 131].

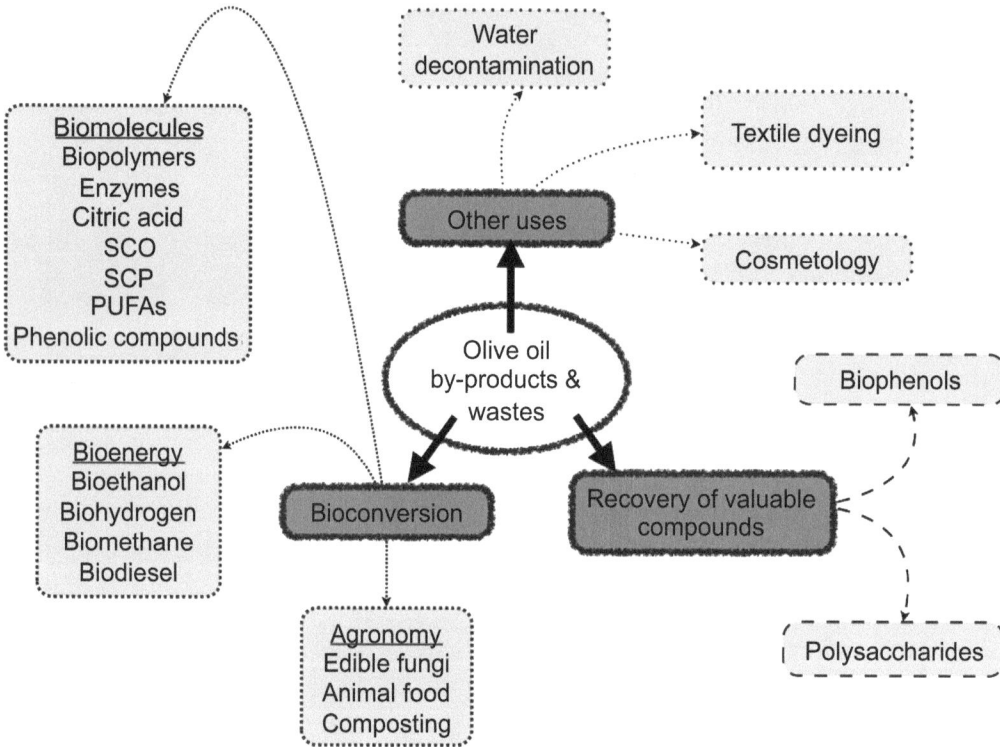

Fig. (9.1). Valorization opportunities for olive mill by-products and wastes [adapted by Demerche *et al.* [120]].

Microbial Products

Various biotechnological processes have been used for the treatment of OMW. Such processes can lead to the simultaneous reduction of COD values, the degradation of phenolic compounds, and the production of value-added compounds, such as biomass, citric acid, ethanol, and enzymes, such as extracellular laccase, manganese peroxidase (MnP), lipases (applicable in the dairy, pharmaceutical, detergent, and other industries), pectinases, and phenol oxidases (used in olive oil extraction process to improve olive oil yield, quality, turbidity, oxidation induction time, and content of aromatic compounds, and to reduce the toxicity of many aromatic compounds)] [17, 19, 20, 97, 99, 101, 105, 107, 115, 122, 125, 128, 132 - 135].

MOLASSES

Molasses, which contains sugars (44-60% *w/w*) and various minerals, is the viscous by-product of the sugar cane or sugar beet processing into sugar [136, 137]. Due to the high content in sugars, molasses has been used as growth medium for producing various (high-) value-added products through microbial fermentation. The water remaining after these bioprocesses, called molasses wastewater (MWW), is characterized by high biochemical oxygen demand (BOD) and chemical oxygen demand (COD) values, strong odor, and dark brown color [138]. The high molecular weight polymer melanoidin is the dark brown pigment found in MWW [139, 140]. The composition of this effluent and its release into the environment without appropriate treatment, may lead to eutrophication phenomena in water. Its dark color hinders photosynthesis by blocking sunlight, labeling molasses deleterious to aquatic life [141].

The simultaneous decolorization of molasses, reduction of its content in melanoidins, and its use as microorganism's substrate for the production of various biotechnological products are basic research goals. Low international prices of molasses as well as its high content in assimilable sugars render this residue as a suitable substrate for the production of numerous biotechnological products. In contrast with various chemical [142] and physicochemical processes [143], biotechnological treatment using fungi (*e.g.Coriolus*, *Aspergillus*, *Phanerochaete*) and bacteria (*e.g.Bacillus*, *Alkaligenes*, *Lactobacillus*) [139, 144 - 146] is more financially feasible on large scale. The biological degradation system includes mainly oxidases and peroxidases, such as glucose oxidase, dependent manganese peroxidases (MnP), and independent manganese peroxidases (MIP) [144, 145]. Boer *et al.* [147] have proved that both MnP and MIP possess the ability to decolorize molasses in presence of H_2O_2; however, one must also remember that the decolorization ability of oxidases and peroxidases reaches a maximum at certain pH and temperature values and is also dependent on their substrate specificity.

Fungi, bacteria, and yeasts have been cultivated on molasses either for melanoidin degradation and reduction of color, BOD and COD values [139, 146], or for the production of value-added metabolites, such as ethanol, gluconic acid, citric acid, fruto-oligosaccharides (FOS), pullulan, succinic acid, single cell oil (SCO), erythromycin, and bacteriocins [17, 148 - 159]. In a limited number of reports, the production of value-added compounds and the detoxification-decolorization of molasses were simultaneously studied [17, 158, 159].

Biological Processes

Various reports in literature suggest using bacteria, yeasts [17, 160], and fungal strains [161] for decolorization of molasses, MWW and melanoidins. Specifically, the fungi *Cunninghamella echinulata* and *Mortierella isabellina* were grown on molasses, showing non-negligible substrate decolorization up to ~75% for *C. echinulata* (400 h of culture) and ~20% for *M. isabellina* (200 h after inoculation) [158]. Moreover, waste molasses was used as growth medium for *Leuconostoc mesenteroides* to produce bacteriocin. Simultaneous decolorization of up to ~27% of this residue was performed by the same species [159]. Ohmomo *et al.* [145] used the fungus *Coriolus versicolour* Ps4a for the decolorization of melanoidins, and achieved a decolorization rate of up to ~80%. Following, Ohmomo *et al.* [146] used *Aspergillus oryzae* strain Y-2-32 which absorbed in its mycelia low molecular weight melanoidins. *A. niger*, used by Miranda *et al.* [161] led to 83% decolorization of MWW. Raghukumar and Rivonkar [162] studied the decolorization of molasses spent wash by white-rot fungus *Flavodon flavus*, isolated from a marine habitat, that was able to quickly degrade the high molecular weight fraction. Tondee *et al.* [160] cultivated *Issatchenkia orientalis* strain No. SF9-246 (isolated from rotten banana) in a malt extract-glucos--peptone broth containing melanoidins, and a decolorization rate of 60.2% was obtained within 7 days.

The capability of bacteria of the genus *Pseudomonas*, *Enterobacter*, *Stenotrophomonas*, *Aeromonas*, *Acinetobacter*, and *Klebsiella* to reduce COD of anaerobically treated molasses spent wash was tested by Ghosh *et al.* [163] resulting in various final decolorization values. Following, Ghosh *et al.* [164] cultivated *Pseudomonas putinda* on the same substrate achieving 24% decolorization whereas when immobilized cells were used, a two-fold increase in the decolorization yield was achieved. Sirianuntapiboon *et al.* [165] cultivated strain No. BP103 of acetogenic bacteria on molasses wastewaters, achieving a 76.4% decolorization yield. Tondee and Sirianuntapiboon [166] cultivated *Lactobacillus plantarum* strain No. PV71-1861 (isolated from pickle samples) under anaerobic and facultative (static) conditions, showing a high potential for use in decolorization of molasses wastewater (maximum yield 68.12% within 7 days). Kumar and Chandra [139] used *Bacillus thuringiensis* MTCC 4714, *Bacillus brevis* MTCC 4716, and *Bacillus* sp. MTCC 6506 on substrates containing synthetic melanoidins. The use of individual cultures did not present significant affection on melanoidins whereas mixed cultures resulted in maximum decolorization of 50%. Kalavathi *et al.* [167] achieved degradation of the melanoidins (decolorization up to 96%) in distillery effluents by the marine cyanobacterium *Oscillatoria boryana* BDU 92181. The cyanobacteria *Lyngbya* sp. and *Synechocystis* sp. managed to decolorize distillery effluents by 81% and 26% respectively [168].

Biotechnological Applications of Molasses

As mentioned above, molasses becomes a highly attractive substrate to produce various biotechnological compounds, due to its high sugar content and low international prices. Literature reports the use of various microorganisms grown on molasses for the production of (high-) value-added products, such as ethanol, citric acid, gluconic acid, fruto-oligosaccharides (FOS), pullulan, succinic acid, single cell oil, erythromycin, bacteriocins, etc [148, 154 - 156, 158].

Ethanol Production

Molasses and molasses waste-water have been used as substrates in various configurations (immobilized yeast strains, fed batch, repeated fed batch and continuous bioreactor trials, mixed yeasts fermentations, non-aseptic conditions *etc.*) for biotechnological production of ethanol [149 - 153].

Other Products

Citric Acid

Citrate, the intermediate metabolite of TCA cycle, may be produced by biotechnological means when molasses is used as substrate, through the of strains of the fungus *Aspergillus niger*. It could be produced either by submerged fermentation (SmF) or liquid surface fermentation (LSF) [169]. The production of citric acid by *A. niger* is highly affected by molasses metal content (such as ferrum, zinc, copper, manganese). According to Majolli and Aguirre [170], the concentration of such heavy metals should be significantly reduced before mycelial growth. Pera and Callieri [171] reported that the production of citric acid, when molasses is used as substrate, is highly affected by the presence of ferrum ions in concentrations above 0.2 ppm. Adham [172] attempted to improve citric acid fermentation (Cit_{max}~73 g/L against ~30 g/L) by *A. niger* grown in beet-molasses medium by adding natural oils [at concentrations of 2% and 4% (v/v)] with high content of unsaturated fatty acids. Ikram-ul *et al.* [173] produced citric acid by selected mutants of *A. niger* [improved by chemical mutagenesis using N-methyl, and N-nitro-N-nitroso- guanidine (MNNG)] when grown on cane molasses. The best selected mutant produced ~96 g/L citric acid 168 h after fermentation in Vogel's medium, using blackstrap molasses previously treated by potassium ferrocyanide and H_2SO_4 as substrate. The production of citric acid from cane molasses by *A. niger* in a pilot study using surface or submerged fermentation, studied by Hamissa and Radwan [174] and Qazi *et al.* [175], resulted in a maximum concentration of citric acid of 60.8 and 67.0 g/L, respectively. Likewise, cell recycling of *A. niger* in surface fermentation of cane molasses was performed significantly, which reduced the fermentation time compared to the normal single cycle batch submerged or surface fermentation process. About 80% of the sugar was converted to citric acid in 5 days of batch fermentation, and three batches were carried out with the same fungal mat without any significant loss of productivity [176]. The production of citric acid from beet molasses by *A. niger* was improved with the addition of phytate (plant constituent that can be found in the seeds of cereals and legumes) to the medium. The effect of phytate addition was found to be dependent on its concentration and the stage of fermentation at which it was added. When added at the beginning of incubation, the optimal concentration of phytate in the medium for citric acid production was 10.0 g/L and resulted in a ~3.1-fold increase in citric acid accumulation. Addition of 16.0 g/L phytate to the medium, after 3 days of incubation, resulted in the maximum citric acid concentration, *i.e.* ~2.4-fold higher than the control experiment [177]. A novel method of citric acid production from beet molasses in which an anion exchange resin packed column was connected to the bioreactor for separation of citric acid from fermentation broth was developed by Wang *et al.* [178]. In comparison with a conventional batch, the new fermentation technique increased citric acid productivity and sugar conversion from 0.34 g/L h^{-1} and 82.2% to more than 0.5 g/L h^{-1} and 94.8%, respectively.

Gluconic Acid

Gluconic acid (pentahydroxycaproic acid) is a mild organic acid derived from glucose by a simple oxidation reaction facilitated by the enzyme glucose oxidase (from fungi) and glucose dehydrogenase (from bacteria such as *Gluconobacter*). Oxidation of the aldehyde group on the C-1 of β-D-glucose to a carboxyl group results in the production of glucono-d-lactone ($C_6H_{10}O_6$) and hydrogen peroxide. Glucono-d-lactone is further hydrolysed to gluconic acid either spontaneously or by lactone hydrolysing enzyme, while hydrogen peroxide is decomposed to water and oxygen by peroxidase or catalase [179].

Gluconic acid and its derivatives (principally sodium gluconate) can be applied in pharmaceutical, detergent, food and leather industry. It is produced by genetically modified strains of *A. niger* in solid-state fermentation using molasses as substrate. The components used for medium hardening should be carefully chosen as they do not only offer the necessary nutrients for microorganism's growth but can also be considered as suspending factors. Sharma *et al.* [156] employed tea waste as solid support (to molasses). The fungus growth was enhanced by the tea components and the maximum yield of gluconic acid production was 80.5%.

Fructooligosaccharides (FOS)

FOS are classified as prebiotics, *i.e.* non-digestible food ingredients that stimulate the growth and/or activity of bacteria (indigenous bifidobacteria) in the digestive system in ways claimed to be beneficial to health. FOS were initially produced on industrial scale using pure sucrose as substrate with the use of enzymes from *Aspergillus* (β-fructofuranosidase) [180, 181] or *Aureobasidium* (fructosyltrans-ferase) [182]. Shin *et al.* [183] cultivated *Aureobasidium pullulans* on molasses (initial sucrose concentration 360 g/L) and achieved total FOS concentration of 166 g/L, 24 h after inoculation with 46% product yield.

Succinic Acid

Liu *et al.* [155] cultivated *Actinobacillus succinogenes* for the production of succinic acid which is a precursor of numerous products [*e.g.* chemicals, pharmaceuticals, food additives, solvents, and biodegradable plastic; see: Willke and Vorlop [184]. They have reported the production of ~44 g/L of succinic acid 60 h after inoculation; acetic acid and formic acid were the by-products of the process. A process in which previously treated molasses, for the recapture of heavy metals, was used as microbial substrate led to a maximum succinic acid concentration of ~51 g/L (consuming 95% of initial sugar concentration) [155].

Single Cell Oil

The yeast *Trichosporon fermentans* was cultivated on molasses under nitrogen limited conditions to produce single cell oil (SCO) that could potentially be used as precursor for biodiesel production. SCO consisted of fatty acids having composition similar to vegetable fatty acids (containing mainly palmitic acid, stearic acid, oleic acid, and linoleic acid) [157]. The factors affecting these fermentations are various, mostly related to substrate composition (*e.g.* C/N ratio) and culture conditions (*e.g.* incubation temperature, dissolved oxygen, and pH) [158, 185 - 190]. Finally, in recent developments, the Zygomycete fungi *Cunninghamella echinulata* and *Mortierella isabellina* were cultivated on sugar-based media including molasses, and it has been demonstrated that the assimilation rate of the sugars used as substrates played a crucial role on the lipid accumulation process. Both fungi presented satisfactory dry cell weight yield on media composed of molasses, while equally remarkable quantities of SCO and the medically and nutritionally important γ-linolenic acid (GLA) were synthesized [158].

Pullulan

Roukas and Liakopoulou-Kyriakides [191] produced the water-soluble polysac-charide pullulan from beet molasses by *Aureobasidium pullulans* in a stirred tank fermenter. Pullulan is consisted of molecules of maltotriose united by α 1,6 glucosidic bonds. This compound has the ability to form films, which have tolerance to oils and are not oxygen permeable. Aeration was an important factor to produce pullulan (the maximum concentration was 23 g/L under 1 vmm aeration, while 0.5 vvm and 0.0 vvm gave maximum pullulan concentrations of 14 g/L and 12 g/L, respectively).

Erythromycin

El-Enshasy *et al.* [154] cultivated *Saccharopolyspora erythraea* in molasses based medium under submerged fermentation for the production of erythromycin. Erythromycin is a secondary metabolite and its production depends on substrate composition and culture conditions. Through cultivation medium optimization (by adding n-propanol and ammonium sulphate) they managed to produce 800 mg/L erythromycin.

Bacteriocins

Bacteriocins, synthesized ribosomically by lactic acid bacteria, are peptides that may act as biopreservatives, since they exert antimicrobial activity against a range of microorganisms. They constitute a naturally produced eco-friendly components amendable for controlling and inhibiting the growth of various undesirable microorganisms, particularly Gram-positive bacteria (including food spoilage and food-borne pathogen microorganisms). Metsoviti *et al.* [159] used waste molasses (with TS_0 concentrations 20 g/L and 30 g/L) as growth medium for *Leuconostoc mesenteroides* to produce bacteriocin, while simultaneous decolorization of the residue up to ~27% was also observed.

CHEESE WHEY

During cheese making and after the precipitation and removal of milk casein, cheese whey (CW; a green-yellowish liquid) remains. Cheese whey is composed of lactose (4.5-5% w/v), soluble proteins (0.6-0.8% w/v), lipids (0.4-0.5% w/v), and mineral salts (0.4-0.5% w/v). Due to its high BOD and COD values and the fact that cheese whey is produced in high volumes, this effluent represents an important environmental problem [192 - 194]. It is of great importance to demonstrate efficient and cost-effective methods of simultaneous treatment (reduction of polluting load) and valorization (conversion of the various compou-nds of the waste into value-added products produced by microbes) of CW [195].

Single Cell Protein (SCP)

SCP can be used as livestock feed, starter culture, or food additive. Yeasts (*e.g. Kluyveromyces marxianus, K. fragilis, Candida pseudotropicalis, C. versati- lis,* and *Torulopsis bovina*) are the most promising candidates for the production of

SCP by valorization means of whey [192, 195]. Schultz *et al.* [196] have used deproteinized sweet and sour cheese whey concentrates as substrates for the production of SCP by *Kluyveromyces marxianus* CBS 6556. After process improvement (culture supplementation with trace elements, such as ammonium, calcium, and vitamins), biomass dry concentrations of up to 50 g/L were obtained ($Y_{X/S}$ values of 0.52 g/g) from sweet whey, and up to 65 g/L ($Y_{X/S}$ values of 0.48 g/g) from sour whey concentrates. Moreover, COD was reduced by 80%. The yeast strain *K. bulgaricus* ATCC 1605 reached a biomass yield of 13.5 mg/mL when cultivated in concentrated whey permeate under aerobic condition [197]. *K. frafilis* strain NRS 5790 reached a cell yield of 0.74 g cell/g lactose when used for a continuous aerobic submerged fermentation of cheese whey [198]. Vamvakaki *et al.* [195] cultivated *Thamnidium elegans* and *Mucor* sp. on whey for the production of biomass. Both fungi seemed incapable of consuming lactose after protein exhaustion. Thus, a supplementary quantity of ammonium sulfate was added in order to favor the consumption of lactose and enhance the production of biomass.

Single Cell Oil (SCO)

Oleaginous microorganisms have been used to convert agro-industrial wastes or raw materials into lipids, rarely found in nature (*e.g.* cocoa-butter), and plant oil equivalents containing poly-unsaturated fatty acids (PUFAs) (*e.g.* γ-linolenic acid). Moreover, a potential use of the oil derived from such microorganisms is as nonconventional substitute for the production of "2nd generation" biodiesel [195].

Cheese whey pretreated by hydrodynamic cavitation (HC) under alkaline conditions was used as substrate for the growth of *Cryptococcus curvatus*, resulting in a growth rate of 7.2 g/L/day, lipid content of 65%, and lipid productivity of 4.68 g/L/day (among the highest reported) [199]. When *C. curvatus* strain TYC-19 was cultivated on cheese whey medium, it produced a notable amount of fatty acid methyl esters (FAMEs; which can be used as biodiesel fuel) with the major ones being linoleic and oleic acid methyl esters [200]. Vamvakaki et al [195] cultivated *Mortierella isabellina* on whey for the production of SCO rich in GLA. This strain consumed all available lactose of the medium and a significant amount of protein. A supplementary quantity of lactose was added into the medium in order to increase the C/N ratio (thus increasing the production of lipids). Growth of *M. isabellina* on lactose- supplemented whey resulted in a maximum GLA production of 301 mg/L.

Ethanol

Literature also discusses the treatment of whey by fermenting lactose to ethanol and the simultaneous reduction of whey pollution [192, 193]. CW, Cheese whey powder (CWP) solution, CW permeate from ultrafiltration and deproteinized CW were used. This treatment presents some limitations like the use of certain microorganisms (which are able to metabolize lactose directly and are not inhibited by moderate sugars and ethanol concentrations) and the use of concentrated whey (significantly reduced costs with lactose concentration up to ~100-120 g/L) [192]. On the other hand, extremely high substrate concentrations may stress the strain used (due to osmotic pressure) and affect ethanol yield and substrate consumption. Also, an increase in the aeration can shift the metabolism of lactose towards biomass rather than ethanol production.

CWP solution was used for ethanol fermentation (yield of ethanol produced on substrate consumed was ~0.35-0.54 g/g) by *K. marxianus* NRRL-1195 in batch experiments, studying also the effects of initial pH, CWP concentration, and external nutrient (N, P) supplementation. The optimum conditions included pH 5 and no nutrient supplementation. The ethanol concentration increased with increasing CWP concentration, indicating no substrate or product inhibition [201]. Crude whey was used as substrate for the cultivation of *K. marxianus* strain MTCC1288 for the production of ethanol and biomass; the yeast was able to metabolize most of the lactose within 22 h to produce 2.10 g/L ethanol and 8.9 g/L biomass [202]. The yeast strain *K. frafilis* NRRL Y 2415 produced the highest ethanol yield (9.1% v/v) when cultivated in concentrated whey permeate under aerobic condition [197].

Biogas

Methane produced by anaerobic digestion of whey can be used as energy source directly in the treatment plant [192]. Specific microorganisms should be used in order to overcome some process drawbacks, such as resistance to degradation by proteins (*e.g.* casein) and lipids. Even if hydrocarbons (*e.g.* lactose) are easily biodegradable compounds, the products obtained from their degradation can cause partial inhibition in the methanogenesis phase. Moreover, despite the fact that anaerobic digestion presents high organic removal efficiency, failure in anaerobic digesters can be caused by low values of alkalinity [193].

Other Products

Different microorganisms and processes may lead to the formation of various products, such as organic acids (acetic, propionic, lactic, butyric acid *etc.*), amino acids, 2,3-butanediol, glycerol, xanthan gum, volatile flavouring compounds, carotenoids, hydrogen, and direct production of electricity through microbial fuel cells [192, 193, 203 - 206].

FAT/OIL/GREASE (FOG) CONTAINING WASTES

FOG could be described as the layer of lipid-rich material from wastewater, generated during industrial cooking and food processing. Its presence on sewage systems could lead to potential forming of hardened deposits through a chemical reaction or a physical aggregation process [207]. The FOG category contains the liquid and solid wastes of slaughter houses, the liquid wastes from restaurants or ready-meal industries that are rich in fat, fried-cooked fats, agro-industrial lipids (stearin, *etc.*), and volatile fatty acids (VFA; obtained from agro industrial lignocellulosic wastes and sludge) [189, 208]. These types of waste are difficult to treat and valorize. Besides their potential use as raw materials for bio-diesel production, they could be used in many biotechnological applications aiming at the synthesis of high value-added metabolites (of great environmental and financial importance), such as SCP, "new" fatty acids that did not exist previously in the substrate, organic acids, bio-surfactants, and lipases [189].

A two-stage fed-batch strategy was designed to enable *Yarrowia lipolytica* strain MUCL 28849 to convert volatile fatty acids (VFAs) into microbial lipids with similar composition to vegetable oils. At the first stage, glucose or glycerol was used as carbon source. After substrate exhaustion, acetic acid was added under nitrogen limiting conditions. This strategy resulted in a lipid content close to 40% w/w of dry cell weight (DCW). The process efficiency was satisfactory also when butyric and propionic acids or a mixture of the aforementioned three acids was used [208].

The ability of *Y. lipolytica* strain ACA-DC 50109 to grow on industrial lipids composed of saturated free fatty acids (stearin) in order to produce SCO was assessed. The process was critically influenced by the medium pH and the incubation temperature, but was independent of nitrogen concentration in the culture medium. Nevertheless, it was favored at high carbon substrate levels and low aeration rates. Significant quantities of lipids were accumulated by yeast cells (44-54% w/w of DCW). Lipids [principally composed of triacylglycerols (55% w/w of total lipids) and free fatty acids (35% w/w)] were rich in stearic acid (80% w/w). Negligible amounts of unsaturated fatty acids were detected too. The reserve lipid content increased when raw glycerol was used as co-substrate [21]. The same strain (*Y. lipolytica* ACA-DC 50109) was used to valorize industrial derivatives of tallow (stearin) in order to produce SCP, SCO and extra-cellular lipases in shake flasks and bioreactor experiments. In shake flasks, high quantities of biomass were produced regardless of the concentration of extra-cellular nitrogen. SCO (maximum quantity of lipids produced was 7.9 mg/ml corresponding to 52.0% (w/w) of lipid in DCW) was accumulated in notable quantities. The process was favored at high initial fat and low initial nitrogen concentrations. Lipase production was critically affected by the medium composition and its concentration clearly increased with increasing concentrations of fat and extracellular nitrogen concentration reaching a maximum of 2.50 IU/mL. Lipase concentration decreased in the stationary growth phase. Significantly higher quantities of biomass were also produced while remarkably lower quantities of cellular lipids and extra-cellular lipase were synthesized in bioreactor trials when higher agitation and aeration conditions were employed [189].

Fickers *et al.* [209] presented examples of wild-type and genetically engineered strains of *Y. lipolytica* able to convert fatty-acid substrates into aroma, SCP, SCO, and citric acid. Moreover, more examples were given with the use of these strains in bioremediation, fine chemistry, steroid biotransformation, and food industry.

FLOUR-RICH WASTE (FRW) STREAMS

FRW streams are produced in tremendous quantities by various industrial sectors (including the manufacture of bread, fresh pastry goods, cakes, rusks, biscuits, preserved pastry goods, and food preparations for infants) and returns from the consumers (such as products with due expiration date or unacceptable for consuming) [22]. FRW contain significant quantities of starch and proteins as well as various micro-nutrients. Such substrates can be used for enzyme production (*e.g.* amylolytic and proteolytic enzymes) *via* solid state fermentation (SSF) using appropriate fungal strains. These types of enzymes can be used to produce nutrient-rich hydrolysates of FRW. Subsequently, the hydrolysates can be used as raw materials for the production of various compounds *via* fermentation processes, amongst others for microbial oil synthesis [22].

Tsouko *et al.* [210] studied the production of bacterial cellulose using sole sunflower meal hydrolysates or blends with crude glycerol as fermentation media. Specifically, the use of the bacterial strain *Komagataeibacter sucrofermentans* DSM 15973 resulted in a cellulose yield of ~13 g/L. The same substrate was used in fed-batch cultures of the oleaginous yeast strains (*Rhodosporidium toruloides*, *Lipomyces starkeyi* and *Cryptococcus curvatus*) to produce SCO conformed with the limits set by certain standards of biodiesel properties. *R. toruloides* led to the production of 37.4 g/L of total dry weight with a microbial oil content of 51.3% (w/w) [211]. Tsakona *et al.* [22] used wheat milling by-products in (SSF) of *Aspergillus awamori* for the production of glucoamylase and protease. These enzyme rich solids were used for hydrolysis of FRW streams and resulted in higher than 90% (w/w) starch to glucose conversion yields and 40% (w/w) total Kjeldahl nitrogen to free amino nitrogen conversion yields (initial FRW concentration ~205 g/L). Following, crude hydrolysates were used as fermentation media in shake flask and fed batch bioreactor cultures using *L. starkeyi* strain DSM 70296. The DCW produced in flask trials reached the amount of 30.5 g/L with a microbial oil content of 40.4% (w/w) whereas in bioreactor cultures the DCW was ~110.0 g/L with a microbial oil content of 57.8% (w/w).

The aforementioned information, regarding the valorization of "simple sugars" residues, are presented and briefly gathered in Table **9.3**.

Table 9.3. Raw material used as substrate, microorganism used for growth and final product resulting from the valorization of "simple sugars" residues by biotechnological methods.

Raw Material / Substrate	Microorganism	Product	Reference
Olive Mill Wastewaters (OMW)	*Cryptococcus albidus* var. *albidus* IMAT 4735	Pectinase (endopolygalacturonase)	Federici *et al.* [212]
	Yeast strains	Bioethanol	Bambalov *et al.* [126]
	Pleurotus pulmonarius LGAM P46	Biomass (edible fungi)	Zervakis *et al.* [125]
	Pleurotus eryngii LGAM P63		
	Yarrowia lipolytica ATCC 20255	Biomass	Scioli and Vollaro [122]
		Lipase	
	Pleurotus ostreatus LGAM P113	Laccase	Aggelis *et al.* [105]
	Pleurotus ostreatus LGAM P115		
	Botryosphaeria rhodina DABAC-P82	Biomass	Crognale *et al.* [213]
		Exopolysaccharide (EPS, β-glucan)	
	Panus tigrinus CBS 577.79	Laccase	D'Annibale *et al.* [214]
		Manganese peroxidase	
	Candida cylindracea NRRL Y-17 506	Lipase	D'Annibale *et al.* [107]
	Saccharomyces cerevisiae sp.	Bioethanol	Zanichelli *et al.* [127]
Olive Mill Wastewaters (OMW)	*Candida holstii*	Biomass	Ben Sassi *et al.* [123]
	Saccharomyces cerevisiae L-6	Bioethanol	Massadeh and Modallal [119]
	Yarrowia lipolytica ATCC 20255	Biomass	Papanikolaou *et al.* [115]
		Citric acid	
		Lipids rich in oleic & palmitoleic acids	

(Table 9.3) cont.....

Raw Material / Substrate	Microorganism	Product	Reference
OMW & Cheese whey (blend)	*Geotrichum candidum* sp.	Biomass (edible fungi)	Aouidi *et al.* [129]
OMW	*Lentinula edodes* strains	Biomass (edible fungi)	Lakhtar *et al.* [117]
		Laccase	
	Yarrowia lipolytica strains	Biomass	Sarris *et al.* [20]
		Lipids with increased concentration of oleic acid	
		Citric acid	
	Saccharomyces cerevisiae MAK-1	Biomass	Sarris *et al.* [19]
		Lipids rich in oleic & linoleic acids	
		Bioethanol	
	Thamnidium elegans CCF-1465	Lipids rich in PUFAs	Bellou *et al.* [124]
	Zygorhynchus moelleri MUCL 1430	Cell mass	
OMW & Molasses (blend)	*Saccharomyces cerevisiae* MAK-1	Biomass	Sarris *et al.* [17]
		Bioethanol	
OMW	*Lipomyces starkeyi* NRRL Y-11557	Lipids	Dourou *et al.* [215]
	Yarrowia lipolytica strains		
	Yarrowia lipolytica A6	Mannitol	
	Yarrowia lipolytica LGAM S (7)	Citric acid	
	Candida tropicalis LFMB 16	Bioethanol	
	Saccharomyces cerevisiae MAK-1		
OMW & Molasses (blend)	*S. cerevisiae* MAK-1	Biomass	Sarris *et al.* [17]
		Bioethanol	

(Table 9.3) cont.....

Raw Material / Substrate	Microorganism	Product	Reference
Molasses	*Aspergillus niger* NRRL 599	Citric acid	Hamissa and Radwan [174]
	Aspergillus niger sp.	Citric acid	Qazi *et al.* [175]
	Aspergillus niger KCU520	Citric acid	Garg *et al.* [176]
	Aspergillus niger W1-2	Citric acid	Wang *et al.* [177]
	Aureobasidium pullulans P56	Pullulans	Roukas and Liakopoulou-Kyriakedes [191]
	Aspergillus niger W1-2	Citric acid	Wang *et al.* [178]
	Aspergillus niger A20	Citric acid	Adham *et al.* [172]
	Saccharomyces cerevisiae sp.	Bioethanol	Nahvi *et al.* [149]
	Aspergillus niger strains	Citric acid	Ikram-ul *et al.* [173]
	Aureobasidium pullulans strains	FOS	Shin *et al.* [183]
	Saccharomyces cerevisiae sp.	Bioethanol	Baptista *et al.* [150]
	Saccharomyces cerevisiae NCYC 1119		
Molasses & henequen (blend)	Kluyveromyces marxianus sp.	Bioethanol	Cáceres-Farfán *et al.* [153]
	Saccharomyces cerevisiae (commercial strain)		
Molasses	*Saccharopolyspora erythraea* NCIMB 8594	Erythromycin	El-Enshasy *et al.* [154]
	Actinobacillus succinogenes CGMCC1593	Succinic acid	Liu *et al.* [155]
	Aspergillus niger ARNU-4	Gluconic acid	Sharma *et al.* [156]
	Saccharomyces cerevisiae AXAZ-1	Bioethanol	Kopsahelis *et al.* [151]
	Trichosporon fermentans CICC 1368	Lipids (mainly palmitic, stearic, oleic, linoleic acids)	Zhu *et al.* [157]
	Cunninghamella echinulata ATHUM4411	Biomass	Chatzifragkou *et al.* [158]
	Mortierella isabellina ATHUM2935	Lipids (containing γ-linolenic (GLA) acid)	
	Leuconostoc mesenteroides	Bacteriocin	Metsoviti *et al.* [159]
	Saccharomyces cerevisiae AXAZ-1	Bioethanol	Kopsahelis *et al.* [152]

(Table 9.3) cont.....

Raw Material / Substrate	Microorganism	Product	Reference
Cheese whey	*Kluyveromyces marxianus* CBS 6556	Biomass (SCP)	Schultz *et al.* [196]
	K. bulgaricus ATCC 1605	Biomass (SCP)	Mahmoud and Kosikowski [197]
	K. frafilis NRRL Y 2415	Ethanol	
	K. frafilis NRS 5790	Biomass (SCP)	Ghaly *et al.* [198]
	Thamnidium elegans CCF 1465	Biomass (SCP)	Vamvakaki *et al.* [195]
	Mucor sp. LGAM 366		
	Mortierella isabellina ATHUM 2935	Lipids rich in GLA	
	K. marxianus NRRL-1195	Ethanol	Kargi and Ozmihci [201]
	K. marxianus MTCC1288	Biomass (SCP)	Zafar *et al.* [202]
		Ethanol	
	Propionibacterium acidipropionici ATCC 4875	Propionic acid	Morales *et al.* [203]
	Clostridium beijerinckii	Butyric acid	Alam *et al.* [206]
	Propionibacterium freudenreichii spp *shermani* NCFB 853	Acetic acid and propionic acid	Haddadin *et al.* [205]
	Propionibacterium acidipropionici NCFB 563		
Volatile Fatty Acids	*Yarrowia lipolytica* MUCL 28849	Lipids	Fontanille *et al.* [208]
Stearin	*Y. lipolytica* ACA-DC 50109	Lipids	Papanikolaou *et al.* [21]
Stearin and glycerol blends			
Stearin	*Y. lipolytica* ACA-DC 50109	Biomass (SCP)	Papanikolaou *et al.* [189]
		Lipids	
		Lipases	
Sunflower meal hydrolysates (also blends with crude glycerol)	*Komagataeibacter sucrofermentans* DSM 15973	Bacterial cellulose	Tsouko *et al.* [210]

(Table 9.3) cont.....

Raw Material / Substrate	Microorganism	Product	Reference
Sunflower meal hydrolysates	*Rhodosporidium toruloides* DSM 4444	Lipids (used for biodiesl production)	Leiva Candia *et al.* [211]
	Lipomyces starkeyi DSM 70296		
	Cryptococcus curvatus ATCC 20509		
Crude hydrolysates of wheat milling by-products	*Lipomyces starkeyi DSM 70296*	Biomass	Tsakona *et al.* [22]
		Lipids	

CONCLUDING REMARKS

There are numerous studies and, therefore, applications presented in the literature for the management, treatment, and valorization of a plethora of food wastes. The role of biotechnology in this crucial matter is mainly focused on both the treatment of such wastes and their valorization, simultaneously, by producing (high-)added value compounds. SSF and SmF have been successfully applied to utilize low-cost materials for biotechnological production of valuable bio-products. The main advantages of SSF is the low contamination risk of substrate, due to the absence of water and the high compounds productivities. However, efforts should be made for suitable reactor design in order to overcome scale-up difficulties. On the other hand, SmF offers controlled operating conditions for scale-up processes. Besides, pretreatment methods should be improved for effective utilization of substrates. In all cases, biotechnological treatment and valorization methods are considered to be more environmentally and financially friendly.

CONSENT FOR PUBLICATION

Not applicable.

CONFLICT OF INTEREST

The authors declare no conflict of interest, financial or otherwise.

ACKNOWLEDGEMENTS

Declared none.

ABBREVIATIONS

FOG   fat/oil/grease

FOS   fructooligosaccharides

FRW   flour-rich waste

OMW   olive mill wastewater

SCO   single cell oil

SCP   single cell Protein

SmF   submerged fermentation

SMS   spent mushroom substrate

SSF   solid-state fermentation

REFERENCES

[1]   Xue Z, Wang L, Ju J, Yu B, Xu P, Ma Y. Efficient production of polymer-grade L-lactic acid from corn stover hydrolyzate by thermophilic *Bacillus* sp. strain XZL4. Springerplus 2012; 1(1): 43.
[http://dx.doi.org/10.1186/2193-1801-1-43] [PMID: 23961368]

[2]   Sindhu R, Kuttiraja M, Prabisha TP, Binod P, Sukumaran RK, Pandey A. Development of a combined pretreatment and hydrolysis strategy of rice straw for the production of bioethanol and biopolymer. Bioresour Technol 2016; 215: 110-6.
[http://dx.doi.org/10.1016/j.biortech.2016.02.080] [PMID: 26949053]

[3]   Torrado AM, Cortés S, Manuel Salgado J, *et al.* Citric Acid production from orange peel wastes by solid-state fermentation. Braz J Microbiol 2011; 42(1): 394-409.
[http://dx.doi.org/10.1590/S1517-83822011000100049] [PMID: 24031646]

[4]   Díaz AB, de Ory I, Caro I, Blandino A. Enhance hydrolytic enzymes production by *Aspergillus awamori* on supplemented grape pomace. Food Bioprod Process 2012; 90(1): 72-8.
[http://dx.doi.org/10.1016/j.fbp.2010.12.003]

[5]   Iqbal HMN, Kyazze G, Keshavarz T. Advances in the valorization of lignocellulosic materials by biotechnology: an overview. BioResources 2013; 8(2): 3157-76.
[http://dx.doi.org/10.15376/biores.8.2.3157-3176]

[6]   Li L, Wang Y, Zhang Q, Li J, Yang X, Jin J. Wheat straw burning and its associated impacts on Beijing air quality. Sci China Ser D Earth Sci 2008; 51(3): 403-14.
[http://dx.doi.org/10.1007/s11430-008-0021-8]

[7]   Corbin KR, Hsieh YS, Betts NS, *et al.* Grape marc as a source of carbohydrates for bioethanol: Chemical composition, pre-treatment and saccharification. Bioresour Technol 2015; 193: 76-83.
[http://dx.doi.org/10.1016/j.biortech.2015.06.030] [PMID: 26117238]

[8]   Phan C-W, Sabaratnam V. Potential uses of spent mushroom substrate and its associated lignocellulosic enzymes. Appl Microbiol Biotechnol 2012; 96(4): 863-73.
[http://dx.doi.org/10.1007/s00253-012-4446-9] [PMID: 23053096]

[9]   Sarkar N, Ghosh SK, Bannerjee S, Aikat K. Bioethanol production from agricultural wastes: An overview. Renew Energy 2012; 37(1): 19-27.
[http://dx.doi.org/10.1016/j.renene.2011.06.045]

[10] De Gregorio A, Mandalari G, Arena N, Nucita F, Tripodo MM, Lo Curto RB. SCP and crude pectinase production by slurry-state fermentation of lemon pulps. Bioresour Technol 2002; 83(2): 89-94.
[http://dx.doi.org/10.1016/S0960-8524(01)00209-7] [PMID: 12056496]

[11]   Binod P, Sindhu R, Singhania RR, *et al.* Bioethanol production from rice straw: An overview. Bioresour Technol 2010; 101(13): 4767-74.
[http://dx.doi.org/10.1016/j.biortech.2009.10.079] [PMID: 19944601]

[12]   Economou ChN, Aggelis G, Pavlou S, Vayenas DV. Single cell oil production from rice hulls hydrolysate. Bioresour Technol 2011; 102(20): 9737-42.
[http://dx.doi.org/10.1016/j.biortech.2011.08.025] [PMID: 21875786]

[13]   Mamma D, Christakopoulos P. Citrus peels: An excellent raw material for the bioconversion into value-added products. WORLD 2008; 94793(59041.4): 19224.9.

[14]   Talebnia F, Karakashev D, Angelidaki I. Production of bioethanol from wheat straw: An overview on pretreatment, hydrolysis and fermentation. Bioresour Technol 2010; 101(13): 4744-53.
[http://dx.doi.org/10.1016/j.biortech.2009.11.080] [PMID: 20031394]

[15]   Zahari NI, Shah UKM, Asa'ari AZM, Mohamad R. Selection of potential fungi for production of cellulase-poor xylanase from rice straw. BioResources 2015; 11(1): 1162-75.
[http://dx.doi.org/10.15376/biores.11.1.1162-1175]

[16]   Couto SR, Sanromán MA. Application of solid-state fermentation to food industry—a review. J Food Eng 2006; 76(3): 291-302.
[http://dx.doi.org/10.1016/j.jfoodeng.2005.05.022]

[17]   Sarris D, Matsakas L, Aggelis G, Koutinas AA, Papanikolaou S. Aerated vs non-aerated conversions of molasses and olive mill wastewaters blends into bioethanol by *Saccharomyces cerevisiae* under non-aseptic conditions. Ind Crops Prod 2014; 56: 83-93.
[http://dx.doi.org/10.1016/j.indcrop.2014.02.040]

[18]   Sarris D, Kotseridis Y, Linga M, Galiotou-Panayotou M, Papanikolaou S. Enhanced ethanol production, volatile compound biosynthesis and fungicide removal during growth of a newly isolated *Saccharomyces cerevisiae* strain on enriched pasteurized grape musts. Eng Life Sci 2009; 9(1): 29-37.
[http://dx.doi.org/10.1002/elsc.200800059]

[19]   Sarris D, Giannakis M, Philippoussis A, Komaitis M, Koutinas AA, Papanikolaou S. Conversions of olive mill wastewater-based media by *Saccharomyces cerevisiae* through sterile and non-sterile bioprocesses. J Chem Technol Biotechnol 2013; 88: 958-69.
[http://dx.doi.org/10.1002/jctb.3931]

[20]   Sarris D, Galiotou-Panayotou M, Koutinas AA, Komaitis M, Papanikolaou S. Citric acid, biomass and cellular lipid production by *Yarrowia lipolytica* strains cultivated on olive mill wastewater-based media. J Chem Technol Biotechnol 2011; 86: 1439-48.
[http://dx.doi.org/10.1002/jctb.2658]

[21]   Papanikolaou S, Chevalot I, Komaitis M, Aggelis G, Marc I. Kinetic profile of the cellular lipid composition in an oleaginous *Yarrowia lipolytica* capable of producing a cocoa-butter substitute from industrial fats. Antonie van Leeuwenhoek 2001; 80(3-4): 215-24.
[http://dx.doi.org/10.1023/A:1013083211405] [PMID: 11827207]

[22]   Tsakona S, Kopsahelis N, Chatzifragkou A, Papanikolaou S, Kookos IK, Koutinas AA. Formulation of fermentation media from flour-rich waste streams for microbial lipid production by *Lipomyces starkeyi*. J Biotechnol 2014; 189: 36-45.
[http://dx.doi.org/10.1016/j.jbiotec.2014.08.011] [PMID: 25150217]

[23]   [accessed on 07 March 2016]; Available online: http://www.faostat.org

[24]   Khan TS, Mubeen U. Wheat straw: A pragmatic overview. Curr Res J Biol Sci 2012; 4(6): 673-5.

[25]   Sastre C, González-Arechavala Y, Santos A. Global warming and energy yield evaluation of Spanish wheat straw electricity generation–A LCA that takes into account parameter uncertainty and variability. Appl Energy 2015; 154: 900-11.
[http://dx.doi.org/10.1016/j.apenergy.2015.05.108]

[26] Nguyen TLT, Hermansen JE, Mogensen L. Environmental performance of crop residues as an energy source for electricity production: the case of wheat straw in Denmark. Appl Energy 2013; 104: 633-41. [http://dx.doi.org/10.1016/j.apenergy.2012.11.057]

[27] Tabka M, Herpoël-Gimbert I, Monod F, Asther M, Sigoillot J. Enzymatic saccharification of wheat straw for bioethanol production by a combined cellulase xylanase and feruloyl esterase treatment. Enzyme Microb Technol 2006; 39(4): 897-902. [http://dx.doi.org/10.1016/j.enzmictec.2006.01.021]

[28] Kootstra AMJ, Beeftink HH, Scott EL, Sanders JP. Comparison of dilute mineral and organic acid pretreatment for enzymatic hydrolysis of wheat straw. Biochem Eng J 2009; 46(2): 126-31. [http://dx.doi.org/10.1016/j.bej.2009.04.020]

[29] Sun Y, Cheng J. Hydrolysis of lignocellulosic materials for ethanol production: a review. Bioresour Technol 2002; 83(1): 1-11. [http://dx.doi.org/10.1016/S0960-8524(01)00212-7] [PMID: 12058826]

[30] Jørgensen H. Effect of nutrients on fermentation of pretreated wheat straw at very high dry matter content by *Saccharomyces cerevisiae*. Appl Biochem Biotechnol 2009; 153(1-3): 44-57. [http://dx.doi.org/10.1007/s12010-008-8456-0] [PMID: 19093228]

[31] Kuhad RC, Gupta R, Singh A. Microbial cellulases and their industrial applications. Enzyme research 2011; 2011 [http://dx.doi.org/10.4061/2011/280696]

[32] Sukumaran RK, Singhania RR, Pandey A. Microbial cellulases-production, applications and challenges. J Sci Ind Res (India) 2005; 64(11): 832.

[33] Kalogeris E, Iniotaki F, Topakas E, Christakopoulos P, Kekos D, Macris BJ. Performance of an intermittent agitation rotating drum type bioreactor for solid-state fermentation of wheat straw. Bioresour Technol 2003; 86(3): 207-13. [http://dx.doi.org/10.1016/S0960-8524(02)00175-X] [PMID: 12688461]

[34] Singh A, Singh N, Bishnoi NR. Production of cellulases by Aspergillus heteromorphus from wheat straw under submerged fermentation. International Journal of Civil and Environmental Engineering 2009; 1(1): 23-6.

[35] Ahmed I, Zia MA, Iqbal HMN. Bioprocessing of proximally analyzed wheat straw for enhanced cellulase production through process optimization with *Trichoderma viride* under SSF. Cellulose 2010; 2(W3): 100.

[36] Hafiz Muhammad Nasir I, Ishtiaq A, Muhammad Anjum Z, Muhammad I. Purification and characterization of the kinetic parameters of cellulase produced from wheat straw by Trichoderma viride under SSF and its detergent compatibility. Adv Biosci Biotechnol 2011; 2011

[37] Philippoussis A, Diamantopoulou P, Papadopoulou K, Lakhtar H, Roussos S, Parissopoulos G, *et al.* Biomass, laccase and endoglucanase production by Lentinula edodes during solid state fermentation of reed grass, bean stalks and wheat straw residues. World J Microbiol Biotechnol 2011; 27(2): 285-97. [http://dx.doi.org/10.1007/s11274-010-0458-8]

[38] Gaitán-Hernández R, Esqueda M, Gutiérrez A, Sánchez A, Beltrán-García M, Mata G. Bioconversion of agrowastes by *Lentinula edodes*: the high potential of viticulture residues. Appl Microbiol Biotechnol 2006; 71(4): 432-9. [http://dx.doi.org/10.1007/s00253-005-0241-1] [PMID: 16331453]

[39] Koutrotsios G, Mountzouris KC, Chatzipavlidis I, Zervakis GI. Bioconversion of lignocellulosic residues by *Agrocybe cylindracea* and *Pleurotus ostreatus* mushroom fungi--assessment of their effect on the final product and spent substrate properties. Food Chem 2014; 161: 127-35. [http://dx.doi.org/10.1016/j.foodchem.2014.03.121] [PMID: 24837930]

[40] Öhgren K, Bura R, Saddler J, Zacchi G. Effect of hemicellulose and lignin removal on enzymatic hydrolysis of steam pretreated corn stover. Bioresour Technol 2007; 98(13): 2503-10.

[http://dx.doi.org/10.1016/j.biortech.2006.09.003] [PMID: 17113771]

[41]   Chundawat SP, Venkatesh B, Dale BE. Effect of particle size based separation of milled corn stover on AFEX pretreatment and enzymatic digestibility. Biotechnol Bioeng 2007; 96(2): 219-31.
[http://dx.doi.org/10.1002/bit.21132] [PMID: 16903002]

[42]   Kadam KL, McMillan JD. Availability of corn stover as a sustainable feedstock for bioethanol production. Bioresour Technol 2003; 88(1): 17-25.
[http://dx.doi.org/10.1016/S0960-8524(02)00269-9] [PMID: 12573559]

[43]   Shinners KJ, Binversie BN, Muck RE, Weimer PJ. Comparison of wet and dry corn stover harvest and storage. Biomass Bioenergy 2007; 31(4): 211-21.
[http://dx.doi.org/10.1016/j.biombioe.2006.04.007]

[44]   Minh NP. Investigation of lactic acid fermentation from corn by-product using *L. casei* and *L. plantarum* strain. Int J Multidisciplinary Res Development 2014; 1(3): 92-100.

[45]   Öhgren K, Bura R, Lesnicki G, Saddler J, Zacchi G. A comparison between simultaneous saccharification and fermentation and separate hydrolysis and fermentation using steam-pretreated corn stover. Process Biochem 2007; 42(5): 834-9.
[http://dx.doi.org/10.1016/j.procbio.2007.02.003]

[46]   Kumar R, Wyman CE. Effect of enzyme supplementation at moderate cellulase loadings on initial glucose and xylose release from corn stover solids pretreated by leading technologies. Biotechnol Bioeng 2009; 102(2): 457-67.
[http://dx.doi.org/10.1002/bit.22068] [PMID: 18781688]

[47]   Schell DJ, Farmer J, Newman M. McMILLAN JD Dilute-sulfuric acid pretreatment of corn stover in pilot-scale reactor Biotechnology for Fuels and Chemicals. Springer 2003; pp. 69-85.

[48]   Zulfiqar A, Anjum F, Zahoor T. Production of lactic acid from corn cobs hydrolysate through fermentation by *Lactobaccillus delbrukii.* Afr J Biotechnol 2009; 8(17)

[49]   Ruengruglikit C, Hang YL. (+)-lactic acid production from corncobs by *Rhizopus oryzae* NRRL-395. Lebensm Wiss Technol 2003; 36(6): 573-5.
[http://dx.doi.org/10.1016/S0023-6438(03)00062-8]

[50]   Cui F, Li Y, Wan C. Lactic acid production from corn stover using mixed cultures of *Lactobacillus rhamnosus* and *Lactobacillus brevis.* Bioresour Technol 2011; 102(2): 1831-6.
[http://dx.doi.org/10.1016/j.biortech.2010.09.063] [PMID: 20943382]

[51]   Ashour A, El-Sharkawy S, Amer M, *et al.* Production of citric acid from corncobs with its biological Evaluation. Journal of cosmetics, Dermatological sciences and applications 2014. 2014
[http://dx.doi.org/10.4236/jcdsa.2014.43020]

[52]   Hang Y, Woodams E. Production of citric acid from corncobs by *Aspergillus niger.* Bioresour Technol 1998; 65(3): 251-3.
[http://dx.doi.org/10.1016/S0960-8524(98)00015-7]

[53]   Hang Y, Woodams E. Corn husks: a potential substrate for production of citric acid by *Aspergillus niger.* Lebensm Wiss Technol 2000; 33(7): 520-1.
[http://dx.doi.org/10.1006/fstl.2000.0711]

[54]   Yang R, Xu S, Wang Z, Yang W. Aqueous extraction of corncob xylan and production of xylooligosaccharides. Lebensm Wiss Technol 2005; 38(6): 677-82.
[http://dx.doi.org/10.1016/j.lwt.2004.07.023]

[55]   Aachary AA, Prapulla SG. Value addition to corncob: production and characterization of xylooligosaccharides from alkali pretreated lignin-saccharide complex using *Aspergillus oryzae* MTCC 5154. Bioresour Technol 2009; 100(2): 991-5.
[http://dx.doi.org/10.1016/j.biortech.2008.06.050] [PMID: 18703333]

[56]   Swain MR, Krishnan C. Improved conversion of rice straw to ethanol and xylitol by combination of

moderate temperature ammonia pretreatment and sequential fermentation using *Candida tropicalis.* Ind Crops Prod 2015; 77: 1039-46.
[http://dx.doi.org/10.1016/j.indcrop.2015.10.013]

[57]   Li W, Li M, Zheng L, *et al.* Simultaneous utilization of glucose and xylose for lipid accumulation in black soldier fly. Biotechnol Biofuels 2015; 8(1): 117.
[http://dx.doi.org/10.1186/s13068-015-0306-z] [PMID: 26273321]

[58]   Yen H-W, Chang J-T. Growth of oleaginous *Rhodotorula glutinis* in an internal-loop airlift bioreactor by using lignocellulosic biomass hydrolysate as the carbon source. J Biosci Bioeng 2015; 119(5): 580-4.
[http://dx.doi.org/10.1016/j.jbiosc.2014.10.001] [PMID: 25454603]

[59]   Zhang T, Liu L, Song Z, *et al.* Biogas production by co-digestion of goat manure with three crop residues. PLoS One 2013; 8(6): e66845.
[http://dx.doi.org/10.1371/journal.pone.0066845] [PMID: 23825574]

[60]   Zhang T, Yang Y, Liu L, Han Y, Ren G, Yang G. Improved biogas production from chicken manure anaerobic digestion using cereal residues as co-substrates. Energy Fuels 2014; 28(4): 2490-5.
[http://dx.doi.org/10.1021/ef500262m]

[61]   Zhu N. Effect of low initial C/N ratio on aerobic composting of swine manure with rice straw. Bioresour Technol 2007; 98(1): 9-13.
[http://dx.doi.org/10.1016/j.biortech.2005.12.003] [PMID: 16427276]

[62]   Chen X, Zhang Y, Gu Y, *et al.* Enhancing methane production from rice straw by extrusion pretreatment. Appl Energy 2014; 122: 34-41.
[http://dx.doi.org/10.1016/j.apenergy.2014.01.076]

[63]   Niladevi KN, Sukumaran RK, Prema P. Utilization of rice straw for laccase production by *Streptomyces psammoticus* in solid-state fermentation. J Ind Microbiol Biotechnol 2007; 34(10): 665-74.
[http://dx.doi.org/10.1007/s10295-007-0239-z] [PMID: 17665235]

[64]   Kang SW, Park YS, Lee JS, Hong SI, Kim SW. Production of cellulases and hemicellulases by *Aspergillus niger* KK2 from lignocellulosic biomass. Bioresour Technol 2004; 91(2): 153-6.
[http://dx.doi.org/10.1016/S0960-8524(03)00172-X] [PMID: 14592744]

[65]   Marín FR, Soler-Rivas C, Benavente-García O, Castillo J, Pérez-Alvarez JA. By-products from different citrus processes as a source of customized functional fibres. Food Chem 2007; 100(2): 736-41.
[http://dx.doi.org/10.1016/j.foodchem.2005.04.040]

[66]   Srivastava P, Malviya R. Extraction, characterization and evaluation of orange peel waste derived pectin as a pharmaceutical excipient. Nat Prod J 2011; 1(1): 65-70.

[67]   Hamdy HS. Citric acid production by *Aspergillus niger* grown on orange peel medium fortified with cane molasses. Ann Microbiol 2013; 63(1): 267-78.
[http://dx.doi.org/10.1007/s13213-012-0470-3]

[68]   TAZE BH, Demir H, Tari C, ÜNLÜTÜRK S, Lahore MF. Evaluation of orange peel, an industrial waste, for the production of *Aspergillus sojae* polygalacturonase considering both morphology and rheology effects. Turk J Biol 2014; 38(4): 537-48.

[69]   Maller A, Damásio ARL, Silva TMd, Jorge JA, Terenzi HF, Polizeli MdLTdM. Biotechnological potential of agro-industrial wastes as a carbon source to thermostable polygalacturonase production in Aspergillus niveus. Enzyme research 2011. 2011

[70]   Irshad M, Anwar Z, Afroz A. Characterization of Exo 1, 4-[Beta] glucanase produced from *Tricoderma viridi* through solid-state bio-processing of orange peel waste. Adv Biosci Biotechnol 2012; 3(5): 580.
[http://dx.doi.org/10.4236/abb.2012.35075]

[71] Mrudula S, Anitharaj R. Pectinase production in solid state fermentation by *Aspergillus niger* using orange peel as substrate. Glob J Biotechnol Biochem 2011; 6(2): 64-71.

[72] Seyis I, Aksoz N. Xylanase production from *Trichoderma harzianum* 1073 D 3 with alternative carbon and nitrogen sources. Food Technol Biotechnol 2005; 43(1): 37-40.

[73] Gema H, Kavadia A, Dimou D, Tsagou V, Komaitis M, Aggelis G. Production of γ-linolenic acid by *Cunninghamella echinulata* cultivated on glucose and orange peel. Appl Microbiol Biotechnol 2002; 58(3): 303-7.
[http://dx.doi.org/10.1007/s00253-001-0910-7] [PMID: 11935180]

[74] Choi IS, Kim J-H, Wi SG, Kim KH, Bae H-J. Bioethanol production from mandarin (*Citrus unshiu*) peel waste using popping pretreatment. Appl Energy 2013; 102: 204-10.
[http://dx.doi.org/10.1016/j.apenergy.2012.03.066]

[75] Wilkins MR, Widmer WW, Grohmann K. Simultaneous saccharification and fermentation of citrus peel waste by *Saccharomyces cerevisiae* to produce ethanol. Process Biochem 2007; 42(12): 1614-9.
[http://dx.doi.org/10.1016/j.procbio.2007.09.006]

[76] Oberoi HS, Vadlani PV, Nanjundaswamy A, *et al.* Enhanced ethanol production from Kinnow mandarin (*Citrus reticulata*) waste via a statistically optimized simultaneous saccharification and fermentation process. Bioresour Technol 2011; 102(2): 1593-601.
[http://dx.doi.org/10.1016/j.biortech.2010.08.111] [PMID: 20863699]

[77] Widmer W, Zhou W, Grohmann K. Pretreatment effects on orange processing waste for making ethanol by simultaneous saccharification and fermentation. Bioresour Technol 2010; 101(14): 5242-9.
[http://dx.doi.org/10.1016/j.biortech.2009.12.038] [PMID: 20189803]

[78] Wikandari R, Nguyen H, Millati R, Niklasson C, Taherzadeh MJ. Improvement of biogas production from orange peel waste by leaching of limonene. BioMed research international 2015. 2015
[http://dx.doi.org/10.1155/2015/494182]

[79] Wikandari R, Millati R, Cahyanto MN, Taherzadeh MJ. Biogas production from citrus waste by membrane bioreactor. Membranes (Basel) 2014; 4(3): 596-607.
[http://dx.doi.org/10.3390/membranes4030596] [PMID: 25167328]

[80] Botella C, Diaz A, De Ory I, Webb C, Blandino A. Xylanase and pectinase production by *Aspergillus awamori* on grape pomace in solid state fermentation. Process Biochem 2007; 42(1): 98-101.
[http://dx.doi.org/10.1016/j.procbio.2006.06.025]

[81] Rodríguez L, Toro M, Vazquez F, Correa-Daneri M, Gouiric S, Vallejo M. Bioethanol production from grape and sugar beet pomaces by solid-state fermentation. Int J Hydrogen Energy 2010; 35(11): 5914-7.
[http://dx.doi.org/10.1016/j.ijhydene.2009.12.112]

[82] Hang Y, Woodams E. Grape pomace: A novel substrate for microbial production of citric acid. Biotechnol Lett 1985; 7(4): 253-4.
[http://dx.doi.org/10.1007/BF01042372]

[83] Stredansky M, Conti E. Xanthan production by solid state fermentation. Process Biochem 1999; 34(6): 581-7.
[http://dx.doi.org/10.1016/S0032-9592(98)00131-9]

[84] Medina E, Paredes C, Pérez-Murcia MD, Bustamante MA, Moral R. Spent mushroom substrates as component of growing media for germination and growth of horticultural plants. Bioresour Technol 2009; 100(18): 4227-32.
[http://dx.doi.org/10.1016/j.biortech.2009.03.055] [PMID: 19409775]

[85] Trejo-Hernandez M, Lopez-Munguia A, Ramirez RQ. Residual compost of *Agaricus bisporus* as a source of crude laccase for enzymic oxidation of phenolic compounds. Process Biochem 2001; 36(7): 635-9.
[http://dx.doi.org/10.1016/S0032-9592(00)00257-0]

[86]   Hublik G, Schinner F. Characterization and immobilization of the laccase from *Pleurotus ostreatus* and its use for the continuous elimination of phenolic pollutants. Enzyme Microb Technol 2000; 27(3-5): 330-6.
[http://dx.doi.org/10.1016/S0141-0229(00)00220-9] [PMID: 10899561]

[87]   Ayotamuno JM, Okparanma RN, Davis DD, Allagoa M. PAH removal from Nigerian oil-based drill-cuttings with spent oyster mushroom (*Pleurotus ostreatus*) substrate. J Food Agric Environ 2010; 8: 914-9.

[88]   Juárez RAC, Dorry LLG, Bello-Mendoza R, Sánchez JE. Use of spent substrate after *Pleurotus pulmonarius* cultivation for the treatment of chlorothalonil containing wastewater. J Environ Manage 2011; 92(3): 948-52.
[http://dx.doi.org/10.1016/j.jenvman.2010.10.047] [PMID: 21078538]

[89]   Chen GQ, Zeng GM, Tu X, Huang GH, Chen YN. A novel biosorbent: characterization of the spent mushroom compost and its application for removal of heavy metals. J Environ Sci (China) 2005; 17(5): 756-60.
[PMID: 16312997]

[90]   Singh AD, Sabaratnam V, Abdullah N, Annuar M, Ramachandran K. Decolourisation of chemically different dyes by enzymes from spent compost of *Pleurotus sajor-caju* and their kinetics. Afr J Biotechnol 2010; 9(1)

[91]   Royse D. Recycling of spent shiitake substrate for production of the oyster mushroom, *Pleurotus sajor-caju*. Appl Microbiol Biotechnol 1992; 38(2): 179-82.
[http://dx.doi.org/10.1007/BF00174464]

[92]   Ashrafi R, Mian M, Rahman M, Jahiruddin M. Recycling of Spent Mushroom Substrate for the Production of Oyster Mushroom. Res Biotechnol 2014; 5(2)

[93]   González Matute R, Figlas D, Curvetto N. *Agaricus blazei* production on non-composted substrates based on sunflower seed hulls and spent oyster mushroom substrate. World J Microbiol Biotechnol 2011; 27(6): 1331-9.
[http://dx.doi.org/10.1007/s11274-010-0582-5] [PMID: 25187132]

[94]   Wang S, Xu F, Li Z, *et al.* The spent mushroom substrates of *Hypsizigus marmoreus* can be an effective component for growing the oyster mushroom *Pleurotus ostreatus*. Sci Hortic (Amsterdam) 2015; 186: 217-22.
[http://dx.doi.org/10.1016/j.scienta.2015.02.028]

[95]   Montoya S, Sanchez OJ, Levin L. Polysaccharide production by submerged and solid-state cultures from several medicinal higher Basidiomycetes. Int J Med Mushrooms 2013; 15(1): 71-9.
[http://dx.doi.org/10.1615/IntJMedMushr.v15.i1.80] [PMID: 23510286]

[96]   Zhu H, Sheng K, Yan E, Qiao J, Lv F. Extraction, purification and antibacterial activities of a polysaccharide from spent mushroom substrate. Int J Biol Macromol 2012; 50(3): 840-3.
[http://dx.doi.org/10.1016/j.ijbiomac.2011.11.016] [PMID: 22138450]

[97]   Lanciotti R, Gianotti A, Baldi D, *et al.* Use of *Yarrowia lipolytica* strains for the treatment of olive mill wastewater. Bioresour Technol 2005; 96(3): 317-22.
[http://dx.doi.org/10.1016/j.biortech.2004.04.009] [PMID: 15474932]

[98]   Arvanitoyannis IS, Kassaveti A. Olive oil waste management: treatment methods and potential uses of treated waste Waste Manage Food Ind. Academic Press 2008; pp. 453-568.

[99]   Niaounakis M, Halvadakis C. Olive processing waste management; literature review and patent survey. Elsevier Ltd. 2006.

[100] Mantzavinos D, Kalogerakis N. Treatment of olive mill effluents Part I. Organic matter degradation by chemical and biological processes--an overview. Environ Int 2005; 31(2): 289-95.
[http://dx.doi.org/10.1016/j.envint.2004.10.005] [PMID: 15661297]

[101] Crognale S, D' Annibale A, Federici F, Fenice M, Quaratino D, Petruccioli M. Olive oil mill wastewater valorization by fungi. J Chem Technol Biotechnol 2006; 81: 1547-55.
[http://dx.doi.org/10.1002/jctb.1564]

[102] Amaral C, Lucas MS, Coutinho J, Crespí AL, do Rosário Anjos M, Pais C. Microbiological and physicochemical characterization of olive mill wastewaters from a continuous olive mill in Northeastern Portugal. Bioresour Technol 2008; 99(15): 7215-23.
[http://dx.doi.org/10.1016/j.biortech.2007.12.058] [PMID: 18261900]

[103] Kissi M, Mountadar M, Assobhei O, *et al.* Roles of two white-rot basidiomycete fungi in decolorisation and detoxification of olive mill waste water. Appl Microbiol Biotechnol 2001; 57(1-2): 221-6.
[http://dx.doi.org/10.1007/s002530100712] [PMID: 11693925]

[104] Tsioulpas A, Dimou D, Iconomou D, Aggelis G. Phenolic removal in olive oil mill wastewater by strains of *Pleurotus* spp. in respect to their phenol oxidase (laccase) activity. Bioresour Technol 2002; 84(3): 251-7.
[http://dx.doi.org/10.1016/S0960-8524(02)00043-3] [PMID: 12118702]

[105] Aggelis G, Iconomou D, Christou M, *et al.* Phenolic removal in a model olive oil mill wastewater using *Pleurotus ostreatus* in bioreactor cultures and biological evaluation of the process. Water Res 2003; 37(16): 3897-904.
[http://dx.doi.org/10.1016/S0043-1354(03)00313-0] [PMID: 12909108]

[106] Ahmadi M, Vahabzadeh F, Bonakdarpour B, Mehranian M. Empirical modeling of olive oil mill wastewater treatment using loofa-immobilized *Phanerochaete chrysosporium*. Process Biochem 2006; 41: 1148-54.
[http://dx.doi.org/10.1016/j.procbio.2005.12.012]

[107] D'Annibale A, Sermanni GG, Federici F, Petruccioli M. Olive-mill wastewaters: a promising substrate for microbial lipase production. Bioresour Technol 2006; 97(15): 1828-33.
[http://dx.doi.org/10.1016/j.biortech.2005.09.001] [PMID: 16236495]

[108] Dermeche S, Nadour M, Larroche C, Moulti-Mati F, Michaud P. Olive mill wastes: Biochemical characterizations and valorization strategies. Process Biochem 2013.
[http://dx.doi.org/10.1016/j.procbio.2013.07.010]

[109] Hamdi M. Future prospects and constraints of olive mill wastewaters use and treatment: A review. Bioprocess Eng 1993; 8: 209-14.
[http://dx.doi.org/10.1007/BF00369831]

[110] Sayadi S, Ellouz R. Roles of Lignin Peroxidase and Manganese Peroxidase from *Phanerochaete chrysosporium* in the Decolorization of Olive Mill Wastewaters. Appl Environ Microbiol 1995; 61(3): 1098-103.
[PMID: 16534959]

[111] Benitez J, Beltran-Heredia J, Torregrosa J, Acero JL, Cercas V. Aerobic degradation of olive mill wastewaters. Appl Microbiol Biotechnol 1997; 47(2): 185-8.
[http://dx.doi.org/10.1007/s002530050910] [PMID: 9077005]

[112] Ayed L, Assas N, Sayadi S, Hamdi M. Involvement of lignin peroxidase in the decolourization of black olive mill wastewaters by *Geotrichum candidum*. Lett Appl Microbiol 2005; 40(1): 7-11.
[http://dx.doi.org/10.1111/j.1472-765X.2004.01626.x] [PMID: 15612995]

[113] Di Serio MG, Lanza B, Murcciarella MR, Russi F, Iannucci E, Marfisi P, *et al.* Effects of olive mill wastewater spreading on the physico-chemical and microbiological characteristics of soil. Int Biodeterior Biodegradation 2008; 1-5.

[114] Paraskeva P, Diamadopoulos E. Review technologies for olive mill wastewater (OMW): a review. J Chem Technol Biotechnol 2006; 81: 1475-85.
[http://dx.doi.org/10.1002/jctb.1553]

[115] Papanikolaou S, Galiotou-Panayotou M, Fakas S, Komaitis M, Aggelis G. Citric acid production by *Yarrowia lipolytica* cultivated on olive-mill wastewater-based media. Bioresour Technol 2008; 99(7): 2419-28.
[http://dx.doi.org/10.1016/j.biortech.2007.05.005] [PMID: 17604163]

[116] Fountoulakis MS, Dokianakis SN, Kornaros ME, Aggelis GG, Lyberatos G. Removal of phenolics in olive mill wastewaters using the white-rot fungus *Pleurotus ostreatus*. Water Res 2002; 36(19): 4735-44.
[http://dx.doi.org/10.1016/S0043-1354(02)00184-7] [PMID: 12448515]

[117] Lakhtar H, Ismaili-Alaoui M, Philippoussis A, Perraud-Gaime I, Roussos S. Screening of strains of *Lentinula edodes* grown on model olive mill wastewater in solid and liquid state culture for polyphenol biodegradation. Int Biodeterior Biodegradation 2010; 64(3): 167-72.
[http://dx.doi.org/10.1016/j.ibiod.2009.10.006]

[118] Sayadi S, Ellouz R. Decolourization of olive mill waste-waters by the white-rot fungus *Phanerochaete chrysosporium*: involvement of the lignin-degrading system. Appl Microbiol Biotechnol 1992; 37: 813-7.
[http://dx.doi.org/10.1007/BF00174851]

[119] Massadeh MI, Modallal N. Ethanol production from olive mill wastewater (OMW) pretreated with *Pleurotus sajor-caju*. Energy Fuels 2008; 22(1): 150-4.
[http://dx.doi.org/10.1021/ef7004145]

[120] Rizzo M, Ventrice D, Varone MA, Sidari R, Caridi A. HPLC determination of phenolics adsorbed on yeasts. J Pharm Biomed Anal 2006; 42(1): 46-55.
[http://dx.doi.org/10.1016/j.jpba.2006.02.058] [PMID: 16631336]

[121] Chtourou M, Ammar E, Nasri M, Medhioub K. Isolation of a yeast, *Trichosporon cutaneum*, able to use low molecular weight phenolic compounds: application to olive mill waste water treatment. J Chem Technol Biotechnol 2004; 79(8): 869-78.
[http://dx.doi.org/10.1002/jctb.1062]

[122] Scioli C, Vollaro L. The use of *Yarrowia lipolytica* to reduce pollution in olive mill wastewaters. Water Res 1997; 31(10): 2520-4.
[http://dx.doi.org/10.1016/S0043-1354(97)00083-3]

[123] Ben Sassi A, Ouazzani N, Walker GM, Ibnsouda S, El Mzibri M, Boussaid A. Detoxification of olive mill wastewaters by Moroccan yeast isolates. Biodegradation 2008; 19(3): 337-46.
[http://dx.doi.org/10.1007/s10532-007-9140-8] [PMID: 18034315]

[124] Bellou S, Makri A, Sarris D, *et al.* The olive mill wastewater as substrate for single cell oil production by *Zygomycetes*. J Biotechnol 2014; 170: 50-9.
[http://dx.doi.org/10.1016/j.jbiotec.2013.11.015] [PMID: 24316440]

[125] Zervakis G, Yiatras P, Balis C. Edible mushrooms from olive oil mill wastes. Int Biodeter Biodegr 1996; 38(3): 237-43.
[http://dx.doi.org/10.1016/S0964-8305(96)00056-X]

[126] Bambalov G, Israilides C, Tanchev S. Alcohol fermentation in olive oil extraction effluents. Biol Wastes 1989; 27(11): 71-5.
[http://dx.doi.org/10.1016/0269-7483(89)90032-3]

[127] Zanichelli D, Carloni F, Hasanaji E, D'Andrea N, Filippini A, Setti L. Production of ethanol by an integrated valorization of olive oil byproducts the role of phenolic inhibition. Environ Sci Pollut Res Int 2007; 14(1): 5-6.
[http://dx.doi.org/10.1065/espr2006.06.316] [PMID: 17352121]

[128] Demerche S, Nadour M, Larroche C, Moulti-Mati F, Michaud P. Olive mill wastes: Biochemical characterizations and valorization. Process Biochem 2013; 48: 1532-52.
[http://dx.doi.org/10.1016/j.procbio.2013.07.010]

[129] Aouidi F, Khelifi E, Asses N, Ayed L, Hamdi M. Use of cheese whey to enhance *Geotrichum candidum* biomass production in olive mill wastewater. J Ind Microbiol Biotechnol 2010; 37(8): 877-82.
[http://dx.doi.org/10.1007/s10295-010-0752-3] [PMID: 20526856]

[130] Rittmann S, Herwig C. A comprehensive and quantitative review of dark fermentative biohydrogen production. Microb Cell Fact 2012; 11(1): 115.
[http://dx.doi.org/10.1186/1475-2859-11-115] [PMID: 22925149]

[131] Aouidi F, Gannoun H, Ben Othman N, Ayed L, Hamdi M. Improvement of fermentative decolorization of olive mill wastewater by *Lactobacillus paracasei* by cheese whey's addition. Process Biochem 2009; 44(5): 597-601.
[http://dx.doi.org/10.1016/j.procbio.2009.02.014]

[132] De Felice B, Pontecorvo G, Carfagna M. Degradation of waste waters from olive oil mills by *Yarrowia lipolytica* ATCC 20255 and *Pseudomonas putida.* Acta Biotechnol 1997; 17: 231-9.
[http://dx.doi.org/10.1002/abio.370170306]

[133] Lopes M, Araújo C, Aguedo M, Gomes N, Gonçalves C, Teixeira JA, *et al.* The use of olive mill wastewater by wild type *Yarrowia lipolytica* strains: medium supplementation and surfactant presence effect. J Chem Technol Biotechnol 2009; 84(4): 533-7.
[http://dx.doi.org/10.1002/jctb.2075]

[134] Yousuf A, Pirozzi D. Prospect of agro-industrial residues as feedstock of biodiesel. 1st International Conference on the Developments in Renewable Energy Technology (ICDRET). 1-5.
[http://dx.doi.org/10.1109/ICDRET.2009.5454233]

[135] Yousuf A, Sannino F, Addorisio V, Pirozzi D. Microbial conversion of olive oil mill wastewaters into lipids suitable for biodiesel production. J Agric Food Chem 2010; 58(15): 8630-5.
[http://dx.doi.org/10.1021/jf101282t] [PMID: 20681652]

[136] Curtin LV, Ed. Molasses—general considerations. West Des Moines, Iowa: National Feed Ingredients Association 1983.

[137] Chen JCP. C.C. C, editors. Cane Sugar Handbook. New York: Wiley; 1993.

[138] Satyawali Y, Balakrishnan M. Wastewater treatment in molasses-based alcohol distilleries for COD and color removal: a review. J Environ Manage 2008; 86(3): 481-97.
[http://dx.doi.org/10.1016/j.jenvman.2006.12.024] [PMID: 17293023]

[139] Kumar P, Chandra R. Decolourisation and detoxification of synthetic molasses melanoidins by individual and mixed cultures of *Bacillus* spp. Bioresour Technol 2006; 97(16): 2096-102.
[http://dx.doi.org/10.1016/j.biortech.2005.10.012] [PMID: 16321521]

[140] Plavsić M, Cosović B, Lee C. Copper complexing properties of melanoidins and marine humic material. Sci Total Environ 2006; 366(1): 310-9.
[http://dx.doi.org/10.1016/j.scitotenv.2005.07.011] [PMID: 16139328]

[141] FitzGibbon F, Singh D, McMullan G, Marchant R. The effect of phenolic acids and molasses spent wash concentration on distillery wastewater remediation by fungi. Process Biochem 1998; 33(8): 799-803.
[http://dx.doi.org/10.1016/S0032-9592(98)00050-8]

[142] Chandra R, Singh H. Chemical decolorization of anaerobically treated distillery effluent. Indian J Environ Prot 1999; 19: 833-7.

[143] Kim SB, Hayase F, Kato H. Decolourisation and degradation products of melanoidins on ozonolysis. Agric Biol Chem 1985; 49: 785-92.

[144] Aoshima I, Tozawa Y, Ohmomo S, Ueda K. Production of decolorizing activity for molasses pigment by *Coriolus versicolor* Ps4a. Agric Biol Chem 1985; 49(7): 2041-5.

[145] Ohmomo S, Aoshima I, Tozawa Y, Sakurada N, Ueda K. Purification and some properties of

melanoidin decolorizing enzymes, P-III and P-IV, from mycelia of *Coriolus versicolor* Ps4a. Agric Biol Chem 1985; 49(7): 2047-53.

[146] Ohmomo S, Kainuma M, Kamimura K, Sirianuntapiboon S, Oshima I, Atthasumpunna P. Adsorption of melanoidin to the mycelia of *Aspergillus oryzae* Y-2-32. Agric Biol Chem 1988; 52: 381-6.

[147] Boer CG, Obici L, De Souza CGM, Peralta RM. Purification and some properties of Mn peroxidase from *Lentinula edodes*. Process Biochem 2006; 41(5): 1203-7.
[http://dx.doi.org/10.1016/j.procbio.2005.11.025]

[148] Roukas T. Continuous ethanol production from nonsterilized carob pod extract by immobilized *Saccharomyces cerevisiae* on mineral kissiris using a two-reactor system. Appl Biochem Biotechnol 1996; 59(3): 299-307.
[http://dx.doi.org/10.1007/BF02783571] [PMID: 8702257]

[149] Nahvi I, Emtiazi G, Alkabi L. Isolation of a flocculating *Saccharomyces cerevisiae* and investigation of its performance in the fermentation of beet molasses to ethanol. Biomass Bioenergy 2002; 23(6): 481-6.
[http://dx.doi.org/10.1016/S0961-9534(02)00070-3]

[150] Baptista CMSG, Cóias JMA, Oliveira ACM, Oliveira NMC, Rocha JMS, Dempsey MJ, *et al.* Natural immobilisation of microorganisms for continuous ethanol production. Enzyme Microb Technol 2006; 40(1): 127-31.
[http://dx.doi.org/10.1016/j.enzmictec.2005.12.025]

[151] Kopsahelis N, Agouridis N, Bekatorou A, Kanellaki M. Comparative study of spent grains and delignified spent grains as yeast supports for alcohol production from molasses. Bioresour Technol 2007; 98(7): 1440-7.
[http://dx.doi.org/10.1016/j.biortech.2006.03.030] [PMID: 17157001]

[152] Kopsahelis N, Bosnea L, Bekatorou A, Tzia C, Kanellaki M. Alcohol production from sterilized and non-sterilized molasses by *Saccharomyces cerevisiae* immobilized on brewer's spent grains in two types of continuous bioreactor systems. Biomass Bioenergy 2012; 45: 87-94.
[http://dx.doi.org/10.1016/j.biombioe.2012.05.015]

[153] Cáceres-Farfán M, Lappe P, Larqué-Saavedra A, Magdub-Méndez A, Barahona-Pérez L. Ethanol production from henequen (*Agave fourcroydes* Lem.) juice and molasses by a mixture of two yeasts. Bioresour Technol 2008; 99(18): 9036-9.
[http://dx.doi.org/10.1016/j.biortech.2008.04.063] [PMID: 18524573]

[154] El-Enshasy HA, Mohamed NA, Farid MA, El-Diwany AI. Improvement of erythromycin production by *Saccharopolyspora erythraea* in molasses based medium through cultivation medium optimization. Bioresour Technol 2008; 99(10): 4263-8.
[http://dx.doi.org/10.1016/j.biortech.2007.08.050] [PMID: 17936622]

[155] Liu Y-P, Zheng P, Sun Z-H, Ni Y, Dong J-J, Zhu L-L. Economical succinic acid production from cane molasses by *Actinobacillus succinogenes*. Bioresour Technol 2008; 99(6): 1736-42.
[http://dx.doi.org/10.1016/j.biortech.2007.03.044] [PMID: 17532626]

[156] Sharma A, Vivekanand V, Singh RP. Solid-state fermentation for gluconic acid production from sugarcane molasses by *Aspergillus niger* ARNU-4 employing tea waste as the novel solid support. Bioresour Technol 2008; 99(9): 3444-50.
[http://dx.doi.org/10.1016/j.biortech.2007.08.006] [PMID: 17881224]

[157] Zhu LY, Zong MH, Wu H. Efficient lipid production with Trichosporon fermentans and its use for biodiesel preparation. Bioresour Technol 2008; 99(16): 7881-5.
[http://dx.doi.org/10.1016/j.biortech.2008.02.033] [PMID: 18394882]

[158] Chatzifragkou A, Fakas S, Galiotou-Panayotou M, Komaitis M, Aggelis G, Papanikolaou S. Commercial sugars as substrates for lipid accumulation in *Cunninghamella echinulata* and *Mortierella isabellina* fungi. Eur J Lipid Sci Technol 2010; 112(9): 1048-57.
[http://dx.doi.org/10.1002/ejlt.201000027]

[159] Metsoviti M, Paramithiotis S, Drosinos EH, Skandamis PN, Galiotou-Panayotou M, Papanikolaou S. Biotechnological valorization of low-cost sugar-based media for bacteriocin production by *Leuconostoc mesenteroides* E131. N Biotechnol 2011; 28(6): 600-9.
[http://dx.doi.org/10.1016/j.nbt.2011.03.004] [PMID: 21419881]

[160] Tondee T, Sirianuntapiboon S, Ohmomo S. Decolorization of molasses wastewater by yeast strain, *Issatchenkia orientalis* No. SF9-246. Bioresour Technol 2008; 99(13): 5511-9.
[http://dx.doi.org/10.1016/j.biortech.2007.10.050] [PMID: 18068358]

[161] Miranda MP, Benito GG, Cristobal NS, Nieto CH. Color elimination from molasses wastewater by *Aspergillus niger.* Bioresour Technol 1996; 57(3): 229-35.
[http://dx.doi.org/10.1016/S0960-8524(96)00048-X]

[162] Raghukumar C, Rivonkar G. Decolorization of molasses spent wash by the white-rot fungus Flavodon flavus, isolated from a marine habitat. Appl Microbiol Biotechnol 2001; 55(4): 510-4.
[http://dx.doi.org/10.1007/s002530000579] [PMID: 11398935]

[163] Ghosh M, Verma SC, Mengoni A, Tripathi AK. Enrichment and identification of bacteria capable of reducing chemical oxygen demand of anaerobically treated molasses spent wash. J Appl Microbiol 2004; 96(6): 1278-86.
[http://dx.doi.org/10.1111/j.1365-2672.2004.02289.x] [PMID: 15139920]

[164] Ghosh M, Ganguli A, Tripathi AK. Decolorization of anaerobically digested molasses spent wash by *Pseudomonas putida.* Prikl Biokhim Mikrobiol 2009; 45(1): 78-83.
[PMID: 19235513]

[165] Sirianuntapiboon S, Phothilangka P, Ohmomo S. Decolorization of molasses wastewater by a strain No.BP103 of acetogenic bacteria. Bioresour Technol 2004; 92(1): 31-9.
[http://dx.doi.org/10.1016/j.biortech.2003.07.010] [PMID: 14643983]

[166] Tondee T, Sirianuntapiboon S. Decolorization of molasses wastewater by *Lactobacillus plantarum* No. PV71-1861. Bioresour Technol 2008; 99(14): 6258-65. b
[http://dx.doi.org/10.1016/j.biortech.2007.12.028] [PMID: 18207387]

[167] Kalavathi DF, Uma L, Subramanian G. Degradation and metabolization of the pigment—melanoidin in distillery effluent by the marine cyanobacterium Oscillator*a boryana BDU 92181.* Enzyme Microb Technol 2001; 29(4–5): 246-51.
[http://dx.doi.org/10.1016/S0141-0229(01)00383-0]

[168] Patel A, Pawar P, Mishra S, Tewari A. Exploitation of marine cyanobacteria for removal of colour from distillery effluent. Ind J Environ Prot 2001; 21: 1118-21.

[169] Milson PT, Meers JI. Citric acid. Comprehensive Biotechnology The Practice of Biotechnology, Current Commodity Products 3. Oxford: Pergamon Press 1985; pp. 665-80.

[170] Majolli MV, Aguirre SN. Effect of trace metals on cell morphology, enzyme activation, and production of citric acid in a strain of *Aspergillus wentii..* Rev Argent Microbiol 1999; 31(2): 65-71.
[PMID: 10425661]

[171] Pera LM, Callieri DA. Influence of calcium on fungal growth, hyphal morphology and citric acid production in *Aspergillus niger.* Folia Microbiol (Praha) 1997; 42(6): 551-6.
[http://dx.doi.org/10.1007/BF02815463] [PMID: 9438355]

[172] Adham NZ. Attempts at improving citric acid fermentation by *Aspergillus niger* in beet-molasses medium. Bioresour Technol 2002; 84(1): 97-100.
[http://dx.doi.org/10.1016/S0960-8524(02)00007-X] [PMID: 12137276]

[173] Ikram-Ul H, Ali S, Qadeer MA, Iqbal J. Citric acid production by selected mutants of *Aspergillus niger* from cane molasses. Bioresour Technol 2004; 93(2): 125-30.
[http://dx.doi.org/10.1016/j.biortech.2003.10.018] [PMID: 15051073]

[174] Hamissa FA, Radwan A. Production of citric acid from cane molasses on a semi-pilot scale. J Gen

Appl Microbiol 1977; 23(6): 325-9.
[http://dx.doi.org/10.2323/jgam.23.325]

[175] Qazi GN, Gaind CN, Chaturvedi SK, Chopra CL, Träger M, Onken U. Pilot-scale citric acid production with *Aspergillus niger* under several conditions. J Ferment Bioeng 1990; 69(1): 72-4.
[http://dx.doi.org/10.1016/0922-338X(90)90170-2]

[176] Garg K, Sharma CB. Repeated batch production of citric acid from sugarcane molasses using recycled solid-state surface culture of *Aspergillus niger.* Biotechnol Lett 1991; 13(12): 913-6.
[http://dx.doi.org/10.1007/BF01022098]

[177] Wang J. Improvement of citric acid production by *Aspergillus niger* with addition of phytate to beet molasses. Bioresour Technol 1998; 65(3): 243-5.
[http://dx.doi.org/10.1016/S0960-8524(98)00026-1]

[178] Wang J, Wen X, Zhou D. Production of citric acid from molasses integrated with in-situ product separation by ion-exchange resin adsorption. Bioresour Technol 2000; 75(3): 231-4.
[http://dx.doi.org/10.1016/S0960-8524(00)00067-5]

[179] Ramachandran S, Fontanille P, Pandey A, Larroche C. Gluconic acid: properties, applications and microbial production. Food Technol Biotechnol 2006; 44(2): 185-95.

[180] Hidaka H, Hirayama M, Sumi N. A Fructooligosaccharide-producing Enzyme from *Aspergillus niger* ATCC 20611. Agric Biol Chem 1988; 52(5): 1181-7.

[181] Hirayama M, Sumi N, Hidaka H. Purification and properties of a fructooligosaccharide-producing β-fructofuranosidase from Asper*gillus niger ATCC* 20611. Agric Biol Chem 1989; 53(3): 667-73.

[182] Yun J, Jung K, Jeon Y, Lee J. Continuous production of fructo-oligosaccharides from sucrose by immobilized cells of *Aureobasidium pullulans.* J Microbiol Biotechnol 1992; 2: 98-101.

[183] Shin HT, Baig SY, Lee SW, *et al.* Production of fructo-oligosaccharides from molasses by *Aureobasidium pullulans* cells. Bioresour Technol 2004; 93(1): 59-62.
[http://dx.doi.org/10.1016/j.biortech.2003.10.008] [PMID: 14987721]

[184] Willke T, Vorlop K-D. Industrial bioconversion of renewable resources as an alternative to conventional chemistry. Appl Microbiol Biotechnol 2004; 66(2): 131-42.
[http://dx.doi.org/10.1007/s00253-004-1733-0] [PMID: 15372215]

[185] Suutari M, Laakso S. Effect of growth temperature on the fatty acid composition of *Mycobacterium phlei.* Arch Microbiol 1993; 159(2): 119-23.
[http://dx.doi.org/10.1007/BF00250270] [PMID: 8439233]

[186] Suutari M, Priha P, Laakso S. Temperature shifts in regulation of lipids accumulated by *Lipomyces starkeyi.* J Am Oil Chem Soc 1993; 70(9): 891-4.
[http://dx.doi.org/10.1007/BF02545349]

[187] Koike Y, Cai HJ, Higashiyama K, Fujikawa S, Park EY. Effect of consumed carbon to nitrogen ratio of mycelial morphology and arachidonic acid production in cultures of *Mortierella alpina.* J Biosci Bioeng 2001; 91(4): 382-9.
[http://dx.doi.org/10.1016/S1389-1723(01)80156-0] [PMID: 16233009]

[188] Fakas S, Papanikolaou S, Galiotou-Panayotou M, Komaitis M, Aggelis G. Lipids of *Cunninghamella echinulata* with emphasis to gamma-linolenic acid distribution among lipid classes. Appl Microbiol Biotechnol 2006; 73(3): 676-83.
[http://dx.doi.org/10.1007/s00253-006-0506-3] [PMID: 16850299]

[189] Papanikolaou S, Chevalot I, Galiotou-Panayotou M, Komaitis M, Marc I, Aggelis G. Industrial derivative of tallow: a promising renewable substrate for microbial lipid, single-cell protein and lipase production by *Yarrowia lipolytica.* Electron J Biotechnol 2007; 10(3): 425-35.
[http://dx.doi.org/10.2225/vol10-issue3-fulltext-8]

[190] Papanikolaou S, Aggelis G. Lipids of oleaginous yeasts. Part II: Technology and potential

applications. Eur J Lipid Sci Technol 2011; 113(8): 1052-73. b
[http://dx.doi.org/10.1002/ejlt.201100015]

[191] Roukas T, Liakopoulou-Kyriakides M. Production of pullulan from beet molasses by *Aureobasidium pullulans* in a stirred tank fermentor. J Food Eng 1999; 40(1): 89-94.
[http://dx.doi.org/10.1016/S0260-8774(99)00043-6]

[192] Siso MG. The biotechnological utilization of cheese whey: a review. Bioresour Technol 1996; 57(1): 1-11.
[http://dx.doi.org/10.1016/0960-8524(96)00036-3]

[193] Prazeres AR, Carvalho F, Rivas J. Cheese whey management: a review. J Environ Manage 2012; 110: 48-68.
[http://dx.doi.org/10.1016/j.jenvman.2012.05.018] [PMID: 22721610]

[194] Carvalho F, Prazeres AR, Rivas J. Cheese whey wastewater: characterization and treatment. Sci Total Environ 2013; 445-446: 385-96.
[http://dx.doi.org/10.1016/j.scitotenv.2012.12.038] [PMID: 23376111]

[195] Vamvakaki A-N, Kandarakis I, Kaminarides S, Komaitis M, Papanikolaou S. Cheese whey as a renewable substrate for microbial lipid and biomass production by *Zygomycetes*. Eng Life Sci 2010; 10(4): 348-60.
[http://dx.doi.org/10.1002/elsc.201000063]

[196] Schultz N, Chang L, Hauck A, Reuss M, Syldatk C. Microbial production of single-cell protein from deproteinized whey concentrates. Appl Microbiol Biotechnol 2006; 69(5): 515-20.
[http://dx.doi.org/10.1007/s00253-005-0012-z] [PMID: 16133331]

[197] Mahmoud M, Kosikowski F. Alcohol and single cell protein production by *Kluyveromyces* in concentrated whey permeates with reduced ash. J Dairy Sci 1982; 65(11): 2082-7.
[http://dx.doi.org/10.3168/jds.S0022-0302(82)82465-X]

[198] Ghaly AE, Kamal M, Correia LR. Kinetic modelling of continuous submerged fermentation of cheese whey for single cell protein production. Bioresour Technol 2005; 96(10): 1143-52.
[http://dx.doi.org/10.1016/j.biortech.2004.09.027] [PMID: 15683905]

[199] Seo YH, Lee I, Jeon SH, Han J-I. Efficient conversion from cheese whey to lipid using *Cryptococcus curvatus*. Biochem Eng J 2014; 90: 149-53.
[http://dx.doi.org/10.1016/j.bej.2014.05.018]

[200] Takakuwa N, Saito K. Conversion of beet molasses and cheese whey into fatty acid methyl esters by the yeast *Cryptococcus curvatus*. J Oleo Sci 2010; 59(5): 255-60.
[http://dx.doi.org/10.5650/jos.59.255] [PMID: 20431242]

[201] Kargi F, Ozmıhcı S. Utilization of cheese whey powder (CWP) for ethanol fermentations: Effects of operating parameters. Enzyme Microb Technol 2006; 38(5): 711-8.
[http://dx.doi.org/10.1016/j.enzmictec.2005.11.006]

[202] Zafar S, Owais M. Ethanol production from crude whey by *Kluyveromyces marxianus*. Biochem Eng J 2006; 27(3): 295-8.
[http://dx.doi.org/10.1016/j.bej.2005.05.009]

[203] Morales J, Choi JS, Kim DS. Production rate of propionic acid in fermentation of cheese whey with enzyme inhibitors. Environ Prog 2006; 25(3): 228-34.
[http://dx.doi.org/10.1002/ep.10153]

[204] Rosa PRF, Santos SC, Sakamoto IK, Varesche MBA, Silva EL. Hydrogen production from cheese whey with ethanol-type fermentation: effect of hydraulic retention time on the microbial community composition. Bioresour Technol 2014; 161: 10-9.
[http://dx.doi.org/10.1016/j.biortech.2014.03.020] [PMID: 24681681]

[205] Haddadin M, Al-Muhirat S, Batayneh N, Robinson R. Production of acetic and propionic acids from labneh whey by fermentation with propionibacteria. Int J Dairy Technol 1996; 49(3): 79-81.

[http://dx.doi.org/10.1111/j.1471-0307.1996.tb02495.x]

[206] Alam S, Stevens D, Bajpai R. Production of butyric acid by batch fermentation of cheese whey with *Clostridium beijerinckii.* J Ind Microbiol 1988; 2(6): 359-64.
[http://dx.doi.org/10.1007/BF01569574]

[207] Long JH, Aziz TN, Francis L, Ducoste JJ. Anaerobic co-digestion of fat, oil, and grease (FOG): a review of gas production and process limitations. Process Saf Environ Prot 2012; 90(3): 231-45.
[http://dx.doi.org/10.1016/j.psep.2011.10.001]

[208] Fontanille P, Kumar V, Christophe G, Nouaille R, Larroche C. Bioconversion of volatile fatty acids into lipids by the oleaginous yeast *Yarrowia lipolytica.* Bioresour Technol 2012; 114: 443-9.
[http://dx.doi.org/10.1016/j.biortech.2012.02.091] [PMID: 22464419]

[209] Fickers P, Benetti P-H, Waché Y, *et al.* Hydrophobic substrate utilisation by the yeast *Yarrowia lipolytica*, and its potential applications. FEMS Yeast Res 2005; 5(6-7): 527-43.
[http://dx.doi.org/10.1016/j.femsyr.2004.09.004] [PMID: 15780653]

[210] Tsouko E, Kourmentza C, Ladakis D, *et al.* Bacterial cellulose production from industrial waste and by-product streams. Int J Mol Sci 2015; 16(7): 14832-49.
[http://dx.doi.org/10.3390/ijms160714832] [PMID: 26140376]

[211] Leiva-Candia DE, Tsakona S, Kopsahelis N, *et al.* Biorefining of by-product streams from sunflower-based biodiesel production plants for integrated synthesis of microbial oil and value-added co-products. Bioresour Technol 2015; 190: 57-65.
[http://dx.doi.org/10.1016/j.biortech.2015.03.114] [PMID: 25930941]

[212] Federici F, Montedoro G, Servili M, Petruccioli M. Pectic enzyme production by Cryptococcus albidus var. albidus on olive vegetation waters enriched with sunflower calathide meal. Biol Wastes 1988; 25(4): 291-301.
[http://dx.doi.org/10.1016/0269-7483(88)90090-0]

[213] Crognale S, Federici F, Petruccioli M. β-Glucan production by *Botryosphaeria rhodina* on undiluted olive-mill wastewaters. Biotechnol Lett 2003; 25(23): 2013-5.
[http://dx.doi.org/10.1023/B:BILE.0000004394.66478.05] [PMID: 14719815]

[214] D'Annibale A, Ricci M, Quaratino D, Federici F, Fenice M. *Panus tigrinus* efficiently removes phenols, color and organic load from olive-mill wastewater. Res Microbiol 2004; 155(7): 596-603.
[http://dx.doi.org/10.1016/j.resmic.2004.04.009] [PMID: 15313262]

[215] Dourou M, Kancelista A, Juszczyk P, Sarris D, Bellou S, Triantaphyllidou I-E, *et al.* Bioconversion of olive mill wastewater into high-added value products. J Clean Prod 2016; 139: 957-69.
[http://dx.doi.org/10.1016/j.jclepro.2016.08.133]

CHAPTER 10

Plant Epigenetics: Basic Research and Expectations for Crop Applications

Amr R.A. Kataya*

University of Stavanger, Centre for Organelle Research, Faculty of Science and Technology, N-4036 Stavanger, Norway

Abstract: Genetics is the science of studying genes that are organized and compacted into nucleosomes and chromatin. Several cases, involving the incorporation of important genes or knocking out others, have been used over the last decades in crop and food applications. Because genetically modified crops are banned and/or ethically under debate, epigenetics that studies gene regulation is thought to have an impact on crop production in future applications. Epigenetics refers to the study of heritable changes in gene expressions without changes in the gene sequence. Epigenetics regulation of gene expression is applied by histone modifications, DNA methylation, histone variants, chromatin remodeling, and small RNAs. This chapter describes what becomes known in plant epigenetics, and future expectations that can employ epigenetics in order to improve crops and produce higher levels of vitamins and proteins that are important for food production and human health.

Keywords: DNA methylation, Epigenetics, Epimutation, Genetics, Genome editing technology, GM crops, Histone modifications, Nutritional epigenetics, RNA dependent DNA methylation, RNA epigenetics.

INTRODUCTION

Currently, around one billion or one out of every 7 humans do not have sufficient daily food. It is estimated that 40% of humans suffer from malnutrition [1], which is a state in which a deficiency or imbalance of essential micronutrients causes measurable adverse effects on tissue/body forms. In most cases, people who suffer from malnutrition consume meals that are mainly limited to specific crops, such as rice, maize, or cassava and lack unavailable or unaffordable other food varieties. Malnutrition and deficiency of vitamin A, zinc, and iron lead to almost two-thirds of childhood deaths worldwide. The world's population is expected to reach 10 billion by 2050, and climate change is another challenge for achieving

* **Corresponding Author Amr R.A. Kataya:** University of Stavanger, Centre for Organelle Research, Faculty of Science and Technology, N-4036 Stavanger, Norway; Tel: +47 5183 2297; Fax: +47 51831750; Email: amr.kataya@uis.no

Ali Osman (Ed.)

global food security [1]. The food security concept (WHO 2015) is to access food that meets people's dietary needs and their food preferences. Tremendous advances in food production and distribution will be required in order to meet future needs.

The development and use of plant biotechnology offer an important contribution to food security and sustainability. Crucial contributions are represented by 30 traits in GM plants, which have been introduced including improved yield, pathogen and herbivore resistance, increased nutrient content and improved product quality [2, 3]. GM crops, for example, gained a rising share of agricultural production mostly in the United States, Brazil, Argentina, Canada and India. Most GM crop production involves soybean, cotton, corn, and canola. However, GM crops usually undergo a series of risk assessments before their approval for food production in order to determine their impact on human health and the environment. These regulations include addressing the GM crop molecular characterization, potential toxicity and/or allergenicity, and nutritional comparison with the available conventional crops. Political opposition and specific regulatory formats in Europe also led to GM crop hindrance [1].

GENETICALLY ENHANCED FOOD CROPS

The central dogma of molecular biology describes the flow of genetic information from DNA to RNA to proteins, where each protein performs a function in the organism. Advances in genetics related sciences and their manipulation boosted the technological methods in plant genome editing. A wide variety of approaches can be achieved; for example, knockout, knockdown, or overexpression of specific genes that can lead to the optimization of a specific trait. The genetic modification using *Agrobacterium tumefacians* that was discovered in 1977 allowed easier manipulation of plant genetics [4]. Moreover, the ability to introduce inter-organismal genes in order to obtain a new trait now is widely used and started after the first reported GMO: a sunflower (*Helianthus annuus*) that acquired a gene from bean (*Phaseolus vulgaris*) [5, 6].

Rice has been modified by introducing multiple genes in order to complete the β-carotene biosynthetic pathway [7]. This example of rice genetic modification (named golden rice) is aimed to minimize the serious problems of vitamin A deficiency, especially in at least 26 countries in Asia, Africa, and Latin America. It is estimated that vitamin A deficiency is responsible for up to 700,000 mortalities worldwide among children annually and causes about 300,000 cases of blindness globally each year. So far, golden rice shows an effective vitamin A source as tested in the United States, Philippines and Taiwan, and several related studies show its positive effect in eliminating vitamin A malnutrition. However, public controversies about using GMOs and the measures posed by developing countries still control its widespread usage [2]. In particular, the humanitarian group 'Greenpeace' used public pressure to block Golden rice [1]. Another example, in this regard, is a new genetically modified "super" banana, which is now under human trials in the USA and is planned to be tried on the fields of Uganda by 2020 as aimed by the Australian and Ugandan scientists who develop it. These trials are held in the US because the safety of new GM crops is determined by showing that they have similar content to their natural counterparts, while Uganda and several African countries have signed the Cartagena Protocol on Biosafety, part of an agreement that applies the precautionary principle which includes "taking into account risks to human health". Like golden rice, the engineered bananas are rich in beta-carotene, which is converted into vitamin A. In East Africa, vitamin A deficiency is common and bananas are a staple crop [8]. As golden rice, super banana receives different critics. Additionally, public critics of the research argue that the long-term implications of the technique are unknown.

Eggplant was also engineered to obtain resistance against the attack of pests, such as the fruit and shoot borer (FSB) *Leucinodes orbonalis*. This type of modification dramatically decreased the usage of pesticides and allowed yield improvement of eggplants. Simplot potato is another useful example of the usage of genetic engineering, where a 20-fold reduction in asparagine has been set in this transgenic line. Starchy foods have asparagine that is converted to acrylamide when products are baked, roasted, or fried. Hence, Simplot potato is an alternative source that leads to less acrylamide accumulation during heat processing [2]. A number of GM crops including corn, soybean, canola, rice, potato and various other crops have been modified to acquire different traits, such as virus resistance, drought resistance, wilt resistance, insect resistance, increased provitamin A, modified oil content, and food improving traits [2, 5, 6]. Besides, several GMOs are now commercialized in a number of countries; these GMO's have shown potential nutritional and economical values, but their use still remains a controversial issue. Most of this controversy comes from the genetic manipulation techniques as well as the introduction of several foreign genes that naturally do not occur in plants, such as antibiotic resistance genes. Therefore, a new number of techniques for plant genetic manipulation are recently developed, which may probably not come under the GMO banner and therefore under GM regulations. These techniques will be discussed briefly at the end of this book chapter.

EPIGENETICS INTRODUCTION AND DIFFERENCE WITH GENETICS

Epigenetics has evolved as a distinct branch in biology, which focuses on studying the changes in gene expression without gene code alteration, and is frequently mitotically inherited [9]. The term epigenetics was derived from the Greek word "epigenesis" which describes the effect of genetics on development [10]. An epigenetically acquired trait is transferable from cell to cell mitotically, and meiotically under distinct circumstances as observed in plants [11]. For example, flax plants (*Linum usitatissimum*) exhibited inherited varieties of phenotypes after fertilizer treatment [12, 13]. It was also reported that heritable phenotypes were developed in *Nicotiana rustica* after growth under specified nutritional conditions that are thought to be related to specific DNA alterations [14 - 16]. Toadflax (*Linaria vulgaris*) has two natural variations, common and pleoric (radially symmetrical) flowers; the pleoric toadflax *LCYC* (a homologue of the cycloidea gene that controls dorsoventral asymmetry) has extensive methylation and is transcriptionally silent and can be passed to offsprings [17]. The gene that is important for fruit ripening and pigment production in tomato, colorless non-ripening (Cnr) locus, was found to be epigenetically affected. This was revealed when an SBP-box (SQUAMOSA promoter binding protein –like) was located in the Cnr locus, and is epigenetically changed to lead to the Cnr phenotype [18]. However, this epigenetic mutation is unstable and occasionally sectors of the mutant fruit can restore their normal pigment production [18].

In general, each organism, in a unicellular or multicellular form, has to cope with the constantly changing environmental and external factors by introducing new alterations. Only the non-mutational alterations that are transmitted from one cell to its daughters or between generations of an organism are considered to be epigenetic changes. In other words, epigenetics refers to "above or beyond genetics" and involves studying heritable changes in response to several external factors [19]. Recently, there has been a rapid advancement in understanding epigenetic mechanisms, which include histone modifications, DNA methylation, small and non-coding RNAs, *etc.*

DNA Methylation

DNA methylation is one of the major epigenetic mechanisms, and is common in plants, mammals, and most fungi, while it is absent or rare in yeast, flies, and nematodes. DNA methylation refers to the methylation of cytosine, which is catalyzed by methyl transferases after DNA replication; this leads to gene activity repression. Methylated cytosines account for 3-8% in mammalian DNA, and 25-30% in plants' DNA [20, 21]. The methylated cytosines in mammals are located on CG, while in plants they are located on CG, CHG and CHH (H is A, T, or C). The symmetric sequences CG and CHG are transmitted through meiosis, while the asymmetrical sequence CHH is not heritable because it has to be re-established in every generation [21]. DNA methyltransferases in plants include DRM2 (domain rearranged methyltransferase), MET1 (methyltransferase 1) and

DNM2 (DNA-(cytosine-5)-methyltransferase 2). DRM2 is involved in *de novo* methylation, MET1 is involved in the maintenance of CG methylation and DNM2 is conserved in eukaryotes but its function is unknown [22]. This mechanism was identified in *Arabidopsis thaliana*, where MET1 primarily maintains DNA methylation at CG sites. CMT3 (chromomethyltransferase) interacts with the H3 Lys9 dimethylation (H3K9me2) to maintain DNA methylation at CHG sites and the DNMT3a/3b (homologs of DRM2) maintains DNA methylation at CHH sites, which requires siRNAs [23].

The functions of DNA methylation include (1) heritable modulation of transcription that includes defense against invasive DNA (transposons), regulation of gene expression, and parental imprinting, (2) modulation of chromatin structure through histone modification and formation of heterochromatin, and through regulation of DNA replication, and (3) regulation of tissue specific genes through promoter methylation [24]. It was reported, from studying chromosome epigenetic landscape in *Arabidopsis thaliana*, that the promoters of endogenous genes were rarely methylated. In contrast, the heterochromatin, transposons, and other repeats were found to harbor high cytosine methylation, which implies the role of DNA methylation in inactivating genes [23]. This was also approved when severe loss of DNA methylation in the quadruple mutant *drm1 drm2 cmt3 met1* led to a genome-wide transcriptional reactivation of transposons and caused embryo lethality [23]. DNA methylation in promoter regions usually leads to repression, while body-methylated genes are usually transcribed at moderate to high levels [23].

Histone Modifications of Chromatin

Chromatin remodeling comprised of histone modifications, such as methylation, phosphorylation, acetylation, s-nitrosoylation, ubiquitination, SUMOylation, proline isomerization, ADP-ribosylation, deamination, and non-covalent histone modifications [25]. Eight histones are included to wrap approximately 147 DNA base pairs to form the nucleosome, which are two copies for each of H2A, H2B, H3 and H4. The N-peptide tails protruding from H3 and H4 are readily modified compared to other histone parts. Through variable histone modifications, chromatin remodeling ATPases that use adenosine-tri-phosphate (ATP)-derived energy are recruited in order to perform nucleosome arrangement and chromatin remodeling, where activators or inhibitors can then modify DNA sequences [26]. Reports indicate that histone modifications by phosphorylation, ubiquitination, and acetylation lead to upregulation of gene expression. In contrast, modifications by dimethylation of H3K9 (Histone 3 Lysine 9) and H3K27, SUMOylation, and biotinylation lead to downregulation of gene expression [26]. In plants, different reports referred to the importance of histone modifications in plant development and in plant defense mechanisms [27 - 30].

Histone Methylation

Histone methylation depends on histone methyl transferases (HMTs). Lysine and arginine residues are the targets of HMTs and these modifications have a variable effect on transcription regulation that includes activation or repression. Recent reports also identified histone demethylation enzymes, such as FLOWERING LOCUS D, LSD1-like 1 and 2 that are involved in plant flowering regulation [19, 26]. Residues can be mono-, di-, or trimethylated where each state refers to different biological effect. For example, H3K9me2 and H3K27me3 downregulate gene expression, while H3K4me1, H3K4me2, H3K4me3, H3K4me2, and H3K4me3 upregulate gene expression [26, 31].

Histone Phosphorylation

Kinases are also involved in histone modifications through the addition of phosphate (PO_4) group to specific histone residues. For example, histone H3 serine 10 (H3S10) is phosphorylated by Aurora-B Kinase during mitosis [19, 32]. Also, H2A (S129, S141, S145), H2BS15, H3 (T3, T11), and H3 (S10, S28) have been reported to be phosphorylated [26, 29, 33]. The function of histone modification by phosphorylation is so far linked to cell mitosis and apoptosis [19, 26, 33].

Histone Acetylation

Histone acetylation usually occurs on the lysine ε-amino groups on the N-terminal tails. This reaction is catalyzed by histone acetyl transferases (HATs), and leads to relaxation of the condensed chromatin and is thought to co-occur with transcription activation [34]. Different studies reported that nucleosomes including K-acetylated residues usually rearrange during plant development. Moreover, this type of modification is linked to flowering, root elongation, and cold tolerance [26, 35 - 38]. Deacetylation of histones, which leads to chromatin re-condensation and transcription inhibition, is carried out by histone deacetylases (HDACs) and NAD+-dependent sirtuin family deacetylases [19]. Acetylation and deacetylation of specific lysine residues are important for a variety of light-induced developmental processes in plants [39].

Histone S-nitrosylation

Nitric oxide (NO) plays an important role in multiple physiological processes in plants. Modification of cysteine residues, through the addition of an NO group, termed as S-nitrosylation, is believed to be important for NO transduction.

Reports indicate that S-nitrosylation of nuclear plant proteins (for instance transcription factors) probably participates in regulation of transcription [40]. Studies of chromatin remodeling in neurons showed that NO alters chromatin acetylation state, where NO is suggested to influence the interaction of transcription factors with chromatin. It was also shown that NO affects histone acetylation by regulating human HDAC2 [40, 41].

Histone Ubiquitination

Ubiquitin is a protein of 76 residues, which is covalently attached to the histone lysine residues. This mechanism is described as monoubiquitination, while polyubiquitination is the mechanism when the histones are covalently attached to one or more ubiquitin monomers, respectively [19]. Ubiquitination on H2B Lysine143 has been reported and is catalyzed by two RING E3 ligases, and is deleted by deubiquitinases. This modification is reported to help plants become immune against necrotrophic fungal pathogens, and regulate seed development [26, 42]. Indicative of its importance for plants, enzymes responsible for the monoubiquitination of H2B were shown to be involved in flowering time regulation in *Arabidopsis thaliana*, where the mutants *histone monoubiquitination (hub)1* and *hub2* lost monoubiquitination of H2B (*H2Bub1*) and flowered early [43].

Other Histone Modifications

Modification of histones by binding to small ubiquitin-related modified protein (SUMO, contains around 100 residues) was named as SUMOylation, and is involved in transcription repression. Histones can also be modified by ADP-ribosyl transferases (ARTs) and poly (ADP-ribose) polymerases (PARPs) to be mono or poly ADP-ribosylated, respectively. Mono-ADP-ribosylation has important roles in cell response to genotoxic stress, and poly-ADP-ribosylation is connected to DNA repair mechanism [19, 25]. Another histone modification includes proline isomerization. This was reported through the discovery of Frp4, a proline isomerase that switches between *cis* and *trans* conformations of prolines 30 and 38 on H3. The change in conformation was shown to lead to regulation of histone lysine methylation [19, 25, 44]. Another histone modification includes the conversion of H3 and H4 arginines to citrullines by PAD14 enzyme in a process called deamination, which has antagonistic effects on arginine methylation status. Histone modifications also include intra/inter-chromosomal interactions, histone exchanges, chromatin repair, and binding with siRNAs and miRNAs [19].

RNA-Mediated Gene Silencing Pathways

As described earlier, DNA methylation and post-translational histone modifications are important epigenetic regulators that affect switching gene transcriptions on and/or off. Epigenetic regulation can also control post-transcriptional products through targeting mRNA degradation or halt. This mechanism, RNA interference (RNAi), is important to control endogenous mRNAs levels and serve as a vital defense against infecting viruses, microbial pathogens, and transgenes [45]. RNAi was discovered in 1998 by Fire and Mello when they investigated the effects of dsRNA, sense, and antisense injection into c-elegans, where they found that dsRNA strongly inhibited gene expression [46]. Because of this discovery, they were awarded the Nobel Prize in Medicine and Physiology in 2007. However, RNAi was explained earlier in different organisms, for example oligodeoxynucleotide was shown to inhibit replication of Rous sarcoma virus [47], quelling in fungi [48], and the description of post-transcriptional silencing (PTGS) in the pigmentation of petunias [49]. The observation of RNAi in petunias was developed when scientists were trying to deepen the purple color of these flowers through the introduction of a pigment-producing gene under the control of a strong promoter. Instead of getting a deep purple color, several flowers appeared variegated or white. This phenomenon was named as co-suppression because both the introduced and endogenous genes were suppressed [49].

RNAi is used by many different organisms to regulate gene activity, and includes two main players: small interfering RNAs (siRNAs) and microRNAs (miRNAs). miRNAs and siRNAs are produced from dsRNA precursors by RNAaseIII-related dicer (DCL) endonuclease. The small RNAs are then bound to a multiprotein RNA-induced silencing complex (RISC), which has in its core a member of the Argonaute (AGO) protein family. The AGO protein binds to the 3` end of the small RNA and uses it to bind with its complementary targets. As a result, the targeted mRNAs can be cleaved by AGO protein's PIWI domain by which translation can be stopped, or chromatin-modifying machinery can be recruited to silence the locus of binding [45, 50]. miRNAs are encoded by endogenous genes where DNA-dependent RNA polymerase II (RNA Pol II) will produce pre-miRNAs. Pre-miRNAs are hairpins with imperfect complementarity in their stems and frequent bulges, mismatches and G:U wobble base pairings. In the case of siRNAs, the dsRNA precursors can be generated through bidirectional transcription by dependent RNA polymerase, such as RNA Pol II, as well as RNA transcripts which can work as templates for RNA-dependent RNA polymerase (RdRP) to generate a complementary strand [45].

Loci encoding miRNAs were primarily discovered as genes not encoding a protein, and it is well known now that the transcripts of these non-coding loci are the sources of small RNAs. miRNAs also evolved another pathway where they induce secondary siRNAs production, which is named trans-acting siRNA (tasiRNA) pathway [45, 51]. In plants, 24-nt siRNAs are used to silence heterochromatin (includes *e.g.* TEs) through recruiting *de novo* cytosine methylation of corresponding genomic sequences. RNA dependent directed methylation (RdDM) was firstly described in tobacco plants infected with viroids, where replicating viroids was found to trigger *de-novo* methylation of viroids cDNA that had been inserted into the genome of infected cells. Moreover, RNA viruses and transgenes engineered to express dsRNAs against promoter sequences were found to induce cytosine methylation, and resulted in viral/gene transcription silencing. RdDM is catalyzed by the DRM class of DNA methyltrasferases [45, 52].

RNA Epigenetics

Similarly, it was shown that RNA is decorated with epigenetic marks that affect its biological functions. RNA molecules such as mRNA, rRNA, tRNA, miRNA and others are engaged in diverse activities such as catalytic, regulatory and translator activities. RNA building blocks contain pyrimidine or purine rings that can be chemically altered. So far, methyl group is the most frequently detected modification as well as acetyl, isopentyl and other groups. Methylation of adenosine at the N6 position (m6A) is reported to be the famous epigenetic mark in eukaryotic mRNA, which was also identified in bacterial rRNAs and tRNAs in 1950s, and was also found in mRNAs of animal and plant cells [53]. Analyzing human transcriptome led to the identification of more than 12000 methylated sites in mRNA molecules derived from 7000 protein-coding genes. The importance of m6A is thought to cover diverse roles of gene expression, translation and alternative splicing. M6A also decorates other RNA species; for example, it is abundant on rRNAs, tRNAs and small nuclear RNAs (snRNAs), which mediate splicing and other RNA processing reactions. Very recently, it was also shown that epigenetic decoration also could regulate miRNAs functions. Researchers found that a protein which binds to m6A marks in a subset of primary miRNA transcripts, and helps to promote primary miRNA processing [54]. Additionally, miRNAs themselves appear to participate in the placement of the m6A epigenetic marks [55].

Epigenetic Genome Regulation in Plants

Plants can be easily mutagenized by physical and chemical treatment, random insertion of transgenes, or transposable elements mobilization. Plant epigenetic research has been boosted in the past decades through the usage of forward and reverse genetics to identify or generate a mutation in epigenetic modifier [45]. Self-pollinated plants as *Arabidopsis thaliana* is an important model plant for epigenetic research because of easy selection of a single locus mutagenized plant.

Moreover, *Arabidopsis thaliana* has extensive natural variation, complete genome sequencing, and an enormous genome wide expression and chromatin modification data. More than 130 genes encoding epigenetic modifiers are known to date and were classified into five groups: DNA modification, Histone modification, RNA silencing, Polycomb-group proteins and interacting components, and chromatin formation or chromatin remodeling (reviewed and summarized in table **1** in [45]).

Selecting Beneficial Epimutation for Successful Breeding

Understanding epigenetics by basic research can supply a basis for possible new sources of beneficial traits for plant breeding. Because epigenetics is responsible for heritable gene variants (epiallels), plant epigenetic and phenotypic variations (Fig. **10.1**) can be employed broadly in order to improve long-term plant productivity and adaptation to environmental challenges [21], such as stress and exposure to different chemicals, which are perceived by plants through signal transduction pathways that affect transcription factors and epigenetic mapping. Recently, it becomes well known that epigenetic changes in chromatin characteristics and small RNAs biogenesis contribute to transcriptional and post-transcriptional regulation of gene expression. Epigenetic induced alleles could thus be transferred to the progeny.

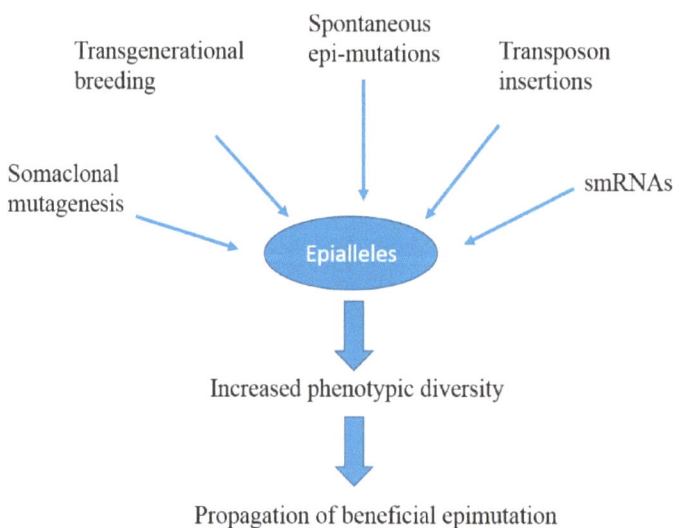

Fig. (10.1). Factors that can develop epigenetic variations in plants. Spontaneous epi-mutations, transposon insertions, small RNAs, tissue culture somaclonal mutagenesis and transgenerational breeding are factors that can contribute to the generation of epiallels. Heritable epiallels can accumulate over generations and increase phenotypic diversity. Modified from Fujimoto *et al.* [56].

Selecting the stable transgenerational inherited epiallel or TEs re-organization under environmental stress are important areas for studying, because epigenetic regulations are known to be relaxed under stress conditions, and this might result in activation or inactivation of genes [57]. For example, hybridization between wheat-rye 2R and 5R monosomic addition lines show wide genetic and epigenetic diversity and is suggested to broaden the genetic diversity of common wheat [58]. As well as, rice and *Zizania latifolia* Griseb hybridization ended up with rice lines different in its DNA methylation patterns [1, 59]. Different reviews have discussed selected aspects of transgenerational epigenetic inheritance in plants [60 - 62], and collected information about different types of stress that led to developing heritable resistance of various plants to different biotic stresses [61]. For example, rice seedlings were treated with azacytidine, an inhibitor of DNA methylation, and their progeny were used to initiate different lines for propagation over several generations. One of these progenies developed acquired resistance to *Xanthomonas oryzae*, a bacterial pathogen that is linked to hypomethylation of the promoter *Xa21G* and hence its constitutive expression. This was a clear example of a heritable epigenetic trait because it was stably transformed [21, 63].

Employing RdDM for Generating Heritable Epimutation

RNA-directed transcriptional gene silencing in plants can be also used for the downregulation of specific gene expression, and can be maintained as an epigenetic trait. For instance, PTGS was initiated in 35S-GFP transgenic plants through infecting with plant RNA viruses, which were modified to have a portion of coding GFP sequence. The result of this infection led to non-heritable silencing through production of siRNAs against GFP. However, RdDM has been developed with the infection by plant viruses harbored a portion of the 35S-promoter (the promoter used to generate GFP transgenic plants). In this case, siRNAs against the 35S promoter led to heritable TGS of GFP and promoter methylation. This later mechanism was found to be inherited and did not require RNAi trigger in the progenies, and depended on MET1 for maintenance [64]. Hence, such an idea can be mobilized in biotechnology, where endogenous siRNAs can be developed to lead to RdDM and PTGS of target locus that will lead to heritable epimutation.

However, the usage of modified plant dsRNA is under concern because of the possible exposure routes to humans or non-target organisms. Studies showed that miRNAs from plant foods can survive digestion and can possibly regulate homologues genes in the body following digestion [65, 66]. In a quiet recent study, scientists analyzed global miRNA levels in humans and 5 other mammalian species following rice consumption, and found a selective uptake of 30 miRNAs. By following the non-digested miRNAs in mice models, they found that one miRNA that is called MIR168a went on to regulate the expression of the liver gene human/mouse low-density lipoprotein receptor adapter protein 1 (*LDRAP1*) [67]. This finding made the authors speculate that plant miRNAs in food can regulate target genes in mammals. Other studies also confirmed the presence of food RNA in human serum from rice, corn, barely, tomato, soybean, wheat, cabbage, grapes, and carrot [68]. In addition, miRNAs were also detected in other human body fluids, such as human breast milk, and were stable in harsh conditions that are known to degrade RNA as boiling, freeze thaw cycle, and high and low acid conditions [65].

Employing Somaclonal Mutagenesis

Genetic variations in plants that have been produced by plant tissue culture can develop somaclonal mutagenesis, which is a general phenomenon of all plant regeneration systems that involve a callus phase. Somaclonal variation is either somatic or meiotic. Somatic variation is often not inherited into subsequent generations, and is mostly used with the amplification of ornamental plants or trees. Meiotic somaclonal mutagenesis is inherited; it is important for the propagation for seeds, and includes two types of variations either genetically or epigenetically [69 - 71]. Tissue culture can be used as a starting material in order to select single cells "encountering beneficial epiallel" and hence can be propagated into multicellular culture. Induction mutagenesis by physical or chemical mutagens can also be applied to tissue culture. Subsequently, preferable epimutations are selected to regenerate whole plants from surviving cells.

Nutritional Epigenetics (Food)

Epigenetic modifications can occur in response to environmental stimuli including the diet. For example, during the winter of 1944-1945, the Netherlands suffered from a famine and the population's intake dropped fewer than 1000 calories per day. Recent studies show that the children who were born during the period of the famine have a higher rate to chronic diseases, such as diabetes, cardiovascular disease, and obesity. Other similar famine problems that show an effect on next generations were reported as seven-year famine in the 19th century from the northern Sweden city "Norrbotten", and potato famine in 1856 in Ireland. From such examples, it is obvious that diet is also a player in epigenetic modifications. The exact epigenetic modifications are not clear, however it was reported that people who were exposed to famine in utero have lower methylation of a gene that is involved in insulin metabolism (insulin-like growth factor II) [72]. Another important example about the influence of diet on epigenetics is shown in mice. Mice population that have an active agouti gene have a yellow coat and a probability of getting obese. This gene can be switched off through DNA methylation. It was reported that if a pregnant agouti mouse eats food that releases methyl groups, such as folic acid or choline, its progeny genes become methylated and thus inactive; consequently, they lose the agouti phenotype [73]. Folic acid uptake is also important for humans, as its insufficient uptake is implicated in developmental problems, such as spina bifida, and other neural tube defects. This is why folic acid is recommended as supplements for pregnant women [74].

The accumulated evidence of food impact on epigenetics led to the emerging of a crucial field of science that is nutritional epigenetics [75]. Different types of food components are thought to affect epigenetics in humans. For instance, broccoli and other cruciferous vegetables have isothiocyanates that can increase histone acetylation. The isoflavone genistein and the polyphenol compound epigallocatechin-3-gallate are implicated in decreasing DNA methylation, and are found in soya and green tea, respectively. Curcumin that is found in turmeric is implicated in both DNA methylation inhibition and histone acetylation modulation [76]. Several other food components, such as folate, biotin, niacin, tea catechin, resveratrol, butyrate, sulforaphane, pantothenic acid and betaine have an impact on various epigenetic mechanisms too. Two different ways of actions were suggested for these food components, where the component can work as either a substrate or an inhibitor of the epigenetic related enzyme. For example, biotin works as a substrate for histone biotinylation, while curcumin is an inhibitor for histone acetyltransferases [19, 76 - 78].

In 2012, researchers from the Institute of Food Research from Newcastle University examined the cells lining the gut wall from volunteers and showed that epigenetic modifications by DNA methylation are affected by ageing and by food intake. Volunteers that have higher vitamin D and selenium were shown to have lower levels of DNA methylation, while higher folate intake was associated with higher levels of DNA methylation [79]. Another research study linked peanut allergy with epigenetics. Around 20% of babies with eczema or milk and egg allergies are prone to develop peanut allergy. The number of children who develop this type of allergy has doubled in U.S. between 1998 and 2003. Researchers from Johns Hopkins University analyzed DNA samples from 2759 participants including children and their biological parents who have this allergy. From this study, 1 million genetic markers across the human genome were investigated to search for clues to what causes food allergy. From this study, the researchers identified an epigenetic variation by DNA methylation in the HLA-DR and -DQ gene region, which they linked to peanut allergy [80]. HLA-DR and –DQ molecules are expressed in a range of cells including B and T cells and are known to play a critical role in allergy development [80].

The role of epigenetics and its impact on human health is also an evolving subject. Current knowledge of epigenetic mechanisms allowed the usage of different strategies against human disorders that are non-related to genomic sequence changes. In this context, the term "Epigenetic Therapy" evolved as a promising field in order to use different drugs to reverse the phenotypic effects and to alleviate related disorders. Representative drugs, such as 5-Azacytidine, 5-Aza-2'-deoxycytidine, 5-fluoro-2'-deoxycytidine, Zebularine, Procainamide, Epigallo-catechin-3-gallate, Psammaplin A, and Antisense oligomers target DNA methylation mechanism, while other drugs target histone deacetylase, such as Phenylbutyric acid, Suberoylanilide hydroxamic acid, Depsipeptide, and Valproic acid [19, 81]

Optimizing Genome-Editing Technology for Acceptable GM Crops

Programmable nucleases, such as zinc finger nucleases (ZFNs), transcription activator-like effector nucleases (TALENs), and RNA guided endonucleases (RGENs) can facilitate genome-editing by increasing the efficiency of homologous recombination. Recently, the newly discovered RGENs: clustered regularly interspaced short palindromic repeats (CRISPRs)/Cas is overcoming ZFNs and TALENs, which depend on time-consuming designs and difficulties related to their optimization [82]. Type II CRISPR-Cas systems have been engineered to affect robust RNA-guided genome modifications in multiple eukaryotic systems [83]. Different Cas9 variants have been engineered to increase the applicability of the CRISPR-Cas9 system. Cas9 has two active sites, and each site is responsible for cutting one of the complementary DNA strands. The double-strand break generated by the guided endonucleases "mentioned above" can either be repaired by the non-homologous end-joining (NHEJ) or homology directed repair (HDR). The error-prone NHEJ is the common pathway and usually induces small deletion or insertion mutations, or point mutation. The balance between NHEJ and HDR can be slightly shifted towards HDR by providing a donor template. By including a transgene within the donor template, gene targeting can be achieved [83 - 85].

CRISPR-Cas9 system has been used in plants and reports show its ability to modify specific genomic targets efficiently. To be able to use this technology, the bacterial Cas9 protein and guide RNAs should be transferred to the plant transiently or stably using the known delivery methods. This can be accomplished through incorporating the coding sequence of Cas9 and guide RNAs into a binary vector, which can be transformed to plants through *Agrobacterium tumefacians* as has been done for example in Arabidopsis, Tomato and Rice [84, 86 - 95]. However, these delivery methods run the risk of creating additional mutations either from the plasmids integration into plant genome, or from the persistence of encoded genome editing players, which can continue to make mutations. This will still face most of the regulatory measures that GM crops face.

Lately, an ambitious technique was adapted by scientists of Seoul National University that avoids the use of plasmids. They preassembled the Cas9 protein and guide RNA complex *in vitro* and they used polyethylene glycol method to directly transfer them by endocytosis into protoplasts (plant cells that have had their cell walls removed). The edited protoplasts can be cultured into small calli, from which a mature plant can be regenerated. They succeeded to transform 4 different plants including Arabidopsis and Lettuce [96]. This new delivery method can completely revolutionize the way we modify GM crops because by this we can avoid the incorporation of any additional unwanted DNA damage, as well as probably allow the GM crops to skip regulatory oversight. Combining this adaptive method and possibly targeting promoters to edit its epigenetic makeup will be beneficial, and can probably make the GM crops not fall under the same set of regulations that currently exist for transgenic plants. The usage of genome editing technology for the epigenome is still in its embryonic stage, however a successful example from human research shows the successful editing of epigenome when the nuclease null dCas9 protein was fused with a human acetyltransferase catalytic subunit, which specifically targeted H3K27 in order to activate genes from promoters and enhancers [97].

CONCLUSION

The debate of using GM crops and their impact on human health and environment should be treated carefully. The increasing population and problems of malnutrition in several places around the world demand more research ideas and applications. However, biotechnology and genetic manipulation helped and there is still hope for supplying better GM crop varieties, but they face tremendous risk assessment regulations that hinder their final use. Epigenetics manipulation can be an alternative source for optimizing crops without affecting genes codes, and instead mostly focuses on selecting beneficial epimutations. Selecting novel traits can be through the generation of heritable epiallels and TEs regulation through generating epimutation by variable methods, such as chemical mutagenesis, RNAi, somaclonal mutagenesis (tissue culture induced *e.g.* banana), and applying different types of stress.

In early 2016, the plant epigenetics community met for the Plant Epigenetics: From Genotype to Phenotype Keystone Symposium [98]. In this meeting, they covered different topics of plant epigenetics including crop improvement. For example, Robert Martienssen (Cold Spring Harbor Laboratory) identified the epiallele that is responsible for poor fruit production in oil palm, and his group developed an assay to early genotype oil palms in order to select the best quality individuals. Moreover, scientists discussed how to use epigenomic markers for agricultural improvement, for example Jon Reinders (DuPont Pioneer) used DNA methylation patterns as markers to predict key phenotypes in maize breeding programs [98]. Interesting to mention, Agro Epigenetics Crop that is a Canadian biotechnology company is specialized in the breeding and propagation of plant species through epigenetic principles. The company is mostly trying to develop faster growing, higher yielding, disease resistant and adaptable plants (http://epigeneticscorp.com/). Finally, employing epigenetics in plant biotechnology, the usage of newly developed gene editing technologies, and understanding the nutritional epigenetics are important for food science applications.

CONSENT FOR PUBLICATION

Not applicable.

CONFLICT OF INTEREST

The author declares no conflict of interest, financial or otherwise.

ACKNOWLEDGEMENTS

Declared none.

ABBREVIATIONS

AGO	Argonaute
ARTs	ADP-ribosyl transferases
CMT3	Chromomethyltransferase
Cnr	Colorless non-ripening
CRISPRs	Clustered regularly interspaced short palindromic repeats
DCL	RNAaseIII-related dicer
DNM	DNA-(cytosine-5)-methyltransferase
DRM	Domain rearranged methyltransferase
FSB	Fruit and shoot borer
GMO	Genetically modified organisms
GM	Genetically modified
H	Histone
HDACs	Histone deacetylases
HDR	Homology directed repair
HMT	Histone methyl transferases
Hub	Histone monoubiquitination
K	Lysine

LDRAP1	Low-density lipoprotein receptor adapter protein 1
me	Methyl
MET	Methyltransferase
miRNAs	MicroRNAs
NHEJ	Non-homologous end-joining
NO	Nitric oxide
PARPs	Poly (ADP-ribose) polymerases
PTGS	Post-transcriptional silencing
RdDM	RNA dependent directed methylation
RDRP	RNA-dependent RNA polymerase
RGENs	RNA guided endonucleases
RISC	RNA-induced silencing complex
RNAi	RNA interference
siRNAs	Small interfering RNAs
snRNAs	Small nuclear RNAs
TALENs	Transcription activator-like effector
tasiRNA	Trans-acting siRNA
TEs	Transposon elements
ZFNs	Zinc finger nucleases

REFERENCES

[1]   Hefferon KL, Makhzoum A. Biotechnological Approaches for Nutritionally Enhanced Food Crop Production Advances in Food Biotechnology. John Wiley & Sons Ltd 2015; pp. 1-12.

[2]   Hallerman E, Grabau E. Crop biotechnology: a pivotal moment for global acceptance. Food Energy Secur 2016; 5(1): 3-17.
[http://dx.doi.org/10.1002/fes3.76]

[3]   GM crops: A story in numbers. Nature 2013; 497(7447): 22-3.
[http://dx.doi.org/10.1038/497022a] [PMID: 23636377]

[4]   Chilton MD, Drummond MH, Merio DJ, *et al.* Stable incorporation of plasmid DNA into higher plant cells: the molecular basis of crown gall tumorigenesis. Cell 1977; 11(2): 263-71.
[http://dx.doi.org/10.1016/0092-8674(77)90043-5] [PMID: 890735]

[5]   Halford NG. Toward two decades of plant biotechnology: successes, failures, and prospects. Food Energy Secur 2012; 1(1): 9-28.
[http://dx.doi.org/10.1002/fes3.3]

[6]   Murai N, Kemp JD, Sutton DW, *et al.* Phaseolin gene from bean is expressed after transfer to sunflower via tumor-inducing plasmid vectors. Science 1983; 222(4623): 476-82.
[http://dx.doi.org/10.1126/science.222.4623.476] [PMID: 17746179]

[7]   Guerinot ML. Perspectives: plant biology. The green revolution strikes gold. Science 2000; 287(5451): 241-3.
[http://dx.doi.org/10.1126/science.287.5451.241] [PMID: 10660423]

[8]   Waltz E, Vitamin A. Super Banana in human trials. Nat Biotechnol 2014; 32(9): 857.
        [http://dx.doi.org/10.1038/nbt0914-857] [PMID: 25203025]

[9]   Riggs AD, Porter TN. Overview of Epigenetic Mechanisms1996 1996.

[10]   Waddington CH. The epigenotype. 1942. Int J Epidemiol 2012; 41(1): 10-3.
        [http://dx.doi.org/10.1093/ije/dyr184] [PMID: 22186258]

[11]   Jablonka E, Lamb MJ. The inheritance of acquired epigenetic variations. J Theor Biol 1989; 139(1): 69-83.
        [http://dx.doi.org/10.1016/S0022-5193(89)80058-X] [PMID: 2593687]

[12]   Cullis CA, Kolodynska K. Variation in the isozymes of flax (Linum usitatissimum) genotrophs. Biochem Genet 1975; 13(9-10): 687-97.
        [http://dx.doi.org/10.1007/BF00484926] [PMID: 1203059]

[13]   Durrant A. The environmental induction of heritable change in Linum. Heredity 1962; 17(1): 27-61.
        [http://dx.doi.org/10.1038/hdy.1962.2]

[14]   Hill J. The environmental induction of heritable changes in NICOTIANA RUSTICA parental and selection lines. Genetics 1967; 55(4): 735-54.
        [PMID: 17248380]

[15]   Hill J, Perkins JM. The environmental induction of heritable changes in NICOTIANA RUSTICA. Effects of genotype-environment interactions. Genetics 1969; 61(3): 661-75.
        [PMID: 17248433]

[16]   Schneeberger RG, Cullis CA. Specific DNA alterations associated with the environmental induction of heritable changes in flax. Genetics 1991; 128(3): 619-30.
        [PMID: 1678726]

[17]   Cubas P, Vincent C, Coen E. An epigenetic mutation responsible for natural variation in floral symmetry. Nature 1999; 401(6749): 157-61.
        [http://dx.doi.org/10.1038/43657] [PMID: 10490023]

[18]   Manning K, Tör M, Poole M, *et al.* A naturally occurring epigenetic mutation in a gene encoding an SBP-box transcription factor inhibits tomato fruit ripening. Nat Genet 2006; 38(8): 948-52.
        [http://dx.doi.org/10.1038/ng1841] [PMID: 16832354]

[19]   Gulluce M, Alaylar B, Koc TY, Karadayi M. Epigenetics: An Innovative Approach for Biotechnology and Food Science. Int J Biosci Biochem Bioinform 2014; 4(3): 195-9. [IJBBB].
        [http://dx.doi.org/10.7763/IJBBB.2014.V4.338]

[20]   Finnegan EJ, Genger RK, Peacock WJ, Dennis ES. DNA Methylation in Plants. Annu Rev Plant Physiol Plant Mol Biol 1998; 49: 223-47.
        [http://dx.doi.org/10.1146/annurev.arplant.49.1.223] [PMID: 15012234]

[21]   Mirouze M, Paszkowski J. Epigenetic contribution to stress adaptation in plants. Curr Opin Plant Biol 2011; 14(3): 267-74.
        [http://dx.doi.org/10.1016/j.pbi.2011.03.004] [PMID: 21450514]

[22]   Chan SWL, Henderson IR, Jacobsen SE. Gardening the genome: DNA methylation in Arabidopsis thaliana. Nat Rev Genet 2005; 6(5): 351-60.
        [http://dx.doi.org/10.1038/nrg1601] [PMID: 15861207]

[23]   Zhang X. The epigenetic landscape of plants. Science 2008; 320(5875): 489-92.
        [http://dx.doi.org/10.1126/science.1153996] [PMID: 18436779]

[24]   Vanyushin BF. DNA methylation in plants. Curr Top Microbiol Immunol 2006; 301: 67-122.
        [http://dx.doi.org/10.1007/3-540-31390-7_4] [PMID: 16570846]

[25]   Herceg Z, Murr R. Mechanisms of histone modifications. Handbook of Epigenetics: The New Molecular and Medical Genetics. Philadelphia: Academic Press 2010; pp. 25-45.

[26]   Chen M, Lv S, Meng Y. Epigenetic performers in plants. Dev Growth Differ 2010; 52(6): 555-66.
[http://dx.doi.org/10.1111/j.1440-169X.2010.01192.x] [PMID: 20646028]

[27]   Lafos M, Schubert D. Balance of power-dynamic regulation of chromatin in plant development. Biol Chem 2009; 390(11): 1113-23.
[http://dx.doi.org/10.1515/BC.2009.129] [PMID: 19747083]

[28]   Wagner D. Chromatin regulation of plant development. Curr Opin Plant Biol 2003; 6(1): 20-8.
[http://dx.doi.org/10.1016/S1369526602000079] [PMID: 12495747]

[29]   Kim JM, To TK, Ishida J, *et al.* Alterations of lysine modifications on the histone H3 N-tail under drought stress conditions in Arabidopsis thaliana. Plant Cell Physiol 2008; 49(10): 1580-8.
[http://dx.doi.org/10.1093/pcp/pcn133] [PMID: 18779215]

[30]   Sokol A, Kwiatkowska A, Jerzmanowski A, Prymakowska-Bosak M. Up-regulation of stress-inducible genes in tobacco and Arabidopsis cells in response to abiotic stresses and ABA treatment correlates with dynamic changes in histone H3 and H4 modifications. Planta 2007; 227(1): 245-54.
[http://dx.doi.org/10.1007/s00425-007-0612-1] [PMID: 17721787]

[31]   Zhou DX. Regulatory mechanism of histone epigenetic modifications in plants. Epigenetics 2009; 4(1): 15-8.
[http://dx.doi.org/10.4161/epi.4.1.7371] [PMID: 19164930]

[32]   Hirota T, Lipp JJ, Toh B-H, Peters J-M. Histone H3 serine 10 phosphorylation by Aurora B causes HP1 dissociation from heterochromatin. Nature 2005; 438(7071): 1176-80.
[http://dx.doi.org/10.1038/nature04254] [PMID: 16222244]

[33]   Zhang K, Sridhar VV, Zhu J, Kapoor A, Zhu JK. Distinctive core histone post-translational modification patterns in Arabidopsis thaliana. PLoS One 2007; 2(11): e1210.
[http://dx.doi.org/10.1371/journal.pone.0001210] [PMID: 18030344]

[34]   Benhamed M, Martin-Magniette M-L, Taconnat L, *et al.* Genome-scale Arabidopsis promoter array identifies targets of the histone acetyltransferase GCN5. Plant J 2008; 56(3): 493-504.
[http://dx.doi.org/10.1111/j.1365-313X.2008.03606.x] [PMID: 18644002]

[35]   Bond DM, Dennis ES, Pogson BJ, Finnegan EJ. Histone acetylation, VERNALIZATION INSENSITIVE 3, FLOWERING LOCUS C, and the vernalization response. Mol Plant 2009; 2(4): 724-37.
[http://dx.doi.org/10.1093/mp/ssp021] [PMID: 19825652]

[36]   He Y, Michaels SD, Amasino RM. Regulation of flowering time by histone acetylation in Arabidopsis. Science 2003; 302(5651): 1751-4.
[http://dx.doi.org/10.1126/science.1091109] [PMID: 14593187]

[37]   Krichevsky A, Zaltsman A, Kozlovsky SV, Tian GW, Citovsky V. Regulation of root elongation by histone acetylation in Arabidopsis. J Mol Biol 2009; 385(1): 45-50.
[http://dx.doi.org/10.1016/j.jmb.2008.09.040] [PMID: 18835563]

[38]   Li C, Wu K, Fu G, *et al.* Regulation of oleosin expression in developing peanut (Arachis hypogaea L.) embryos through nucleosome loss and histone modifications. J Exp Bot 2009; 60(15): 4371-82.
[http://dx.doi.org/10.1093/jxb/erp275] [PMID: 19737778]

[39]   Jiao Y, Lau OS, Deng XW. Light-regulated transcriptional networks in higher plants. Nat Rev Genet 2007; 8(3): 217-30.
[http://dx.doi.org/10.1038/nrg2049] [PMID: 17304247]

[40]   Mengel A, Chaki M, Shekariesfahlan A, Lindermayr C. Effect of nitric oxide on gene transcription - S-nitrosylation of nuclear proteins. Front Plant Sci 2013; 4: 293.
[http://dx.doi.org/10.3389/fpls.2013.00293] [PMID: 23914201]

[41]   Nott A, Watson PM, Robinson JD, Crepaldi L, Riccio A. S-Nitrosylation of histone deacetylase 2 induces chromatin remodelling in neurons. Nature 2008; 455(7211): 411-5.

[http://dx.doi.org/10.1038/nature07238] [PMID: 18754010]

[42] Liu Y, Koornneef M, Soppe WJ. The absence of histone H2B monoubiquitination in the Arabidopsis hub1 (rdo4) mutant reveals a role for chromatin remodeling in seed dormancy. Plant Cell 2007; 19(2): 433-44.
[http://dx.doi.org/10.1105/tpc.106.049221] [PMID: 17329563]

[43] Cao Y, Dai Y, Cui S, Ma L. Histone H2B monoubiquitination in the chromatin of FLOWERING LOCUS C regulates flowering time in Arabidopsis. Plant Cell 2008; 20(10): 2586-602.
[http://dx.doi.org/10.1105/tpc.108.062760] [PMID: 18849490]

[44] Kaufmann K, Pajoro A, Angenent GC. Regulation of transcription in plants: mechanisms controlling developmental switches. Nat Rev Genet 2010; 11(12): 830-42.
[http://dx.doi.org/10.1038/nrg2885] [PMID: 21063441]

[45] Pikaard CS, Mittelsten Scheid O. Epigenetic regulation in plants. Cold Spring Harb Perspect Biol 2014; 6(12): a019315.
[http://dx.doi.org/10.1101/cshperspect.a019315] [PMID: 25452385]

[46] Fire A, Xu S, Montgomery MK, Kostas SA, Driver SE, Mello CC. Potent and specific genetic interference by double-stranded RNA in Caenorhabditis elegans. Nature 1998; 391(6669): 806-11.
[http://dx.doi.org/10.1038/35888] [PMID: 9486653]

[47] Zamecnik PC, Stephenson ML. Inhibition of Rous sarcoma virus replication and cell transformation by a specific oligodeoxynucleotide. Proc Natl Acad Sci USA 1978; 75(1): 280-4.
[http://dx.doi.org/10.1073/pnas.75.1.280] [PMID: 75545]

[48] Cogoni C, Macino G. Isolation of quelling-defective (qde) mutants impaired in posttranscriptional transgene-induced gene silencing in Neurospora crassa. Proc Natl Acad Sci USA 1997; 94(19): 10233-8.
[http://dx.doi.org/10.1073/pnas.94.19.10233] [PMID: 9294193]

[49] Jorgensen R. Altered gene expression in plants due to trans interactions between homologous genes. Trends Biotechnol 1990; 8(12): 340-4.
[http://dx.doi.org/10.1016/0167-7799(90)90220-R] [PMID: 1366894]

[50] Djupedal I, Ekwall K. Epigenetics: heterochromatin meets RNAi. Cell Res 2009; 19(3): 282-95.
[http://dx.doi.org/10.1038/cr.2009.13] [PMID: 19188930]

[51] Allen E, Howell MD. miRNAs in the biogenesis of trans-acting siRNAs in higher plants. Semin Cell Dev Biol 2010; 21(8): 798-804.
[http://dx.doi.org/10.1016/j.semcdb.2010.03.008] [PMID: 20359543]

[52] Wassenegger M, Heimes S, Riedel L, Sänger HL. RNA-directed de novo methylation of genomic sequences in plants. Cell 1994; 76(3): 567-76.
[http://dx.doi.org/10.1016/0092-8674(94)90119-8] [PMID: 8313476]

[53] Roundtree IA, He C. RNA epigenetics-chemical messages for posttranscriptional gene regulation. Curr Opin Chem Biol 2016; 30: 46-51.
[http://dx.doi.org/10.1016/j.cbpa.2015.10.024] [PMID: 26625014]

[54] Alarcón CR, Goodarzi H, Lee H, Liu X, Tavazoie S, Tavazoie SF. HNRNPA2B1 Is a Mediator of m(6)A-Dependent Nuclear RNA Processing Events. Cell 2015; 162(6): 1299-308.
[http://dx.doi.org/10.1016/j.cell.2015.08.011] [PMID: 26321680]

[55] Chen T, Hao Y-J, Zhang Y, et al. m(6)A RNA methylation is regulated by microRNAs and promotes reprogramming to pluripotency. Cell Stem Cell 2015; 16(3): 289-301.
[http://dx.doi.org/10.1016/j.stem.2015.01.016] [PMID: 25683224]

[56] Fujimoto R, Sasaki T, Ishikawa R, Osabe K, Kawanabe T, Dennis ES. Molecular mechanisms of epigenetic variation in plants. Int J Mol Sci 2012; 13(8): 9900-22.
[http://dx.doi.org/10.3390/ijms13089900] [PMID: 22949838]

[57]   Madlung A, Comai L. The effect of stress on genome regulation and structure. Ann Bot 2004; 94(4): 481-95.
[http://dx.doi.org/10.1093/aob/mch172] [PMID: 15319229]

[58]   Fu S, Sun C, Yang M, *et al.* Genetic and epigenetic variations induced by wheat-rye 2R and 5R monosomic addition lines. PLoS One 2013; 8(1): e54057.
[http://dx.doi.org/10.1371/journal.pone.0054057] [PMID: 23342073]

[59]   Dong ZY, Wang YM, Zhang ZJ, *et al.* Extent and pattern of DNA methylation alteration in rice lines derived from introgressive hybridization of rice and Zizania latifolia Griseb. Theor Appl Genet 2006; 113(2): 196-205.
[http://dx.doi.org/10.1007/s00122-006-0286-2] [PMID: 16791687]

[60]   Paszkowski J, Grossniklaus U. Selected aspects of transgenerational epigenetic inheritance and resetting in plants. Curr Opin Plant Biol 2011; 14(2): 195-203.
[http://dx.doi.org/10.1016/j.pbi.2011.01.002] [PMID: 21333585]

[61]   Holeski LM, Jander G, Agrawal AA. Transgenerational defense induction and epigenetic inheritance in plants. Trends Ecol Evol (Amst) 2012; 27(11): 618-26.
[http://dx.doi.org/10.1016/j.tree.2012.07.011] [PMID: 22940222]

[62]   Sharma A. Transgenerational epigenetic inheritance: focus on soma to germline information transfer. Prog Biophys Mol Biol 2013; 113(3): 439-46.
[http://dx.doi.org/10.1016/j.pbiomolbio.2012.12.003] [PMID: 23257323]

[63]   Akimoto K, Katakami H, Kim HJ, *et al.* Epigenetic inheritance in rice plants. Ann Bot 2007; 100(2): 205-17.
[http://dx.doi.org/10.1093/aob/mcm110] [PMID: 17576658]

[64]   Jones L, Ratcliff F, Baulcombe DC. RNA-directed transcriptional gene silencing in plants can be inherited independently of the RNA trigger and requires Met1 for maintenance. Curr Biol 2001; 11(10): 747-57.
[http://dx.doi.org/10.1016/S0960-9822(01)00226-3] [PMID: 11378384]

[65]   Sirinathsinghji E. Epigenetics and Implications for GM Crops Using RNAi: Institute of Science in Society 2014. Available from: http://www.i-sis.org.uk/How_Food_Affects_Genes.php

[66]   Ho MW. New GM Nightmares with RNA: Institute of Science in Society 2013. Available from: http://www.i-sis.org.uk/New_GM_nightmares_with_RNA.php

[67]   Zhang L, Hou D, Chen X, *et al.* Exogenous plant MIR168a specifically targets mammalian LDLRAP1: evidence of cross-kingdom regulation by microRNA. Cell Res 2012; 22(1): 107-26.
[http://dx.doi.org/10.1038/cr.2011.158] [PMID: 21931358]

[68]   Wang K, Li H, Yuan Y, *et al.* The complex exogenous RNA spectra in human plasma: an interface with human gut biota? PLoS One 2012; 7(12): e51009.
[http://dx.doi.org/10.1371/journal.pone.0051009] [PMID: 23251414]

[69]   Miguel C, Marum L. An epigenetic view of plant cells cultured in vitro: somaclonal variation and beyond. J Exp Bot 2011; 62(11): 3713-25.
[http://dx.doi.org/10.1093/jxb/err155] [PMID: 21617249]

[70]   Kaeppler SM, Kaeppler HF, Rhee Y. Epigenetic aspects of somaclonal variation in plants. Plant Mol Biol 2000; 43(2-3): 179-88.
[http://dx.doi.org/10.1023/A:1006423110134] [PMID: 10999403]

[71]   Krishna H, Alizadeh M, Singh D, Singh U, Chauhan N, Eftekhari M, *et al.* Somaclonal variations and their applications in horticultural crops improvement. 3 Biotech 2016; 6(1): 54.

[72]   Heijmans BT, Tobi EW, Stein AD, *et al.* Persistent epigenetic differences associated with prenatal exposure to famine in humans. Proc Natl Acad Sci USA 2008; 105(44): 17046-9.
[http://dx.doi.org/10.1073/pnas.0806560105] [PMID: 18955703]

[73] Dolinoy DC. The agouti mouse model: an epigenetic biosensor for nutritional and environmental alterations on the fetal epigenome. Nutr Rev 2008; 66 (Suppl. 1): S7-S11.
[http://dx.doi.org/10.1111/j.1753-4887.2008.00056.x] [PMID: 18673496]

[74] de Benoist B. Conclusions of a WHO Technical Consultation on folate and vitamin B12 deficiencies. Food Nutr Bull 2008; 29(2) (Suppl.): S238-44.
[http://dx.doi.org/10.1177/15648265080292S129] [PMID: 18709899]

[75] Landecker H. Food as exposure: Nutritional epigenetics and the new metabolism. Biosocieties 2011; 6(2): 167-94.
[http://dx.doi.org/10.1057/biosoc.2011.1] [PMID: 23227106]

[76] Flabouraris G, Karikas GA. Nutri-epigenetics and synthetic analogs in cancer chemoprevention. J BUON 2016; 21(1): 4-16.
[PMID: 27061524]

[77] Choi SW, Friso S. Epigenetics: A new bridge between nutrition and health. Adv Nutr 2010; 1(1): 8-16.
[http://dx.doi.org/10.3945/an.110.1004] [PMID: 22043447]

[78] Su LJ, Mahabir S, Ellison GL, McGuinn LA, Reid BC. Epigenetic Contributions to the Relationship between Cancer and Dietary Intake of Nutrients, Bioactive Food Components, and Environmental Toxicants. Front Genet 2012; 2: 91.
[http://dx.doi.org/10.3389/fgene.2011.00091] [PMID: 22303385]

[79] Tapp HS, Commane DM, Bradburn DM, *et al.* Nutritional factors and gender influence age-related DNA methylation in the human rectal mucosa. Aging Cell 2013; 12(1): 148-55.
[http://dx.doi.org/10.1111/acel.12030] [PMID: 23157586]

[80] Hong X, Hao K, Ladd-Acosta C, *et al.* Genome-wide association study identifies peanut allergy-specific loci and evidence of epigenetic mediation in US children. Nat Commun 2015; 6: 6304.
[http://dx.doi.org/10.1038/ncomms7304] [PMID: 25710614]

[81] Egger G, Liang G, Aparicio A, Jones PA. Epigenetics in human disease and prospects for epigenetic therapy. Nature 2004; 429(6990): 457-63.
[http://dx.doi.org/10.1038/nature02625] [PMID: 15164071]

[82] Gaj T, Gersbach CA, Barbas CF III. ZFN, TALEN, and CRISPR/Cas-based methods for genome engineering. Trends Biotechnol 2013; 31(7): 397-405.
[http://dx.doi.org/10.1016/j.tibtech.2013.04.004] [PMID: 23664777]

[83] Horvath P, Barrangou R. CRISPR/Cas, the immune system of bacteria and archaea. Science 2010; 327(5962): 167-70.
[http://dx.doi.org/10.1126/science.1179555] [PMID: 20056882]

[84] Mao Y, Zhang H, Xu N, Zhang B, Gou F, Zhu JK. Application of the CRISPR-Cas system for efficient genome engineering in plants. Mol Plant 2013; 6(6): 2008-11.
[http://dx.doi.org/10.1093/mp/sst121] [PMID: 23963532]

[85] Pennisi E. The CRISPR craze. Science 2013; 341(6148): 833-6.
[http://dx.doi.org/10.1126/science.341.6148.833] [PMID: 23970676]

[86] Jia H, Wang N. Targeted genome editing of sweet orange using Cas9/sgRNA. PLoS One 2014; 9(4): e93806.
[http://dx.doi.org/10.1371/journal.pone.0093806] [PMID: 24710347]

[87] Jiang W, Zhou H, Bi H, Fromm M, Yang B, Weeks DP. Demonstration of CRISPR/Cas9/sgRNA-mediated targeted gene modification in Arabidopsis, tobacco, sorghum and rice. Nucleic Acids Res 2013; 41(20): e188.
[http://dx.doi.org/10.1093/nar/gkt780] [PMID: 23999092]

[88] Shan Q, Wang Y, Li J, *et al.* Targeted genome modification of crop plants using a CRISPR-Cas system. Nat Biotechnol 2013; 31(8): 686-8.

[http://dx.doi.org/10.1038/nbt.2650] [PMID: 23929338]

[89] Xie K, Yang Y. RNA-guided genome editing in plants using a CRISPR-Cas system. Mol Plant 2013; 6(6): 1975-83.
[http://dx.doi.org/10.1093/mp/sst119] [PMID: 23956122]

[90] Xie K, Zhang J, Yang Y. Genome-wide prediction of highly specific guide RNA spacers for CRISPR-Cas9-mediated genome editing in model plants and major crops. Mol Plant 2014; 7(5): 923-6.
[http://dx.doi.org/10.1093/mp/ssu009] [PMID: 24482433]

[91] Xing H-L, Dong L, Wang Z-P, *et al.* A CRISPR/Cas9 toolkit for multiplex genome editing in plants. BMC Plant Biol 2014; 14(1): 327.
[http://dx.doi.org/10.1186/s12870-014-0327-y] [PMID: 25432517]

[92] Fauser F, Schiml S, Puchta H. Both CRISPR/Cas-based nucleases and nickases can be used efficiently for genome engineering in Arabidopsis thaliana. Plant J 2014; 79(2): 348-59.
[http://dx.doi.org/10.1111/tpj.12554] [PMID: 24836556]

[93] Feng Z, Mao Y, Xu N, *et al.* Multigeneration analysis reveals the inheritance, specificity, and patterns of CRISPR/Cas-induced gene modifications in Arabidopsis. Proc Natl Acad Sci USA 2014; 111(12): 4632-7.
[http://dx.doi.org/10.1073/pnas.1400822111] [PMID: 24550464]

[94] Feng Z, Zhang B, Ding W, *et al.* Efficient genome editing in plants using a CRISPR/Cas system. Cell Res 2013; 23(10): 1229-32.
[http://dx.doi.org/10.1038/cr.2013.114] [PMID: 23958582]

[95] Belhaj K, Chaparro-Garcia A, Kamoun S, Nekrasov V. Plant genome editing made easy: targeted mutagenesis in model and crop plants using the CRISPR/Cas system. Plant Methods 2013; 9(1): 39.
[http://dx.doi.org/10.1186/1746-4811-9-39] [PMID: 24112467]

[96] Woo JW, Kim J, Kwon SI, *et al.* DNA-free genome editing in plants with preassembled CRISPR-Cas9 ribonucleoproteins. Nat Biotechnol 2015; 33(11): 1162-4.
[http://dx.doi.org/10.1038/nbt.3389] [PMID: 26479191]

[97] Hilton IB, D'Ippolito AM, Vockley CM, *et al.* Epigenome editing by a CRISPR-Cas9-based acetyltransferase activates genes from promoters and enhancers. Nat Biotechnol 2015; 33(5): 510-7.
[http://dx.doi.org/10.1038/nbt.3199] [PMID: 25849900]

[98] Slotkin RK. Plant epigenetics: from genotype to phenotype and back again. Genome Biol 2016; 17(1): 57.
[http://dx.doi.org/10.1186/s13059-016-0920-5] [PMID: 27025922]

SUBJECT INDEX

A

ACE 5, 7, 9, 12
  inhibition 5, 7, 9, 12
  inhibitory activity 5, 7
Acetylation, histone 436, 437, 443
Acid(s) 12, 21, 32, 53, 54, 55, 81, 82, 93, 106,
    107, 111, 112, 113, 131, 135, 137, 139,
    140, 150, 153, 155, 156, 158, 159, 160,
    194, 224, 227, 228, 232, 235, 240, 241,
    242, 243, 244, 245, 246, 247, 258, 259,
    284, 288, 290, 291, 296, 298, 299, 301,
    302, 303, 305, 307, 309, 310, 368, 386,
    388, 396, 401, 403, 404, 405, 409, 413,
    414, 443
  acetic 131, 139, 240, 241, 244, 284, 386,
    388, 405, 409, 414
  ascorbic 158, 241, 258, 259, 291
  aspartic 12, 368
  caffeic 298, 299, 301
  carminic 232, 235, 246
  carnosic 246, 298
  ellagic 291, 298, 299, 302
  ferulic 242, 246, 298, 299, 305, 372
  folic 155, 156, 158, 160, 443
  formic 137, 155, 156, 159, 405
  gallic 241, 246, 288, 290, 291, 296, 298,
    299, 303, 307, 310
  gluconic 194, 241, 401, 403, 404
  glucuronic 53, 54, 55, 305
  lactobionic 242, 246, 247, 372
  L-glutamic 224, 227, 228
  L-lactic 245
  phosphatidic (PA) 61, 81, 82, 93, 106, 107,
    111, 112, 113
  propionic 131, 409, 414
  protocatechuic 290, 296, 298, 299, 309
  succinic 137, 139, 160, 401, 403, 405, 413
Acidified milk 134
Acidolysis 86, 88, 89, 91
Activity 10, 11, 66, 111, 140, 145, 147, 162,
    185, 186, 187, 205, 223, 234, 238, 240,
    243, 246, 259, 260, 292, 297, 356, 359,
    387, 406
  analgesic 11
  antimicrobial 162, 185, 234, 238, 243, 292,
    297, 406
  antitumor 10, 66
  biological 223, 234, 238, 240, 246, 259, 260
  maximum 111, 359, 387
  metabolic 140, 145, 147, 186, 187, 205, 356
  opioid 11
Acyl group modification 81, 82, 83, 85, 91
Agrobacterium aurantiacum 240
Alcalase 13, 29, 30
Alcohol production 150
Amino acid 224, 225, 226, 227, 370, 371
  production 224, 225, 226
  purification 227
  large hydrophobic 370, 371
Anaerobic digestion 390, 399, 408
Anthocyanins 234, 278, 290, 305, 398
Antihypertensive activity 3, 4, 7, 12
Antioxidant activity 8, 10, 29, 52, 67, 234,
    238, 288, 291, 309
Antitumor properties 52, 65, 66
Apoptotic effect 301, 302, 303, 304
Applications 50, 51, 356, 359, 360, 372, 384,
    391, 409
  biotechnological 384, 391, 409
  commercial 356, 359, 360, 372
  non-food 50, 51
Arabidopsis thaliana 435, 437, 439, 440
Archaea 93, 346, 350, 351, 352
Aspergillus niger 244, 293, 371, 389, 390,
    391, 393, 396, 413
Aspergillus oryzae 13, 133, 358, 362, 363,
    366, 368
Aspergillus oryzae β-galactosidase 363, 364
Assay 289, 290, 301, 302, 304, 308
  acid phosphatase 301, 302, 304
  lipid peroxidation 289, 290, 308
  metal chelating activity 289, 290
Autotaxin 96, 111, 112

Proteolysis 1, 2, 12, 23, 28, 29, 31, 140, 155, 157, 354, 369
Proteolytic enzymes 29, 31, 140, 154, 229, 410
Pseudomonas aeruginosa 293, 294, 295, 296
Psychrophilic enzymes 32, 345, 347, 348

Q

Quercetin-3-O-rutinoside 298, 299

R

Reactive oxygen species (ROS) 8, 150, 281, 307, 311
Refractive index (RI) 59
Renin angiotensin system (RAS) 3, 6
Residues 365, 411
  phenylalanine 365
  simple sugars 411
Rhizomucor miehei lipase 88, 371, 372
RNA dependent DNA methylation 431
RNA epigenetics 431, 439
RNA interference (RNAi) 438, 441, 445
RSM optimization 87, 88
Rubropunctatin 237, 238, 239

S

Saccharomyces cerevisiae 133, 134, 147, 152, 349, 355, 386, 392, 396, 413
Salmonella typhimurium 151, 293, 295, 296, 297
Scavenging activity 10, 67, 289, 290, 291
Short chain fatty acids (SCFAs) 144, 180, 192, 196, 197, 198, 199
Single cell oil (SCO) 384, 401, 403, 405, 407, 409, 410
Single cell protein (SCP) 384, 392, 396, 406, 407, 409, 410, 414
Solid-lipid nanoparticles 277
Solid-lipid NPs 283

Solid-state fermentation (SSF) 384, 390, 391, 392, 393, 404, 410, 415
Soy protein hydrolysates 10
Soy protein isolates (SPI) 19
Sphingophospholipids 81
Stabilize enzymes 25
Staphylococcus aureus 151, 293, 294, 295, 296, 297, 395
Steviol glycosides 248, 254, 255
Strains, glutamicum 226
Streptococcus thermophilus 134, 137, 142, 146, 151, 154
Sulfur AA Bacteriocin 159
Sweet plant proteins 256
Synbiotic treatment 202, 203, 204
Synthesis of nanoparticles 283, 284
Systolic blood pressure (SBP) 5, 6

T

Thaumatin 248, 256, 257
Therapeutic Modulation 177, 184
Thermophilic enzymes 32, 348
Thermophilic proteins 352
Thermophilus 53, 148, 155, 156, 157, 158, 159, 202, 204, 352
Transcription activator-like effector nucleases (TALENs) 444
Transcription factors 437, 440
Transesterification 81, 86, 95
Transgalactosylation 341, 348, 361, 362, 363, 364, 366, 367
Transgalactosylation activity 195, 365, 366
Transglutaminase cross-linking 1, 2, 23, 28
Transphosphatidylation 92, 93, 94, 95, 101, 107, 111, 112

U

UDP-galactose 56, 57
UDP-glucose 56

UHT milk, free 358
Utilization and applications of OMW 399

V

Varidose systems 357, 358, 359
Viscoelastic properties 19, 20, 22, 26, 28
Vitamin and cofactor pool 159

W

Whey protein isolates (WPI) 10, 17, 18, 19,
    23, 24, 30
Whey proteins 7, 10, 11, 17, 18, 24, 27, 138,
    157, 190, 372

X

Xylanase 26, 390, 391, 392, 393, 397
    production of 391, 393
Xylitol production 253

Y

Yarrowia lipolytica strains 412
Yeast extract 58, 148, 230, 251, 353
Yersinia enterocolitica 295, 296, 297
Yoghurt bacteria 156, 157, 158
Yoghurt consortium 154, 155, 157, 160

Z

Zinc finger nucleases (ZFNs) 444

www.ingramcontent.com/pod-product-compliance
Lightning Source LLC
Chambersburg PA
CBHW041933220326
41598CB00058BA/771